VOLUME FIVE HUNDRED AND FOURTEEN

METHODS IN ENZYMOLOGY

Ghrelin

METHODS IN ENZYMOLOGY

VOLUME FIVE HUNDRED AND FOURTEEN

METHODS IN ENZYMOLOGY

Ghrelin

Edited by

MASAYASU KOJIMA

Professor of Molecular Genetics
Institute of Life Science
Kurume University
Kurume, Fukuoka, Japan

KENJI KANGAWA

Director
National Cerebral and Cardiovascular
Center Research Institute
Suita, Osaka, Japan

ELSEVIER

AMSTERDAM • BOSTON • HEIDELBERG • LONDON
NEW YORK • OXFORD • PARIS • SAN DIEGO
SAN FRANCISCO • SINGAPORE • SYDNEY • TOKYO
Academic Press is an imprint of Elsevier

Academic Press is an imprint of Elsevier
525 B Street, Suite 1900, San Diego, CA 92101-4495, USA
225 Wyman Street, Waltham, MA 02451, USA
The Boulevard, Langford Lane, Kidlington, Oxford, OX51GB, UK
32, Jamestown Road, London NW1 7BY, UK
Radarweg 29, PO Box 211, 1000 AE Amsterdam, The Netherlands

First edition 2012

For information on all Academic Press publications
visit our website at store.elsevier.com

ISBN: 978-0-12-381272-8
ISSN: 0076-6879

Printed and bound in United States of America
12 13 14 15 11 10 9 8 7 6 5 4 3 2 1

CONTENTS

Section 2
Molecular Forms of Ghrelin and Measuring of the Concentrations

Section 3
Enzymes for Processing of Ghrelin

Section 4
Synthesis of Ghrelin Agonists and Antagonists

Section 5
Action of Ghrelin

Section 6
Tg and KO Mice of Ghrelin

Section 7
Clinical Application of Ghrelin

CONTRIBUTORS

Takashi Akamizu
Ghrelin Research Project of Translational Research Center, Kyoto University Graduate School of Medicine, Kyoto; The First Department of Medicine, Wakayama Medical University, Wakayama, Japan

Haruka Amitani
Department of Psychosomatic Internal Medicine, Kagoshima University Graduate School of Medical and Dental Sciences, Kagoshima, Japan

Hiroyuki Ariyasu
Department of Endocrinology and Metabolism, and Ghrelin Research Project of Translational Research Center, Kyoto University Graduate School of Medicine, Kyoto, Japan

Masahiro Asaka
Cancer Preventive Medicine, Hokkaido University Graduate School of Medicine, Sapporo, Hokkaido, Japan

Akihiro Asakawa
Department of Psychosomatic Internal Medicine, Kagoshima University Graduate School of Medical and Dental Sciences, Kagoshima, Japan

Jef D. Boeke
Department of Molecular Biology & Genetics and High Throughput Biology Center, The Johns Hopkins University School of Medicine, Baltimore, Maryland, USA

Cyril Y. Bowers
Section of Endocrinology and Metabolism, Department of Medicine, Tulane University Health Sciences Center, New Orleans, Louisiana, USA

Philip A. Cole
Department of Pharmacology & Molecular Sciences, The Johns Hopkins University School of Medicine, Baltimore, Maryland, USA

Tamer Coskun
Endocrinology, Eli Lilly and Company, Indianapolis, Indiana, USA

Yukari Date
Frontier Science Research Center, University of Miyazaki, Kiyotake, Miyazaki, Japan

Katsuya Dezaki
Division of Integrative Physiology, Department of Physiology, Jichi Medical University School of Medicine, Shimotsuke, Japan

Yuichiro Doki
Division of Gastroenterological Surgery, Department of Surgery, Graduate School of Medicine, Osaka University, Osaka, Japan

Mineko Fujimiya
Department of Anatomy, Sapporo Medical University School of Medicine, Hokkaido, Japan

Naoki Fujitsuka
Department of Psychosomatic Internal Medicine, Kagoshima University Graduate School of Medical and Dental Sciences, Kagoshima, and Tsumura Research Laboratories, Ibaraki, Japan

Jesus A. Gutierrez
Translational Science and Technologies, Eli Lilly and Company, Indianapolis, Indiana, USA

John E. Hale
Hale Biochemical Consulting LLC, Klamath Falls, Oregon, USA

Tomohisa Hattori
Tsumura & Co., Tsumura Research Laboratories, Ibaraki, Japan

Kristy M. Heppner
Metabolic Diseases Institute, Department of Medicine, University of Cincinnati, Cincinnati, Ohio, USA

Richard E. Higgs
Global Statistics and Advanced Analytics, Eli Lilly and Company, Indianapolis, Indiana, USA

Yuichiro Hiura
Division of Gastroenterological Surgery, Department of Surgery, Graduate School of Medicine, Osaka University, Osaka, Japan

Hiroshi Hosoda
Department of Biochemistry, and Department of Regenerative Medicine and Tissue Engineering, National Cerebral and Cardiovascular Center Research Institute, Suita, Osaka, Japan

Mari Hotta
Health Services Center, National Graduate Institute for Policy Studies, and Department of Medicine, Tokyo Women's Medical University, Tokyo, Japan

Po-Yuan Hsiao
Department of Pharmacology & Molecular Sciences, The Johns Hopkins University School of Medicine, Baltimore, Maryland, USA

Yousang Hwang
Department of Pharmacology & Molecular Sciences, The Johns Hopkins University School of Medicine, Baltimore, Maryland, USA

Takanori Ida
Interdisciplinary Research Organization, University of Miyazaki, Kiyotake, Miyazaki, Japan

Akio Inui
Department of Psychosomatic Internal Medicine, Kagoshima University Graduate School of Medical and Dental Sciences, Kagoshima, Japan

Hiroshi Iwakura
Department of Endocrinology and Metabolism; Ghrelin Research Project of Translational Research Center, and Medial Innovation Center, Kyoto University Graduate School of Medicine, Kyoto, Japan

Haruaki Kageyama
Department of Anatomy, Showa University School of Medicine, Tokyo, Japan

Hiroyuki Kaiya
Department of Biochemistry, National Cerebral and Cardiovascular Center Research Institute, Suita, Osaka, Japan

Kenji Kangawa
National Cerebral and Cardiovascular Center Research Institute, Suita, Osaka, and Ghrelin Research Project of Translational Research Center, Kyoto University Graduate School of Medicine, Kyoto, Japan

Daniel M. Keenan
Department of Statistics, University of Virginia, Charlottesville, Virginia, USA

Michael D. Knierman
Translational Science and Technologies, Eli Lilly and Company, Indianapolis, Indiana, USA

Masayasu Kojima
Molecular Genetics, Institute of Life Science, Kurume University, Hyakunenkouen, Kurume, Fukuoka, Japan

Blerina Kola
Centre for Endocrinology, William Harvey Research Institute, Barts and The London School of Medicine and Dentistry, Queen Mary University of London, London, United Kingdom

Márta Korbonits
Centre for Endocrinology, William Harvey Research Institute, Barts and The London School of Medicine and Dentistry, Queen Mary University of London, London, United Kingdom

Chung Thong Lim
Centre for Endocrinology, William Harvey Research Institute, Barts and The London School of Medicine and Dentistry, Queen Mary University of London, London, United Kingdom

Ligen Lin
Department of Pediatrics, USDA/ARS Children's Nutrition Research Center, Baylor College of Medicine, Houston, Texas, USA

Francesca Lolli
Centre for Endocrinology, William Harvey Research Institute, Barts and The London School of Medicine and Dentistry, Queen Mary University of London, London, United Kingdom

Timo D. Müller
Institute for Diabetes and Obesity, Helmholtz Center and Technical University Munich, Munich, Germany

Tomohiro Makino
Faculty of Pharmacology II, Asubio Pharma Co. Ltd., Chuo-ku, Kobe, Japan

Masaru Matsumoto
R&D Group Vaccine Business Planning Department Business Intelligence Division, Daiichi Sankyo Co. Ltd., Edogawa-ku, Tokyo, Japan

Nobuhiro Matsumoto
Division of Neurology, Respirology, Endocrinology, and Metabolism, Department of Internal Medicine, Faculty of Medicine, University of Miyazaki, Miyazaki, Japan

Hiroharu Mifune
Institute of Animal Experimentation, Kurume University School of Medicine, Asahi-machi, Kurume, Japan

Yoshiharu Minamitake
Board Director, Asubio Pharma Co. Ltd., Chuo-ku, Kobe, Japan

Yasuhiro Miyazaki
Division of Gastroenterological Surgery, Department of Surgery, Graduate School of Medicine, Osaka University, Osaka, Japan

Mikiya Miyazato
Department of Biochemistry, National Cerebral and Cardiovascular Center Research Institute, Suita, Osaka, Japan

Kohei Murakami
Division of Gastroenterological Surgery, Department of Surgery, Graduate School of Medicine, Osaka University, Osaka, Japan

Shuichi Muto
Pathophysiology and Therapeutics, Hokkaido University Faculty of Pharmaceutical Sciences, Sapporo, and Gastroenterology, Tomakomai City General Hospital, Tomakomai, Hokkaido, Japan

Miwa Nahata
Tsumura & Co., Tsumura Research Laboratories, Ibaraki, Japan

Koji Nakagawa
Pathophysiology and Therapeutics, Hokkaido University Faculty of Pharmaceutical Sciences, Sapporo, Hokkaido, Japan

Kazuwa Nakao
Department of Endocrinology and Metabolism, Kyoto University Graduate School of Medicine, Kyoto, Japan

Masamitsu Nakazato
Division of Neurology, Respirology, Endocrinology, and Metabolism, Department of Internal Medicine, Faculty of Medicine, University of Miyazaki, Miyazaki, Japan

Yoshihiro Nishi
Department of Physiology, Kurume University School of Medicine, Asahi-machi, Kurume, Japan

Hideko Ohgusu
Molecular Genetics, Institute of Life Science, Kurume University, Kurume, Fukuoka, Japan

Shunsuke Ohnishi
Gastroenterology, Tomakomai City General Hospital, Tomakomai, Hokkaido, Japan

Rina Ohwada
Department of Medicine, Tokyo Women's Medical University, Tokyo, Japan

Doug R. Perkins
Biotechnology Discovery Research, Eli Lilly and Company, Indianapolis, Indiana, USA

Chiharu Sadakane
Pathophysiology and Therapeutics, Hokkaido University Faculty of Pharmaceutical Sciences, Sapporo, Hokkaido, and Tsumura & Co., Tsumura Research Laboratories, Ibaraki, Japan

Yayoi Saegusa
Pathophysiology and Therapeutics, Hokkaido University Faculty of Pharmaceutical Sciences, Sapporo, Hokkaido, and Tsumura & Co., Tsumura Research Laboratories, Ibaraki, Japan

Takahiro Sato
Institute of Life Science, Kurume University, Kurume, Japan

Motoyasu Satou
Department of Biochemistry, Dokkyo Medical University School of Medicine, Mibu, Tochigi, Japan

Tamotsu Shibasaki
Department of Physiology, Nippon Medical School, Tokyo, Japan

Seiji Shioda
Department of Anatomy, Showa University School of Medicine, Tokyo, Japan

Patricia J. Solenberg
BioVenio LLC, Indianapolis, Indiana, USA

Hiroyuki Sugimoto
Department of Biochemistry, Dokkyo Medical University School of Medicine, Mibu, Tochigi, Japan

Yuxiang Sun
Department of Pediatrics, USDA/ARS Children's Nutrition Research Center, Baylor College of Medicine, Houston, Texas, USA

Tomoko Takahashi
Molecular Genetics, Institute of Life Science, Kurume University, Kurume, Fukuoka, Japan

Akihiro Takata
Division of Gastroenterological Surgery, Department of Surgery, Graduate School of Medicine, Osaka University, Osaka, Japan

Hiroshi Takeda
Pathophysiology and Therapeutics, Hokkaido University Faculty of Pharmaceutical Sciences, and Gastroenterology and Hematology, Hokkaido University Graduate School of Medicine, Sapporo, Hokkaido, Japan

Fumiko Takenoya
Department of Anatomy, Showa University School of Medicine, and Department of Physical Education, Hoshi University School of Pharmacy and Pharmaceutical Science, Tokyo, Japan

Shuji Takiguchi
Division of Gastroenterological Surgery, Department of Surgery, Graduate School of Medicine, Osaka University, Osaka, Japan

Martin S. Taylor
Department of Pharmacology & Molecular Sciences, and Department of Molecular Biology & Genetics and High Throughput Biology Center, The Johns Hopkins University School of Medicine, Baltimore, Maryland, USA

Julia D. Thomas
Centre for Endocrinology, William Harvey Research Institute, Barts and The London School of Medicine and Dentistry, Queen Mary University of London, London, United Kingdom

Jenny Tong
Metabolic Diseases Institute, Department of Medicine, University of Cincinnati, Cincinnati, Ohio, USA

Matthias H. Tschöp
Metabolic Diseases Institute, Department of Medicine, University of Cincinnati, Cincinnati, Ohio, USA, and Institute for Diabetes and Obesity, Helmholtz Center and Technical University Munich, Munich, Germany

Johannes D. Veldhuis
Department of Medicine, and Department of Physiology, Endocrine Research Unit and Biophysics Section, Mayo School of Graduate Medical Education, Clinical Translational Science Center, Mayo Clinic, Rochester, Minnesota, USA

Jill A. Willency
Translational Science and Technologies, Eli Lilly and Company, Indianapolis, Indiana, USA

Toshihiko Yada
Division of Integrative Physiology, Department of Physiology, Jichi Medical University School of Medicine, Shimotsuke, Japan

Go Yamada
Department of Endocrinology and Metabolism, Kyoto University Graduate School of Medicine, Kyoto, Japan

Morikatsu Yoshida
Department of Biochemistry, National Cerebral and Cardiovascular Center Research Institute, Suita, Osaka, Japan

PREFACE

Thirteen years have passed since the discovery of ghrelin. During these years, many research have been done to elucidate the physiological functions of ghrelin, not only a mere growth hormone-releasing hormone but also an important appetite regulator, energy conservator, and sympathetic nerve suppressor. At present, ghrelin is the only circulating orexigenic hormone that is secreted from the peripheral organ and acts on the hypothalamic arcuate nucleus, the regulatory region of appetite.

Although the discovery of ghrelin is dated back to 1999, it has a long history since 1950s when Dr. Davis reported the gastric cells similar to the pancreatic alpha cells. These A-like cells turned out to be ghrelin cells. The first GHS (growth hormone secretagogue), a synthetic ghrelin mimetic, was discovered in 1976 by Dr. Bowers and led to the identification of the GHS receptor, which was the key strategic molecule for the discovery of ghrelin. Among the authors in this volume, we sincerely thank Dr. Bowers because he is the father of ghrelin and contributes the history before ghrelin discovery.

This volume provides descriptions of several aspects of ghrelin, from its structure to clinical applications. Authors were selected based on the research contributions on ghrelin and encouraged to open their protocols and guides in a clear and reproducible way to make it possible to adapt the methods to other peptide hormones.

Last, let us express our personal pleasure. We are very happy and honored to edit this volume of the *Methods in Enzymology*. We, of course, referred the several volumes of *Methods in Enzymology* in our research on peptides.

Masayasu Kojima
Kenji Kangawa

METHODS IN ENZYMOLOGY

VOLUME I. Preparation and Assay of Enzymes
Edited by SIDNEY P. COLOWICK AND NATHAN O. KAPLAN

VOLUME II. Preparation and Assay of Enzymes
Edited by SIDNEY P. COLOWICK AND NATHAN O. KAPLAN

VOLUME III. Preparation and Assay of Substrates
Edited by SIDNEY P. COLOWICK AND NATHAN O. KAPLAN

VOLUME IV. Special Techniques for the Enzymologist
Edited by SIDNEY P. COLOWICK AND NATHAN O. KAPLAN

VOLUME V. Preparation and Assay of Enzymes
Edited by SIDNEY P. COLOWICK AND NATHAN O. KAPLAN

VOLUME VI. Preparation and Assay of Enzymes *(Continued)*
Preparation and Assay of Substrates
Special Techniques
Edited by SIDNEY P. COLOWICK AND NATHAN O. KAPLAN

VOLUME VII. Cumulative Subject Index
Edited by SIDNEY P. COLOWICK AND NATHAN O. KAPLAN

VOLUME VIII. Complex Carbohydrates
Edited by ELIZABETH F. NEUFELD AND VICTOR GINSBURG

VOLUME IX. Carbohydrate Metabolism
Edited by WILLIS A. WOOD

VOLUME X. Oxidation and Phosphorylation
Edited by RONALD W. ESTABROOK AND MAYNARD E. PULLMAN

VOLUME XI. Enzyme Structure
Edited by C. H. W. HIRS

VOLUME XII. Nucleic Acids (Parts A and B)
Edited by LAWRENCE GROSSMAN AND KIVIE MOLDAVE

VOLUME XIII. Citric Acid Cycle
Edited by J. M. LOWENSTEIN

VOLUME XIV. Lipids
Edited by J. M. LOWENSTEIN

VOLUME XV. Steroids and Terpenoids
Edited by RAYMOND B. CLAYTON

VOLUME XVI. Fast Reactions
Edited by KENNETH KUSTIN

VOLUME XVII. Metabolism of Amino Acids and Amines (Parts A and B)
Edited by HERBERT TABOR AND CELIA WHITE TABOR

VOLUME XVIII. Vitamins and Coenzymes (Parts A, B, and C)
Edited by DONALD B. MCCORMICK AND LEMUEL D. WRIGHT

VOLUME XIX. Proteolytic Enzymes
Edited by GERTRUDE E. PERLMANN AND LASZLO LORAND

VOLUME XX. Nucleic Acids and Protein Synthesis (Part C)
Edited by KIVIE MOLDAVE AND LAWRENCE GROSSMAN

VOLUME XXI. Nucleic Acids (Part D)
Edited by LAWRENCE GROSSMAN AND KIVIE MOLDAVE

VOLUME XXII. Enzyme Purification and Related Techniques
Edited by WILLIAM B. JAKOBY

VOLUME XXIII. Photosynthesis (Part A)
Edited by ANTHONY SAN PIETRO

VOLUME XXIV. Photosynthesis and Nitrogen Fixation (Part B)
Edited by ANTHONY SAN PIETRO

VOLUME XXV. Enzyme Structure (Part B)
Edited by C. H. W. HIRS AND SERGE N. TIMASHEFF

VOLUME XXVI. Enzyme Structure (Part C)
Edited by C. H. W. HIRS AND SERGE N. TIMASHEFF

VOLUME XXVII. Enzyme Structure (Part D)
Edited by C. H. W. HIRS AND SERGE N. TIMASHEFF

VOLUME XXVIII. Complex Carbohydrates (Part B)
Edited by VICTOR GINSBURG

VOLUME XXIX. Nucleic Acids and Protein Synthesis (Part E)
Edited by LAWRENCE GROSSMAN AND KIVIE MOLDAVE

VOLUME XXX. Nucleic Acids and Protein Synthesis (Part F)
Edited by KIVIE MOLDAVE AND LAWRENCE GROSSMAN

VOLUME XXXI. Biomembranes (Part A)
Edited by SIDNEY FLEISCHER AND LESTER PACKER

Volume XLVIII. Enzyme Structure (Part F)
Edited by C. H. W. Hirs and Serge N. Timasheff

Volume XLIX. Enzyme Structure (Part G)
Edited by C. H. W. Hirs and Serge N. Timasheff

Volume L. Complex Carbohydrates (Part C)
Edited by Victor Ginsburg

Volume LI. Purine and Pyrimidine Nucleotide Metabolism
Edited by Patricia A. Hoffee and Mary Ellen Jones

Volume LII. Biomembranes (Part C: Biological Oxidations)
Edited by Sidney Fleischer and Lester Packer

Volume LIII. Biomembranes (Part D: Biological Oxidations)
Edited by Sidney Fleischer and Lester Packer

Volume LIV. Biomembranes (Part E: Biological Oxidations)
Edited by Sidney Fleischer and Lester Packer

Volume LV. Biomembranes (Part F: Bioenergetics)
Edited by Sidney Fleischer and Lester Packer

Volume LVI. Biomembranes (Part G: Bioenergetics)
Edited by Sidney Fleischer and Lester Packer

Volume LVII. Bioluminescence and Chemiluminescence
Edited by Marlene A. DeLuca

Volume LVIII. Cell Culture
Edited by William B. Jakoby and Ira Pastan

Volume LIX. Nucleic Acids and Protein Synthesis (Part G)
Edited by Kivie Moldave and Lawrence Grossman

Volume LX. Nucleic Acids and Protein Synthesis (Part H)
Edited by Kivie Moldave and Lawrence Grossman

Volume 61. Enzyme Structure (Part H)
Edited by C. H. W. Hirs and Serge N. Timasheff

Volume 62. Vitamins and Coenzymes (Part D)
Edited by Donald B. McCormick and Lemuel D. Wright

Volume 63. Enzyme Kinetics and Mechanism (Part A: Initial Rate and Inhibitor Methods)
Edited by Daniel L. Purich

Volume 79. Interferons (Part B)
Edited by Sidney Pestka

Volume 80. Proteolytic Enzymes (Part C)
Edited by Laszlo Lorand

Volume 81. Biomembranes (Part H: Visual Pigments and Purple Membranes, I)
Edited by Lester Packer

Volume 82. Structural and Contractile Proteins (Part A: Extracellular Matrix)
Edited by Leon W. Cunningham and Dixie W. Frederiksen

Volume 83. Complex Carbohydrates (Part D)
Edited by Victor Ginsburg

Volume 84. Immunochemical Techniques (Part D: Selected Immunoassays)
Edited by John J. Langone and Helen Van Vunakis

Volume 85. Structural and Contractile Proteins (Part B: The Contractile Apparatus and the Cytoskeleton)
Edited by Dixie W. Frederiksen and Leon W. Cunningham

Volume 86. Prostaglandins and Arachidonate Metabolites
Edited by William E. M. Lands and William L. Smith

Volume 87. Enzyme Kinetics and Mechanism (Part C: Intermediates, Stereo-chemistry, and Rate Studies)
Edited by Daniel L. Purich

Volume 88. Biomembranes (Part I: Visual Pigments and Purple Membranes, II)
Edited by Lester Packer

Volume 89. Carbohydrate Metabolism (Part D)
Edited by Willis A. Wood

Volume 90. Carbohydrate Metabolism (Part E)
Edited by Willis A. Wood

Volume 91. Enzyme Structure (Part I)
Edited by C. H. W. Hirs and Serge N. Timasheff

Volume 92. Immunochemical Techniques (Part E: Monoclonal Antibodies and General Immunoassay Methods)
Edited by John J. Langone and Helen Van Vunakis

VOLUME 93. Immunochemical Techniques (Part F: Conventional Antibodies, Fc Receptors, and Cytotoxicity)
Edited by JOHN J. LANGONE AND HELEN VAN VUNAKIS

VOLUME 94. Polyamines
Edited by HERBERT TABOR AND CELIA WHITE TABOR

VOLUME 95. Cumulative Subject Index Volumes 61–74, 76–80
Edited by EDWARD A. DENNIS AND MARTHA G. DENNIS

VOLUME 96. Biomembranes [Part J: Membrane Biogenesis: Assembly and Targeting (General Methods; Eukaryotes)]
Edited by SIDNEY FLEISCHER AND BECCA FLEISCHER

VOLUME 97. Biomembranes [Part K: Membrane Biogenesis: Assembly and Targeting (Prokaryotes, Mitochondria, and Chloroplasts)]
Edited by SIDNEY FLEISCHER AND BECCA FLEISCHER

VOLUME 98. Biomembranes (Part L: Membrane Biogenesis: Processing and Recycling)
Edited by SIDNEY FLEISCHER AND BECCA FLEISCHER

VOLUME 99. Hormone Action (Part F: Protein Kinases)
Edited by JACKIE D. CORBIN AND JOEL G. HARDMAN

VOLUME 100. Recombinant DNA (Part B)
Edited by RAY WU, LAWRENCE GROSSMAN, AND KIVIE MOLDAVE

VOLUME 101. Recombinant DNA (Part C)
Edited by RAY WU, LAWRENCE GROSSMAN, AND KIVIE MOLDAVE

VOLUME 102. Hormone Action (Part G: Calmodulin and Calcium-Binding Proteins)
Edited by ANTHONY R. MEANS AND BERT W. O'MALLEY

VOLUME 103. Hormone Action (Part H: Neuroendocrine Peptides)
Edited by P. MICHAEL CONN

VOLUME 104. Enzyme Purification and Related Techniques (Part C)
Edited by WILLIAM B. JAKOBY

VOLUME 105. Oxygen Radicals in Biological Systems
Edited by LESTER PACKER

VOLUME 106. Posttranslational Modifications (Part A)
Edited by FINN WOLD AND KIVIE MOLDAVE

Volume 107. Posttranslational Modifications (Part B)
Edited by Finn Wold and Kivie Moldave

Volume 108. Immunochemical Techniques (Part G: Separation and Characterization of Lymphoid Cells)
Edited by Giovanni Di Sabato, John J. Langone, and Helen Van Vunakis

Volume 109. Hormone Action (Part I: Peptide Hormones)
Edited by Lutz Birnbaumer and Bert W. O'Malley

Volume 110. Steroids and Isoprenoids (Part A)
Edited by John H. Law and Hans C. Rilling

Volume 111. Steroids and Isoprenoids (Part B)
Edited by John H. Law and Hans C. Rilling

Volume 112. Drug and Enzyme Targeting (Part A)
Edited by Kenneth J. Widder and Ralph Green

Volume 113. Glutamate, Glutamine, Glutathione, and Related Compounds
Edited by Alton Meister

Volume 114. Diffraction Methods for Biological Macromolecules (Part A)
Edited by Harold W. Wyckoff, C. H. W. Hirs, and Serge N. Timasheff

Volume 115. Diffraction Methods for Biological Macromolecules (Part B)
Edited by Harold W. Wyckoff, C. H. W. Hirs, and Serge N. Timasheff

Volume 116. Immunochemical Techniques (Part H: Effectors and Mediators of Lymphoid Cell Functions)
Edited by Giovanni Di Sabato, John J. Langone, and Helen Van Vunakis

Volume 117. Enzyme Structure (Part J)
Edited by C. H. W. Hirs and Serge N. Timasheff

Volume 118. Plant Molecular Biology
Edited by Arthur Weissbach and Herbert Weissbach

Volume 119. Interferons (Part C)
Edited by Sidney Pestka

Volume 120. Cumulative Subject Index Volumes 81–94, 96–101

Volume 121. Immunochemical Techniques (Part I: Hybridoma Technology and Monoclonal Antibodies)
Edited by John J. Langone and Helen Van Vunakis

VOLUME 122. Vitamins and Coenzymes (Part G)
Edited by FRANK CHYTIL AND DONALD B. MCCORMICK

VOLUME 123. Vitamins and Coenzymes (Part H)
Edited by FRANK CHYTIL AND DONALD B. MCCORMICK

VOLUME 124. Hormone Action (Part J: Neuroendocrine Peptides)
Edited by P. MICHAEL CONN

VOLUME 125. Biomembranes (Part M: Transport in Bacteria, Mitochondria, and Chloroplasts: General Approaches and Transport Systems)
Edited by SIDNEY FLEISCHER AND BECCA FLEISCHER

VOLUME 126. Biomembranes (Part N: Transport in Bacteria, Mitochondria, and Chloroplasts: Protonmotive Force)
Edited by SIDNEY FLEISCHER AND BECCA FLEISCHER

VOLUME 127. Biomembranes (Part O: Protons and Water: Structure and Translocation)
Edited by LESTER PACKER

VOLUME 128. Plasma Lipoproteins (Part A: Preparation, Structure, and Molecular Biology)
Edited by JERE P. SEGREST AND JOHN J. ALBERS

VOLUME 129. Plasma Lipoproteins (Part B: Characterization, Cell Biology, and Metabolism)
Edited by JOHN J. ALBERS AND JERE P. SEGREST

VOLUME 130. Enzyme Structure (Part K)
Edited by C. H. W. HIRS AND SERGE N. TIMASHEFF

VOLUME 131. Enzyme Structure (Part L)
Edited by C. H. W. HIRS AND SERGE N. TIMASHEFF

VOLUME 132. Immunochemical Techniques (Part J: Phagocytosis and Cell-Mediated Cytotoxicity)
Edited by GIOVANNI DI SABATO AND JOHANNES EVERSE

VOLUME 133. Bioluminescence and Chemiluminescence (Part B)
Edited by MARLENE DELUCA AND WILLIAM D. MCELROY

VOLUME 134. Structural and Contractile Proteins (Part C: The Contractile Apparatus and the Cytoskeleton)
Edited by RICHARD B. VALLEE

VOLUME 135. Immobilized Enzymes and Cells (Part B)
Edited by KLAUS MOSBACH

VOLUME 136. Immobilized Enzymes and Cells (Part C)
Edited by KLAUS MOSBACH

VOLUME 137. Immobilized Enzymes and Cells (Part D)
Edited by KLAUS MOSBACH

VOLUME 138. Complex Carbohydrates (Part E)
Edited by VICTOR GINSBURG

VOLUME 139. Cellular Regulators (Part A: Calcium- and Calmodulin-Binding Proteins)
Edited by ANTHONY R. MEANS AND P. MICHAEL CONN

VOLUME 140. Cumulative Subject Index Volumes 102–119, 121–134

VOLUME 141. Cellular Regulators (Part B: Calcium and Lipids)
Edited by P. MICHAEL CONN AND ANTHONY R. MEANS

VOLUME 142. Metabolism of Aromatic Amino Acids and Amines
Edited by SEYMOUR KAUFMAN

VOLUME 143. Sulfur and Sulfur Amino Acids
Edited by WILLIAM B. JAKOBY AND OWEN GRIFFITH

VOLUME 144. Structural and Contractile Proteins (Part D: Extracellular Matrix)
Edited by LEON W. CUNNINGHAM

VOLUME 145. Structural and Contractile Proteins (Part E: Extracellular Matrix)
Edited by LEON W. CUNNINGHAM

VOLUME 146. Peptide Growth Factors (Part A)
Edited by DAVID BARNES AND DAVID A. SIRBASKU

VOLUME 147. Peptide Growth Factors (Part B)
Edited by DAVID BARNES AND DAVID A. SIRBASKU

VOLUME 148. Plant Cell Membranes
Edited by LESTER PACKER AND ROLAND DOUCE

VOLUME 149. Drug and Enzyme Targeting (Part B)
Edited by RALPH GREEN AND KENNETH J. WIDDER

VOLUME 150. Immunochemical Techniques (Part K: *In Vitro* Models of B and T Cell Functions and Lymphoid Cell Receptors)
Edited by GIOVANNI DI SABATO

VOLUME 166. Branched-Chain Amino Acids
Edited by ROBERT HARRIS AND JOHN R. SOKATCH

VOLUME 167. Cyanobacteria
Edited by LESTER PACKER AND ALEXANDER N. GLAZER

VOLUME 168. Hormone Action (Part K: Neuroendocrine Peptides)
Edited by P. MICHAEL CONN

VOLUME 169. Platelets: Receptors, Adhesion, Secretion (Part A)
Edited by JACEK HAWIGER

VOLUME 170. Nucleosomes
Edited by PAUL M. WASSARMAN AND ROGER D. KORNBERG

VOLUME 171. Biomembranes (Part R: Transport Theory: Cells and Model Membranes)
Edited by SIDNEY FLEISCHER AND BECCA FLEISCHER

VOLUME 172. Biomembranes (Part S: Transport: Membrane Isolation and Characterization)
Edited by SIDNEY FLEISCHER AND BECCA FLEISCHER

VOLUME 173. Biomembranes [Part T: Cellular and Subcellular Transport: Eukaryotic (Nonepithelial) Cells]
Edited by SIDNEY FLEISCHER AND BECCA FLEISCHER

VOLUME 174. Biomembranes [Part U: Cellular and Subcellular Transport: Eukaryotic (Nonepithelial) Cells]
Edited by SIDNEY FLEISCHER AND BECCA FLEISCHER

VOLUME 175. Cumulative Subject Index Volumes 135–139, 141–167

VOLUME 176. Nuclear Magnetic Resonance (Part A: Spectral Techniques and Dynamics)
Edited by NORMAN J. OPPENHEIMER AND THOMAS L. JAMES

VOLUME 177. Nuclear Magnetic Resonance (Part B: Structure and Mechanism)
Edited by NORMAN J. OPPENHEIMER AND THOMAS L. JAMES

VOLUME 178. Antibodies, Antigens, and Molecular Mimicry
Edited by JOHN J. LANGONE

VOLUME 179. Complex Carbohydrates (Part F)
Edited by VICTOR GINSBURG

Volume 180. RNA Processing (Part A: General Methods)
Edited by James E. Dahlberg and John N. Abelson

Volume 181. RNA Processing (Part B: Specific Methods)
Edited by James E. Dahlberg and John N. Abelson

Volume 182. Guide to Protein Purification
Edited by Murray P. Deutscher

Volume 183. Molecular Evolution: Computer Analysis of Protein and Nucleic Acid Sequences
Edited by Russell F. Doolittle

Volume 184. Avidin-Biotin Technology
Edited by Meir Wilchek and Edward A. Bayer

Volume 185. Gene Expression Technology
Edited by David V. Goeddel

Volume 186. Oxygen Radicals in Biological Systems (Part B: Oxygen Radicals and Antioxidants)
Edited by Lester Packer and Alexander N. Glazer

Volume 187. Arachidonate Related Lipid Mediators
Edited by Robert C. Murphy and Frank A. Fitzpatrick

Volume 188. Hydrocarbons and Methylotrophy
Edited by Mary E. Lidstrom

Volume 189. Retinoids (Part A: Molecular and Metabolic Aspects)
Edited by Lester Packer

Volume 190. Retinoids (Part B: Cell Differentiation and Clinical Applications)
Edited by Lester Packer

Volume 191. Biomembranes (Part V: Cellular and Subcellular Transport: Epithelial Cells)
Edited by Sidney Fleischer and Becca Fleischer

Volume 192. Biomembranes (Part W: Cellular and Subcellular Transport: Epithelial Cells)
Edited by Sidney Fleischer and Becca Fleischer

Volume 193. Mass Spectrometry
Edited by James A. McCloskey

Volume 194. Guide to Yeast Genetics and Molecular Biology
Edited by Christine Guthrie and Gerald R. Fink

Volume 195. Adenylyl Cyclase, G Proteins, and Guanylyl Cyclase
Edited by Roger A. Johnson and Jackie D. Corbin

Volume 196. Molecular Motors and the Cytoskeleton
Edited by Richard B. Vallee

Volume 197. Phospholipases
Edited by Edward A. Dennis

Volume 198. Peptide Growth Factors (Part C)
Edited by David Barnes, J. P. Mather, and Gordon H. Sato

Volume 199. Cumulative Subject Index Volumes 168–174, 176–194

Volume 200. Protein Phosphorylation (Part A: Protein Kinases: Assays, Purification, Antibodies, Functional Analysis, Cloning, and Expression)
Edited by Tony Hunter and Bartholomew M. Sefton

Volume 201. Protein Phosphorylation (Part B: Analysis of Protein Phosphorylation, Protein Kinase Inhibitors, and Protein Phosphatases)
Edited by Tony Hunter and Bartholomew M. Sefton

Volume 202. Molecular Design and Modeling: Concepts and Applications (Part A: Proteins, Peptides, and Enzymes)
Edited by John J. Langone

Volume 203. Molecular Design and Modeling: Concepts and Applications (Part B: Antibodies and Antigens, Nucleic Acids, Polysaccharides, and Drugs)
Edited by John J. Langone

Volume 204. Bacterial Genetic Systems
Edited by Jeffrey H. Miller

Volume 205. Metallobiochemistry (Part B: Metallothionein and Related Molecules)
Edited by James F. Riordan and Bert L. Vallee

Volume 206. Cytochrome P450
Edited by Michael R. Waterman and Eric F. Johnson

Volume 207. Ion Channels
Edited by Bernardo Rudy and Linda E. Iverson

Factors and Inhibitors)
Edited by Laszlo Lorand and Kenneth G. Mann

Volume 223. Proteolytic Enzymes in Coagulation, Fibrinolysis, and Complement Activation (Part B: Complement Activation, Fibrinolysis, and Nonmammalian Blood Coagulation Factors)
Edited by Laszlo Lorand and Kenneth G. Mann

Volume 224. Molecular Evolution: Producing the Biochemical Data
Edited by Elizabeth Anne Zimmer, Thomas J. White, Rebecca L. Cann, and Allan C. Wilson

Volume 225. Guide to Techniques in Mouse Development
Edited by Paul M. Wassarman and Melvin L. DePamphilis

Volume 226. Metallobiochemistry (Part C: Spectroscopic and Physical Methods for Probing Metal Ion Environments in Metalloenzymes and Metalloproteins)
Edited by James F. Riordan and Bert L. Vallee

Volume 227. Metallobiochemistry (Part D: Physical and Spectroscopic Methods for Probing Metal Ion Environments in Metalloproteins)
Edited by James F. Riordan and Bert L. Vallee

Volume 228. Aqueous Two-Phase Systems
Edited by Harry Walter and Göte Johansson

Volume 229. Cumulative Subject Index Volumes 195–198, 200–227

Volume 230. Guide to Techniques in Glycobiology
Edited by William J. Lennarz and Gerald W. Hart

Volume 231. Hemoglobins (Part B: Biochemical and Analytical Methods)
Edited by Johannes Everse, Kim D. Vandegriff, and Robert M. Winslow

Volume 232. Hemoglobins (Part C: Biophysical Methods)
Edited by Johannes Everse, Kim D. Vandegriff, and Robert M. Winslow

Volume 233. Oxygen Radicals in Biological Systems (Part C)
Edited by Lester Packer

Volume 234. Oxygen Radicals in Biological Systems (Part D)
Edited by Lester Packer

Volume 235. Bacterial Pathogenesis (Part A: Identification and Regulation of Virulence Factors)
Edited by Virginia L. Clark and Patrik M. Bavoil

Volume 236. Bacterial Pathogenesis (Part B: Integration of Pathogenic Bacteria with Host Cells)
Edited by Virginia L. Clark and Patrik M. Bavoil

Volume 237. Heterotrimeric G Proteins
Edited by Ravi Iyengar

Volume 238. Heterotrimeric G-Protein Effectors
Edited by Ravi Iyengar

Volume 239. Nuclear Magnetic Resonance (Part C)
Edited by Thomas L. James and Norman J. Oppenheimer

Volume 240. Numerical Computer Methods (Part B)
Edited by Michael L. Johnson and Ludwig Brand

Volume 241. Retroviral Proteases
Edited by Lawrence C. Kuo and Jules A. Shafer

Volume 242. Neoglycoconjugates (Part A)
Edited by Y. C. Lee and Reiko T. Lee

Volume 243. Inorganic Microbial Sulfur Metabolism
Edited by Harry D. Peck, Jr., and Jean LeGall

Volume 244. Proteolytic Enzymes: Serine and Cysteine Peptidases
Edited by Alan J. Barrett

Volume 245. Extracellular Matrix Components
Edited by E. Ruoslahti and E. Engvall

Volume 246. Biochemical Spectroscopy
Edited by Kenneth Sauer

Volume 247. Neoglycoconjugates (Part B: Biomedical Applications)
Edited by Y. C. Lee and Reiko T. Lee

Volume 248. Proteolytic Enzymes: Aspartic and Metallo Peptidases
Edited by Alan J. Barrett

Volume 249. Enzyme Kinetics and Mechanism (Part D: Developments in Enzyme Dynamics)
Edited by Daniel L. Purich

Volume 250. Lipid Modifications of Proteins
Edited by Patrick J. Casey and Janice E. Buss

Volume 281. Vitamins and Coenzymes (Part K)
Edited by Donald B. McCormick, John W. Suttie, and Conrad Wagner

Volume 282. Vitamins and Coenzymes (Part L)
Edited by Donald B. McCormick, John W. Suttie, and Conrad Wagner

Volume 283. Cell Cycle Control
Edited by William G. Dunphy

Volume 284. Lipases (Part A: Biotechnology)
Edited by Byron Rubin and Edward A. Dennis

Volume 285. Cumulative Subject Index Volumes 263, 264, 266–284, 286–289

Volume 286. Lipases (Part B: Enzyme Characterization and Utilization)
Edited by Byron Rubin and Edward A. Dennis

Volume 287. Chemokines
Edited by Richard Horuk

Volume 288. Chemokine Receptors
Edited by Richard Horuk

Volume 289. Solid Phase Peptide Synthesis
Edited by Gregg B. Fields

Volume 290. Molecular Chaperones
Edited by George H. Lorimer and Thomas Baldwin

Volume 291. Caged Compounds
Edited by Gerard Marriott

Volume 292. ABC Transporters: Biochemical, Cellular, and Molecular Aspects
Edited by Suresh V. Ambudkar and Michael M. Gottesman

Volume 293. Ion Channels (Part B)
Edited by P. Michael Conn

Volume 294. Ion Channels (Part C)
Edited by P. Michael Conn

Volume 295. Energetics of Biological Macromolecules (Part B)
Edited by Gary K. Ackers and Michael L. Johnson

Volume 296. Neurotransmitter Transporters
Edited by Susan G. Amara

Volume 313. Antisense Technology
(Part A: General Methods, Methods of Delivery, and RNA Studies)
Edited by M. Ian Phillips

Volume 314. Antisense Technology (Part B: Applications)
Edited by M. Ian Phillips

Volume 315. Vertebrate Phototransduction and the Visual Cycle
(Part A)
Edited by Krzysztof Palczewski

Volume 316. Vertebrate Phototransduction and the Visual Cycle
(Part B)
Edited by Krzysztof Palczewski

Volume 317. RNA–Ligand Interactions (Part A: Structural Biology
Methods)
Edited by Daniel W. Celander and John N. Abelson

Volume 318. RNA–Ligand Interactions (Part B: Molecular Biology
Methods)
Edited by Daniel W. Celander and John N. Abelson

Volume 319. Singlet Oxygen, UV-A, and Ozone
Edited by Lester Packer and Helmut Sies

Volume 320. Cumulative Subject Index Volumes 290–319

Volume 321. Numerical Computer Methods (Part C)
Edited by Michael L. Johnson and Ludwig Brand

Volume 322. Apoptosis
Edited by John C. Reed

Volume 323. Energetics of Biological Macromolecules (Part C)
Edited by Michael L. Johnson and Gary K. Ackers

Volume 324. Branched-Chain Amino Acids (Part B)
Edited by Robert A. Harris and John R. Sokatch

Volume 325. Regulators and Effectors of Small GTPases
(Part D: Rho Family)
Edited by W. E. Balch, Channing J. Der, and Alan Hall

Volume 326. Applications of Chimeric Genes and Hybrid Proteins
(Part A: Gene Expression and Protein Purification)
Edited by Jeremy Thorner, Scott D. Emr, and John N. Abelson

VOLUME 327. Applications of Chimeric Genes and Hybrid Proteins (Part B: Cell Biology and Physiology)
Edited by JEREMY THORNER, SCOTT D. EMR, AND JOHN N. ABELSON

VOLUME 328. Applications of Chimeric Genes and Hybrid Proteins (Part C: Protein–Protein Interactions and Genomics)
Edited by JEREMY THORNER, SCOTT D. EMR, AND JOHN N. ABELSON

VOLUME 329. Regulators and Effectors of Small GTPases (Part E: GTPases Involved in Vesicular Traffic)
Edited by W. E. BALCH, CHANNING J. DER, AND ALAN HALL

VOLUME 330. Hyperthermophilic Enzymes (Part A)
Edited by MICHAEL W. W. ADAMS AND ROBERT M. KELLY

VOLUME 331. Hyperthermophilic Enzymes (Part B)
Edited by MICHAEL W. W. ADAMS AND ROBERT M. KELLY

VOLUME 332. Regulators and Effectors of Small GTPases (Part F: Ras Family I)
Edited by W. E. BALCH, CHANNING J. DER, AND ALAN HALL

VOLUME 333. Regulators and Effectors of Small GTPases (Part G: Ras Family II)
Edited by W. E. BALCH, CHANNING J. DER, AND ALAN HALL

VOLUME 334. Hyperthermophilic Enzymes (Part C)
Edited by MICHAEL W. W. ADAMS AND ROBERT M. KELLY

VOLUME 335. Flavonoids and Other Polyphenols
Edited by LESTER PACKER

VOLUME 336. Microbial Growth in Biofilms (Part A: Developmental and Molecular Biological Aspects)
Edited by RON J. DOYLE

VOLUME 337. Microbial Growth in Biofilms (Part B: Special Environments and Physicochemical Aspects)
Edited by RON J. DOYLE

VOLUME 338. Nuclear Magnetic Resonance of Biological Macromolecules (Part A)
Edited by THOMAS L. JAMES, VOLKER DÖTSCH, AND ULI SCHMITZ

VOLUME 339. Nuclear Magnetic Resonance of Biological Macromolecules (Part B)
Edited by THOMAS L. JAMES, VOLKER DÖTSCH, AND ULI SCHMITZ

VOLUME 340. Drug–Nucleic Acid Interactions
Edited by JONATHAN B. CHAIRES AND MICHAEL J. WARING

VOLUME 341. Ribonucleases (Part A)
Edited by ALLEN W. NICHOLSON

VOLUME 342. Ribonucleases (Part B)
Edited by ALLEN W. NICHOLSON

VOLUME 343. G Protein Pathways (Part A: Receptors)
Edited by RAVI IYENGAR AND JOHN D. HILDEBRANDT

VOLUME 344. G Protein Pathways (Part B: G Proteins and Their Regulators)
Edited by RAVI IYENGAR AND JOHN D. HILDEBRANDT

VOLUME 345. G Protein Pathways (Part C: Effector Mechanisms)
Edited by RAVI IYENGAR AND JOHN D. HILDEBRANDT

VOLUME 346. Gene Therapy Methods
Edited by M. IAN PHILLIPS

VOLUME 347. Protein Sensors and Reactive Oxygen Species (Part A: Selenoproteins and Thioredoxin)
Edited by HELMUT SIES AND LESTER PACKER

VOLUME 348. Protein Sensors and Reactive Oxygen Species (Part B: Thiol Enzymes and Proteins)
Edited by HELMUT SIES AND LESTER PACKER

VOLUME 349. Superoxide Dismutase
Edited by LESTER PACKER

VOLUME 350. Guide to Yeast Genetics and Molecular and Cell Biology (Part B)
Edited by CHRISTINE GUTHRIE AND GERALD R. FINK

VOLUME 351. Guide to Yeast Genetics and Molecular and Cell Biology (Part C)
Edited by CHRISTINE GUTHRIE AND GERALD R. FINK

VOLUME 352. Redox Cell Biology and Genetics (Part A)
Edited by CHANDAN K. SEN AND LESTER PACKER

VOLUME 353. Redox Cell Biology and Genetics (Part B)
Edited by CHANDAN K. SEN AND LESTER PACKER

Volume 354. Enzyme Kinetics and Mechanisms (Part F: Detection and Characterization of Enzyme Reaction Intermediates)
Edited by Daniel L. Purich

Volume 355. Cumulative Subject Index Volumes 321–354

Volume 356. Laser Capture Microscopy and Microdissection
Edited by P. Michael Conn

Volume 357. Cytochrome P450, Part C
Edited by Eric F. Johnson and Michael R. Waterman

Volume 358. Bacterial Pathogenesis (Part C: Identification, Regulation, and Function of Virulence Factors)
Edited by Virginia L. Clark and Patrik M. Bavoil

Volume 359. Nitric Oxide (Part D)
Edited by Enrique Cadenas and Lester Packer

Volume 360. Biophotonics (Part A)
Edited by Gerard Marriott and Ian Parker

Volume 361. Biophotonics (Part B)
Edited by Gerard Marriott and Ian Parker

Volume 362. Recognition of Carbohydrates in Biological Systems (Part A)
Edited by Yuan C. Lee and Reiko T. Lee

Volume 363. Recognition of Carbohydrates in Biological Systems (Part B)
Edited by Yuan C. Lee and Reiko T. Lee

Volume 364. Nuclear Receptors
Edited by David W. Russell and David J. Mangelsdorf

Volume 365. Differentiation of Embryonic Stem Cells
Edited by Paul M. Wassauman and Gordon M. Keller

Volume 366. Protein Phosphatases
Edited by Susanne Klumpp and Josef Krieglstein

Volume 367. Liposomes (Part A)
Edited by Nejat Düzgüneş

Volume 368. Macromolecular Crystallography (Part C)
Edited by Charles W. Carter, Jr., and Robert M. Sweet

Volume 369. Combinational Chemistry (Part B)
Edited by Guillermo A. Morales and Barry A. Bunin

Volume 370. RNA Polymerases and Associated Factors (Part C)
Edited by Sankar L. Adhya and Susan Garges

Volume 371. RNA Polymerases and Associated Factors (Part D)
Edited by Sankar L. Adhya and Susan Garges

Volume 372. Liposomes (Part B)
Edited by Nejat Düzgüneş

Volume 373. Liposomes (Part C)
Edited by Nejat Düzgüneş

Volume 374. Macromolecular Crystallography (Part D)
Edited by Charles W. Carter, Jr., and Robert W. Sweet

Volume 375. Chromatin and Chromatin Remodeling Enzymes (Part A)
Edited by C. David Allis and Carl Wu

Volume 376. Chromatin and Chromatin Remodeling Enzymes (Part B)
Edited by C. David Allis and Carl Wu

Volume 377. Chromatin and Chromatin Remodeling Enzymes (Part C)
Edited by C. David Allis and Carl Wu

Volume 378. Quinones and Quinone Enzymes (Part A)
Edited by Helmut Sies and Lester Packer

Volume 379. Energetics of Biological Macromolecules (Part D)
Edited by Jo M. Holt, Michael L. Johnson, and Gary K. Ackers

Volume 380. Energetics of Biological Macromolecules (Part E)
Edited by Jo M. Holt, Michael L. Johnson, and Gary K. Ackers

Volume 381. Oxygen Sensing
Edited by Chandan K. Sen and Gregg L. Semenza

Volume 382. Quinones and Quinone Enzymes (Part B)
Edited by Helmut Sies and Lester Packer

Volume 383. Numerical Computer Methods (Part D)
Edited by Ludwig Brand and Michael L. Johnson

Volume 384. Numerical Computer Methods (Part E)
Edited by Ludwig Brand and Michael L. Johnson

Volume 385. Imaging in Biological Research (Part A)
Edited by P. Michael Conn

Volume 386. Imaging in Biological Research (Part B)
Edited by P. Michael Conn

VOLUME 403. GTPases Regulating Membrane Targeting and Fusion
Edited by WILLIAM E. BALCH, CHANNING J. DER, AND ALAN HALL

VOLUME 404. GTPases Regulating Membrane Dynamics
Edited by WILLIAM E. BALCH, CHANNING J. DER, AND ALAN HALL

VOLUME 405. Mass Spectrometry: Modified Proteins and Glycoconjugates
Edited by A. L. BURLINGAME

VOLUME 406. Regulators and Effectors of Small GTPases: Rho Family
Edited by WILLIAM E. BALCH, CHANNING J. DER, AND ALAN HALL

VOLUME 407. Regulators and Effectors of Small GTPases: Ras Family
Edited by WILLIAM E. BALCH, CHANNING J. DER, AND ALAN HALL

VOLUME 408. DNA Repair (Part A)
Edited by JUDITH L. CAMPBELL AND PAUL MODRICH

VOLUME 409. DNA Repair (Part B)
Edited by JUDITH L. CAMPBELL AND PAUL MODRICH

VOLUME 410. DNA Microarrays (Part A: Array Platforms and Web-Bench Protocols)
Edited by ALAN KIMMEL AND BRIAN OLIVER

VOLUME 411. DNA Microarrays (Part B: Databases and Statistics)
Edited by ALAN KIMMEL AND BRIAN OLIVER

VOLUME 412. Amyloid, Prions, and Other Protein Aggregates (Part B)
Edited by INDU KHETERPAL AND RONALD WETZEL

VOLUME 413. Amyloid, Prions, and Other Protein Aggregates (Part C)
Edited by INDU KHETERPAL AND RONALD WETZEL

VOLUME 414. Measuring Biological Responses with Automated Microscopy
Edited by JAMES INGLESE

VOLUME 415. Glycobiology
Edited by MINORU FUKUDA

VOLUME 416. Glycomics
Edited by MINORU FUKUDA

VOLUME 417. Functional Glycomics
Edited by MINORU FUKUDA

VOLUME 418. Embryonic Stem Cells
Edited by IRINA KLIMANSKAYA AND ROBERT LANZA

Volume 419. Adult Stem Cells
Edited by Irina Klimanskaya and Robert Lanza

Volume 420. Stem Cell Tools and Other Experimental Protocols
Edited by Irina Klimanskaya and Robert Lanza

Volume 421. Advanced Bacterial Genetics: Use of Transposons and Phage for Genomic Engineering
Edited by Kelly T. Hughes

Volume 422. Two-Component Signaling Systems, Part A
Edited by Melvin I. Simon, Brian R. Crane, and Alexandrine Crane

Volume 423. Two-Component Signaling Systems, Part B
Edited by Melvin I. Simon, Brian R. Crane, and Alexandrine Crane

Volume 424. RNA Editing
Edited by Jonatha M. Gott

Volume 425. RNA Modification
Edited by Jonatha M. Gott

Volume 426. Integrins
Edited by David Cheresh

Volume 427. MicroRNA Methods
Edited by John J. Rossi

Volume 428. Osmosensing and Osmosignaling
Edited by Helmut Sies and Dieter Haussinger

Volume 429. Translation Initiation: Extract Systems and Molecular Genetics
Edited by Jon Lorsch

Volume 430. Translation Initiation: Reconstituted Systems and Biophysical Methods
Edited by Jon Lorsch

Volume 431. Translation Initiation: Cell Biology, High-Throughput and Chemical-Based Approaches
Edited by Jon Lorsch

Volume 432. Lipidomics and Bioactive Lipids: Mass-Spectrometry–Based Lipid Analysis
Edited by H. Alex Brown

VOLUME 433. Lipidomics and Bioactive Lipids: Specialized Analytical Methods and Lipids in Disease
Edited by H. ALEX BROWN

VOLUME 434. Lipidomics and Bioactive Lipids: Lipids and Cell Signaling
Edited by H. ALEX BROWN

VOLUME 435. Oxygen Biology and Hypoxia
Edited by HELMUT SIES AND BERNHARD BRÜNE

VOLUME 436. Globins and Other Nitric Oxide-Reactive Protiens (Part A)
Edited by ROBERT K. POOLE

VOLUME 437. Globins and Other Nitric Oxide-Reactive Protiens (Part B)
Edited by ROBERT K. POOLE

VOLUME 438. Small GTPases in Disease (Part A)
Edited by WILLIAM E. BALCH, CHANNING J. DER, AND ALAN HALL

VOLUME 439. Small GTPases in Disease (Part B)
Edited by WILLIAM E. BALCH, CHANNING J. DER, AND ALAN HALL

VOLUME 440. Nitric Oxide, Part F Oxidative and Nitrosative Stress in Redox Regulation of Cell Signaling
Edited by ENRIQUE CADENAS AND LESTER PACKER

VOLUME 441. Nitric Oxide, Part G Oxidative and Nitrosative Stress in Redox Regulation of Cell Signaling
Edited by ENRIQUE CADENAS AND LESTER PACKER

VOLUME 442. Programmed Cell Death, General Principles for Studying Cell Death (Part A)
Edited by ROYA KHOSRAVI-FAR, ZAHRA ZAKERI, RICHARD A. LOCKSHIN, AND MAURO PIACENTINI

VOLUME 443. Angiogenesis: *In Vitro* Systems
Edited by DAVID A. CHERESH

VOLUME 444. Angiogenesis: *In Vivo* Systems (Part A)
Edited by DAVID A. CHERESH

VOLUME 445. Angiogenesis: *In Vivo* Systems (Part B)
Edited by DAVID A. CHERESH

VOLUME 446. Programmed Cell Death, The Biology and Therapeutic Implications of Cell Death (Part B)
Edited by ROYA KHOSRAVI-FAR, ZAHRA ZAKERI, RICHARD A. LOCKSHIN, AND MAURO PIACENTINI

VOLUME 447. RNA Turnover in Bacteria, Archaea and Organelles
Edited by LYNNE E. MAQUAT AND CECILIA M. ARRAIANO

VOLUME 448. RNA Turnover in Eukaryotes: Nucleases, Pathways and Analysis of mRNA Decay
Edited by LYNNE E. MAQUAT AND MEGERDITCH KILEDJIAN

VOLUME 449. RNA Turnover in Eukaryotes: Analysis of Specialized and Quality Control RNA Decay Pathways
Edited by LYNNE E. MAQUAT AND MEGERDITCH KILEDJIAN

VOLUME 450. Fluorescence Spectroscopy
Edited by LUDWIG BRAND AND MICHAEL L. JOHNSON

VOLUME 451. Autophagy: Lower Eukaryotes and Non-Mammalian Systems (Part A)
Edited by DANIEL J. KLIONSKY

VOLUME 452. Autophagy in Mammalian Systems (Part B)
Edited by DANIEL J. KLIONSKY

VOLUME 453. Autophagy in Disease and Clinical Applications (Part C)
Edited by DANIEL J. KLIONSKY

VOLUME 454. Computer Methods (Part A)
Edited by MICHAEL L. JOHNSON AND LUDWIG BRAND

VOLUME 455. Biothermodynamics (Part A)
Edited by MICHAEL L. JOHNSON, JO M. HOLT, AND GARY K. ACKERS (RETIRED)

VOLUME 456. Mitochondrial Function, Part A: Mitochondrial Electron Transport Complexes and Reactive Oxygen Species
Edited by WILLIAM S. ALLISON AND IMMO E. SCHEFFLER

VOLUME 457. Mitochondrial Function, Part B: Mitochondrial Protein Kinases, Protein Phosphatases and Mitochondrial Diseases
Edited by WILLIAM S. ALLISON AND ANNE N. MURPHY

VOLUME 458. Complex Enzymes in Microbial Natural Product Biosynthesis, Part A: Overview Articles and Peptides
Edited by DAVID A. HOPWOOD

VOLUME 459. Complex Enzymes in Microbial Natural Product Biosynthesis, Part B: Polyketides, Aminocoumarins and Carbohydrates
Edited by DAVID A. HOPWOOD

VOLUME 460. Chemokines, Part A
Edited by TRACY M. HANDEL AND DAMON J. HAMEL

VOLUME 461. Chemokines, Part B
Edited by TRACY M. HANDEL AND DAMON J. HAMEL

VOLUME 462. Non-Natural Amino Acids
Edited by TOM W. MUIR AND JOHN N. ABELSON

VOLUME 463. Guide to Protein Purification, 2nd Edition
Edited by RICHARD R. BURGESS AND MURRAY P. DEUTSCHER

VOLUME 464. Liposomes, Part F
Edited by NEJAT DÜZGÜNEŞ

VOLUME 465. Liposomes, Part G
Edited by NEJAT DÜZGÜNEŞ

VOLUME 466. Biothermodynamics, Part B
Edited by MICHAEL L. JOHNSON, GARY K. ACKERS, AND JO M. HOLT

VOLUME 467. Computer Methods Part B
Edited by MICHAEL L. JOHNSON AND LUDWIG BRAND

VOLUME 468. Biophysical, Chemical, and Functional Probes of RNA Structure, Interactions and Folding: Part A
Edited by DANIEL HERSCHLAG

VOLUME 469. Biophysical, Chemical, and Functional Probes of RNA Structure, Interactions and Folding: Part B
Edited by DANIEL HERSCHLAG

VOLUME 470. Guide to Yeast Genetics: Functional Genomics, Proteomics, and Other Systems Analysis, 2nd Edition
Edited by GERALD FINK, JONATHAN WEISSMAN, AND CHRISTINE GUTHRIE

VOLUME 471. Two-Component Signaling Systems, Part C
Edited by MELVIN I. SIMON, BRIAN R. CRANE, AND ALEXANDRINE CRANE

VOLUME 472. Single Molecule Tools, Part A: Fluorescence Based Approaches
Edited by NILS G. WALTER

VOLUME 473. Thiol Redox Transitions in Cell Signaling, Part A Chemistry and Biochemistry of Low Molecular Weight and Protein Thiols
Edited by ENRIQUE CADENAS AND LESTER PACKER

Volume 474. Thiol Redox Transitions in Cell Signaling, Part B Cellular Localization and Signaling
Edited by Enrique Cadenas and Lester Packer

Volume 475. Single Molecule Tools, Part B: Super-Resolution, Particle Tracking, Multiparameter, and Force Based Methods
Edited by Nils G. Walter

Volume 476. Guide to Techniques in Mouse Development, Part A Mice, Embryos, and Cells, 2nd Edition
Edited by Paul M. Wassarman and Philippe M. Soriano

Volume 477. Guide to Techniques in Mouse Development, Part B Mouse Molecular Genetics, 2nd Edition
Edited by Paul M. Wassarman and Philippe M. Soriano

Volume 478. Glycomics
Edited by Minoru Fukuda

Volume 479. Functional Glycomics
Edited by Minoru Fukuda

Volume 480. Glycobiology
Edited by Minoru Fukuda

Volume 481. Cryo-EM, Part A: Sample Preparation and Data Collection
Edited by Grant J. Jensen

Volume 482. Cryo-EM, Part B: 3-D Reconstruction
Edited by Grant J. Jensen

Volume 483. Cryo-EM, Part C: Analyses, Interpretation, and Case Studies
Edited by Grant J. Jensen

Volume 484. Constitutive Activity in Receptors and Other Proteins, Part A
Edited by P. Michael Conn

Volume 485. Constitutive Activity in Receptors and Other Proteins, Part B
Edited by P. Michael Conn

Volume 486. Research on Nitrification and Related Processes, Part A
Edited by Martin G. Klotz

Volume 487. Computer Methods, Part C
Edited by Michael L. Johnson and Ludwig Brand

Volume 488. Biothermodynamics, Part C
Edited by Michael L. Johnson, Jo M. Holt, and Gary K. Ackers

VOLUME 489. The Unfolded Protein Response and Cellular Stress, Part A
Edited by P. MICHAEL CONN

VOLUME 490. The Unfolded Protein Response and Cellular Stress, Part B
Edited by P. MICHAEL CONN

VOLUME 491. The Unfolded Protein Response and Cellular Stress, Part C
Edited by P. MICHAEL CONN

VOLUME 492. Biothermodynamics, Part D
Edited by MICHAEL L. JOHNSON, JO M. HOLT, AND GARY K. ACKERS

VOLUME 493. Fragment-Based Drug Design Tools,
Practical Approaches, and Examples
Edited by LAWRENCE C. KUO

VOLUME 494. Methods in Methane Metabolism, Part A
Methanogenesis
Edited by AMY C. ROSENZWEIG AND STEPHEN W. RAGSDALE

VOLUME 495. Methods in Methane Metabolism, Part B
Methanotrophy
Edited by AMY C. ROSENZWEIG AND STEPHEN W. RAGSDALE

VOLUME 496. Research on Nitrification and Related Processes, Part B
Edited by MARTIN G. KLOTZ AND LISA Y. STEIN

VOLUME 497. Synthetic Biology, Part A
Methods for Part/Device Characterization and Chassis Engineering
Edited by CHRISTOPHER VOIGT

VOLUME 498. Synthetic Biology, Part B
Computer Aided Design and DNA Assembly
Edited by CHRISTOPHER VOIGT

VOLUME 499. Biology of Serpins
Edited by JAMES C. WHISSTOCK AND PHILLIP I. BIRD

VOLUME 500. Methods in Systems Biology
Edited by DANIEL JAMESON, MALKHEY VERMA, AND HANS V. WESTERHOFF

VOLUME 501. Serpin Structure and Evolution
Edited by JAMES C. WHISSTOCK AND PHILLIP I. BIRD

VOLUME 502. Protein Engineering for Therapeutics, Part A
Edited by K. DANE WITTRUP AND GREGORY L. VERDINE

Volume 503. Protein Engineering for Therapeutics, Part B
Edited by K. Dane Wittrup and Gregory L. Verdine

Volume 504. Imaging and Spectroscopic Analysis of Living Cells
Optical and Spectroscopic Techniques
Edited by P. Michael Conn

Volume 505. Imaging and Spectroscopic Analysis of Living Cells
Live Cell Imaging of Cellular Elements and Functions
Edited by P. Michael Conn

Volume 506. Imaging and Spectroscopic Analysis of Living Cells
Imaging Live Cells in Health and Disease
Edited by P. Michael Conn

Volume 507. Gene Transfer Vectors for Clinical Application
Edited by Theodore Friedmann

Volume 508. Nanomedicine
Cancer, Diabetes, and Cardiovascular, Central Nervous System, Pulmonary and Inflammatory Diseases
Edited by Nejat Düzgüneş

Volume 509. Nanomedicine
Infectious Diseases, Immunotherapy, Diagnostics, Antifibrotics, Toxicology and Gene Medicine
Edited by Nejat Düzgüneş

Volume 510. Cellulases
Edited by Harry J. Gilbert

Volume 511. RNA Helicases
Edited by Eckhard Jankowsky

Volume 512. Nucleosomes, Histones & Chromatin, Part A
Edited by Carl Wu and C. David Allis

Volume 513. Nucleosomes, Histones & Chromatin, Part B
Edited by Carl Wu and C. David Allis

Volume 514. Ghrelin
Edited by Masayasu Kojima and Kenji Kangawa

SECTION 1

History of GHS and Ghrelin

CHAPTER ONE

History to the Discovery of Ghrelin

Cyril Y. Bowers[1]

Section of Endocrinology and Metabolism, Department of Medicine, Tulane University Health Sciences Center, New Orleans, Louisiana, USA

[1]Corresponding author: e-mail address: cybowers@tulane.edu

Contents

Abstract

The most important initial historical time points in the development of the enlarging ghrelin system were 1973, 1976, 1982, 1984, 1990, 1996, 1998, and 1999. At these respective times, the following occurred sequentially: isolation of somatostatin, discovery of unnatural growth-hormone-releasing peptides (GHRPs), isolation of growth-hormone-releasing hormone (GHRH), hypothesis of a new natural GHRP different from GHRH, GHRP + GHRH synergism in humans, discovery of the growth hormone secretagogue GHS/GHRP receptor, cloning of the receptor, and finally, isolation and identification of the new natural endogenous GHRP ghrelin.

To understand the pharmacology and probably also the physiological regulation of growth hormone (GH) secretion, an important finding was that GHRP increased pulsatile GH secretion in children as well as normal younger and older men and women. This requires endogenous GHRH secretion, even though GHRP alone substantially releases GH from the pituitary *in vitro* without the addition of GHRH.

Methods in Enzymology, Volume 514
ISSN 0076-6879
http://dx.doi.org/10.1016/B978-0-12-381272-8.00001-5

Unnatural GHRP gave rise to natural GHRP ghrelin because of many talented researchers worldwide. GHRP was first envisioned to be an analog of GHRH but, from comparison of the activity of GHRH and GHRPs between 1982 and1984, it was hypothesized to reflect the activity of a new hormone regulator of GH secretion yet to be isolated and identified. Intravenous bolus GHRP releases more GH than GHRH in humans, but the reverse occurs *in vitro*. GHRPs are pleiotropic peptides with major effects on GH, nutrition, and metabolism, especially as an additional hormone in combination with GHRH as a new regulator of pulsatile GH secretion. The first indication of pleiotropism was an increase of food intake by GHRP. A major reason for the prolonged initial interest in the GHRPs has been its similar, yet different and complementary, action with GHRH on GH regulation and secretion. Particularly noteworthy is the variable chemistry of the GHRPs. They consist of three major chemical classes, including peptides, partial peptides, and nonpeptides, and all probably act via the same receptor and cellular mechanisms. Generally, most GHRPs have been active by all routes of administration, intravenously (iv), subcutaneously (sc), orally, intranasally, and intracerebroventricularly (icv), which supports their possible broad future clinical utility. From evolutionary studies starting with the zebrafish, the natural receptor and hormone have been present for hundreds of years, underscoring the fundamental evolutionary and functional importance of the ghrelin system. GHRPs were well established to act directly on both the hypothalamus and pituitary several years before the GHS receptor assay (Howard et al., 1996; Smith et al., 1996; Van der Ploeg et al., 1998). Finally, the ghrelin chemical isolation and identification was accomplished surprisingly from the stomach, which is the major site but not the only site, for example, the hypothalamus (Bowers, 2005; Kojima et al., 1999; Sato et al., 2005). Ghrelin was isolated and identified by Kojima and Kangawa et al. in 1999. A primary action of GHRPs continues to concern GH secretion and regulation, but increasingly this has included direct and indirect effects on nutrition and metabolism as well as a variety of other actions which may be pharmacological and/or physiological. Possible continuing and expanding roles of this new hormonal receptor include the central nervous system as well as the cardiovascular, renal, gastrointestinal, pancreatic, immunological, and anti-inflammatory systems. Our basic and clinical studies have mainly involved effects on GH regulation and secretion and this relationship to metabolism. So far in our studies, the actions of GHRPs and ghrelin on GH secretion and regulation in rats and probably in humans have generally been the same. A current objective is the incorporation of ghrelin into the diffuse endocrine hormonal system especially via GH.

1. INTRODUCTION/PRE-GHRP STUDIES

Initially, our GHRP studies were derived, interrelated, and rooted in the objective of isolating various hypothalamic hypophysiotropic hormones (HHHs), with Schally starting the work in 1960. Andrew had been involved in this objective for the previous 5 years, and I had developed a tadpole and subsequently a mouse T_3–TRH bioassay for measuring thyroid-stimulating hormone (TSH) in order to isolate, identify, and characterize TRH presumably present in multiple fractions of porcine hypothalami. Between 1960 and 1965, various pituitary hormonal bioassays and eventually radioimmunoassays (RIAs) were established in order to determine the HHH activity of a number of different HHHs including the growth-hormone-releasing hormone (GHRH) and LHRH (Schally and Bowers, 1964). They were envisioned to be small peptides. Essentially, pure thyrotropin-releasing factor or thyrotropin-releasing hormone (TRH) was isolated in 1966. By 1968, Andrew and I started working with Karl Folkers at the University of Texas in Austin on the final structural determination of TRH.

Notably, the synthetic TRH structure was determined 6 months before the determination of the natural porcine TRH structure, and unnatural GHRP appeared 20 years before natural ghrelin. Several publications on the isolation and structure of TRH were published, which consisted of the chemical–biological properties of synthetic and natural porcine TRH and eventually the identity of the synthetic and natural TRH molecules (Boler et al., 1969; Bowers and Schally, 1969; Bowers et al., 1970; Folkers et al., 1969; Schally and Bowers, 1970; Schally et al., 1966, 1969). Also, TRH was the same molecule in sheep hypothalami (Burgus et al., 1969) as well as in human hypothalami (Bowers et al., 1971).

Another group of relevant pre-GHRP studies consisted of the development of synthetic analogs of TRH and the natural decapeptide luteinizing-hormone-releasing hormone (LHRH) between 1970 and 1973. Notable was that the C-terminal amidation and also the N-terminal cyclization of Gln were absolute requirements for biological activity of both TRH and LHRH. C-terminal amidation also has been required for almost all of our growth-hormone-releasing peptides (GHRPs). Some special analogs of TRH and LHRH consisted of substitution of unnatural amino acids (aa) at various positions in these two hormones which enhanced or inhibited the TSH or luteinizing hormone/follicle-stimulating hormone (LH/FSH) activity *in vitro* and *in vivo*. Of note is that 7 of 10 aa of LHRH were substituted with unnatural aa in one of our receptor antagonists, namely, Antide. Also, after many studies it was determined that TRH released TSH as well as prolactin (PRL) in animals and humans (Bowers et al., 1973a,b, 1975). Thus, during this pre-GHRP time, we were already aware of unpredicted functional hormonal activity of small peptides and the possible role of substituted unnatural aa, as well as the fact that experimental conditions and reagents might affect these activities. Between 1971 and 1976, one of our projects concerned whether there was a single one or two individual

HHHs that regulated LH and FSH release from the pituitary, that is, LHRH versus FSHRH. Unlike GHRP/ghrelin and GHRH, the single HHH GnRH decapeptide physiologically regulates both LH and FSH.

Multiple combined *in vitro* and *in vivo* strategies and approaches were developed for assessing the actions of TRH as well as LHRH and similarly have been utilized for the development of GHRPs (Bowers et al., 1973a,b, 1983). For example, the pituitary actions of TRH were determined *in vitro* by the direct addition of the somatostatin-releasing-inhibiting factor (SRIF), triiodothyronine (T3), prostaglandin E1 (PGE1), theophylline, and dibutyryl cyclic adenylyl monophosphate (cAMP), and the subsequent measurement of TSH release and pituitary cAMP levels (Bowers et al., 1971, 1983).

2. GHRP BEGINNING

In 1974, John Hughes isolated the Met and Leu enkephalin opiate peptides. Shortly thereafter, a former collaborator on TRH and LHRH, Jaw-Kang Chang, sent me some enkephalin analogs which we assayed for possible HHH activity by determining their effects on rat pituitary hormonal release *in vitro*. Because of our previous work on structure–activity studies of HHHs and the fact that opiates are well established to release growth hormone (GH) *in vivo*, the newly synthesized peptides were assayed for GH first and then other pituitary hormones. The first enkephalin analog to release GH was DTrp^2-Met enkephalin amide. As recorded in Fig. 1.1,

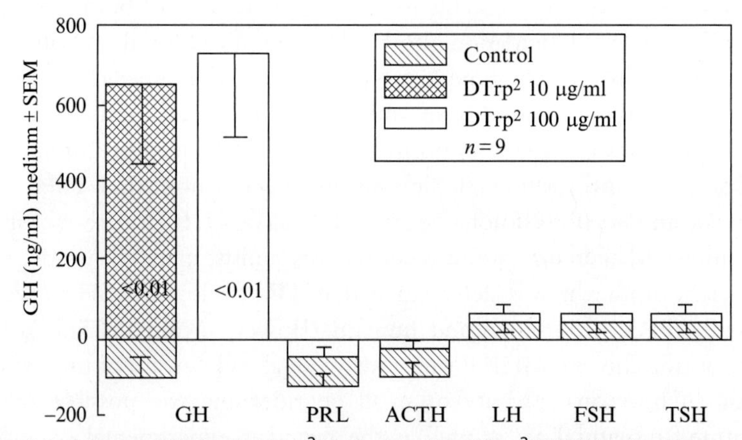

Figure 1.1 Release of GH by TyrDTrp^{2}GlyPheMetNH$_2$ (DTrp^{2}GHRP) *in vitro*. There was no release of any of the other hormones measured. Pituitaries of 20-day-old female rats. Mean of 6 ± SEM (reprinted with permission from Bowers et al., 1977).

$DTrp^2$ had low potency *in vitro* but did not release any other pituitary hormones. In addition, it was only active *in vitro* and it did not have opiate activity (Bowers et al., 1977). Multiple analogs were synthesized and tested both *in vitro* and *in vivo*. Peptides with both *in vitro* and *in vivo* activity eventually were classified into four pentapeptide chemical templates, that is, $DTrp^2$, $DTrp^3$, $DTrp^{2,3}$, and $DTrp^2$, spacer, $LTrp^4$ (*vide infra*). These chemical templates were eventually utilized by other investigators as well as pharmaceutical companies. Thus began the search for a new endogenous GHRP, which resulted in its receptor identification and cloning during 1996–1998 by the Merck group and then the final isolation of the natural endogenous hormone ghrelin in 1999 by Kojima and Kangawa. Since this chapter represents more than 30 years of studies, the details of the methods utilized can be found in the references (Bowers et al., 1973a,b, 1975, 1983) and throughout this chapter.

In addition to our TRH and LHRH analog studies and strategies of incorporating unnatural aa into the GHRPs, the theoretical low-energy conformational peptide strategy of Momany was incorporated as an additional step to improve activity (Momany et al., 1981). This involved integrated sequential solid-phase peptide synthesis, assay, chemical–functional analysis, and design of new analogs. The peptides were quantitatively assayed by RIA methods in multiple dosages for GH, TSH, LH, and FSH, and sometimes PRL and ACTH *in vitro* utilizing intact pituitaries of immature and/or adult male or female rats (Bowers et al., 1977) and eventually were extended to *in vivo* studies in rats and various other animals as well as humans.

Momany very effectively incorporated and integrated his theoretical conformational approach with our chemical linear unnatural aa substitution approach and we performed hormonal release studies weekly *in vitro* and subsequently *in vivo*. This low-energy conformational approach included molecular chemical mechanisms and dynamic computational methods as well as, occasionally, nuclear magnetic resonance solution studies. Conformational analyses of sterically constrained analogs were also performed, that is, disulfide-linked analogs and incorporation of methylated N- or C-alpha atoms of select peptide bonds, as demonstrated in Fig. 1.2 (Momany and Bowers, 1996). The working hypothesis was that, if 3-dimensional peptide conformation in solution was the same as the bound conformation of the peptide receptor, this would support that these two different types of information could be directly interrelated. In addition, conformational computation results were performed on the solvated configuration of GHRP-6 in a water environment (not shown).

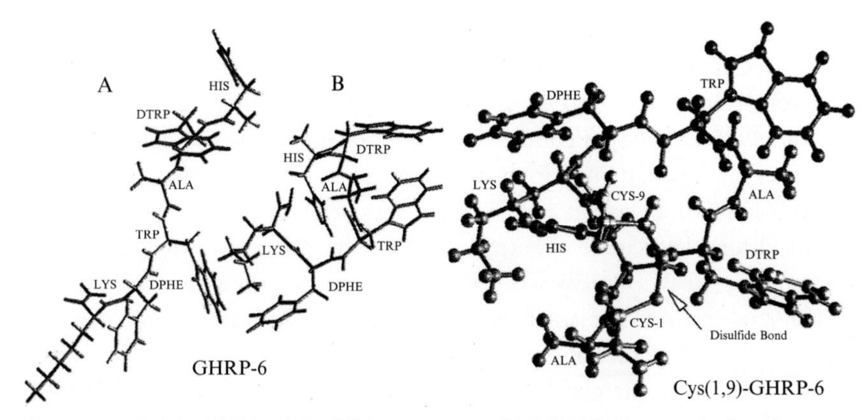

Figure 1.2 Starting (A) and final (B) structures of GHRP-6 before and after dynamics simulation *in vacuo*. The simulation was performed for 200 ps at a temperature of 300 K. The amino acids are labeled (left panel). Conformer C-4 of Cys(1,9)-Ala-GHRP-6. The disulfide bridge is facing the viewer (right panel) (reprinted with permission from Momany and Bowers, 1996).

Assays developed for TRH and LHRH agonists, and sometimes antagonist activity, were utilized for GHRP experimental studies. *In vivo* GH activity was quantitatively determined in 26-day-old female rats without treatment or pretreated with pentobarbital, and the RIAs were performed with reagents obtained from the NIAMD, NIH program (Bowers et al., 1973a,b, 1983). Since the procedures and reagents utilized for these studies are no longer used, they are not discussed in detail but only referenced.

Between 1975 and 1980, although our initial synthetic GHRPs were low in potency, they were of particular value in contrast to hypothalamic extracts because of the peptide purity as well as the known and varied chemistry which readily could be altered in many specific ways and they specifically released GH by a direct pituitary site of action. After considerable efforts of ours as well as other investigators, natural GHRH had defied isolation. Furthermore, unexpectedly, in 1973 an inhibitor of GH release, somatostatin 1–14 as well as 1–28 (SRIF) was isolated from hypothalamic extracts and thus this peptide in part may have been a significant factor for the inability to isolate GHRH earlier since the two SRIFs so effectively inhibit GHRH release (Brazeau et al., 1973; Guillemin et al., 1982) as well as the pentapeptides of GHRP-6 (Bowers et al., 1981, 1983).

Our initial GHRP *in vitro* pituitary approach maximized specificity because of the direct pituitary action of the peptides but minimized the complexity because GHRPs may act at both the hypothalamic and the pituitary sites to release GH. However, concomitant measurement of other pituitary hormone release *in vitro* added another increment of specificity. When GHRPs became active *in vivo*, in 1980 it was possible to consider both the hypothalamus and pituitary as the primary sites of action. Of special note is that the pituitary, in particular, the somatotrophs, was also being considered from an additional more complex regulatory and secretory viewpoint. This concerned the three hypothalamic–pituitary sites selectively responsive to GHRH, GHRP, and SRIF, which were involved more globally in the stimulation and inhibition of GH release.

Pentapeptides developed between 1976 and 1980 generically could be classified into four general types according to the number, position, and type of Trp aa residue (L- or D-). Examples of these peptides were initially derived from the newly discovered natural opiate peptides Met or Leu enkephalin and are as follows: TyrDTrpGlyPheMetNH_2 (DTrp^2), TyrGlyDTrpPheMetNH_2 (DTrp^3), TyrDTrpDTrpPheMetNH_2 (DTrp^{2,3}), and TyrDTrpAlaTrpPheNH_2 (DTrp^2, LTrp^4). Notable is that each of these four classes of pentapeptides became chemical templates to us as well as to other investigators for developing small unnatural peptides and nonpeptides that specifically release GH both *in vitro* and *in vivo* at low dosages. C-terminal amidation also was a similar requirement for GH-releasing activity.

Since DTrp^2 and DTrp^3 pentapeptides released GH by a direct pituitary action, the same strategies and approaches previously described for TRH, LHRH, and SRIF studies were utilized to further characterize their actions *in vitro*. Select findings of these studies were the following: PGE1 and theophylline increased both pituitary cyclic adenosine monophosphate (cAMP) and GH release, while DTrp^2 and DTrp^3 increased release of GH but not pituitary cAMP, suggesting that these pentapeptides do not act via the cAMP intracellular signaling pathway. DTrp^2 and DTrp^3 further increased GH release with lower levels of glucose during pituitary incubation and further decreased GH release with higher glucose levels. On repeated DTrp^2 stimulation, GH was released by 5 min at lower dosages and the effect of GH release was specific. GH release by DTrp^2 and DTrp^3 was more sustained from pituitaries of immature than mature rats. Additionally, SRIF inhibition of GH release of immature rats was more readily reversible. In contrast to these DTrp pentapeptide results, repeated PGE1 or theophylline stimulation *in vitro* of pituitaries of mature rats induced sustained GH release. Some conclusions of these *in vitro* studies were the following: (1) DTrp^2, DTrp^3 may be unique for learning about the function and behavior of the putative GH-releasing receptor of the pituitary somatotrophs; (2) dual effects of SRIF plus DTrp^2 or DTrp^3 rather than the effects of SRIF alone appear to produce a more irreversible inhibition of the release of GH; and (3) studies on the regulation of GH release *in vitro* of immature and mature rats might reveal unique differences which have important physiological implications such as regulation of GH secretion and rate of body growth (Bowers et al., 1981, 1983).

3. GHRP-6 ACTIVE *IN VITRO* AND *IN VIVO* (1982)

In 1980, the most active peptide *in vitro* was reported to be TyrDTrpAlaTrpDPheNH_2, but it was active only at 30 ng/ml medium. In 1981, more active small peptides, such as the hexapeptide HisDTrpAlaTrpDPheLysNH_2 (GHRP-6), were designed, synthesized, and assayed, which specifically released GH both *in vitro* and *in vivo* in low doses (*vide infra*). As reported at the Endocrine Society Annual Meeting (Bowers et al., 1982), "one of these peptides, (GHRP), significantly released GH *in vitro* from the pituitary of immature female rats at 1 ng/ml incubation medium. The control GH value in ng/ml was -167 ± 114 while the stimulated release values were 955 ± 272 ($<.01$) at 1 ng, 1603 ± 305 ($<.001$) at 3 ng, 2244 ± 172 ($<.001$) at 10 ng and 2198 ± 358 ($<.001$) at 30 ng. When this peptide was administered to unanesthetized 21 day old female rats, GH (ng/ml serum) acutely released by 1, 10, and 100 micrograms (μg) was 12 ± 172 ($<.02$), 151 ± 56 ($<.02$) and 381 ± 61 ($<.001$), respectively; control value was 1 ± 0.6. Additionally, the peptide significantly ($<.001$) augmented the body weight gain of 16 day old female rats by a net increase of 17.5–10% after 9 and 25 days of treatment with 30–100 μg once or twice daily; each dosage produced essentially the same effect. At the end of the 9 and 25 day treatment periods the peptide also was found to release GH acutely after administration of 30 μg. Other results of this GHRP included the following. The GHRP did not release LH, FSH, TSH or PRL *in vitro* or *in vivo* (rat), SRIF and SRIF 1-28 inhibited the GHRP induced GH levels (rat) *in vivo* and *in vitro* with SRIF 1–28 being more potent especially *in vivo* and GH levels rose acutely 10–25 fold within 2–10 min in rhesus monkeys, lambs and calves. Also, on repeated administration of low and high dosages of the GHRP within a day to immature rats, it was possible to demonstrate an equivocal potentiation and an unequivocal down regulation of the GHRP response." In addition, our GHRP pentapeptides bind to the same receptor as GHRP-6. Our first GHRP receptor antagonist was derived from HisDTrpAlaTrpDPheLysNH_2 ($DTrp^2$, spacer, $LTrp^4$) or GHRP-6 by substituting $DLys^3$ for Ala^3. GHRP-6 was the first GHRP, in 1980, that was active in pituitary assays *in vitro* and *in vivo* in various animals and, eventually, in 1989, the first GHRP to be administered to humans specifically to release GH (Bowers et al., 1990).

4. GHRH ISOLATION AND IDENTIFICATION (1982)

Most important and timely, as well as fortuitously, the natural linear GHRH 1-44NH_2 hormone at last was isolated from functional pancreatic tumors of patients with acromegaly (Guillemin et al., 1982, 1996) and GHRH 1-40 COOH (Rivier et al., 1982). As previously described, 1982 also was the year in which we published for the first time the *in vitro* and *in vivo* results of GHRP-6 which had been envisioned to have the GH-releasing activity of the elusive natural GHRH (Bowers et al., 1980, 1982). Although the chemistry of GHRH and GHRP-6 was very different, they were both peptides that specifically released GH by a direct pituitary action. Because of our chemistry experience with varied peptides and their pituitary hormonal releasing actions, we considered our GHRPs might possibly be GHRH analogs even though they were so very different chemically. Also, during 1983, the availability of both natural GHRH and unnatural GHRP-6 allowed detailed *in vitro* and *in vivo* comparisons of the GH-releasing activity of the two peptides. These 1983 results revealed that only GHRH increased cAMP while GH was released by GHRH, GHRP pentapeptides ($DTrp^2$, $DTrp^3$), and the hexapeptide GHRP-6, but the GHRPs did not release cAMP (Bowers et al., 1983).

5. GHRP VERSUS GHRH (1981–1984)

Our 1981 *in vitro* $DTrp^2$, $DTrp^3$ pentapeptides and our 1982 GHRP-6 hexapeptide *in vitro* and *in vivo* results together with our 1982–1984 GHRH and GHRP-6 comparative *in vitro* and *in vivo* results led to the conclusion in 1984 that GHRH and GHRP GH effects were not the same (Bowers et al., 1984a,b; Momany et al., 1984). The commonality and differences of GHRH and GHRP-6 actions were equally impressive. In 1984, this led to the conclusion that there are two different natural GHRHs and two different receptors on the pituitary somatotrophs. To appreciate these similarities and differences of GHRH and GHRP-6, various conceptualizations, strategies, and *in vitro* and *in vivo* integrated studies utilizing different approaches and techniques were utilized. The results revealed that the two peptides were independent, dependent, synergistic, additive, and permissive in their specific actions on GH release (Bowers et al., 1984a,b). Some notable initial differences were that GHRH and GHRP-6 released GH synergistically *in vivo* and additively *in vitro*; pituitary cAMP was increased by GHRH but not GHRPs; and the GHRH radioreceptor assay (RRA) showed a lack of competitive binding by GHRP-6 (W. Vale, unpublished data). The parallelism of the *in vitro* GH-releasing actions of GHRP-6 and GHRH again was emphasized by other studies (Sartor et al., 1985a,b). Calcium-blocking agents inhibited the GH release of both peptides, and the two peptides in combination released GH additively rather than synergistically (Sartor et al., 1985a,b).

In two other different types of assay in 1984 and 1985, again there were marked similarities but definite differences in the GH-releasing activity of GHRH 1-44NH_2 and GHRP-6. One of the assays (Badger et al., 1984) involved perifusion of enzymatically dispersed pituitary cells of adult male rats, which demonstrated the effects of GHRH and GHRP-6 infusion on GH secretion. The other assay (McCormick et al., 1985) involved pretreatment of adult male rats with SRIF antiserum and diethyldithiocarbonate (dopamine-β-hydroxylase inhibitor) in order to functionally isolate the pituitary from the hypothalamus. These were open-loop-type assays in order to obviate or minimize feedback effects on the feedforward effects of GHRH and GHRP-6. Both peptides induced dose–response release of GH when administered repeatedly at 1–1.5-h intervals over several hours. However, as shown in Fig. 1.3, in the perifusion assay of Badger, continuous GHRH but not continuous GHRP-6 induced sustained GH release over several hours. During continuous GHRP-6 administration, the GH immediate rise was substantial but only lasted for 1 h, and then returned and remained at the baseline for several hours. Thus, the GH response of GHRP-6 but not GHRH was acutely downregulated at the pituitary level by continuous but not by hourly, intermittent GHRH or GHRP-6 administration. Also, the *in vivo* assay of McCormick revealed that the slopes of the dose–response curves to intermittent hourly GHRP-6 versus GHRH were very different. Thus results of these different types of assays supported the conclusion that functionally GHRP-6 was not a GHRH mimic but rather supported the activity of a new hormone and new receptor.

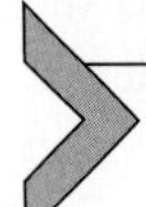

6. GHRP REFLECTS ACTIVITY OF A NEW HORMONE AND NEW RECEPTOR—HYPOTHALAMIC–PITUITARY MODELS (1984–1991)

In 1984, we noted “a fundamental problem still unsolved is the number and type of HHH that normally exist.” Using TRH as an example, the following was stated. “TRH is equally potent in releasing TSH and PRL in

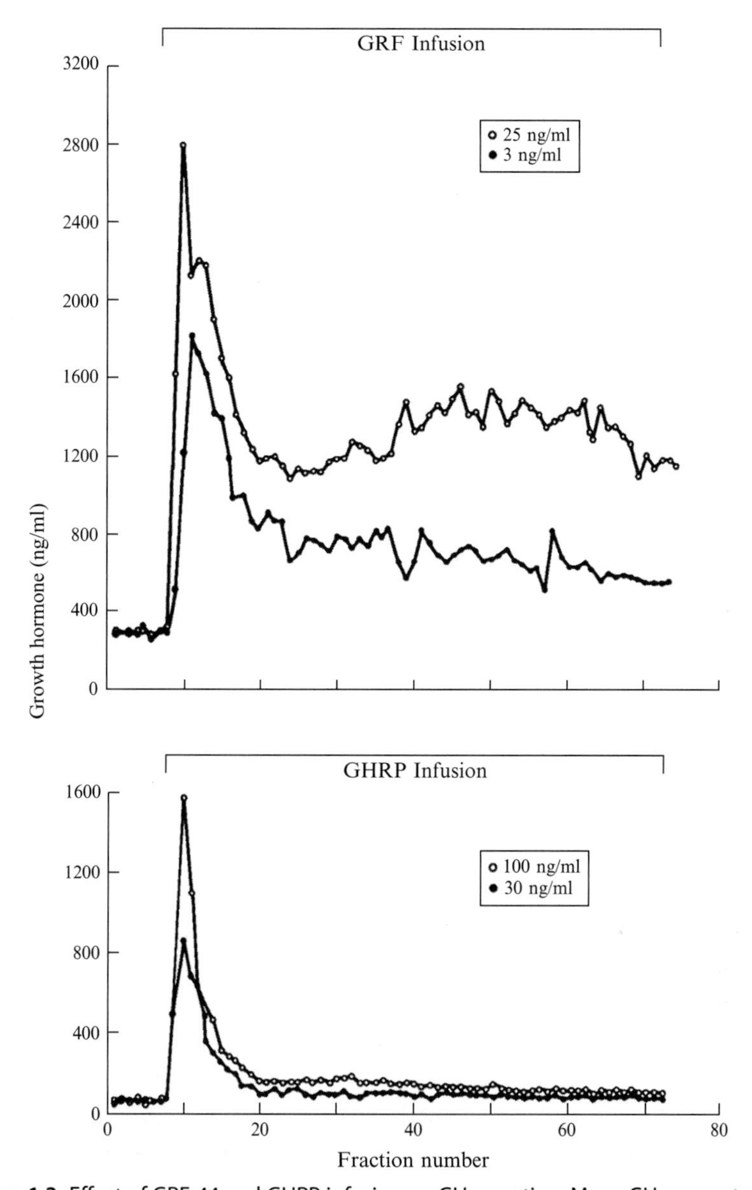

Figure 1.3 Effect of GRF-44 and GHRP infusion on GH secretion. Mean GH concentration for three columns receiving either 25 or 3 ng/ml GRF. There were 6.4×10^6 cells/column (upper panel). Mean GH concentration of three columns receiving either 100 or 30 ng/ml GHRP. There were 5.4×10^6 cells/column (lower panel) (reprinted with permission from Badger et al., 1984).

humans indicating that the designation PTRH in humans may not be unreasonable. However, the species dependency of the action of TRH is emphasized by the finding that TRH releases mainly TSH and very little PRL in rats and thus, in rats it should be designated TRH while in the monkey it stimulated primarily PRL and perhaps should be designated PRH. Added to these complexities is the fact that some investigators still believe PRH (PRF) is yet to be isolated" (Bowers et al., 1984a,b). The above points emphasize some of the issues that arose and still are in the unnatural to natural GHRP sequence of events. Experimentally, this required the determination and consideration of species, age, sex, dose, and unexpected effects, each of which is important and also required various conceptualizations, strategies, approaches, and techniques.

Differences in the effects of GHRP and GHRH on GH release became of increasing value in appreciating and understanding the more complex relationships and actions of GHRP on GH release and possible roles of the GH system. These particular results were the reason for previously emphasizing that, in order to appreciate the full spectrum of GHRP actions on GH release, the hypothalamic–pituitary approach rather than just the pituitary would be required. Results of an *in vivo* assay (McCormick et al., 1985) revealed that much less GH release was induced by GHRP-6 than by GHRH. This was envisioned to be the result of a functional isolation of the pituitary, which in turn indirectly supported the conclusion that GHRP-6 acts on both the hypothalamus and pituitary whereas GHRH acts specifically on the pituitary to release GH. This dual anatomical site of action of GHRP to regulate GH secretion *in vivo* again supported the prediction that the GHRP-6 GH release reflects the activity of a new hypothalamic hormone and a new receptor.

From our composite *in vitro* and *in vivo* studies on GHRP, GHRH, and SRIF on the regulation of GH secretion obtained between 1984 and 1994, results were presented at the Serono Conference in 1994 and discussed in terms of three models, that is, pituitary, hypothalamic, and hypothalamic–pituitary (Bowers et al., 1994). A summary of these results was the following: "Since it has been concluded that the GHRP-like ligand should, rather than could exist, the compelling guiding point seems to be that despite how irrational it may seems to be to include the GHRP-like ligand along with GHRH and SRIF as a physiological regulator of GH secretion, a variety of evidence indicates that the attempt to include this new putative regulator should be diligently pursued. Various aspects and conceptual models of the latter have been discussed and proposed. As a more detailed understanding of the GHRP(s) receptor, molecular intracellular mechanism, and endocrine mechanism by which it releases GH via both the hypothalamus and pituitary is elucidated, the physiological role of the GHRP-like ligand in the regulation of GH release will become more apparent. Evidence is strong that the actions of GHRP(s) and GHRH on release of GH can be dependent and independent as well as complementary to each other. A seemingly important general conclusion that has evolved from these studies is that GH secretion can be markedly increased apparently without involving endogenous GHRH or involving it only in a permissive way. Thus, new results support that new dimensions will be revealed in the near future about the physiological regulation of GH secretion as well as the pharmacological release of GH from the hypothalamic-pituitary unit."

In these studies on the regulation of GH secretion via the hypothalamic–pituitary axis in 26-day-old female rats, a number of different methodological approaches and techniques were utilized. Some of the approaches included administration of GHRH or SRIF antiserum alone and together with and without opiates, GHRH, GHRP, and SRIF in various dosages and combinations, times of administration, and experimental designs. GH was measured by RIA. Concomitant effects on GH release and pituitary cAMP levels were also determined in 26-day-old female rats. Administration of peptide opiates and some kappa opiates in combination with GHRPs or GHRH synergistically released GH. Since pretreatment of rats with SRIF antiserum prevented the synergistic release induced by combined opiates and GHRPs as well as GHRH, it has been concluded that opiates inhibit endogenous SRIF release and, thus by this mechanism, enhance the GH release by inhibition of SRIF release. Also, these results indicate neither GHRP nor GHRH released GH by inhibiting endogenous release of SRIF (Bowers et al., 1994).

7. GHRP/GHS RECEPTOR ASSAYS (1989–2000)

Results of *in vitro* RRA for GHRP were published between 1989 and 1998. These assay results included specific binding of GHRH, GHRPs, as well as partial and nonpeptide GHSs. GHRP-6 did not bind in the GHRH RRA; however, GHRP-6-specific binding was demonstrated in the hypothalamus and pituitary (Bitar et al., 1991; Codd et al., 1989; Howard et al., 1996, *vide infra*; Sethumadhavan et al., 1991; Smith et al., 1996, *vide infra*; Van der Ploeg et al., 1998, *vide infra*; Veeraragavan et al., 1992). In general, receptor-binding results and GH-releasing results parallel each other. Also, specific binding was demonstrated for the first GHRP receptor antagonist we designed and developed, namely, $DLys^3$-GHRP-6, HisDTrpDLysTrpDPheLysNH$_2$, which inhibited the GH-releasing response of GHRP-6 but not GHRH *in vitro* (Bowers et al., 1991). In addition, we identified an analog, [$DArg^1$ $DPhe^5DTrp^{7,9}Leu^{11}$]-Substance-P, that inhibited GHRP-6 receptor binding (Bitar et al., 1991). Subsequently, in a series of notable GHS-R 1a-binding studies (Holst et al., 2003), high constitutive activity of this receptor was demonstrated and that this GHRP Substance-P antagonist at low dosages inhibited the constitutive activity while at higher dosages it inhibited the receptor binding of GHRP-6 and other GHSs.

8. GHS/GHRP RECEPTOR (1996–1998)

The seminal cloned GHS-R 1a-binding assay was published in 1996 (Howard et al., 1996; Van der Ploeg et al., 1998) (Fig. 1.4, left panel). This is a seven-transmembrane-domain G-protein-coupled receptor. It is located anatomically on hypothalamic arcuate GHRH neurons and pituitary somatotroph cells as well as other CNS anatomical sites. High specific binding for the GHS MK-0677 as well as GHRP-6 and GHRP-2 was demonstrated in this receptor assay by Howard et al. The hypothalamus and pituitary receptors have identical sequences and binding properties but not necessarily the same GH-releasing properties. $DLys^3$-GHRP-6 inhibits the binding of the GHS agonist [^{35}S]-MK-0677 to this receptor as well as all the GHRPs.

A series of *in vitro* studies were performed on the effects of GHRH and GHRP on GH-secreting pituitary tumor cells of rats and humans (Adams et al., 1996, 1998). The results in Fig. 1.4, right panel, reveal that GHRP-6 and GHRP-2 dose-dependently release GH and increase the rate of phosphatidylinositol (PI) formation in these GH-releasing tumor cells. With the exception that GHRP-2 is more potent than GHRP-6, the effects of these two GHRPs were the same. Furthermore, in these tumor cells, even when adenylyl cyclase activity was elevated, cAMP production could be further increased by GHRP. This provided evidence for concluding that cross talk existed between the PI and adenylyl cyclase intracellular transductions systems in the tumor pituitary cells. The GH-releasing effect of GHRP-6 and GHRP-2 were reduced or abolished by a protein kinase-C inhibitor, namely, phloretin, and a calmodulin inhibitor. Interestingly, when the basal cAMP levels of the GH-secreting tumor cells were elevated, they were further increased by GHRPs but not GHRH. In

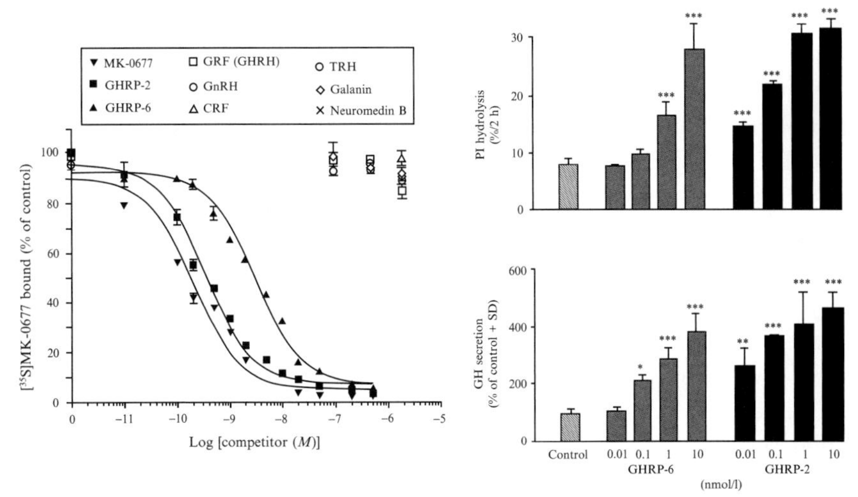

Figure 1.4 Binding of [^{35}S]MK-0677 to crude membranes from COS-7 cells transfected with human type 1a GHS-R cDNA. Competition analysis (0.24 n*M* [^{35}S]MK-0677). Competition binding data analyzed by nonlinear regression curve-fitting program. Results are mean ($\pm$SEM) of triplicate determinations. Competition analysis with GHRP-2 and GHRP-6 as well as MK-0677 identifies GHS-R 1a high-affinity binding sites. Galanin, GHRH, TRH, GnRH, and neurotensin fail to compete for binding of MK-0677 with the GHS-R 1a (left panel) (adapted and reprinted with permission from Howard et al., 1996). Stimulatory effect of GHRP-6 and GHRP-2 on the rate of PI hydrolysis (upper right panel) and GH secretion (lower right panel) by cell culture of a human pituitary somatotropinoma. Values are the mean$\pm$SD. Similar effects of GHRP-2 have been observed in two other experiments (not shown) using cultures from different somatotropinomas. *$P<0.05$; **$P<0.01$; ***$P<0.001$ versus control (reprinted with permission from Adams et al., 1996).

contrast, in tumor cells with low basal cAMP levels, combined GHRP and GHRH increased both PI and cAMP *in vitro*. Thus, it is possible that the effect on GH release by combined GHRP and GHRH on GH secretion will vary as a function of the factors that influence pituitary somatotroph cell levels of cAMP and PI. A general conclusion that evolved from these 1996 studies is the following: "The development of the GHRPs and the many studies on their GH releasing ability through the transduction system and cell surface receptors different from those used by GHRH emphasize the remarkable complexity of intracellular mechanisms involved in controlling GH secretion and have important implications for normal and pathophysiology of pituitary somatotrophs."

The anatomical site of this gene receptor parallels the site of the GH-releasing effects of the GHSs and GHRPs. Overexpression of the GHRP/GHS receptors has been demonstrated in human pituitary tumors of patients with acromegaly as shown in Fig. 1.4, right panel (Adams et al., 1996). Complementary *in vivo* and *in vitro* GHRP-6 studies were performed on the GH responses in patients with acromegaly and also on pituitary tumors removed from patients with acromegaly (Alster et al., 1993). The *in vivo* results revealed high parallel activity of the GH response to GHRP-6, GHRH, and TRH. In 11 acromegalic patients, intravenous (iv) bolus GHRH and GHRP-6 (1 μg/kg) significantly released GH in all patients, while iv bolus TRH (1 μg/kg) released GH in 9 of 10 of the patients.

Increased food intake in conscious male rats was first demonstrated after intracerebroventricular (icv) administration of GHRP-6 (Locke et al., 1995). Later it was demonstrated that GHRP-2 also increased food intake after icv administration to conscious male rats, as shown in Fig. 1.5 (Okada et al., 1996). In addition, icv GHRH + GHRP-2 increased food intake additively but not synergistically. However, we had demonstrated earlier in humans that it is necessary to give a very low dose of GHRP-6 to more readily show the synergistic release of GH by combined GHRP and GHRH (Bowers et al., 1990). Thus, studying different dosages may be necessary to more definitively assess the combined effects of the peptides on food intake.

Normal spontaneous secretion of GH of conscious male rats is recorded in Fig. 1.6A, left panel (Tannenbaum and Bowers, 2001). Also, as recorded, iv GHRH antiserum pretreatment inhibited the GHRP-6-stimulated release of GH (Fig. 1.6B), while somatostatin antiserum increased the GH release by GHRP-6 (Fig. 1.6C). Parallel studies (right panel) were obtained after iv administration without (Fig. 1.6A) and with pretreatment of GHRH antiserum (Fig. 1.6B) on the GH response of ghrelin (Tannenbaum et al., 2003). The results again corroborate the projected parallel effects of GHRP-6 and ghrelin on GH secretion. GHRH antiserum inhibited the induced release of GH by both GHRP-6 and ghrelin, while somatostatin antiserum pretreatment augmented the GH release of GHRP-6.

9. CLINICAL RESULTS IN HUMANS

9.1. GHRP-6 + GHRH synergistic effect in humans (1989)

Our first human GHRP-6 studies in normal young men were performed in collaboration with Michael Thorner (Bowers et al., 1990). These studies (Fig. 1.7, left panel) revealed that iv bolus GHRP-6 released GH and, when given together with GHRH, released GH synergistically. One of the most characteristic and consistent *in vivo* actions of GHRPs in various animals as well as humans of both sexes and all ages is the synergistic release of GH

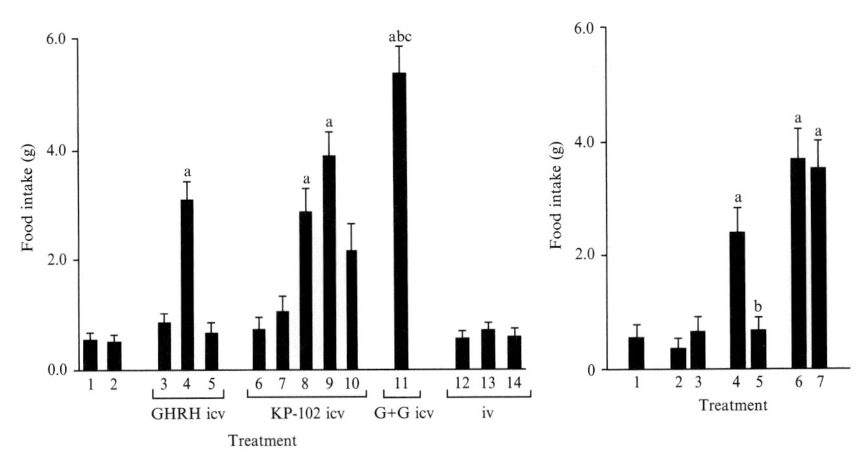

Figure 1.5 Effect of GHRH and GHRP-2 on food intake in free-feeding rats. 1—Control; 2—saline icv; 3—GHRH icv 4 pmol; 4—40 pmol; 5—400 pmol; 6—GHRP-2 icv 0.112 pmol; 7—1.12 pmol; 8—11.2 pmol; 9—112 pmol; 10—1.12 nmol; 11—GHRH + GHRP-2-icv 40 + 112 pmol, respectively; 12—Saline iv; 13—GHRH iv 40 pmol; 14—GHRP-2 iv 112 pmol ($n = 8$). Food intake was measured on two consecutive days from 13:00 to 15:00 h before treatment to obtain control values ± SEM. a. $P < 0.05$ versus icv saline; b. $P < 0.05$ versus 40 pmol icv GHRH; c. $P < 0.05$ versus 112 pmol icv GHRP-2 (left panel). Effect of GHRH antagonist (GHRH-A) on food intake stimulated by GHRH or GHRP-2 in free-feeding rats. 1—Control; 2—saline–saline; 3—GHRH-A 2.87 pmol + saline; 4—saline + GHRH 40 pmol; 5—GHRH-A 2.87 pmol + GHRH 40 pmol; 6—saline + GHRP-2 112 pmol; 7—GHRH-A 2.87 pmol + GHRP-2 112 pmol. Values are mean ± SEM of 2-h food intake (13:00–15:00 h) after icv administration of saline, GHRH, or GHRP-2. GHRH-A was administered centrally 15 min before icv saline, GHRH, or GHRP-2. a. $P < 0.05$ versus saline + saline; b. $P < 0.05$ versus saline + GHRH (right panel) (modified and reprinted with permission from Okada et al., 1996).

when GHRP is administered concomitantly with GHRH by iv bolus. Subsequently, this was also found for continuous 24/7 subcutaneous (sc) infusion. Also recorded in Fig. 1.7, right panel, is the comparative GH-releasing effects of iv bolus GHRP-6, -1, -2, and GHRH in normal young men. The potency of the three GHRPs we developed over several years was increasingly effective in releasing GH, and each released more GH than GHRH in normal young men. In addition, this was also found to occur in normal young women (Bowers, 1996).

Results in Fig. 1.8 of normal young men (left panel) and women (right panel) demonstrate that iv bolus combined GHRP-2 and GHRH at the respective doses of 1 μg/kg GHRH and a subthreshold GH-releasing dose of 0.03 μg/kg GHRP-2 released GH synergistically (Bowers, 1998). From these studies, GHRP is envisioned to act on the hypothalamus to release an unknown factor (U factor) rather than endogenous GHRH which

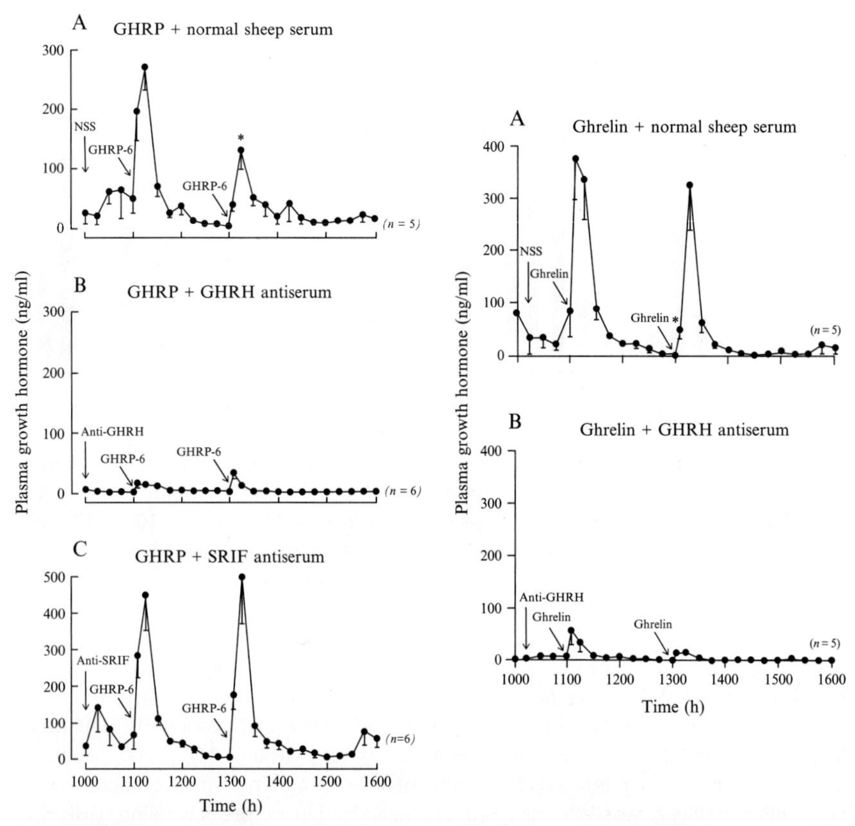

Figure 1.6 Effects of immunoneutralization with GHRH and SRIF antisera on mean GH plasma responses to GHRP-6 (5 μg) administrated during peak (11:00 h) and trough (13:00 h) periods of GH secretion. $^{*}P < 0.02$ compared with GHRP-6 induced GH release at 11:00 h (left panel) (reprinted with permission from Tannenbaum and Bowers, 2001). Effects of passive immunoneutralization with GHRH antiserum on GH responsiveness to ghrelin (5 μg, iv). Arrows indicate the times of injection; vertical lines represent the SE. The number of animals in each group is shown in parentheses (right panel) (reprinted with permission from Tannenbaum et al., 2003).

subsequently acts concomitantly with GHRH on the pituitary somatotroph to release GH synergistically. In this study, the important specific finding is that GHRP-2 augments GHRH release even when GHRH is present in excess amounts, and the concomitant GHRP-2 dose of 0.03 μg/kg is a subthreshold GH-releasing amount. Thus, GHRP + GHRH is not releasing GH in this study by augmenting endogenous GHRH release and, furthermore, GHRP+GHRH release *in vitro* is additive and not synergistic. In addition, from other high-dosage GHRP-2 data, that is, 10 μg/kg sc (not shown), we have postulated that at high doses GHRPs do act on the

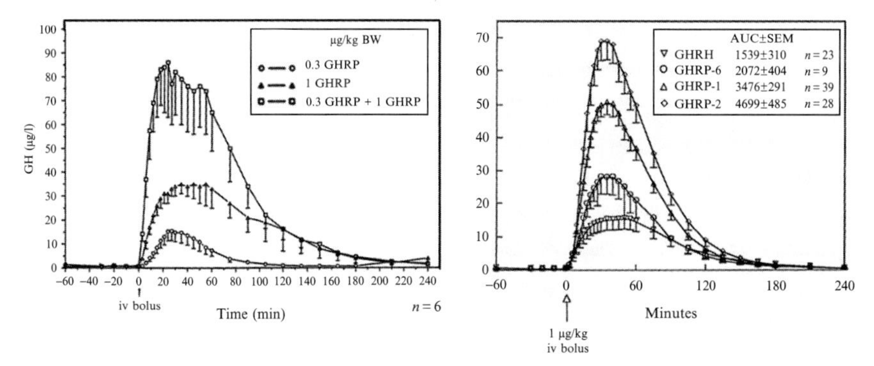

Figure 1.7 GHRP-6 + GHRH synergism in normal young men. Peptides were administered alone as well as in combination to the same subjects. Values are the mean ± SEM (left panel) (reprinted with permission from Bowers et al., 1990). Comparative GH responses to 1 μg/kg GHRH 1-44NH_2, GHRP-6, GHRP-1, and GHRP-2 in normal young men. Values are the mean ± SEM (right panel) (reprinted with permission from Bowers, 1996).

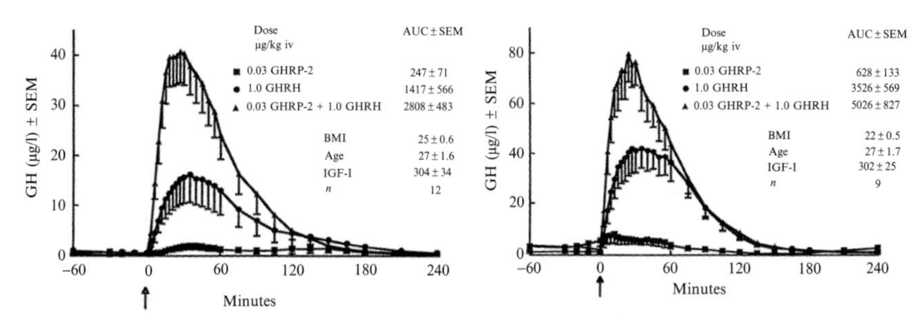

Figure 1.8 Effects of a very low dose of GHRP-2, 0.03 μg/kg, combined with a high dose of GHRH, 1.0 μg/kg, on the synergistic release of GH in normal young men (left panel) and women (right panel). GH values are the mean ± SEM (reprinted with permission from Bowers, 1998a,b).

hypothalamus to release endogenous GHRH because high-dose GHRP-2 (10 μg/kg sc) releases the same large amount of GH released by combined GHRH + GHRP-2 at 1 + 1 μg/kg iv (Bowers, 1998a,b).

9.2. 24-h Continuous infusion

In Fig 1.9, left panel, are results after 24-h continuous infusion of GHRP-6, which demonstrate an increase in normal pulsatile GH secretion (Jaffe et al., 1993). At the end of the infusion, 1 μg/kg iv bolus TRH did not release GH, while iv bolus GHRH increased GH release. An important hallmark of normal GH secretion in animals and humans is its pulsatile secretion.

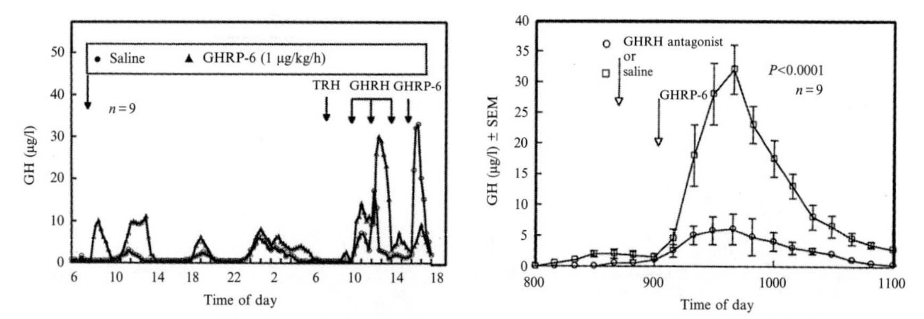

Figure 1.9 Mean (±SE) of plasma GH concentrations in nine normal men during saline or GHRP-6 infusion from 06:00 h on day 1 to 18:00 h on day 2 (left panel) (modified and reproduced with permission from Jaffe et al., 1993). Mean (±SE) GH responses to GHRP-6 after saline and GHRH antagonist, N-Ac-Tyr^1,DArg^2-GHRH 1-29NH$_2$, 400 μg/kg (right panel) (reprinted with permission from Pandya et al., 1998).

Pulsatile GH secretion had been demonstrated to occur in earlier studies of continuous GHRH administration (Bowers and Granda-Ayala, 2001). Normal pulsatile GH secretion also occurred by more prolonged sc GHRP-2 infusion for 1–3 months when administered to normal older men and women with low serum IGF-I levels (not shown). In these studies, serum IGF-I and IGFBP-3 and -5 were increased and the normal circadian secretion of cortisol and PRL was maintained (Bowers et al., 2004). These results again underscore the fundamental involvement and basic importance of endogenous GHRH and ghrelin secretion in the regulation of GH secretion. In Fig. 1.9, right panel, the data reveal that a GHRH receptor antagonist blocked the GHRP-6 GH response, demonstrating that the GHRP-6-induced release of GH is dependent on endogenous GHRH secretion (Pandya et al., 1998). In previous rat studies, it was shown that GHRH antiserum decreased GH release while SRIF antiserum increased GHRP-induced GH release (Bowers et al., 1992), which supports the results of Pandya et al.

An additional clinical attribute of the GHRPs is that they effectively release GH after almost all routes of administration (Bowers, 1999a–c). The pharmacokinetic (PK) and pharmacodynamic (PD) studies summarized by Bowers included iv, sc, oral, and nasal administration. There are special clinical advantages to the various different types of administration, including sc depot administration once a month which results in sustained delivery of GHRP-1 and elevated GH levels in rats and dogs (Bowers et al., 2008). Later, these same effects on the blood levels of GHRP-1, GH, and IGF-I were obtained in normal young men (C.Y. Bowers, unpublished data).

9.3. GHRP administration in children

Results were obtained in 1991 in which small synthetic peptides were developed that specifically released GH after oral administration to humans. Significant serum GH rise occurred after administration of the GHRP-6 hexapeptide to both normal men and seven of nine short-stature children without any adverse clinical or laboratory effects. Serum concentrations of GHRP-6 measured by RIA paralleled the increase of GH release after iv and oral administration (Bowers et al., 1992).

The finding that GHRP-2 given iv intranasally or GHRP-1 SC to children exerted a significant GH response suggested that it could be beneficial to assess pituitary GH function in the outpatient setting. A convenient, well-tolerated GH-releasing agent could represent a possible attractive treatment option for children with hypothalamic GH deficiency (Mericq et al., 1995; Pihoker et al., 1995).

From PK and PD studies in children, the following was concluded: (1) GHRP-2 produced a predictable and significant increase in plasma GH concentrations; (2) the PK and PD link model enabled quantitative assessment of GHRP-2 modulation of serum GH levels; and (3) EC_{50} for GHRP-2 would enable PK and PD evaluations of extravascular dosing regimens for children (Pihoker et al., 1998).

Data of chronic intranasal administration of the methylated derived GHRP-6 hexarelin peptide are shown in Fig. 1.10 (left panel) and GHRP-2 (center panel), both of which demonstrate increased height velocity of short-stature children (Bowers et al., 2008) These same effects also occurred when GHRP-2 was administered chronically sc (right panel) (Klinger et al., 1998; Mericq et al., 1998; Pihoker et al., 1997). The effects on height velocity were less than those reported with rhGH daily administration; nevertheless, the height of the children was increased, and our results indirectly indicate that a more optimum approach would be continuous sc 24-h delivery of GHRPs.

Also, from our combined adult studies with Veldhuis, we found that continuous sc GHRP administration has been more of a physiological approach because of the enhancement of normal pulsatile GH, increased serum IGF-I levels as well as IGFBP-3 and -5, and the maintenance of normal circadian secretion of serum cortisol and PRL (Bowers et al., 2004). Already a once-per-month delivery of GHRP has been developed, with the feasibility of less frequent administration. With the availability of more potent GHRPs and improved delivery systems, it is possible to reasonably forecast GHRP delivery at 6- or 12-month intervals.

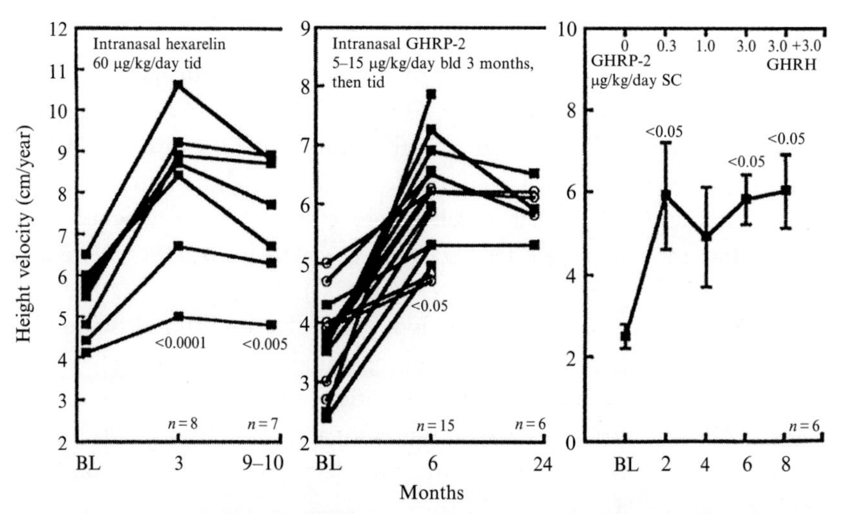

Figure 1.10 Results of three different investigators demonstrating that there is a sustained, increased rate of growth in height velocity of short-stature children with varying degrees of GH deficiency by prolonged administration of hexarelin and GHRP-2. In the left panel are the results of Klinger et al. (1996); in the middle panel, those of Pihoker et al. (1997); and in the right panel, those of Mericq et al. (1998) (adapted and reprinted with permission from Bowers et al., 2008).

9.4. Effect of GHRP-2, GHRH, GHRP-2 + GHRH, ghrelin, and ghrelin + GHRH in humans

In Fig. 1.11, data of two obese subjects are presented with low serum IGF-I levels during continuous 24-h subcutaneous infusion of GHRP-2, ghrelin, GHRH alone, and GHRP-2 or ghrelin combined with GHRH (Bowers et al., 2008). The data show the pulsatile GH secretion over the entire 24 h of each of these peptides alone as well as with GHRH. In addition, serum IGF-I levels were increased. When GHRP-2 (left panel) or ghrelin (right panel) was infused with GHRH, the pulsatile GH secretion was synergistically increased and the IGF-I further increased. These results again support the parallel actions of GHRP-2 and ghrelin; most notable is that a normal physiological pulsatile secretion is induced over the entire 24-h period by the peptides alone and in combination.

After a 5-day infusion of GHRP-2 plus TRH to critically ill patients, the combination of these two peptides reactivated blunted GH and TSH secretion with preserved pulsatility, peripheral responsiveness, and feedback inhibition without affecting serum cortisol with a shift toward anabolic metabolism. This short-term GHRP-2 plus TRH effect provides evidence

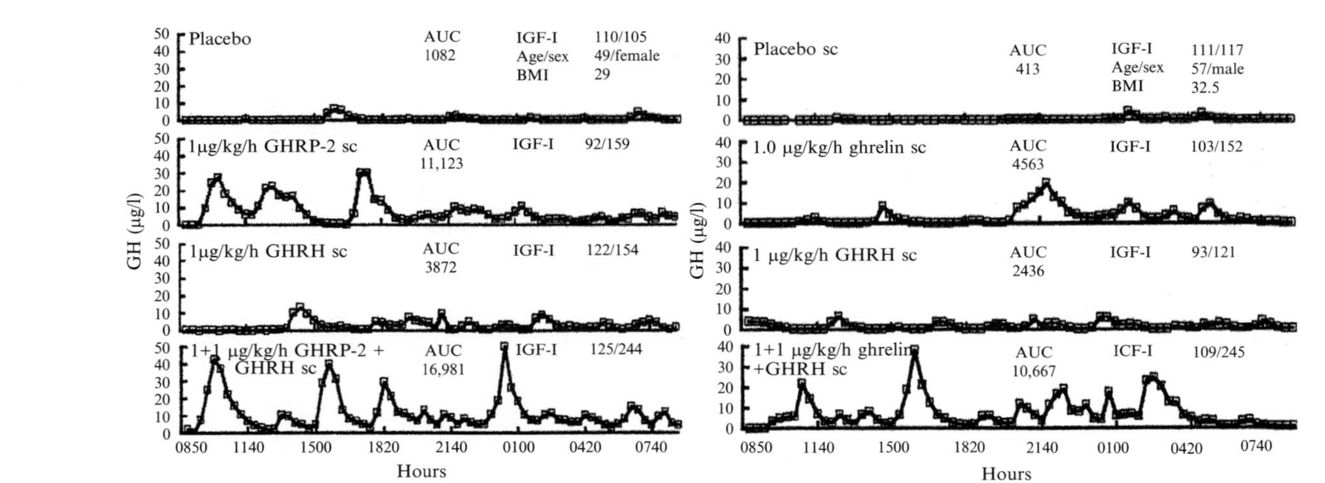

Figure 1.11 Twenty-four-hour continuous subcutaneous infusion of GHRP-2 (left panel) produced the same effect as ghrelin (right panel) infusion at 1 μg/kg/h. When GHRP-2 or ghrelin was infused with GHRH, the pulsatile GH secretion pattern was the same but the amplitude of the GH pulses was further increased. IGF-I levels rose by about the same to both peptides (adapted and reprinted with permission from Bowers et al., 2008).

of the metabolic response of this combination particularly in the wasting condition (Van den Berghe et al., 1999).

10. SUMMARY/CONCLUSION

Between 2000 and 2002, we presented a series of abstracts at the Annual Endocrine Society Meeting in the United States on comparative effects of GHRP-2, GHRH, and the newly discovered ghrelin *in vivo* and *in vitro*. Our objective in these studies was to qualitatively and quantitatively determine the similarities and differences of GHRP-2 and ghrelin administration in animals and humans.

A summary of these results is given below, some of which were previously discussed (Bowers, 1993, 1994, 2001, 2002). In the same *in vivo* and *in vitro* rat assay systems utilized in studies and actions of early GHRP pentapeptides, GHRP-2, and ghrelin, qualitative and quantitative results essentially agreed with each other. Even though GHRP-2 and ghrelin elicited the same GH effect, these effects overlapped with those of GHRH but also were distinguishable from them. Human ghrelin, rat ghrelin 1–28, and GHRP-2 were nearly equipotent in releasing GH, whereas desoctanoyl ghrelin neither released nor inhibited the GH release of these peptides. Combinations of GHRH plus ghrelin or GHRP-2, but not ghrelin plus GHRP-2, synergistically released GH in nonpentobarbital 26-day-old rats. SRIF was equally efficacious in inhibiting the GH response to GHRP-2, GHRH, and ghrelin; however, this SRIF effect on GHRP-2 or ghrelin plus GHRH was attenuated. Also, increasing the dosages of these three individual GHRPs only partially attenuated the inhibitory action of SRIF. In contrast, GHRP-2 or ghrelin in combination with GHRH completely attenuated the inhibitory SRIF effect on GH release even at low dosages in rats. In addition, in pentobarbital-pretreated rats, the GHRP receptor antagonists [DArg^{1}DPhe^{5}DTrp^{7,9}Leu^{11}]-Substance-P as well as the GHRH receptor antagonists [ibuTyrDArg^{2}Phe (4Cl)^{6} Abu^{15}NLeu^{27}] (gift from A.V. Schally) each inhibited the GH-releasing action of GHRP-2, ghrelin, and GHRH. These results underscore the importance of the concomitant GHRH action with GHRPs or ghrelin on the magnitude of GH release. This conclusion was uniquely revealed in normal young men by substantial inhibition of the GHRP-6-induced GH release by concomitant administration of the DArg^{2}-GHRH receptor antagonist. In addition, in 2000, our conclusion of the above *in vivo* rat studies was that the GH-releasing activities of

GHRP-2 and ghrelin closely paralleled each other. Like the findings of Kojima et al., our data strongly support that ghrelin is the natural endogenous GHRP hormone. Furthermore, in my opinion, the amount of data obtained in animals and humans on the GHRPs over the last 20–30 years probably will be applicable to ghrelin.

Most fortunate as well as outstanding the seven-transmembrane G-protein coupled receptor, GHS-R 1a, was isolated, cloned and established the quintessential assay of the ghrelin receptor (Howard et al., 1996; Smith et al., 1996; Van der Ploeg et al., 1998), while Kojima and Kangawa isolated and identified the exotic endogenous putative GHRP hormone designated ghrelin in 1999. These accomplishments not only established the major milestones of another unique new biological system but also became the infrastructure for new and continuing concepts, strategies, approaches, and techniques in the cornucopia of hormonal regulation and metabolism in animals and humans.

Our GHRP studies have focused on the regulation of GH secretion and have been encouraged by the unique interactions of GHRH, GHRP, and SRIF on the regulation of GH secretion. Recently, Veldhuis and I (Bowers et al., 2012) have proposed the GHRH, ghrelin, and SRIF tripartite hormones as the primary regulators of the physiological secretion of GH. The complexity of this decision is definitely curvilinear because of the stomach and hypothalamic origins of ghrelin and because this involves ghrelin in peripheral blood and locally in the hypothalamus and pituitary as well as having a specific regulating role on the ghrelin system via the vagal nerve and secondary modulating effects on adrenal, gonadal, and thyroid hormone secretion. Regarding the role of ghrelin in the physiological regulation of GH, it appears to be how and to what degree and not whether it interacts with GHRH to regulate GH and perhaps food intake, etc. A tentative succinct working hypothesis is that the action of ghrelin, especially in adult women and men, is to interrelate, influence, and modulate physiological GH secretion with metabolism and nutrition. It is particularly relevant to obtain additional information about the possible existence and role of ghrelin subtypes, desacyl ghrelin biological actions, and the comparative effects of ghrelin, GHRPs, and GHSs.

Seemingly poetic, valuable, and instructive was the concomitant publication of GHRP-6 and the isolation and identification of GHRH 1-44NH_2 in 1982. Both peptides released GH, and their effects on GH secretion were very similar to the degree that initially GHRP-6 was considered to be an analog of GHRH. Thus, in 1982, if only unnatural GHRP-6 had been available and natural ghrelin eventually isolated, ghrelin might have been considered and designated to be GHRH. Parenthetically, the eventual isolation of the GHRH 1-44NH_2 molecule would be possibly designated ghrelin. Even clinically possible in the future is a pancreatic tumor that secretes ghrelin and induces acromegaly.

GHRH, GHRPs/GHSs, and ghrelin interrelationships are of definite importance physiologically, pharmacologically, diagnostically, and therapeutically and thus merit continued exploration.

ACKNOWLEDGMENTS

I wish to thank all the past members of the Tulane Endocrine Laboratory who contributed to these studies and especially to George-Ann Reynolds, Lilly Zeller, Kathryn Friedman, and Christine Maier for their unequivocal loyalty and devotion to the basic and clinical studies included in this chapter. Also, I would like to acknowledge the National Institutes of Health, Tulane Clinical Research Center, and Kaken Pharmaceutical Co., Ltd. for the GHRP-2 (KP-102) and support. In addition, I would like to deeply thank my various collaborators as well as all the investigators who contributed greatly to the success of the history of ghrelin via the GHRPs.

REFERENCES

Adams, E.F., Lei, T., Buchfelder, M., Bowers, C.Y., Fahlbusch, R., 1996. Protein kinase C-dependent growth hormone releasing peptides stimulate cyclic adenosine 3',5'-monophosphate production by human pituitary somatotropinomas expressing gsp ncogenes: evidence for cross-talk between transduction pathways. Mol. Endocrinol. 10, 432–438.

Adams, E.F., Huang, B., Buchfelder, M., Howard, A., Smith, R.G., Feighner, S.D., Van der Ploeg, L.H.T., Bowers, C.Y., Fahlbusch, R., 1998. Presence of growth hormone secretagogue receptor messenger ribonucleic acid in human pituitary tumors and rat GH_3 cells. J. Clin. Endocrinol. Metab. 83, 638–642.

Alster, D.K., Bowers, C.Y., Jaffe, C.A., Ho, P.J., Barkan, A.L., 1993. The growth hormone (GH) response to GH releasing peptide (HisDTrpAlaTrpDPheLysNH_2), GH releasing hormone, and thyrotropin releasing hormone in acromegaly. J. Clin. Endocrinol. Metab. 77, 842–845.

Badger, T.M., Millard, W.J., McCormick, G.F., Bowers, C.Y., Martin, J.B., 1984. The effects of growth hormone (GH) releasing peptides on GH secretion in perifused pituitary cells of adult male rats. Endocrinology 115, 1432–1438.

Bitar, K.G., Bowers, C.Y., Coy, D.H., 1991. Effects of substance P/Bombesin antagonists on the release of growth hormone by GHRP and GHRH. Biochem. Biophys. Res. Commun. 180, 156–161.

Boler, J., Enzmann, F., Folkers, K., Bowers, C.Y., Schally, A.V., 1969. The identity of chemical and hormonal properties of the thyrotropin releasing hormone and pyroglutamyl-histidyl-proline amide. Biochem. Biophys. Res. Commun. 37, 705–710.

Bowers, C.Y., 1971. Studies on the role of cyclic AMP in the release of anterior pituitary hormones. In: Robinson, G.A., Nahas, G.G., Triner, L. (Eds.), Annals of the New York Academy of Sciences, vol. 185. New York Academy of Sciences, USA, pp. 263–290.

Bowers, C.Y., 1993. A new dimension on the induced release of growth hormone in obese subjects. J. Clin. Endocrinol. Metab. 76, 817–818.

Bowers, C.Y., 1994. On a peptidomimetic growth hormone releasing peptide (GHRP). J. Clin. Endocrinol. Metab. 79, 940–942.
Bowers, C.Y., 1996. Xenobiotic growth hormone secretagogues: growth hormone releasing peptides. In: Bercu, B.B., Walker, R.F. (Eds.), Growth Hormone Secretagogues. Springer-Verlag, New York, pp. 9–28.
Bowers, C.Y., 1998a. Growth hormone releasing peptide (GHRP). Cell. Mol. Life Sci. 54, 1316–1329.
Bowers, C.Y., 1998b. Synergistic release of growth hormone by GHRP and GHRH: scope and implication. In: Bercu, B.B., Walker, R.F. (Eds.), Growth Hormone Secretagogues in Clinical Practice. Marcel Dekker, New York/Basel/Hong Kong, pp. 1–25.
Bowers, C.Y., 1999a. Growth hormone releasing peptides (GHRPs). In: Kostyo, J., Goodman, H.M. (Eds.), Handbook of Physiology. Oxford University Press, New York/Oxford, pp. 267–297.
Bowers, C.Y., 1999b. Unnatural toward the natural. In: Ghigo, E., Boghen, M., Casanueva, F.F., Dieguez, C. (Eds.), Growth Hormone Secretagogues. Elsevier, New York, pp. 5–18.
Bowers, C.Y., 1999c. GHRP-GH secretagogues. In: Bengtsson, B.-A. (Ed.), Growth Hormone. Kluwer Academic Publishers, Boston/Dordrecht/London, pp. 335–351.
Bowers, C.Y., 2001. Unnatural growth hormone releasing peptide begets natural ghrelin. J. Clin. Endocrinol. Metab. 86, 1464–1469.
Bowers, C.Y., 2002. New insights into the control of growth hormone secretion. In: Kleinberg, D.L., Clemmons, D.R. (Eds.), Central and Peripheral Metabolisms in Pituitary Disease. BioScientifica Ltd., Bristol, pp. 163–176.
Bowers, C.Y., Granda-Ayala, R., 2001. Growth hormone/insulin-like growth factor-1 response to acute and chronic growth hormone-releasing peptide-2, growth hormone - releasing hormone 1-44NH_2 and in combination in older obese men and women with decreased growth hormone secretion. Endocrine 14 (1), 79–86.
Bowers, C.Y., Schally, A.V., 1969. Assay of thyrotropin releasing hormone. In: Meites, J. (Ed.), Proceedings of NIH Conference on Hypothalamic Hypophysiotropic Hormones. The Williams and Wilkins Company, Baltimore, Maryland, pp. 74–80.
Bowers, C.Y., Schally, A.V., Enzmann, F., Boler, J., Folkers, K., 1970. Porcine thyrotropin releasing hormone is (Pyro)GluHisProNH_2. Endocrinology 86, 1143–1153.
Bowers, C.Y., Schally, A.V., Weil, A., Reynolds, G.A., Folkers, K., 1971. Chemical and biological identity of thyrotropin releasing hormone (TRH) of bovine and human origin. In: In: Fellinger, K., Hofer, R. (Eds.), Further Advances in Thyroid Research, vol. 2. Verlag der Wiener Medizinischen Akademie, Vienna, Austria, pp. 1029–1049.
Bowers, C.Y., Friesen, H., Folkers, K., 1973a. Further evidence that TRH is also a physiological regulator of PRL secretion in man. Biochem. Biophys. Res. Commun. 51, 512–521.
Bowers, C.Y., Chang, K., Folkers, K., 1973b. Studies on the LH and FSH releasing activity of the hypothalamic hormone, pGluHisTrpSerTyrGlyLeuArgProGlyNH_2 (Decapeptide), its analogs and other small peptides. In: Gual, C., Rosemberg, E. (Eds.), Hypothalamic Hypophysiotropic Hormones. Excerpta Medica Foundation, Amsterdam, pp. 68–88.
Bowers, C.Y., Friesen, H.G., Malacara, J.M., Folkers, K., 1975. Clinical uses of hypothalamic releasing hormones. In: McMahon, F.G. (Ed.), Endocrine-Metabolic Drugs, Vol. VI. Futura Publishing Company, Inc, Mount Kisco, New York, pp. 115–152.
Bowers, C.Y., Chang, J., Momany, F., Folkers, K., 1977. Effect of the enkephalins and enkephalin analogs on release of pituitary hormones *in vitro*. In: MacIntyre, I., Szelke, M. (Eds.), Proceedings of 6th International Conference on Endocrinology, Molecular Endocrinology. Elsevier/North Holland, Amsterdam, pp. 287–292.
Bowers, C.Y., Momany, F., Chang, D., Hong, A., Chang, K., 1980. Structure-activity relationships of a synthetic pentapeptide that specifically releases GH *in vitro*. Endocrinology 106 (3), 663–667.

Bowers, C.Y., Reynolds, G.A., Chang, D., Hong, A., Chang, K., Momany, F., 1981. A study on the regulation of GH release from the pituitary of rats *in vitro*. Endocrinology 108 (3), 1071–1079.

Bowers, C.Y., Momany, F., Reynolods, G.A., 1982. *In vitro* and *in vivo* activity of a small synthetic peptide with potent GH releasing activity. (Abstract) The Endocrine Society, San Francisco, CA. P. 205.

Bowers, C.Y., Reynolds, G.A., Hong, A., Momany, F., 1983. Studies on pituitary cyclic AMP and GH levels and the release of GH *in vitro*. In: Bhatnager, A.S. (Ed.), The Anterior Pituitary Gland. Raven Press, New York, pp. 165–176.

Bowers, C.Y., Reynolds, G.A., Momany, F., 1984a. New advances on the regulation of growth hormone (GH) secretion. Int. J. Neurol. 18, 188–205.

Bowers, C.Y., Momany, F., Reynolds, G.A., Hong, A., 1984b. On the *in vitro* and *in vivo* activity of a new synthetic hexapeptide that acts on the pituitary to specifically release growth hormone. Endocrinology 114, 1537–1545.

Bowers, C.Y., Reynolds, G.A., Durham, D., Barrera, C.M., Pezzoli, S.S., Thorner, M.O., 1990. Growth hormone releasing peptide stimulates GH release in normal men and acts synergistically with GH releasing hormone. J. Clin. Endocrinol. Metab. 70, 975–982.

Bowers, C.Y., Sartor, A.O., Reynolds, G.A., Badger, T.M., 1991. On the actions of the growth hormone releasing hexapeptide, GHRP. Endocrinology 128, 2027–2035.

Bowers, C.Y., Alster, D.K., Frentz, J.M., 1992. The growth hormone releasing activity of a synthetic hexapeptide in normal men and short statured children after oral administration. J. Clin. Endocrinol. Metab. 74, 292–298.

Bowers, C.Y., Veeraragavan, K., Sethumadhavan, K., 1994. Atypical growth hormone releasing peptides. In: Bercu, B.B., Walker, R.F. (Eds.), Growth Hormone II Basic and Clinical Aspects. Springer-Verlag, New York, pp. 203–222.

Bowers, C.Y., Granda, R., Mohan, S., Kuipers, J., Baylink, D., Veldhuis, J.D., 2004. Sustained elevation of pulsatile growth hormone (GH) secretion and insulin-like growth factor I (IGF-I), IGF-binding protein-3 (IGFBP-3) and IGFBP-5 concentrations during 30 day continuous subcutaneous infusion of GH releasing peptide-2 in older men and women. J. Clin. Endocrinol. Metab. 89, 2290–2300.

Bowers, C.Y., 2005. Octanoyl ghrelin is hypothalamic rooted. Endocrinology 146, 2508–2509.

Bowers, C.Y., Laferrere, B., Hurley, D.L., Veldhuis, J.D., 2008. The role of growth hormone secretagogues and ghrelin in feeding and body composition. In: Donohoue, P.A. (Ed.), Energy Metabolism and Obesity. Humana Press, Totowa, New Jersey, pp. 125–154.

Bowers, C.Y., Reynolds, G.A., Veldhuis, J.D., 2012. Ghrelin: a history of its discovery. In: Smith, R.G., Thorner, M.O. (Eds.), Ghrelin in Health and Disease. Springer Science+Business Media, Inc, New York, pp. 1–35.

Brazeau, P.W., Vale, W., Burgus, R., Ling, N., Butcher, M., Riviere, J., Guillemin, R., 1973. Hypothalamic polypeptide that inhibits the secretion of immunoreactive pituitary growth hormone. Science 179, 77–79.

Burgus, R., Dunn, T.F., Ward, D.N., Vale, W., Amoss, M., Guillemin, R., 1969. Derives poltyeptidiques de synthese doues d'activite hypophysiotrope TRF. C.R. Acad. Sci. Paris 268, 2116–2118.

Codd, E.E., Shu, A., Walker, R.F., 1989. Binding of a growth hormone releasing hexapeptide to specific hypothalamic and pituitary sites. Neuropharmacology 28, 1139–1144.

Folkers, K., Enzmann, F., Boler, J., Bowers, C.Y., Schally, A.V., 1969. Discovery of modification of the synthetic tripeptide sequence of the thyrotropin releasing hormone having activity. Biochem. Biophys. Res. Commun. 37, 123–126.

Guillemin, R., 1996. Growth hormone releasing factor: a brief history of its time. In: Bercu, B.B., Walker, R. (Eds.), Growth Hormone Secretagogues. Springer-Verlag, New York, pp. 3–8.

Guillemin, R., Brazeau, P., Bohlen, P., Esch, F., Ling, N., Wehrenberg, W.B., 1982. Growth hormone releasing factor from a human pancreatic tumor that caused acromegaly. Science 218, 585–587.

Holst, B., Cygankiewicz, A., Jensen, T.H., Ankersen, M., Schwartz, T.W., 2003. High constitutively signaling of the ghrelin receptor-identification of a potent inverse agonist. Mol. Endocrinol. 17, 2201–2210.

Howard, A.D., Feighner, S.D., Cully, D.F., Arena, J.P., Liberator, P.A., Rosenblum, C.I., Hamelin, M., Hreniuk, D.L., Palyha, O.C., Anderson, J., Paress, P.S., Diaz, C., Chou, M., Liu, K.K., McKee, K.K., Pong, S.S., Chaung, L.Y., Elbrecht, A., Dashkevicz, M., Heavens, R., Rigby, M., Sirinathsinghji, D.J.S., Dean, D.D., Melillo, D.G., Patchett, A.A., Nargund, R., Griffin, P.R., DeMartino, J.A., Gupta, S.K., Schaeffer, J.M., Smith, R.G., Van der Ploeg, L.H.T., 1996. A receptor in pituitary and hypothalamus that functions in growth hormone release. Science 273, 974–977.

Jaffe, C.A., Ho, J., Demott-Frieberg, R.D., Bowers, C.Y., Barkan, A.L., 1993. Effects of a prolonged growth hormone (GH) releasing peptide infusion on pulsatile GH secretion in normal men. J. Clin. Endocrinol. Metab. 77, 1641–1647.

Klinger, B., Silbergeld, A., Deghenghi, R., Frenkel, J., Laron, Z., 1996. Desensitization from long term intranasal treatment with hexarelin does not interfere with the biological effects of this growth hormone releasing peptide in short children. Eur. J. Endocrinol. 134, 716–719.

Kojima, M., Hosada, H., Date, Y., Nakazato, M., Matsuo, M., Kangawa, K., 1999. Ghrelin is a growth hormone releasing acylated peptide from stomach. Nature 402, 656–660.

Locke, W., Kirgis, H.D., Bowers, C.Y., Abdoh, A.A., 1995. Effect of intra-cerebroventricular injection of growth hormone releasing peptide on feeding behavior and plasma growth hormone levels of rats. Life Sci. 56, 1347–1352.

McCormick, G.F., Millard, W.J., Badger, T.M., Bowers, C.Y., Martin, J.B., 1985. Dose-response characteristics of various peptides with growth hormone releasing activity in the unanesthetized male rat. Endocrinology 117, 97–105.

Mericq, V., Cassorla, F., Garcia, H., Avila, A., Bowers, C.Y., Merriam, G., 1995. Growth hormone responses to growth hormone releasing peptide (GHRP) in growth hormone deficient children (GHD). J. Clin. Endocrinol. Metab. 80, 1681–1684.

Mericq, V., Cassorla, F., Salazar, T., Avila, A., Inguez, G., Bowers, C.Y., Merriam, G.R., 1998. Effects of eight months treatment with graded doses of a growth hormone releasing peptide in GH-deficient children. J. Clin. Endocrinol. Metab. 83, 2355–2360.

Momany, F.A., Bowers, C.Y., 1996. Computer assisted modeling of xenobiotic growth hormone secretagogues. In: Bercu, B.B., Walker, R.F. (Eds.), Growth Hormone Secretagogues. Springer, New York, pp. 73–83.

Momany, F., Bowers, C.Y., Reynolds, G.A., Chang, D., Hong, A., Newlander, K., 1981. Design, synthesis and biological activity of peptides which release growth hormone *in vitro*. Endocrinology 108 (1), 31–39.

Momany, F., Bowers, C.Y., Reynolds, G.A., Hong, A., Newlander, K., 1984. Conformational energy studies and *in vivo* activity data on active GH releasing peptides. Endocrinology 114, 1531–1536.

Okada, K., Ishii, S., Minami, S., Sugihara, H., Shibasaki, T., Wakabayashi, I., 1996. Intracerebroventricular administration of the growth hormone releasing peptide KP-102 increases food intake in free-feeding rats. Endocrinology 137 (11), 5155–5158.

Pandya, C.A., Demott-Friberg, R., Bowers, C.Y., Barkan, A.L., Jaffe, C.A., 1998. Growth hormone (GH) releasing peptide-6 requires endogenous hypothalamic GH releasing hormone for maximal GH stimulation. J. Clin. Endocrinol. Metab. 83, 1186–1189.

Pihoker, C., Middleton, R., Reynolds, G.A., Bowers, C.Y., Badger, T.M., 1995. Diagnostic studies with intravenous and intranasal growth hormone releasing peptide-2 in children of short stature. J. Clin. Endocrinol. Metab. 80, 2987–2992.

Pihoker, C., Badger, T.M., Reynolds, G.A., Bowers, C.Y., 1997. Treatment effects of intranasal growth hormone releasing peptide-2 in children with short stature. J. Endocrinol. 155, 79–86.

Pihoker, C., Kearns, G.L., French, D., Bowers, C.Y., 1998. Pharmacokinetics and pharmacodynamics of growth hormone releasing peptide-2: a phase I study in children. J. Clin. Endocrinol. Metab. 83, 1168–1172.

Rivier, K.J., Spiess, M., Thorner, M., Vale, W., 1982. Characterization of a growth hormone releasing factor from a human pancreatic islet tumor. Nature 300, 276–278.

Sartor, O., Bowers, C.Y., Chang, K., 1985a. Parallel studies of His DTrpAlaTrpDPheLysNH_2 and hpGRF in rat pituitary primary cell monolayer culture. Endocrinology 116, 952–957.

Sartor, O., Bowers, C.Y., Reynolds, G.A., Momany, F., 1985b. Variables determining the GH response of HisDTrpAlaTrpDPheLysNH_2 (GHRP-6) in the rat. Endocrinology 117, 1441–1447.

Sato, T., Fukue, Y., Teranishi, H., Yoshida, Y., Kojima, M., 2005. Molecular forms of hypothalamic ghrelin and its regulation by fasting and 2-deoxy-D-glucose administration. Endocrinology 146, 2510–2516.

Schally, A.V., Bowers, C.Y., 1964. Purification of luteinizing hormone releasing factor from bovine hypothalami. Endocrinology 75, 608–614.

Schally, A.V., Bowers, C.Y., 1970. The nature of thyrotropin-releasing hormone (TRH). In: Kenny, A.D., Anderson, R.R. (Eds.), Proceedings of Sixth Midwest Conference on the Thyroid and Endocrinology. University of Missouri, Columbia, pp. 25–63.

Schally, A.V., Bowers, C.Y., Redding, T.W., Barrett, J.F., 1966. Isolation of thyrotropin releasing factor (TRF) from porcine hypothalamus. Biochem. Biophys. Res. Commun. 25, 165–169.

Schally, A.V., Redding, T.W., Bowers, C.Y., Barrett, J.F., 1969. Isolation and properties of porcine thyrotropin releasing hormone. J. Biol. Chem. 244, 4077–4088.

Sethumadhavan, K., Veeraragavan, K., Bowers, C.Y., 1991. Demonstration and characterization of the specific binding of growth hormone releasing peptide (GHRP) to rat anterior pituitary and hypothalamic membranes. Biochem. Biophys. Res. Commun. 178, 31–37.

Smith, R.G., Pong, S.S., Hickey, G., Jacks, T., Cheng, K., Leonard, R., Cohen, C.J., Arena, J.P., Chang, C.H., Drisko, J., Wyvratt, M., Fisher, M., Nargund, R., Patchett, A., 1996. Modulation of pulsatile GH release through a novel receptor in hypothalamus and pituitary gland. In: Conn, P.M. (Ed.), Recent Progress in Hormone Research. The Endocrine Society, Bethesda, Maryland, pp. 261–286.

Tannenbaum, G.S., Bowers, C.Y., 2001. Interactions of growth hormone secretagogues and growth hormone releasing hormone/somatostatin. Endocrine 14, 21–27.

Tannenbaum, G.S., Epelbaum, J., Bowers, C.Y., 2003. Interrelationship between the most novel peptide ghrelin and somatostatin/growth hormone releasing hormone in regulation of pulsatile growth hormone secretion. Endocrinology 144, 967–974.

Van den Berghe, G., Wouters, P., Weekers, F., Mohan, S., Baxter, R.C., Veldhuis, J.D., Bowers, C.Y., Bouillon, R., 1999. Reactivation of pituitary hormone release and metabolic improvement by infusion of growth hormone releasing peptide and thyrotropin releasing peptide in patients with protracted critical illness. J. Clin. Endocrinol. Metab. 84, 1311–1323.

Van der Ploeg, L.H.T., Howard, A.D., Smith, R.G., Feighner, S.D., 1998. Molecular cloning and characterization of human, swine, and rat growth hormone secretagogue receptors. In: Bercu, B.B., Walker, R.F. (Eds.), Growth Hormone Secretagogues. Markel Dekker, Inc, New York, pp. 57–76.

Veeraragavan, K., Sethumadhavan, K., Bowers, C.Y., 1992. Growth hormone releasing peptide (GHRP) binding to porcine anterior pituitary and hypothalamic membranes. Life Sci. 50, 1149–1155.

CHAPTER TWO

Orphan GPCRs and Methods for Identifying Their Ligands

Morikatsu Yoshida*, Mikiya Miyazato*, Kenji Kangawa[†,1]
*Department of Biochemistry, National Cerebral and Cardiovascular Center Research Institute, Suita, Osaka, Japan
[†]National Cerebral and Cardiovascular Center Research Institute, Suita, Osaka, Japan
[1]Corresponding author: e-mail address: kangawa@ri.ncvc.go.jp

Contents

Abstract

G protein-coupled receptors (GPCRs) constitute the largest family of cell-surface receptors. These proteins play a crucial role in physiology by facilitating cell communication through recognition of diverse ligands, including bioactive peptides, amines, nucleosides, and lipids. The human genome sequencing project identified more than 100 orphan GPCRs, whose ligands had not yet been discovered. We subsequently identified ghrelin, neuromedin U, and neuromedin S as endogenous ligands of various orphan GPCRs and have proposed various mechanisms through which these peptides regulate physiological functions through their cognate GPCRs. In this chapter, we review methods for identifying novel peptide ligands of orphan GPCRs.

1. INTRODUCTION

Bioactive peptides, such as hormones and neuropeptides, play important roles in the cell-to-cell communication necessary for numerous physiological phenomena. These diverse effects typically are mediated by G protein-coupled receptor (GPCR) signaling pathways.

Methods in Enzymology, Volume 514
ISSN 0076-6879
http://dx.doi.org/10.1016/B978-0-12-381272-8.00002-7

GPCRs constitute the largest family of cell-surface receptors and facilitate cell-to-cell communication through their recognition of diverse ligands, including bioactive peptides, amines, nucleosides, and lipids. The hamster β_2-adrenergic receptor (Dixon et al., 1986) and rat muscarinic acetylcholine receptor (Kubo et al., 1986) were the first GPCRs to be identified. Typical of their protein family, both these receptors share a seven-transmembrane domain topology with an extracellular N terminus and a cytoplasmic C terminus. The amino acid and nucleotide sequences of these domains are highly conserved among GPCRs (Probst et al., 1992). This structural feature led to a search for novel GPCRs. Nucleic acid-based homology screening approaches such as low-stringency hybridization, degenerate polymerase chain reaction techniques, and bioinformatic analysis of genomes have been applied to find novel GPCRs (Kobilka et al., 1987; Libert et al., 1989; Takeda et al., 2002). Since the cloning of the cDNA encoding the substance K receptor and its identification as a GPCR (Masu et al., 1987), numerous other bioactive peptide and neurotransmitter receptor cDNAs have been identified as GPCRs (Vassilatis et al., 2003).

Owing to the successful mapping of the human genome, the genes of approximately 400 nonolfactory GPCRs have been cloned to date. More than 100 GPCRs still remain as "orphan receptors," for which the endogenous ligands are unknown (Vassilatis et al., 2003). Moreover, agonists and antagonists of GPCR function are important pharmaceutical agents and constitute more than 40% of all marketed drugs (Drews, 2000). Therefore, the identification of endogenous ligands of orphan GPCRs likely will promote the discovery of novel regulatory mechanisms of physiological function and new drug targets. Here, we review several methods for identifying novel peptide ligands of orphan GPCRs.

2. STRATEGIES FOR IDENTIFYING LIGANDS OF ORPHAN GPCRs

The conventional approach for identifying bioactive peptides is initiated by the discovery of a biological activity. The novel bioactive peptide responsible for the activity then is isolated. Subsequently, the peptide is used to reveal the corresponding receptor, which is used as a drug target in high-throughput screening.

In contrast, the current typical strategy for identifying ligands of orphan GPCRs is known as "reverse pharmacology" (Libert et al., 1991). In the reverse pharmacology approach (Fig. 2.1), the orphan GPCR of interest first

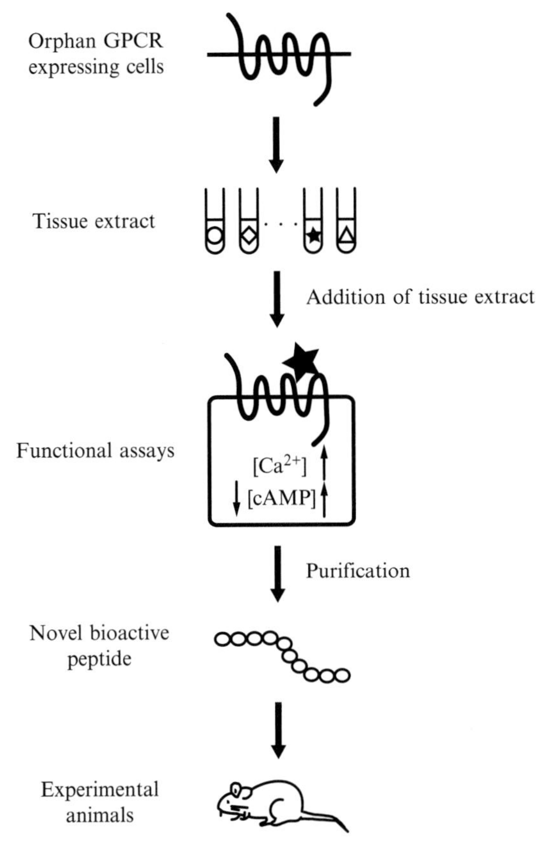

Figure 2.1 Reverse pharmacology approach to identify endogenous ligands of orphan GPCRs. The target orphan GPCR is expressed in a recombinant expression system, such as CHO or HEK293 cells. Candidate ligands in tissue extracts are screened against cell lines expressing target orphan GPCRs by measuring changes in intracellular second messengers. An active peptide is purified from tissue extract with ligand properties as a potential endogenous ligand of an orphan GPCR. This novel peptide can be used to study the physiological role of its cognate receptor.

is expressed in eukaryotic cells for functional analysis. The expressed receptor then is used in functional assays to screen candidate ligands. Once identified, the specific ligand is used to characterize the biological and physiological roles of its receptor. This strategy was employed successfully for the first time in 1995 to identify the neuropeptide nociceptin/orphanin FQ as an endogenous ligand of the opioid receptor-like 1 (ORL-1) receptor (Meunier et al., 1995; Reinscheid et al., 1995). Since the discovery of nociceptin/orphanin FQ, reverse pharmacology methods have yielded several additional bioactive peptides (Table 2.1).

Table 2.1 Bioactive peptides of orphan GPCRs identified using reverse pharmacology approach

Ligand	Orphan GPCR	Reference
Nociceptin/Orphanin FQ	ORL-1	Meunier et al. (1995) and Reinscheid et al. (1995)
Orexin-A, Orexin-B	OX_1R, OX_2R	Sakurai et al. (1998)
Prolactin-releasing peptide	GPR10	Hinuma et al. (1998)
Apelin	APJ	Tatemoto et al. (1998)
Motilin	GPR38	Feighner et al. (1999)
Melanin-concentrating hormone	SLC-1	Saito et al. (1999)
Ghrelin	GHS-R	Kojima et al. (1999)
Neuromedin U	FM-3/GPR66, FM-4/TGR-1	Kojima et al. (2000)
RFamide-related peptide	GPR147	Hinuma et al. (2000)
Metastin	GPR54	Ohtaki et al. (2001)
Neuropeptide W	GPR7, GPR8	Shimomura et al. (2002)
Neuropeptide B	GPR7, GPR8	Fujii et al. (2002)
Pyroglutamylated RFamide peptide	GPR103	Fukusumi et al. (2003)
Neuropeptide S	GPR154	Xu et al. (2004)
Neuromedin S	FM-3/GPR66, FM-4/TGR-1	Mori et al. (2005)

Successful identification of ligands of orphan GPCRs involves at least four crucial aspects: (1) selecting the target orphan GPCR, (2) generating a recombinant expression system, (3) screening the ligand source, and (4) using functional assays.

3. SELECTION OF TARGET ORPHAN GPCRs

Owing to advances in bioinformatics, the current wealth of genomic information can be used to classify orphan GPCRs phylogenetically and thus predict the type of ligand that a target receptor might bind. For example, comparison of deduced amino acid sequences revealed that the orphan

GPCR growth hormone secretagogue receptor (GHS-R) was 35% identical to the neurotensin receptor and 29% identical to the thyrotropin-releasing hormone (TRH) receptor (Howard et al., 1996). Therefore, like neurotensin and TRH, the endogenous ligand of GHS-R was predicted to be a bioactive peptide, in light of the high sequence similarity among the sequences of the cognate receptors. Ghrelin, a peptide isolated from rat stomach, subsequently was identified as the endogenous ligand of GHS-R (Kojima et al., 1999).

Information in genomic databases indicates four additional orphan GPCRs whose structures are similar to that of GHS-R (Fig. 2.2). One of these is GPR38, which has been identified as the motilin receptor (Feighner et al., 1999). Two additional receptors, FM-3/GPR66 and FM-4/TGR-1, are now known to be the neuromedin U and neuromedin S receptors (Kojima et al., 2000; Mori et al., 2005). Because of their high protein sequence homology (51%), FM-3/GPR66 and FM-4/TGR-1 can bind same ligands. Although Zhang et al. (2005) reported that obestatin, a peptide derived from the ghrelin precursor, was the cognate ligand of GPR39 (the fourth GHS-R-like orphan GPCR), recent studies have failed to support this assignment (Chartrel et al., 2007; Holst et al., 2007; Lauwers et al., 2006). Because of the high protein sequence homology between GHS-R and GPR39 (McKee et al., 1997), the endogenous ligand of GPR39 is predicted to be a bioactive peptide.

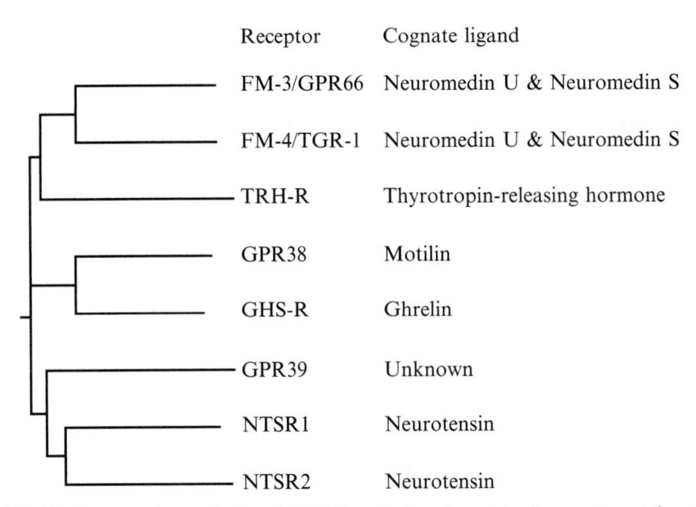

Figure 2.2 Phylogenetic analysis of GHS-R and closely related receptors. The tree was generated from protein sequence alignment of human receptors by using the ClustalW program in the GenomeNet database (http://www.genome.jp).

4. RECOMBINANT EXPRESSION SYSTEMS FOR ORPHAN GPCRs

High-throughput screening of candidate endogenous ligands typically relies on cells that stably express the orphan GPCR protein. For these recombinant expression systems to detect ligand activity with high sensitivity, the target receptors need to be expressed at high levels at the cell surface. To this end, an expression vector is created that links DNA encoding the target GPCR with a strong promoter sequence (e.g., cytomegalovirus promoter). This construct then is introduced into an appropriate cell line (typically CHO [Chinese hamster ovary] or HEK [human embryonic kidney] 293 cells) for protein expression. In the absence of a ligand to confirm target receptor expression, it is important to verify the presence of the target receptor at both the mRNA and protein levels. Although Northern blotting of cell lines readily indicates receptor expression, whether the receptor protein is expressed at the cell surface is unclear. To avoid this problem, several groups have generated receptors that contain short epitope tags at the N terminus, such as FLAG and/or hemagglutinin tags (Koller et al., 1997). Another approach has been to generate the target receptor as a fusion protein with green fluorescent protein (Kallal and Benovic, 2000). Coupled with fluorescence-activated cell sorting, confocal fluorescent microscopy, or immunocytochemistry, both of these approaches have enabled verification of the site of target protein expression.

5. SOURCES OF POTENTIAL LIGANDS OF ORPHAN GPCRs

5.1. Tissue extracts

Tissue extracts are the usual sources for candidate endogenous ligands of orphan GPCRs. Orphan GPCR-expressing cells are used to screen tissue extracts for candidate ligands. Ligands present in the extract bind to and consequently activate GPCRs, which in turn couple heterotrimeric G proteins containing α, β, and γ subunits to alter the concentrations and activities of intracellular second-messenger molecules, such as 3′,5′-cyclic adenosine monophosphate (cAMP) and calcium. After a source tissue extract is identified, a second-messenger assay is used in the chromatography-based purification of endogenous ligands of orphan GPCRs. Using the approach highlighted in Fig. 2.3, we identified ghrelin, neuromedin U, and

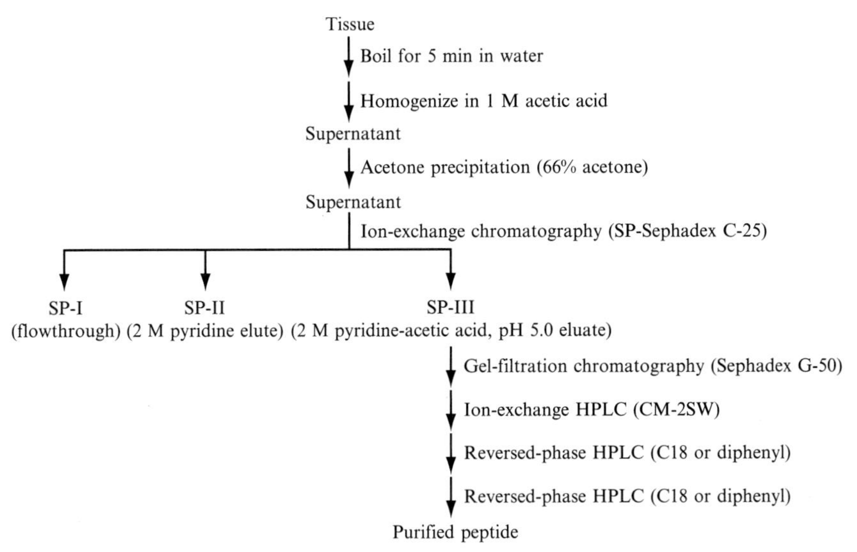

Figure 2.3 The extraction and purification of bioactive peptides. Almost all bioactive peptides are present at low concentrations in normal tissues and are easily hydrolyzed by intrinsic proteases during extraction procedures. We developed the method to boil tissues in water to inactivate intrinsic proteases before peptides extraction from tissues. By combining high-performance liquid chromatography (HPLC) procedures with this method, we have been able to purify bioactive peptides in high yield.

neuromedin S as endogenous ligands of orphan GPCRs (Kojima et al., 1999, 2000; Mori et al., 2005).

5.2. Bioinformatics

A motif that is a common feature among the sequences of bioactive peptide precursors in a genomic database can be used to discover novel genes that encode a preproprotein. To this end, mature peptides predicted from the database are synthesized and screened against cell lines expressing various orphan GPCRs by measuring changes in intracellular second-messenger molecules. Neuropeptide B (Fujii et al., 2002), RF amide-related peptide (Hinuma et al., 2000), and pyroglutamylated RF amide peptide (Fukusumi et al., 2003) were identified as endogenous ligands of orphan GPCRs through this approach. However, this bioinformatics-based technique was unsuccessful for ghrelin, which is modified by *n*-octanoic acid. The acyl modification of ghrelin is necessary for its bioactivity and binding to GHS-R (Kojima et al., 1999); such posttranslational modification cannot currently be predicted from genomic data. Further development of predictive algorithms for novel peptide precursors likely will improve the efficiency of the bioinformatics approach.

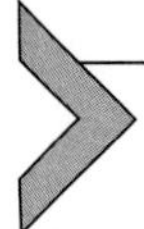

6. FUNCTIONAL ASSAYS FOR SCREENING CANDIDATE LIGANDS OF ORPHAN GPCRs

As mentioned earlier, activated GPCRs couple to heterotrimeric G proteins and alter the amounts and activities of second-messenger molecules. Mammalian G protein α subunits consist of 16 subunits and are organized into four categories depending on their signaling cascades (Cabrera-Vera et al., 2003). Proteins in the $G\alpha_s$ family couple to adenylate cyclase to increase intracellular cAMP, whereas members of the $G\alpha_{i/o}$ family inhibit adenylate cyclase, open potassium channels, and close calcium channels. $G\alpha_{q/11}$ proteins activate phospholipase Cβ (PLCβ) to release intracellular calcium from the endoplasmic reticulum (ER). $G\alpha_{12/13}$ proteins activate the signaling pathway of RhoA, which belongs to the Ras superfamily of GTP-binding proteins (Wettschureck and Offermanns, 2005). In general, functional assays used to screen for activating ligands measure changes in intracellular second messengers, such as cAMP and calcium. These changes are either measured directly or quantitated in reporter gene assays.

One effective way to identify ligands for $G\alpha_q$-coupled orphan GPCRs is to measure changes in intracellular calcium concentration by using a fluorometric imaging plate reader (FLIPR™; Molecular Devices, Sunnyvale, CA). Because a particular benefit of this technology is the ability to add compounds to and detect signals from every well of a 96-well plate simultaneously, this system is used widely for high-throughput screening assays. For example, we used this methodology to identify ghrelin, neuromedin U, and neuromedin S as endogenous ligands of orphan GPCRs (Kojima et al., 1999, 2000; Mori et al., 2005).

One typical method to identify ligands for $G\alpha_s$-coupled orphan GPCRs is to measure ligand-induced cAMP accumulation in whole cells. Although radioimmunoassays previously were used widely to measure intracellular cAMP production (Katafuchi et al., 2003), fluorescence- or luminescence-based assays are more typical choices currently, to avoid the use of radioactivity (Hill et al., 2010). For example, cAMP assay kits based on the amplified luminescent proximity homogeneous assay screen (Alpha Screen™; Perkin Elmer, Waltham, MA) measure cAMP accumulation in response to ligand-associated receptor activation. The high sensitivity and simplicity of these assay protocols makes them ideal for high-throughput screening. For

$G\alpha_i$-coupled orphan GPCRs, the activity of the receptor is monitored through its inhibition of forskolin (a direct activator of adenylate cyclase) and subsequent cAMP accumulation.

Increases in intracellular second-messenger concentrations induce several transcription factors to bind to their response elements and regulate target gene expression. In reporter gene assays, activated target GPCRs lead to transcriptional induction of a second-messenger-responsive reporter gene. For example, receptors coupled to $G\alpha_s$ activation typically are monitored by using reporter constructs containing cAMP response elements. Activation of $G\alpha_q$- and $G\alpha_i$-coupled receptors are monitored by using constructs whose promoters contain either TPA (12-O-tetradecanoylphorbol-13-acetate) response elements, serum response factor response elements, nuclear factor of activated T-cells (NFAT) response elements, or nuclear factor κB response elements (Kotarsky et al., 2003). Luciferase, which is often used as a reporter enzyme, has no internal background in eukaryotic cells (Greer and Szalay, 2002). In these assays, the target GPCR expression vector and second-messenger-responsive luciferase gene plasmid usually are transiently cotransfected into CHO or HEK293 cells.

Which functional assay to use to screen candidate ligands of orphan GPCRs is a difficult choice, because we cannot accurately predict the G proteins likely to couple these receptors. To circumvent this problem, several approaches are helpful. Promiscuous G protein α subunits such as $G\alpha_{15}$ and $G\alpha_{16}$ can bind both $G\alpha_s$- and $G\alpha_i$-coupled receptors to stimulate PLCβ to increase intracellular calcium (Offermanns and Simon, 1995). In chimeric G proteins, the five C-terminal residues of $G\alpha_q$ are replaced with the five C-terminal residues of $G\alpha_s$ or $G\alpha_i$. The five C-terminal residues of Gα proteins are sufficient for receptor contact, and the rest of the subunit interacts with effector molecules (Conklin et al., 1993). For example, a $G\alpha_{q/i3}$ chimera designed to drive $G\alpha_i$-coupled GPCR to $G\alpha_q$ activation contained the five C-terminal residues of $G\alpha_{i3}$ and retained the rest of the $G\alpha_q$ sequence (Saito et al., 1999). Methods based on chimeric or promiscuous G proteins are independent of the coupling specificity of receptors to G proteins and allow screening of all GPCRs in the high-throughput FLIPR system. In these assays, the target GPCR and chimeric or promiscuous G protein are transiently cotransfected into CHO or HEK293 cells.

In mammalian cell assays, candidate ligands have to be screened against both untransfected cells and the transfected cell line containing the orphan GPCR to confirm specific detection of target receptor-expressed cells. The presence of endogenous GPCRs can result in background noise in a functional assay as a result of the activation of the endogenous receptors in a transfected cell line. To circumvent this problem, Jurkat cells cotransfected with the human GPR103 cDNA and a luciferase reporter driven by NFAT were used to monitor ligand-associated activities (Takayasu et al., 2006). Jurkat cells express only a limited number of GPCRs that become activated by substances in crude tissue extracts.

7. SUMMARY

No new bioactive peptides for orphan GPCRs have been identified since 2005, when neuromedin S was identified as an endogenous ligand of both FM-3/GPR66 and FM-4/TGR-1 (Mori et al., 2005). Two main issues currently impede the discovery process.

The first drawback is the low tissue content of these novel peptide ligands under normal physiological conditions. Perhaps the concentrations of some bioactive peptides might be increased in specific tissues under particular pathophysiological conditions. Alternatively, a ligand of a target GPCR may be labile under the conditions of conventional peptide extraction. For example, acyl-modified ghrelin is cleaved easily into des-acyl ghrelin during peptide extraction (Hosoda et al., 2000). If the ligand of a target orphan GPCR has low tissue content and is labile, conventional approaches are unlikely to be successful. The development of animal models of various pathological states and new methods for peptide extraction will help to overcome these problems.

The second issue challenging the discovery of ligands of orphan GPCRs is that some of these proteins are retained in the ER when expressed in recombinant expression system such as CHO and HEK293 cells. Such receptors may require accessory proteins, other GPCRs as with the $GABA_B$ receptor, to translocate to the cell surface (Kaupmann et al., 1998). In addition, even though a target receptor might be expressed at the cell surface, other molecules such as receptor-activity-modifying proteins may have to associate with the receptor so that it can bind cognate ligand (McLatchie et al., 1998). The development of novel recombinant expression systems and discovery of novel accessory molecules that regulate orphan GPCRs likely will yield important insights into the functions and activities of these receptors.

ACKNOWLEDGMENTS

This work was supported in part by Grants-in-Aid for Scientific Research from the Ministry of Education, Science, Sports, and Culture of Japan to M. Y., M. M., and K. K.

REFERENCES

Cabrera-Vera, T.M., et al., 2003. Insights into G protein structure, function, and regulation. Endocr. Rev. 24, 765–781.

Chartrel, N., et al., 2007. Comment on "Obestatin, a peptide encoded by the ghrelin gene, opposes ghrelin's effects on food intake" Science 315, 766.

Conklin, B.R., et al., 1993. Substitution of 3 amino acids switches receptor specificity of $G_q\alpha$ to that of $G_i\alpha$. Nature 363, 274–276.

Dixon, R.A.F., et al., 1986. Cloning of the gene and cDNA for mammalian β-adrenergic receptor and homology with rhodopsin. Nature 321, 75–79.

Drews, J., 2000. Drug discovery: a historical perspective. Science 287, 1960–1964.

Feighner, S.D., et al., 1999. Receptor for motilin identified in the human gastrointestinal system. Science 284, 2184–2188.

Fujii, R., et al., 2002. Identification of a neuropeptide modified with bromine as an endogenous ligand for GPR7. J. Biol. Chem. 277, 34010–34016.

Fukusumi, S., et al., 2003. A new peptidic ligand and its receptor regulating adrenal function in rats. J. Biol. Chem. 278, 46387–46395.

Greer, L.F., Szalay, A.A., 2002. Imaging of light emission from the expression of luciferases in living cells and organisms: a review. Luminescence 17, 43–74.

Hill, S.J., et al., 2010. Insights into GPCR pharmacology from the measurement of changes in intracellular cyclic AMP; advantages and pitfalls of differing methodologies. Br. J. Pharmacol. 161, 1266–1275.

Hinuma, S., et al., 1998. A prolactin-releasing peptide in the brain. Nature 393, 272–276.

Hinuma, S., et al., 2000. New neuropeptides containing carboxy-terminal RFamide and their receptor in mammals. Nat. Cell Biol. 2, 703–708.

Holst, B., et al., 2007. GPR39 signaling is stimulated by zinc ions but not by obestatin. Endocrinology 148, 13–20.

Hosoda, H., et al., 2000. Ghrelin and des-acyl ghrelin: two major forms of rat ghrelin peptide in gastrointestinal tissue. Biochem. Biophys. Res. Commun. 279, 909–913.

Howard, A.D., et al., 1996. A receptor in pituitary and hypothalamus that functions in growth hormone release. Science 273, 974–977.

Kallal, L., Benovic, J.L., 2000. Using green fluorescent proteins to study G-protein-coupled receptor localization and trafficking. Trends Pharmacol. Sci. 21, 175–180.

Katafuchi, T., et al., 2003. Calcitonin receptor-stimulating peptide, a new member of the calcitonin gene-related peptide family. Its isolation from porcine brain, structure, tissue distribution, and biological activity. J. Biol. Chem. 278, 12046–12054.

Kaupmann, K., et al., 1998. $GABA_B$-receptor subtypes assemble into functional heteromeric complexes. Nature 396, 683–687.

Kobilka, B.K., et al., 1987. cDNA for the human beta-2-adrenergic receptor: a protein with multiple membrane-spanning domains and encoded by a gene whose chromosomal location is shared with that of the receptor for platelet-derived growth factor. Proc. Natl. Acad. Sci. U.S.A. 84, 46–50.

Kojima, M., et al., 1999. Ghrelin is a growth-hormone-releasing acylated peptide from stomach. Nature 402, 656–660.

Kojima, M., et al., 2000. Purification and identification of neuromedin U as an endogenous ligand for an orphan receptor GPR66 (FM3). Biochem. Biophys. Res. Commun. 276, 435–438.

Koller, K.J., et al., 1997. A generic method for the production of cell lines expressing high levels of 7-transmembrane receptors. Anal. Biochem. 250, 51–60.

Kotarsky, K., et al., 2003. Improved reporter gene assays used to identify ligands acting on orphan seven-transmembrane receptors. Pharmacol. Toxicol. 93, 249–258.

Kubo, T., et al., 1986. Cloning, sequencing, and expression of complementary DNA encoding the muscarinic acetylcholine receptor. Nature 323, 411–416.

Lauwers, E., et al., 2006. Obestatin does not activate orphan G protein-coupled receptor GPR39. Biochem. Biophys. Res. Commun. 351, 21–25.

Libert, F., et al., 1989. Selective amplification and cloning of four new members of the G protein-coupled receptor family. Science 244, 569–572.

Libert, F., et al., 1991. Current developments in G-protein-coupled receptors. Curr. Opin. Cell Biol. 3, 218–223.

Masu, Y., et al., 1987. cDNA cloning of bovine substance-K receptor through oocyte expression system. Nature 329, 836–838.

McKee, K.K., et al., 1997. Cloning and characterization of two human G protein-coupled receptor genes (GPR38 and GPR39) related to the growth hormone secretagogue and neurotensin receptors. Genomics 46, 426–434.

McLatchie, L.M., et al., 1998. RAMPs regulate the transport and ligand specificity of the calcitonin-receptor-like receptor. Nature 393, 333–339.

Meunier, J.C., et al., 1995. Isolation and structure of the endogenous agonist of opioid receptor-like ORL1 receptor. Nature 377, 532–535.

Mori, K., et al., 2005. Identification of neuromedin S and its possible role in the mammalian circadian oscillator system. EMBO J. 24, 325–335.

Offermanns, S., Simon, M.I., 1995. $G\alpha_{15}$ and $G\alpha_{16}$ couple a wide variety of receptors to phospholipase C. J. Biol. Chem. 270, 15175–15180.

Ohtaki, T., et al., 2001. Metastasis suppressor gene KiSS-1 encodes peptide ligand of a G-protein-coupled receptor. Nature 411, 613–617.

Probst, W.C., et al., 1992. Sequence alignment of the G-protein coupled receptor superfamily. DNA Cell Biol. 11, 1–20.

Reinscheid, R.K., et al., 1995. Orphanin FQ: a neuropeptide that activates an opioidlike G protein-coupled receptor. Science 270, 792–794.

Saito, Y., et al., 1999. Molecular characterization of the melanin-concentrating-hormone receptor. Nature 400, 265–269.

Sakurai, T., et al., 1998. Orexins and orexin receptors: a family of hypothalamic neuropeptides and G protein-coupled receptors that regulate feeding behavior. Cell 92, 573–585.

Shimomura, Y., et al., 2002. Identification of neuropeptide W as the endogenous ligand for orphan G-protein-coupled receptors GPR7 and GPR8. J. Biol. Chem. 277, 35826–35832.

Takayasu, S., et al., 2006. A neuropeptide ligand of the G protein-coupled receptor GPR103 regulates feeding, behavioral arousal, and blood pressure in mice. Proc. Natl. Acad. Sci. U.S.A. 103, 7438–7443.

Takeda, S., et al., 2002. Identification of G protein-coupled receptor genes from the human genome sequence. FEBS Lett. 520, 97–101.

Tatemoto, K., et al., 1998. Isolation and characterization of a novel endogenous peptide ligand for the human APJ receptor. Biochem. Biophys. Res. Commun. 251, 471–476.

Vassilatis, D.K., et al., 2003. The G protein-coupled receptor repertoires of human and mouse. Proc. Natl. Acad. Sci. U.S.A. 100, 4903–4908.

Wettschureck, N., Offermanns, S., 2005. Mammalian G proteins and their cell-type-specific functions. Physiol. Rev. 85, 1159–1204.

Xu, Y.L., et al., 2004. Neuropeptide S: a neuropeptide promoting arousal and anxiolytic-like effects. Neuron 43, 487–497.

Zhang, J.V., et al., 2005. Obestatin, a peptide encoded by the ghrelin gene, opposes ghrelin's effects on food intake. Science 310, 996–999.

CHAPTER THREE

Purification of Rat and Human Ghrelins

Masayasu Kojima*[,1], Hiroshi Hosoda[†], Kenji Kangawa[‡]

*Molecular Genetics, Institute of Life Science, Kurume University, Kurume, Fukuoka, Japan
[†]Department of Biochemistry, National Cerebral and Cardiovascular Center Research Institute, Suita, Osaka, Japan
[‡]National Cerebral and Cardiovascular Center Research Institute, Suita, Osaka, Japan
[1]Corresponding author: e-mail address: kojima_masayasu@kurume-u.ac.jp

Contents

Abstract

Small synthetic molecules called growth hormone secretagogues (GHSs) stimulate the release of growth hormone (GH) from the pituitary. They act through the GHS-R, a G-protein-coupled receptor highly expressed in the hypothalamus and pituitary. Using an orphan receptor strategy with a stable cell line expressing GHS-R, we purified endogenous ligands for GHS-R from rat and human stomach and named it "ghrelin," after a word root (ghre) in Proto-Indo-European languages meaning "grow." Ghrelin is a peptide hormone in which the third amino acid, usually a serine but in some species a threonine, is modified by a fatty acid; this modification is essential for ghrelin's activity.

Methods in Enzymology, Volume 514
ISSN 0076-6879
http://dx.doi.org/10.1016/B978-0-12-381272-8.00003-9

The main active form of rat ghrelin is 28-amino acid peptides with *n*-octanoyl modification. In rat stomach, a second type of ghrelin peptide was purified, identified as des-Gln14-ghrelin. With the exception of the deletion of Gln14, des-Gln14-ghrelin is identical to ghrelin, retaining the *n*-octanoic acid modification. Des-Gln14-ghrelin is encoded by an mRNA created by alternative splicing of the ghrelin gene. As in the rat, the major active form of human ghrelin is a 28-amino acid peptide with an *n*-octanoylated Ser3. However, in human stomach, several minor forms of human ghrelin peptides have been isolated. These can be classified into four groups by the type of acylation observed at Ser3 and into two groups by the amino acids in length. The discovery of ghrelin indicates that the release of GH from the pituitary and appetite stimulation might be regulated by ghrelin derived from the stomach.

1. INTRODUCTION

An orphan receptor strategy has served as a powerful tool to discover novel peptides or peptide hormones (Civelli et al., 2006). This method has led to the identification of several important bioactive peptides, such as nociceptin/orphanin FQ, orexin/hypocretin, prolactin-releasing peptide, apelin, metastin (kisspeptin), neuropeptide B, and neuropeptide W. Similarly, we used an orphan receptor strategy to discover the peptide ghrelin (Kojima and Kangawa, 2005; Kojima et al., 1999). A detailed description of the orphan receptor strategy is provided in another chapter of this book (Chapter 2, Yoshida et al.).

Growth hormone secretagogues (GHSs) are synthetic compounds that are potent stimulators of growth hormone (GH) release (Bowers, 1998; Smith et al., 1997), working through a novel G protein-coupled receptor, the GHS-receptor (GHS-R) (Howard et al., 1996), now called the ghrelin receptor. Because GHSs are a group of artificial compounds that do not exist in nature, the existence of endogenous GHS-R ligands had been postulated until the discovery of ghrelin. Here, we survey the purification and structure of rat and human ghrelins.

2. GHS AND ITS RECEPTOR

2.1. History

In 1976, Bowers et al. found that some opioid peptide derivatives had weak GH-releasing activity, but no opioid activity (Bowers et al., 1980). These factors were dubbed GHSs. The structure of the first GHS was Tyr-D-Trp-Gly-Phe-Met-NH_2, which released GH through direct action on the pituitary. This synthetic peptide was a methionine enkephalin derivative, in which the second Gly was replaced by D-Trp and there was a C-terminal amide. The GH-releasing activity of early GHSs was so weak that the activity was only observed *in vitro*. Subsequently, many peptidyl derivatives were synthesized in search of more potent GHSs (Bowers, 1998).

In 1984, a potent GHS, GHRP-6, was synthesized on the basis of conformational energy calculations in conjunction with peptide chemistry modifications and a biological activity assay (Bowers et al., 1984). Hexapeptide GHRP-6 was shown to be active in both *in vitro* and *in vivo*, suggesting the possibility of clinical applications. Then, in 1993, the first nonpeptide GHS, L-692,429, was synthesized by Smith et al. (1993). This nonpeptidyl GHS represented a milestone in the clinical use of GHSs, since L-692,429 was sufficiently active even when orally administered.

During this period, researchers investigated the mechanisms of GHS action. While GH release from the pituitary was known to be stimulated by the hypothalamic growth hormone-releasing hormone (GHRH), exogenous GHSs were thought to induce GH release through a different pathway. GHRH acts on the GHRH receptor to increase intracellular cAMP, which serves as a second messenger. By contrast, GHSs act on a different receptor, increasing intracellular calcium concentrations via inositol 1,4,5-trisphosphate (IP_3) signal transduction (Akman et al., 1993).

In 1996, GHS-R was identified by expression cloning (Howard et al., 1996). The strategy was based on the finding that GHSs stimulated phospholipase C, leading to increased inositol triphosphate and intracellular calcium. *Xenopus* oocytes were injected with *in vitro* transcribed cRNAs derived from swine pituitary, supplemented simultaneously with mRNAs encoding various G-alpha subunits. MK0677-stimulated calcium increases could be detected by bioluminescence of the jellyfish photoprotein aequorin, which was incorporated in the *Xenopus* oocytes. The identified GHS-R was a typical G protein-coupled seven-transmembrane receptor. *In situ* hybridization analyses showed that GHS-R was expressed in the pituitary, hypothalamus, and hippocampus (Guan et al., 1997). This receptor was for some time an example of an "orphan GPCR," that is, a GPCR with no known natural ligand. After the identification of the GHS-R, a search for its endogenous ligand was actively undertaken, using the "orphan receptor strategy."

2.2. The GHS-R superfamily

Figure 3.1 shows a dendrogram of GHS-R and its related receptor superfamily. In 1996, the GHS-R was identified by expression cloning and shown to be a typical G-protein-coupled seven-transmembrane receptor

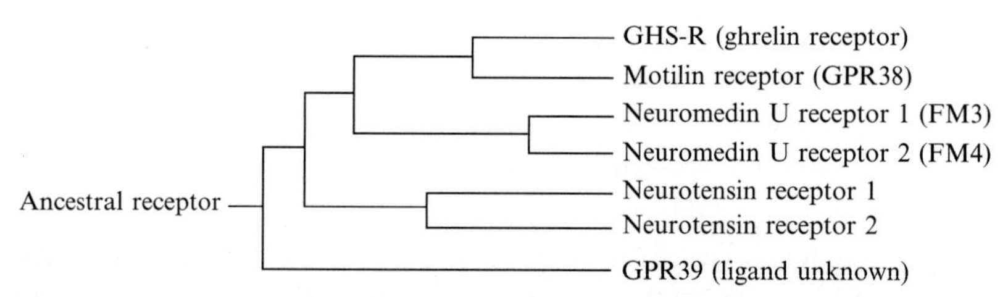

Figure 3.1 Dendrogram alignment of the ghrelin receptor (GHS-R) with homologous G-protein-coupled receptors. The ghrelin receptor is a member of the GPCR superfamily, which includes the receptors for motilin, neuromedin U, and neurotensin. The ghrelin receptor (GHS-R) is most homologous to the motilin receptor. Because their endogenous ligands, ghrelin and motilin, share amino acid sequence homology, the ghrelin and motilin systems may have evolved from a common ancestral peptide system. The GPCR superfamily also contains an orphan receptor, GPR39, for which obestatin is the putative endogenous ligand. However, this claim is highly controversial.

(Howard et al., 1996). The family member most closely related to GHS-R is the motilin receptor (GPR38); the human forms of these receptors share a 52% amino acid homology (Feighner et al., 1999). Neuromedin U receptors (NMU-R1 and -R2), which bind a neuropeptide that promotes smooth-muscle contraction and suppresses food intake, are also homologous to GHS-R (Howard et al., 2000; Kojima et al., 2000). Because motilin and NMU are found mainly in gastrointestinal organs, it was speculated that the endogenous ligand for GHS-R may be another gastrointestinal peptide. This hypothesis was confirmed by the isolation of ghrelin from stomach tissue.

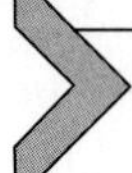

3. PURIFICATION OF RAT GHRELIN

3.1. Construction of GHS-R-expressing cells

As many orphan GPCRs have no ligands for use as positive controls in assay development, the GPCR-expressing cells have to be used without confirming that the receptors are functional. While the mRNA levels of the expressed orphan receptor can be assessed by Northern blotting, high mRNA expression levels do not mean that the expressed orphan receptor is functional. Thus, the search for novel ligands using the orphan receptor strategy is inherently a high-risk challenge. By contrast, GHS-R was known to bind artificial ligands such as GHRP-6 and hexarelin, providing a convenient positive control for constructing the assay used to identify the endogenous ligand. A cultured cell line expressing GHS-R was established and used to identify tissue extracts that could stimulate the GHS-R as monitored by increases in intracellular calcium levels.

3.1.1 Construction of the GHS-R-expressing cell line

1. The full-length cDNA of rat GHS-R was obtained by RT-PCR using a rat brain cDNA as the template. Sense and antisense primers were synthesized on the basis of the reported sequence of rat GHS-R (NCBI Reference Sequence: NM_032075.3). The sense primer was 5′-ATGTGGAACGCGACCCCCAGCGA-3′, and the antisense primer was 5′-ACCCCCAATTGTTTCCAGACCCAT-3′.
2. PCR was conducted using 35 cycles of 94 °C for 1 min, 63 °C for 2 min, and 72 °C for 3 min. After amplification, the PCR products were resolved by electrophoresis in a 1% agarose gel, and the band that matched to the predicted length of GHS-R cDNA was isolated and purified.
3. The purified cDNA of rat GHS-R was ligated into the pcDNA3.1 vector (Invitrogen, Carlsbad, CA).
4. The expression vector, GHS-R-pcDNA3.1, was transfected into Chinese hamster ovary (CHO) cells using FuGENE6 transfection reagent (Roche, Indianapolis, IN).
5. Stable transfectants were selected in 1 mg/ml G418 and isolated as single clonal cells.
6. The isolated cell lines were screened for intracellular calcium concentration changes induced by GHRP-6 (Peninsula Laboratories, Belmont, CA).

3.2. Monitoring intracellular calcium via a fluorometric imaging plate reader-based assay

Figure 3.2 shows changes in intracellular calcium concentrations in several GHS-R-expressing cell lines as detected by fluorometric imaging plate reader (FLIPR)-based assays. Isolated GHS-R-expressing cell lines were activated by GHRP-6, an artificial ligand to GHS-R. The calcium changes varied in each cell line in relation to the expression levels of GHS-R mRNA.

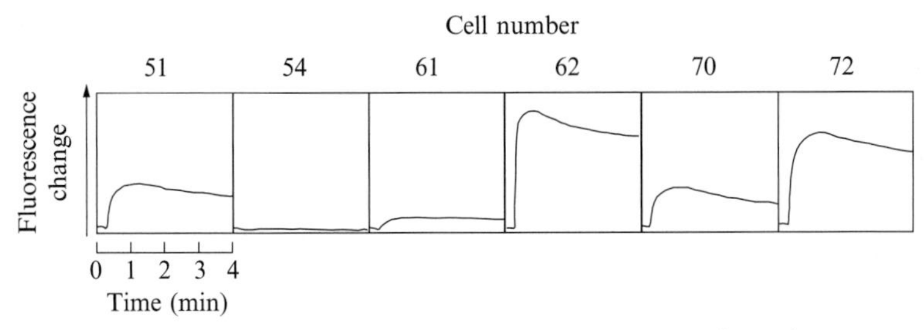

Figure 3.2 Intracellular calcium flux in GHS-R-expressing CHO cells. Each square represents the responses of isolated stably GHS-R-expressing cells as determined by FLIPR. Intracellular calcium concentrations, measured over 4 min, were read as changes in fluorescence of the indicator dye, Fluo-4. GHRP-6 (10^{-9} *M*) was used to activate the cells.

3.2.1 FLIPR assay

1. GHS-R-expressing cells were plated 12–15 h before the assay in flat-bottomed, black-walled, 96-well plates (Corning Costar Corporation, Cambridge, MA) at 4×10^4 cells/well.
2. Cells were incubated for 1 h with 4 m*M* Fluo-4-AM fluorescent indicator dye (Molecular Probes, Inc., Eugene, OR) in assay buffer: Hanks Balanced Salts Solution, 10 m*M* HEPES, 2.5 m*M* probenecid, 1% fetal calf serum (FCS). Then, cells were washed four times in assay buffer without FCS.
3. Changes in intracellular calcium concentrations were measured by a FLIPR (Molecular Devices, Sunnyvale, CA). We used the maximum change in fluorescence over baseline to determine agonist responses.
4. The clone showing the highest response (CHO-GHSR62) was used in the following assays.

3.3. Purification of rat ghrelin

The search for an endogenous GHS-R ligand had been focused on brain and hypothalamic extracts, because GHS-R is primarily expressed in the pituitary and hypothalamus, suggesting that its ligand should be similarly localized. However, numerous attempts failed to find the ligand in the brain, and unexpectedly, we succeeded in purifying and identifying the endogenous GHS-R ligand, ghrelin, from the stomach (Kojima et al., 1999). The name ghrelin is derived from "ghre," a word root in Proto-Indo-European languages meaning "grow," reflecting the peptide's role in stimulating GH release.

Figure 3.3 shows the steps involved in purification of rat ghrelin from the stomach. Cells contain many proteases to digest and process endogenous and exogenous molecules for the protection of the organism. In intact cells, these proteases are compartmentalized into restricted cellular components (e.g., lysosomes, microsomes, mitochondoria, and the Golgi apparatus), and bioactive peptides are sequestered from proteases as undigested propeptides in secretary granules. However, when cells are homogenized to extract peptide fractions, intracellular proteases are released and sometimes inactive forms of proteases become activated. These proteases digest and inactivate endogenous peptides very easily. Therefore, it is important to inactivate intracellular proteases before purification. Our simple method of boiling the tissue in water for 5–10 min before homogenization is a very effective way to maintain intact peptides during purification (Kangawa and Matsuo, 1984).

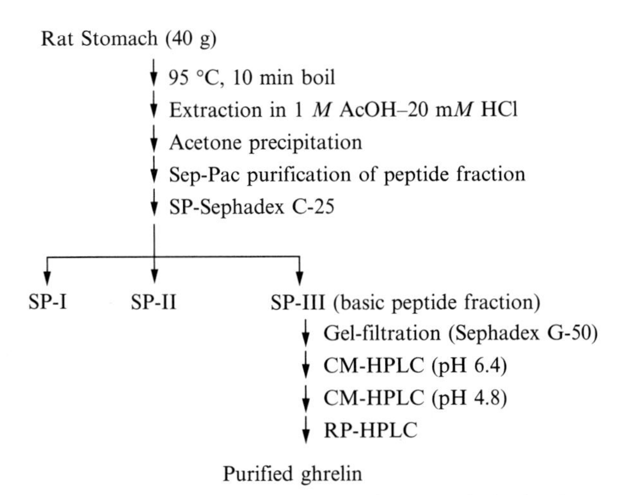

Figure 3.3 Outline of the steps involved in purification of ghrelin from rat stomach. AcOH, acetic acid; CM, carboxymethyl; RP-HPLC, reverse-phase high-performance liquid chromatography.

3.3.1 Purification of ghrelin from rat stomach tissue

1. Fresh rat stomach (40 g) was minced and boiled for 10 min in 5× volumes of water to inactivate intrinsic proteases. Boiling in an acidic solution should be avoided, as it can induce nonspecific peptide cleavage. With respect to ghrelin, boiling in acetic acid cleaves the ester bond between the modified acyl acid and Ser3, inactivating the peptide.
2. The boiled stomach tissue was cooled on ice. Then, the solution was adjusted to 1 *M* acetic acid (AcOH), 20 m*M* HCl.
3. The tissue was homogenized with a Polytron mixer and centrifuged for 30 min at 11,000 rpm.
4. The supernatant was concentrated to approximately 40 ml by evaporation.
5. Acetone (2× volume) was added to the concentrated supernatant to precipitate high-molecular-weight proteins.
6. Acetone was removed from the isolated supernatants by evaporation.
7. To collect the peptide fraction, the supernatant was loaded onto a Sep-Pak C18 cartridge (10 g; Waters, Milford, MA) pre-equilibrated in 0.1% trifluoroacetic acid (TFA). Then, the cartridge was washed with 10% acetonitrile (CH_3CN)/0.1% TFA, and the bound peptides were eluted in 60% CH_3CN/0.1% TFA.
8. The eluate was evaporated and lyophilized.

9. The residual materials were redissolved in 1 *M* AcOH and adsorbed on a SP-Sephadex C-25 column (H^+-form; Pharmacia, Uppsala, Sweden) pre-equilibrated with 1 *M* AcOH.
10. Successive elutions with 1 *M* AcOH, 2 *M* pyridine, and 2 *M* pyridine-AcOH (pH 5.0) yielded three fractions: SP-I, SP-II, and SP-III (basic peptide fraction).
11. The SP-III fraction was lyophilized (dry weight: 39 mg).
12. The lyophilized sample was dissolved in 1 ml of 1 N acetic acid and applied to a Sephadex G50 gel-filtration column (1.8 × 130 cm; Pharmacia).
13. Fractions (5 ml) were collected, and a portion of each fraction was tested for GHS-R ligand activity using the intracellular calcium assay with CHO-GHSR62 cells as targets (Fig. 3.4A).

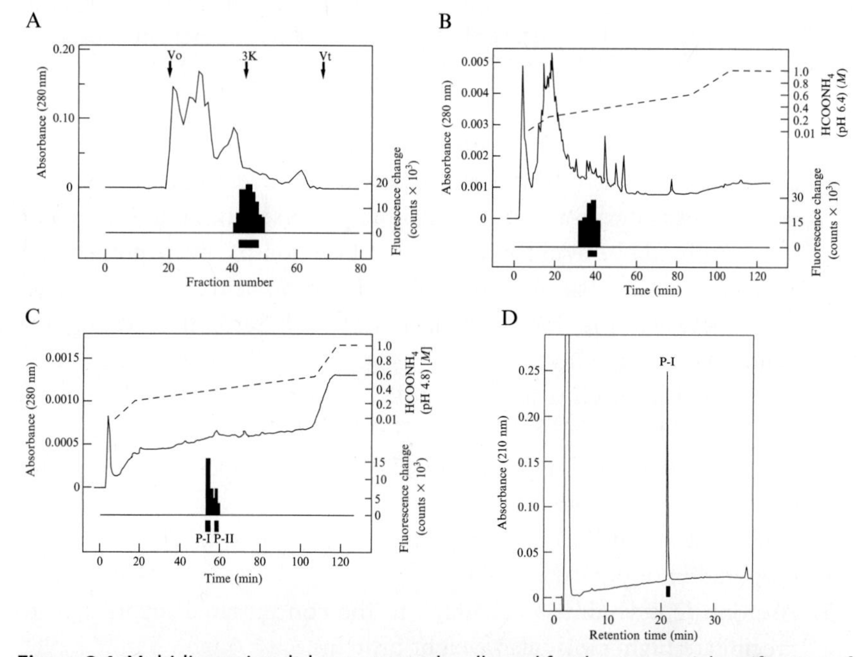

Figure 3.4 Multidimensional chromatography allowed for the sequential purification of ghrelin from rat stomach. Black bars indicate fractions that contained calcium flux activity on CHO-GHSR62 cells as determined by the FLIPR assay. Gradient profiles are indicated by the dotted lines. (A) Gel-filtration chromatography of the basic peptide fraction extracted from 40 g of rat stomach. The active fractions (43–48) eluted at approximately M_r 3000 (3 K). (B) CM ion-exchange HPLC (pH 6.4) of the active fractions derived from gel filtration. (C) A second CM ion-exchange HPLC (pH 4.8) of the active fractions from the first CM-HPLC. Two active fractions, P-I and P-II, were identified. (D) Final purification of P-I by RP-HPLC. Vo, void volume; Vt, total volume.

14. Active fractions were separated by carboxymethyl (CM) ion-exchange high-performance liquid chromatography (HPLC) on a TSK CM-2SW column (4.6×250 mm; Tosoh, Tokyo, Japan) using an ammonium formate gradient ($HCOONH_4$, pH 6.4) from 10 mM to 1 M in the presence of 10% CH_3CN at 1 ml/min for 100 min. Fractions (2 ml) were collected and tested for GHS-R ligand activity using the intracellular calcium assay (Fig. 3.4B).
15. Active fractions were further fractionated by a second round of CM-HPLC on the same column at pH 4.8, yielding two active peaks (P-I and P-II) (Fig. 3.4C). Ghrelin was purified from P-I, and des-Gln14-ghrelin was isolated from P-II as described later (see Section 5.1).
16. P-I was further purified manually using a C18 reverse-phase HPLC (RP-HPLC) column (Symmetry300, 3.9×150 mm; Waters) (Fig. 3.4D).

3.4. Structural determination of rat ghrelin

Amino acid sequence analysis revealed that ghrelin is a 28-amino acid peptide. However, the third amino acid was not determined by Edman degradation. An expressed sequence tag clone containing the coding region of the peptide revealed that the third residue is a Ser. Then, this sequence was confirmed with cDNA clones encoding the peptide precursor isolated from a rat stomach cDNA library (NCBI Reference Sequence: AB029433.1).

A 28-amino acid peptide based on the Ser-containing cDNA sequence was synthesized and its characteristics were compared with those of purified ghrelin. This comparison revealed the following: (1) the synthetic peptide, unlike the purified peptide, did not increase intracellular calcium levels in GHS-R-expressing cells; (2) purified ghrelin was retained longer on RP-HPLC than was the synthetic peptide, indicating that the purified form was more hydrophobic; and (3) by mass spectrometric analysis, the molecular weight (M_r) of purified ghrelin (3315) was 126 Da greater than that of the synthetic peptide (3189).

The most likely modification was the fatty acid *n*-octanoic acid (Kojima et al., 2001). When the hydroxyl group of Ser3 in the synthetic peptide ($M_r = 3189$) was esterified by *n*-octanoic acid ($M_r = 144$), the resulting modified peptide had the same molecular weight as purified ghrelin (3315). Moreover, the fatty acid modification appeared to increase the hydrophobicity of the peptide, resulting in an increased retention time when the sample was subjected to HPLC. The proposed peptide was then synthesized with a Ser3 *n*-octanoyl modification and its characteristics were compared

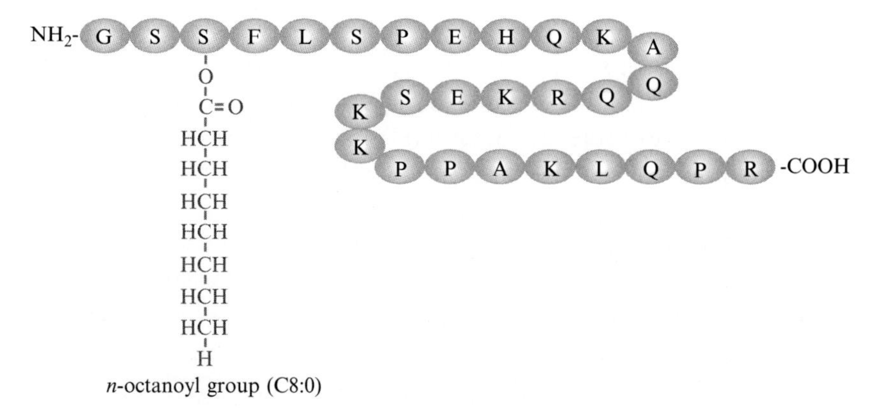

Figure 3.5 Structure of rat ghrelin. The third amino acid, serine, is modified by a fatty acid, *n*-octanoic acid, which is essential for activity. (See Color Insert.)

with those of purified ghrelin. This synthetically derived peptide eluted at the same retention time as did purified ghrelin on RP-HPLC, and the mass spectrometric fragmentation patterns of the two peptides were the same. Moreover, the synthetically derived peptide had the same effect as purified ghrelin on intracellular calcium concentrations in GHS-R-expressing cells. These results confirmed the primary structure of ghrelin, an octanoyl-modified peptide (Fig. 3.5) (Kojima et al., 1999).

4. PURIFICATION OF HUMAN GHRELIN

As compared to ghrelin obtained from the rat stomach, where the *n*-octanoyl and des-acyl varieties constitute the two main species, ghrelin in the human stomach exists in additional forms (Hosoda et al., 2003). The method of peptide preparation from human stomach tissue is almost the same as that followed for purification of rat ghrelin. The multiple forms of human ghrelin peptides are separated by an ion-exchange chromatography and HPLC.

4.1. Purification of human ghrelin from stomach tissue

1. Human stomach mucosa (27 g) was minced and boiled for 5 min in 5 × volumes of water to inactivate intrinsic proteases.
2. After cooling on ice, the solution was adjusted to 1 *M* AcOH, 20 m*M* HCl.
3. The tissue was homogenized with a Polytron mixer and centrifuged for 30 min at 11,000 rpm.

4. The supernatant was concentrated to ~25 ml by evaporation and subjected to acetone precipitation in 66% acetone.
5. After removal of the precipitate, the supernatant was evaporated to remove the acetone. Then, the sample was loaded onto a Sep-Pak C18 cartridge (Waters) pre-equilibrated in 0.1% TFA. The cartridge was washed with 10% CH_3CN/0.1% TFA, and the peptides were eluted in 60% CH_3CN/0.1% TFA. The eluate was evaporated and lyophilized.
6. The lyophilized materials were then redissolved in 1 *M* AcOH and applied to a Sephadex G-50 gel-filtration column (1.8 × 130 cm; Amersham Biosciences, Uppsala, Sweden). Fractions (5 ml) were collected. A portion of each fraction was subjected to ghrelin-specific radioimmunoassay (RIA) and the intracellular calcium flux assay using CHO-GHSR62 cells (Fig. 3.6A).
7. The active fractions (#43–46) were separated by CM ion-exchange HPLC on a TSK CM-2SW column (4.6 × 250 mm; Tosoh) using an ammonium acetate ($HCOONH_4$, pH 4.8) gradient of 10 m*M* to 1 *M* in the presence of 10% CH_3CN at 1 ml/min for 100 min. Fractions (1 ml) were collected and subjected to ghrelin-specific RIAs and intracellular calcium influx assays (Fig. 3.6B).
8. The six active fractions (Fractions A–F) separated by CM-HPLC were individually purified using C18 HPLC columns (Symmetry 300, 3.9 × 150 mm; Waters). The amino acid sequences of the purified peptides were analyzed using a protein sequencer (494; Applied Biosystems, Foster City, CA).

4.2. Mass spectrometric analysis of human ghrelins

1. Electrospray ionization mass spectrometry (ESI-MS) was performed on a quadrupole mass spectrometer SSQ7000 (Finnigan San Jose, CA) equipped with a Finnigan ESI source.
2. A capillary needle was heated to 150 °C to evaporate the samples. Samples (<20 pmol) were dissolved in 50% (v/v) methanol and 1% AcOH and introduced into the +4.5 kV (positive ionization) ion source at a flow rate of 5 μl/min by direct infusion using a syringe pump.
3. The molecular masses of the purified peptides were calculated using the ICIS software Bioworks provided by Finnigan.

4.3. Molecular forms of human ghrelin

The major form of active ghrelin in the human stomach is, like rat ghrelin, a 28-amino acid peptide with an octanoyl modification at its third amino

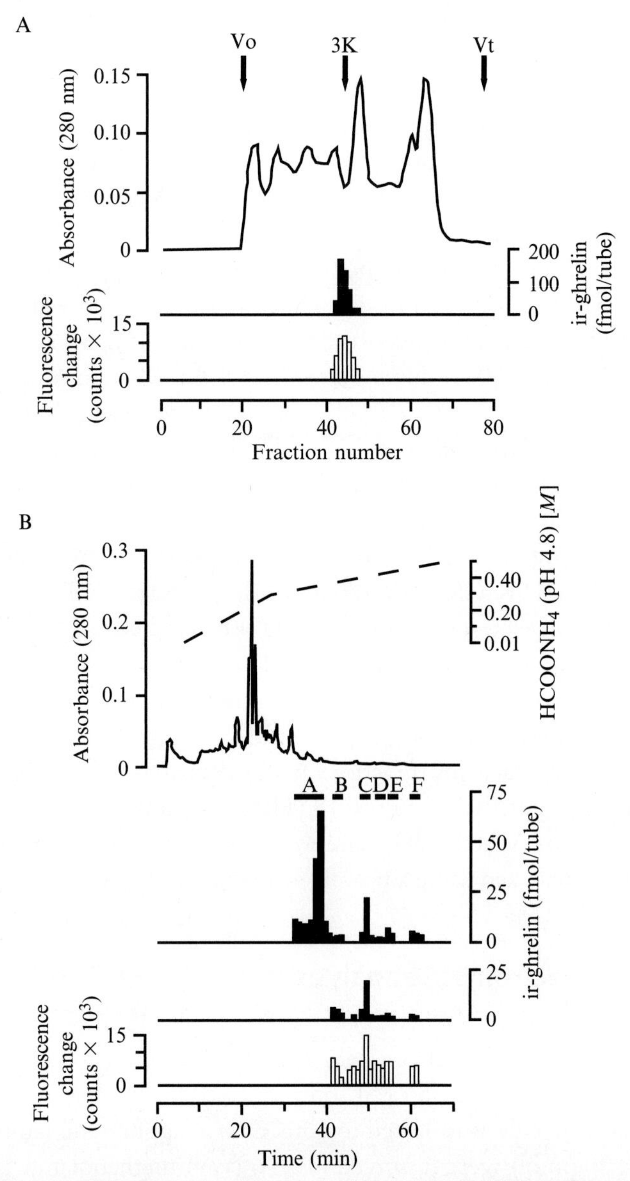

Figure 3.6 Multidimensional chromatography allowed for the sequential purification of ghrelin from human stomach. Black bars indicate ir-ghrelin content, and open bars indicate fractions that contained calcium flux activity on CHO-GHSR62 cells as determined by the FLIPR assay. Gradient profiles are indicated by the dotted lines. (A) Gel-filtration chromatography of extract from 27 g of human stomach. The active fractions (43–46) were eluted at approximately M_r 3000 (3 K). An aliquot from each fraction (5 mg of wet tissue equivalent) was subjected to ghrelin C-RIA. (B) CM ion-exchange HPLC (pH 4.8) of the active fractions derived from gel filtration was monitored by C-RIA (top), N-RIA (middle), and calcium flux assays (bottom). Active fractions, indicated by solid bars, were separated into six fractions (Fractions A–F). A portion of each fraction (5 mg of wet tissue equivalent) was subjected to ghrelin-specific RIA. Vo, void volume; Vt, total volume.

Table 3.1 Structural analyses of human ghrelin and ghrelin-derived peptides

Fraction	Amino acid sequence	Linked fatty acid
A	Gly^1-Arg^{28}	C8:0
B	Gly^1-Pro^{27}	C8:0
C	Gly^1-Arg^{28}	C8:0
D	Gly^1-Pro^{27}	C10:0
E	Gly^1-Arg^{28}	C10:1
F	Gly^1-Arg^{28}	C10:0

acid, serine. During the course of purification, we isolated several minor forms of ghrelin peptides (Table 3.1) (Hosoda et al., 2003). These peptides were classified into four groups according to the type of acylation observed at Ser3: nonacylated, octanoylated (C8:0), decanoylated (C10:0), and possibly decenoylated (C10:1). All peptides found were either 27 or 28 amino acids in length, with the former variety lacking the C-terminal Arg28. All of the observed peptides were derived from the same ghrelin precursor through two alternative pathways. Synthetic octanoylated and decanoylated ghrelins induced intracellular calcium increases in GHS-R-expressing cells and stimulated GH release in rats to similar degrees.

5. PURIFICATION OF RAT DES-GLN14-GHRELIN

In the rat stomach, two isoforms of mRNA encoding proghrelin are produced by an alternative splicing mechanism (Hosoda et al., 2000). One mRNA encodes the ghrelin precursor, and the other encodes a precursor for des-Gln14-ghrelin, a peptide identical to ghrelin, but with a deletion of Gln14. This deletion results from the use of the CAG codon, which encodes Gln14 and also acts as a splicing signal. Thus, two types of active ghrelin peptides are produced in the rat stomach, namely, ghrelin and des-Gln14-ghrelin. However, des-Gln14-ghrelin is only present in low amounts. The full-length ghrelin is the major active form isolated from rat stomach.

5.1. Purification of des-Gln14-ghrelin from rat stomach tissue

1. Des-Gln14-ghrelin was purified using nearly the same method as was employed to isolate rat ghrelin. Steps 1–13 are the same as those described in Section 3.3.1.

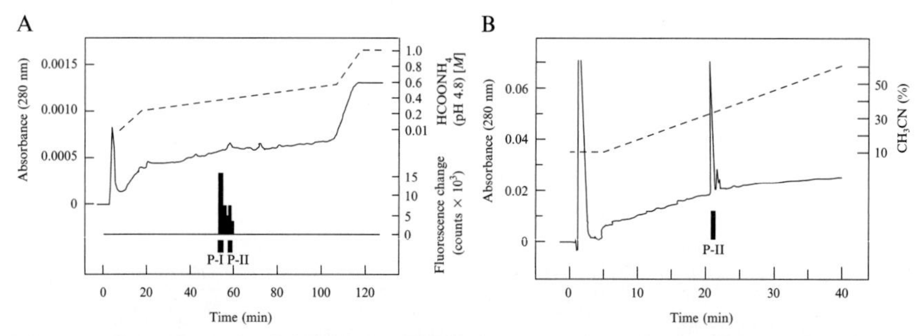

Figure 3.7 Purification of des-Gln14-ghrelin from rat stomach. Black bars indicate fractions that contained calcium flux activity on CHO-GHSR62 cells as determined by the FLIPR assay. Gradient profiles are indicated by the dotted lines. (A) A second CM ion-exchange HPLC (pH 4.8) separation was performed on the active fractions from the first CM-HPLC (Fig. 3.4B). Two active fractions, P-I and P-II, are indicated by solid bars. (B) Final purification of P-II by RP-HPLC.

2. Active fractions from CM-HPLC were further fractionated by a second round of CM-HPLC on the same column at pH 4.8 to give two active peaks (P-I and P-II) (Fig. 3.7A).
3. P-II was purified manually using a C18 RP-HPLC column (Symmetry 300, 3.9 × 150 mm; Waters) and identified as des-Gln14-ghrelin (Fig. 3.7B).

6. OBESTATIN, AN ANOREXIGENIC HORMONE OR A DIGESTED PEPTIDE FRAGMENT FROM A GHRELIN PRECURSOR?

Among the members of the GPCR superfamily to which GHS-R belongs, GPR39 raised controversy with respect to the identity of its endogenous ligand. In November 2005, Zhang et al., a group of researchers from Stanford University, reported a novel peptide hormone "obestatin," named from the Latin "obedere," meaning to devour, and "statin," meaning suppression, because it suppressed food intake (Zhang et al., 2005). Interestingly, obestatin is processed from the ghrelin precursor; this means that the two peptide hormones with opposing action on food intake, orexigenic ghrelin and anorectic obestatin, are derived from the same hormone precursor. Zhang and colleagues proposed that the lack of an obvious phenotype in the ghrelin knockout mouse was because both ghrelin and obestatin were deleted in these animals.

Although these initial results were provocative, several subsequent reports raised objections to the proposed action of obestatin and its putative receptor. A number of independent research groups reported that obestatin did not bind to or activate GPR39 (Chartrel et al., 2007; Holst et al., 2007). Moreover, several groups observed that injection of obestatin did not suppress food intake. The original group later reported that an exact 15 min delay, neither 0 nor 30 min, in food replacement after obestatin injection was essential to demonstrate the food suppressive effect of obestatin. In addition, this group could not repeat their earlier studies showing that obestatin bound to GPR39 (Zhang et al., 2007).

The amino acid sequences of mammalian obestatins are well conserved. However, in nonmammalian species, only the ghrelin portions of the common precursor are well conserved (Fig. 3.8). Moreover, the original paper on obestatin reported that the C-terminal amide structure of obestatin was essential for binding and activating GPR39; however, the Gly residue required to form this structure is missing from the nonmammalian common precursor (Fig. 3.8). Thus, nonmammalian obestatins, if they were present in the stomach, would not exhibit the C-terminal amide structure. Furthermore, general processing sites for prohormone convertases, such as Arg-Arg or Lys-Arg, were not found in the nonmammalian obestatin sequences. In addition, if both ghrelin and obestatin were processed from the same precursor protein, the levels of both peptides should be similar. However, the plasma content of ghrelin is much higher than that of obestatin—plasma ghrelin versus obestatin, rat: 344.4 ± 40.6 versus not detected; human: 132.4 ± 13.1 versus 6.9 ± 0.28 fmol/ml (Mondal et al., 2008). Thus, it is likely that obestatin is not an endogenous hormone derived from physiologically relevant processing of the ghrelin precursor, but rather represents a peptide fragment produced by nonspecific protease digestion during peptide extraction.

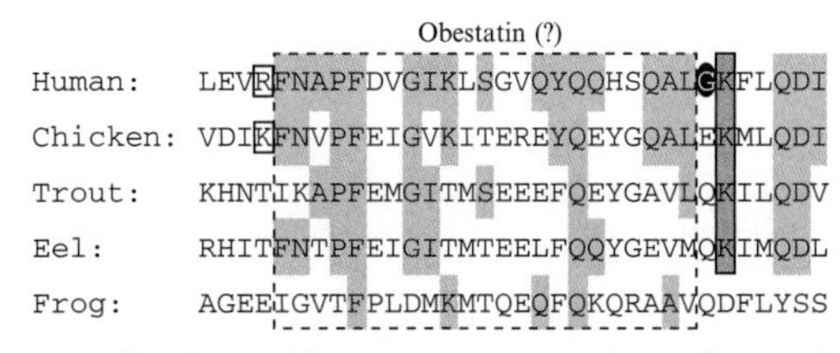

Figure 3.8 Alignment of amino acid sequences around vertebrate obestatin parts. Residues identical in at least two species are shaded. Obestatin parts are boxed by dotted line. Putative processing sites for the prohormone convertase are boxed. The glycine residue required for C-terminal amide processing in human obestatin is indicated by a white letter in black circle. Note that obestatin is not highly conserved, particularly with respect to the functionally important C-terminal amide structure. (For color version of this figure, the reader is referred to the online version of this chapter.)

7. CONCLUSION

In general, we can predict the active forms of bioactive peptides from a precursor sequence based on the existence of prohormone processing sites. However, it was impossible to predict the acyl modification of ghrelin, which is essential for its activity, from the precursor sequence. In this case, we needed to purify the endogenous peptides from tissues using multi-dimensional chromatography to determine the endogenous active form. The methods described here are not novel; however, the details of ghrelin purification are still valuable for researchers looking to discover novel peptide hormones or bioactive peptides.

REFERENCES

Akman, M.S., et al., 1993. Mechanisms of action of a second generation growth hormone-releasing peptide (Ala-His-D-beta Nal-Ala-Trp-D-Phe-Lys-NH2) in rat anterior pituitary cells. Endocrinology 132, 1286–1291.

Bowers, C.Y., 1998. Growth hormone-releasing peptide (GHRP). Cell. Mol. Life Sci. 54, 1316–1329.

Bowers, C.Y., et al., 1980. Structure-activity relationships of a synthetic pentapeptide that specifically releases growth hormone *in vitro*. Endocrinology 106, 663–667.

Bowers, C.Y., et al., 1984. On the *in vitro* and *in vivo* activity of a new synthetic hexapeptide that acts on the pituitary to specifically release growth hormone. Endocrinology 114, 1537–1545.

Chartrel, N., et al., 2007. Comment on "Obestatin, a peptide encoded by the ghrelin gene, opposes ghrelin's effects on food intake" Science 315, 766 author reply 766.

Civelli, O., et al., 2006. Orphan GPCRs and their ligands. Pharmacol. Ther. 110, 525–532.

Feighner, S.D., et al., 1999. Receptor for motilin identified in the human gastrointestinal system. Science 284, 2184–2188.

Guan, X.M., et al., 1997. Distribution of mRNA encoding the growth hormone secretagogue receptor in brain and peripheral tissues. Brain Res. Mol. Brain Res. 48, 23–29.

Holst, B., et al., 2007. GPR39 signaling is stimulated by zinc ions but not by obestatin. Endocrinology 148, 13–20.

Hosoda, H., et al., 2000. Purification and characterization of rat des-Gln14-Ghrelin, a second endogenous ligand for the growth hormone secretagogue receptor. J. Biol. Chem. 275, 21995–22000.

Hosoda, H., et al., 2003. Structural divergence of human ghrelin. Identification of multiple ghrelin-derived molecules produced by post-translational processing. J. Biol. Chem. 278, 64–70.

Howard, A.D., et al., 1996. A receptor in pituitary and hypothalamus that functions in growth hormone release. Science 273, 974–977.

Howard, A.D., et al., 2000. Identification of receptors for neuromedin U and its role in feeding. Nature 406, 70–74.

Kangawa, K., Matsuo, H., 1984. Purification and complete amino acid sequence of alpha-human atrial natriuretic polypeptide (alpha-hANP). Biochem. Biophys. Res. Commun. 118, 131–139.

Kojima, M., Kangawa, K., 2005. Ghrelin: structure and function. Physiol. Rev. 85, 495–522.

Kojima, M., et al., 1999. Ghrelin is a growth-hormone-releasing acylated peptide from stomach. Nature 402, 656–660.

Kojima, M., et al., 2000. Purification and identification of neuromedin U as an endogenous ligand for an orphan receptor GPR66 (FM3). Biochem. Biophys. Res. Commun. 276, 435–438.

Kojima, M., et al., 2001. Ghrelin: discovery of the natural endogenous ligand for the growth hormone secretagogue receptor. Trends Endocrinol. Metab. 12, 118–122.

Mondal, M.S., et al., 2008. Characterization of obestatin in rat and human stomach and plasma, and its lack of acute effect on feeding behavior in rodents. J. Endocrinol. 198, 339–346.

Smith, R.G., et al., 1993. A nonpeptidyl growth hormone secretagogue. Science 260, 1640–1643.

Smith, R.G., et al., 1997. Peptidomimetic regulation of growth hormone secretion. Endocr. Rev. 18, 621–645.

Zhang, J.V., et al., 2005. Obestatin, a peptide encoded by the ghrelin gene, opposes ghrelin's effects on food intake. Science 310, 996–999.

Zhang, J.V., et al., 2007. Response to comment on "obestatin, a peptide encoded by the ghrelin gene, opposes ghrelin's effects on food intake" Science 315, 766.

CHAPTER FOUR

Variety of Acyl Modifications in Mammalian Ghrelins

Takanori Ida[1]
Interdisciplinary Research Organization, University of Miyazaki, Kiyotake, Miyazaki, Japan
[1]Corresponding author: e-mail address: a0d203u@cc.miyazaki-u.ac.jp

Contents

Abstract

Ghrelin, a 28-amino acid-long peptide with an *n*-octanoyl modification at Ser^3, has been isolated from rat and human stomachs as an endogenous ligand for the growth hormone secretagogue receptor. It is very important to study the ghrelin from mammals (especially, domestic animals) that serve as human companions, food resources, and model organisms. We purified feline and caprine ghrelin and observed that the administration of synthetic ghrelin increased plasma growth hormone (GH) levels in cats and goats. Therefore, we believe that ghrelin may play important roles in GH release in mammals.

1. INTRODUCTION

Kojima et al. originally isolated ghrelin, an endogenous growth hormone (GH) secretagogue, from the rat stomach as an endogenous ligand for the GH secretagogue receptor (GHS-R) (Kojima et al., 1999). The third residue in ghrelin, a serine (Ser^3), is modified by *n*-octanoic acid (C8:0). This octanoyl modification is essential for receptor binding and the resulting

Methods in Enzymology, Volume 514
ISSN 0076-6879
http://dx.doi.org/10.1016/B978-0-12-381272-8.00004-0

biological activities such as stimulation of GH release and food intake (Nakazato et al., 2001).

Structural divergence has been observed in rat and human ghrelins. For example, ghrelins lacking Gln^{14} or Arg^{28} have been isolated from rat and human, respectively (Hosoda et al., 2002, 2003). In addition to octanoylated (C8:0) ghrelin, decanoylated (C10:0) and decenoylated (C10:1) ghrelins were also found in the human stomach (Hosoda et al., 2003). This structural divergence in peptide length and fatty acid modification has also been reported in ghrelins of nonmammalian vertebrates such as rainbow trout, chicken, and bullfrog (Kaiya et al., 2001, 2002, 2003). We isolated some mammalian ghrelins and identified variations in ghrelin length and modification (Ida et al., 2007, 2010).

2. PURIFICATION OF MAMMALIAN GHRELIN

2.1. Purification of feline ghrelin

2.1.1 Calcium mobilization assays

1. During the purification process, ghrelin activity was followed by applying fractions to a cell line stably expressing rat GHS-R (Chinese hamster ovary (CHO)-GHSR62) and measuring changes in intracellular calcium concentrations ($[Ca^{2+}]_i$) with a fluorometric imaging plate reader (FLIPR) system (Molecular Devices, CA, USA), as described previously (Kojima et al., 1999).
2. CHO-GHSR62 cells were plated in flat-bottom, black-walled, 96-well plates (Corning Corstar Corporation, Cambridge, MA) at 3×10^4 cells/well for 12–15 h before the assay.
3. Cells were loaded with 4 m*M* Fluo-4-AM fluorescent indicator dye (Molecular Probes, Inc., Eugene, OR) for 1 h in the assay buffer (Hanks' balanced salts solution, 10 m*M* HEPES, 2.5 m*M* probenecid, and 1% fetal calf serum) and washed four times in the assay buffer without serum.
4. Changes in $[Ca^{2+}]_i$ were assayed after loading the samples onto a FLIPR system.

2.1.2 Purification of feline ghrelin

1. The obtained feline stomach (15 g) was quickly frozen on dry ice and used as the starting material.
2. The stomach was minced and boiled for 10 min in 10 volumes of water to inactivate intrinsic proteases.
3. The solution was adjusted to 1 *M* acetic acid (AcOH).
4. The stomach tissues were homogenized with a Polytron mixer.

5. The supernatants of the extracts, obtained after a 30-min centrifugation at 11,000 rpm, were concentrated to ∼15 ml by evaporation.
6. Two volumes of acetone were added to the concentrate for acetone precipitation to a final concentration of 66% acetone.
7. The solution was agitated overnight.
8. The supernatants of the extracts, obtained after a 30-min centrifugation at 11,000 rpm, were filtered by glass filter.
9. Acetone was removed from the isolated supernatants by evaporation.
10. The supernatant was loaded onto a 10-g cartridge of Sep-Pak C18 (Waters, Milford, MA) pre-equilibrated in 0.1% trifluoroacetic acid (TFA). The Sep-Pak cartridge was washed with 10% acetonitrile (CH_3CN) and 0.1% TFA, and the peptide fraction was eluted in a solution of 60% CH_3CN and 0.1% TFA.
11. The eluate was evaporated and lyophilized.
12. The residual materials were redissolved in 1 *M* AcOH and adsorbed on a column of SP-Sephadex C-25 (H^+-form; Amersham Pharmacia Biotech) pre-equilibrated with 1 *M* AcOH. Successive elutions with 1 M AcOH, 2 *M* pyridine, and 2 *M* pyridine–AcOH (pH 5.0) provided three fractions: SP-I, SP-II, and SP-III.
13. The lyophilized SP-III fraction was subjected to carboxymethyl (CM) ion-exchange high-performance liquid chromatography (HPLC) on a column of TSK CM-2SW (4.6 × 250 mm; Tosoh, Tokyo, Japan) with an ammonium acetate ($HCOONH_4$) (pH 6.5) linear gradient of 0.01–0.6 *M* in the presence of 10% CH_3CN, at a flow rate of 1 ml/min, for 10 min up to 130 min.
14. Active fractions were separated by reverse-phase (RP) HPLC with a μBondasphere C18 column (3.9 × 150 mm; Waters, MA, USA) at a flow rate of 1 ml/min with a linear gradient of 10–60% ACN and 0.1% TFA for 80 min.
15. Active fractions were further purified by RP-HPLC using a diphenyl column (2.1 × 150 mm, 219TP5125; Vydac, Hesperia, CA, USA) and a Chemcosorb 3ODS-H column (2.1 × 75 mm; Chemco, Osaka, Japan) for 80 min under a linear gradient from 10–60% ACN in 0.1% TFA, at a flow rate of 0.2 ml/min. Fractions corresponding to the absorption peaks were collected, and an aliquot of each fraction (1-g tissue equivalent) was assayed by the FLIPR system.
16. Approximately 20 pmol of the final purified peptide from the main activity fraction was analyzed with a protein sequencer (model 494; Applied Biosystems, CA, USA).

17. Approximately 1 pmol of each active fraction was used for molecular weight determination by matrix-assisted laser desorption–ionization time of flight mass spectrometry with a Voyager-DE PRO instrument (Applied Biosystems).

2.1.3 Cloning of feline ghrelin cDNA

1. Total RNA was extracted from a feline stomach with TRIzol reagent (Invitrogen, Carlsbad, CA, USA), and poly(A)$^+$ RNA was isolated with an mRNA purification kit (TaKaRa Bio Inc., Kyoto, Japan).
2. For 3′-RACE PCR, first-strand cDNAs were synthesized from 200 ng of poly(A)$^+$ RNA by using an adaptor primer supplied with the 3′-RACE system (Invitrogen) and the SuperScript II reverse transcriptase (RT) (Invitrogen).
3. One-tenth of the cDNA was used as template. Primary PCR was performed as described with four degenerate primers based on the N-terminal 7 amino acids of human ghrelin (GSSFLSP):
   GRL-S7, 5′-GGGTCGAGYTTCTTRTCNCC-3′;
   GRL-S8, 5′-GGGTCGAGYTTCTTRAGYCC-3′;
   GRL-S9, 5′-GGGTCGAGYTTCCTNTCNCC-3′;
   and GRL-S10, 5′-GGGTCGAGYTTCCTNAGYCC-3′.
4. Amplification was performed as follows: 94 °C for 1 min; 35 cycles of 94 °C for 30 s, 58 °C for 30 s, and 72 °C for 1 min; and a final extension for 3 min at 72 °C.
5. Amplified products were purified with the Wizard PCR Preps DNA purification system (Promega, Madison, WI).
6. For second-round nested PCR, a nested sense primer for feline ghrelin-(7–13) (5′-CCNGARCAYCARAARGTNCARC-3′) was used. The amplification reaction was 94 °C for 1 min; 35 cycles of 94 °C for 30 s, 55 °C for 30 s, and 72 °C for 1 min; and a final extension for 3 min at 72 °C.
7. The candidate ghrelin cDNA fragment was subcloned with the TOPO TA cloning kit (pCR II-TOPO vector, Invitrogen) and sequenced.
8. For 5′-RACE PCR, first-strand cDNAs were synthesized from 200 ng of poly(A)$^+$ RNA with oligo-d$T_{12–18}$ primers and SuperScript II RT at 42 °C for 1 h.
9. One-fifth of the purified cDNA was subjected to a TdT-tailing reaction of the 5′-ends of the first-strand cDNA with deoxy CTP, according to the manufacturer's protocol (Invitrogen).

10. The resultant dC-tailed cDNAs were used as template. A gene-specific primer was designed on the basis of the sequence of the feline ghrelin cDNA as determined by 3′-RACE PCR: Fel GRL-AS2, 5′-GTGGATCAAGCCTTCCAGAG-3′; Fel GRL-AS3, 5′-GACA GCTTGATTCCAACATC-3′. Primary PCR was performed with Fel GRL-AS3, an abridged anchor primer supplied with the 5′-RACE kit, and Ex *Taq* DNA polymerase under the following reaction conditions: 94 °C for 1 min; 35 cycles of 94 °C for 30 s, 57 °C for 30 s, and 72 °C for 1 min; and a final extension for 3 min at 72 °C.
11. The resulting product was purified with Wizard PCR Preps, and the second-round nested PCR was performed with Fel GRL-AS2 and an abridged universal amplification primer. The amplification reaction was 94 °C for 1 min; 30 cycles of 94 °C for 30 s, 55 °C for 30 s, and 72 °C for 1 min; and a final extension for 3 min at 72 °C.
12. The candidate PCR product was subcloned into the pCR-II-TOPO vector and sequenced.
13. The nucleotide sequence of the isolated cDNA fragment was determined by automated sequencing (DNA sequencer model 3100, Applied Biosystems) according to the protocol for the BigDye terminator cycle sequencing kit (Applied Biosystems).

2.1.4 Structural determination of feline ghrelin

1. The complete amino acid sequence of the main activity fraction was determined to be GSXFLSPEHQKVQQRKESKKPPAKLQPR (X was unidentified by the sequencer because of acyl modification) by protein sequencing.
2. From cDNA analysis, the unidentified third amino acid was determined to be serine.
3. Figure 4.1 shows sequence comparison of mammalian ghrelins. Table 4.1 shows the actual measured molecular masses of the isolated peptides. The molecular forms are deduced from the molecular masses, in addition to the analyses of the peptide and cDNA sequences.

2.2. Purification of caprine ghrelin

Caprine ghrelin was purified and determined to be structurally similar to feline ghrelin. Table 4.2 shows the actual measured molecular masses of the isolated peptides, the deduced molecular forms, and the yields.

```
          *
Goat      GSSFLSPEHQKLQ-RKEPKKPSGRLKPR
Sheep     GSSFLSPEHQKLQ-RKEPKKPSGRLKPR
Cow       GSSFLSPEHQKLQ-RKEAKKPSGRLKPR
Pig       GSSFLSPEHQKVQQRKESKKPAAKLKPR
Horse     GSSFLSPEHHKVQHRKESKKPPAKLKPR
Dog       GSSFLSPEHQKLQQRKESKKPPAKLQPR
Cat       GSSFLSPEHQKVQQRKESKKPPAKLQPR
Human     GSSFLSPEHQRVQQRKESKKPPAKLQPR
Rat       GSSFLSPEHQKAQQRKESKKPPAKLQPR
Mouse     GSSFLSPEHQKAQQRKESKKPPAKLQPR
```

Figure 4.1 Sequence comparison of mammalian ghrelins. An asterisk indicates Ser^3 modified by fatty acid *(adapted from Ida et al., 2010).*

Table 4.1 Molar yield of purified feline ghrelin and ghrelin-derived molecules

Expected molecular form	Mass [M + H]	Yield (pmol)
Ghrelin-(1–27)(C8:0)	3188.16	20.00
Ghrelin-(1–27)(C10:0)	3216.64	3.13
Ghrelin-(1–27)(C10:1)	3214.35	10.63
Ghrelin-(1–27)(C10:2)	3212.65	5.63
Ghrelin-(1–28)(C8:0)	3344.88	107.50
Ghrelin-(1–28)(C8:1)	3343.21	18.75
Ghrelin-(1–28)(C10:0)	3372.77	11.25
Ghrelin-(1–28)(C10:1)	3371.22	38.75
Ghrelin-(1–28)(C10:2)	3369.50	10.00
Ghrelin-(1–28)(C13:0)	3413.35	1.50
Ghrelin-(1–28)(C13:1)	3412.68	0.63
Des-Gln^{14}-ghrelin-(1–27)(C8:0)	3060.56	8.75
Des-Gln^{14}-ghrelin-(1–28)(C8:0)	3215.90	8.00

Adapted from Ida et al. (2007).

Table 4.2 Molar yield of purified caprine ghrelin and ghrelin-derived molecules

Expected molecular form	Mass [M+H]	Yield (pmol)
Ghrelin-(1–27)(C8:0)	3242.35	387.5
Ghrelin-(1–27)(C8:1)	3241.20	12.4
Ghrelin-(1–27)(C9:0)	3256.57	102.0
Ghrelin-(1–27)(C10:0)	3270.87	87.5
Ghrelin-(1–26)(+142)	3102.82	31.2
Desacyl-ghrelin-(1–27)(+142)	3258.81	195.8
Desacyl-ghrelin-(1–27)(+143)	3259.67	50.0
Desacyl-ghrelin-(1–27)(+156)	3273.02	5.0

Adapted from Ida et al. (2010).

3. GH-RELEASING ACTIVITY OF GHRELIN IN MAMMALS

3.1. GH-releasing activity of ghrelin in cats

3.1.1 Preparation of plasma samples

1. Cephalic vein catheters (19 mm long, Insyte 24 GA; Becton Dickinson, Sandy, UT, USA) were introduced into healthy adult cats without tranquilizer.
2. The experiments were performed on four groups: saline control, and doses of 0.05, 0.5, and 2.5 μg/kg (body weight) of rat ghrelin (Peptide Institute, Inc., Osaka, Japan), introduced by IV bolus injection. Cats were randomly assigned to groups, and each dose of ghrelin was administered two times to five cats; the second injection was administered 1 month after the first.
3. Blood (500 μl per sample) was collected from the cephalic vein—before and 10, 20, 40, and 80 min after injection—into tubes containing EDTA-2Na (1 mg/ml blood) (Sigma, St. Louis, USA).
4. Plasma was separated by centrifugation (3000 rpm, 10 min at 4 °C) and kept at −80 °C until GH measurement.

3.1.2 Radioiodination of GH

Plasma GH concentrations were measured by a heterologous canine radioimmunoassay (RIA) that has been previously validated for feline GH (Peterson et al., 1990). The canine GH RIA kit was supplied by the National

Hormone and Peptide Program (NIDDK, CA, USA). Radioiodination of canine GH was performed by the chloramine-T method.

1. The canine GH (10 μg) was dissolved in 0.25 *M* PBS (20 μl).
2. $Na^{125}I$ (10 μl) and 0.25 *M* PBS (50 μl) were added to the solution.
3. Ten microliters of chloramine-T (1 mg/ml in 0.05 *M* PBS) was added to the solution and incubated for 30 s.
4. Fifty microliters of sodium metabisulfite (3.33 mg/ml in 0.05 *M* PBS) was added to the solution.
5. Two-hundred microliters of 0.05 *M* PBS with 0.5% BSA was immediately added to the solution and loaded onto the gel filtration column (Amersham Pharmacia Biotech) and fractions (500 μl) were collected.

3.1.3 Radioimmunoassay of GH

Day 1: The RIA incubation mixture was composed of 100 μl of standard GH or unknown samples and 100 μl of anticanine GH antiserum monkey diluted with RIA buffer (50 mmol/l sodium phosphate buffer (pH 7.4), 1.8% BSA, 0.5% Triton-X 100, 137 mmol/l NaCl, and 10 mmol/l EDTA-2Na) containing 1% normal monkey serum. The anticanine GH antiserum was used at a final dilution of 1/50,000 in the RIA buffer.

Day 2: ^{125}I-labeled tracer (100 μl, 18,000 cpm) was added.

Day 3: Secondary antibody (250 μl of goat antimonkey IgG serum, 1/60), with 3.5% polyethylene glycol, was added and incubated for 3 h at room temperature. Free and bound tracers were separated by centrifugation at 3000 rpm for 20 min. After aspiration of supernatant, the radioactivity in the pellet was assessed with a gamma counter (ARC-1000M; Aloka, Tokyo, Japan).

Figure 4.2A shows the time course of plasma GH concentration after intravenous injections of synthetic rat ghrelin into cats.

3.2. GH-releasing activity of ghrelin in goats

Goat plasma GH concentrations were measured by an ovine GH RIA kit supplied by the National Hormone and Peptide Program (NIDDK). Radioiodination of ovine GH (NIDDK-oGH-I-5) was performed by the chloramine-T method. The primary and secondary antibodies used were antiovine GH antiserum (rabbit) (NIDDK-anit-oGH-2) and antirabbit IgG antiserum (donkey), respectively. The assay procedure was performed like in the feline study.

Figure 4.2B shows the time course of plasma GH concentration after intravenous injections of synthetic rat ghrelin into goats.

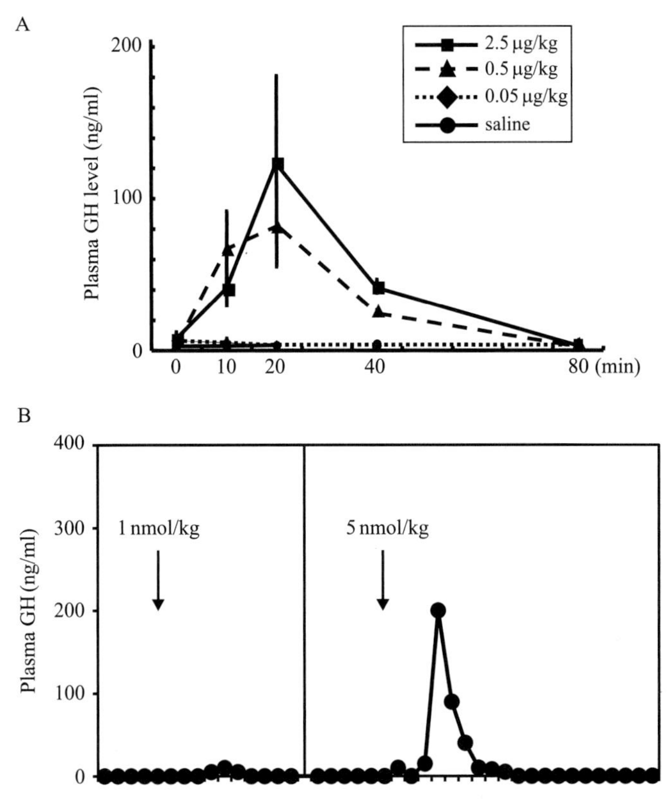

Figure 4.2 Biological activity of ghrelin in cats and goats. (A) Time course of plasma growth hormone (GH) concentration after intravenous injections of synthetic rat ghrelin into cats. Each symbol and vertical line on the graph represents the mean ± SEM of the data from five cats (adapted from Ida et al., 2007). (B) Effect of bovine ghrelin on plasma GH level in goats *(adapted from Ida et al., 2010).*

4. SUMMARY

We report the purification and characterization of the major feline and caprine ghrelins, as well as other minor ghrelin-derived molecules from cat and goat stomachs. In addition, we show that injection of synthetic ghrelin can alter the GH levels in cats and goats. The major active form of feline ghrelin is a 28-amino acid-long peptide with an *n*-octanoyl modification at Ser^3 and that of caprine ghrelin is a 27-amino acid-long peptide with an *n*-octanoyl modification at Ser^3. The major form of acyl modification of feline and caprine ghrelins was *n*-octanoic acid, as in all the known mammalian and nonmammalian ghrelins except rainbow trout ghrelin (Hosoda et al., 2003; Kaiya et al., 2003).

To study feline ghrelin, we observed ghrelin-(1–28)(C8:1), ghrelin-(1–28)(C10:2), and ghrelin-(1–27)(C10:2) which were not observed in other mammalian species, and ghrelin-(1–28)(C13:0) and ghrelin-(1–28)(C13:1) which were not observed in any species. In addition, *n*-nonanoylated (C9:0) forms of ghrelin were identified in goat. An *n*-nonanoylated form has not been found in other species aside from *Suncus* (Ishida et al., 2009). The most notable difference between caprine ghrelin and ghrelin from other species was the presence of several unknown acyl-modified forms (in ~33% of all purified ghrelins).

Recently, two groups demonstrated that acyl modification of ghrelin with *n*-octanoic acid is catalyzed by a specific acyltransferase named ghrelin *O*-acyltransferase (GOAT) (Gutierrez et al., 2008; Yang et al., 2008). Most likely, ghrelins are modified by GOAT in cats and goats, but the reasons underlying the modification of ghrelin with various fatty acids remain unknown. The mechanisms governing the acylation of ghrelin are also unknown, but feeding conditions or food composition may influence the type and extent of acyl modification of ghrelin.

In cats and goats, ghrelin stimulated the release of GH. Further analysis will be required to determine the physiological significance of the various forms of feline ghrelin. Furthermore, many similarities exist in the pathophysiology between humans, cats, goats, and other domestic animals. Therefore, we anticipate that studies of the physiological functions of ghrelin in domestic animals, including the effects on GH release, will help us understand the role of ghrelin in human pathophysiology.

ACKNOWLEDGMENTS

This work was financially supported in part by the Program to Disseminate Tenure Tracking System for Promoting Science and Technology from the Japanese Ministry of Education, Culture, Sports, Science and Technology; by a grant for Scientific Research on Priority Areas from the University of Miyazaki; and by grants-in-aid from the Ministry of Education, Science, Sports, and Culture, Japan.

REFERENCES

Gutierrez, J.A., et al., 2008. Ghrelin octanoylation mediated by an orphan lipid transferase. Proc. Natl. Acad. Sci. U.S.A. 105, 6320–6325.

Hosoda, H., et al., 2002. Purification and characterization of rat des-Gln14-Ghrelin, a second endogenous ligand for the growth hormone secretagogue receptor. J. Biol. Chem. 275, 21995–22000.

Hosoda, H., et al., 2003. Structural divergence of human ghrelin. Identification of multiple ghrelin-derived molecules produced by post-translational processing. J. Biol. Chem. 278, 64–70.
Ida, T., et al., 2007. Purification and characterization of feline ghrelin and its possible role. Domest. Anim. Endocrinol. 32, 93–105.
Ida, T., et al., 2010. Purification and characterization of caprine ghrelin and its effect on growth hormone release. J. Mol. Neurosci. 42, 99–105.
Ishida, Y., et al., 2009. Identification of ghrelin in the house musk shrew (*Suncus murinus*): cDNA cloning, peptide purification and tissue distribution. Peptides 30, 982–990.
Kaiya, H., et al., 2001. Bullfrog ghrelin is modified by n-octanoic acid at its third threonine residue. J. Biol. Chem. 276, 40441–40448.
Kaiya, H., et al., 2002. Chicken ghrelin: purification, cDNA cloning and biological activity. Endocrinology 143, 3454–3463.
Kaiya, H., et al., 2003. Peptide purification, complementary deoxyribonucleic acid (DNA) and genomic DNA cloning, and functional characterization of ghrelin in rainbow trout. Endocrinology 144, 5215–5226.
Kojima, M., et al., 1999. Ghrelin is a growth hormone-releasing acylated peptide from the stomach. Nature 402, 656–660.
Nakazato, M., et al., 2001. A role for ghrelin in the central regulation of feeding. Nature 409, 194–198.
Peterson, M., et al., 1990. Acromegaly in 14 cats. J. Vet. Intern. Med. 4, 192–201.
Yang, J., et al., 2008. Identification of the acyltransferase that octanoylates ghrelin, an appetite-stimulating peptide hormone. Cell 132, 387–396.

CHAPTER FIVE

Determination of Nonmammalian Ghrelin

Hiroyuki Kaiya*[,1], Hiroshi Hosoda[†], Kenji Kangawa[‡], Mikiya Miyazato*

*Department of Biochemistry, National Cerebral and Cardiovascular Center Research Institute, Suita, Osaka, Japan

[†]Department of Regenerative Medicine and Tissue Engineering, National Cerebral and Cardiovascular Center Research Institute, Suita, Osaka, Japan

[‡]National Cerebral and Cardiovascular Center Research Institute, Suita, Osaka, Japan

[1]Corresponding author: e-mail address: kaiya@ri.ncvc.go.jp

Contents

Abstract

Ghrelin is a peptide with a unique molecular modification by a fatty acid such as *n*-octanoic acid. The acyl modification is necessary for ghrelin to bind to its receptor (growth hormone secretagogue-receptor 1a, GHS-R1a) and to induce subsequent intracellular Ca^{2+} signaling. Ghrelin is widely expressed in mammals as well as in nonmammalian vertebrates. In our laboratory, a method for efficient purification of ghrelin

Methods in Enzymology, Volume 514
ISSN 0076-6879
http://dx.doi.org/10.1016/B978-0-12-381272-8.00005-2

from a small amount of tissues has been established. Here, we introduce the identification process of ghrelin in nonmammalian vertebrates.

1. INTRODUCTION

Ghrelin, a peptide discovered in rat stomach, is the endogenous ligand for GHS-R1a (Kojima et al., 1999). This novel 28-amino acid peptide possesses a unique serine residue at the N-terminal position 3 (Ser^3), which is modified by *n*-octanoic acid. The acylation is essential for ghrelin binding to the receptor and for subsequent ghrelin signaling and activity. Ghrelin is present not only in mammals but also in nonmammalian vertebrates (Kaiya et al., 2008), and we have succeeded in identifying ghrelin in cartilaginous fish (red stingray, Kaiya et al., 2009a,b; sharks, Kawakoshi et al., 2007), bony fish (catfish, Kaiya et al., 2005; eel, Kaiya et al., 2003c; goldfish, Miura et al., 2009; tilapia, Kaiya et al., 2003b; rainbow trout, Kaiya et al., 2003a), amphibians (bullfrog, Kaiya et al., 2001), reptiles (red-eared slider turtle, Kaiya et al., 2004), and birds (chicken, Kaiya et al., 2002). Various molecular forms of ghrelin have been determined by peptide purification, followed by cDNA cloning and mass spectrometric analysis. Here, we summarize the process of determining primary structure of ghrelin in nonmammalian vertebrates. The most different point, when compared to the purification of mammalian ghrelin (Kojima et al., 1999), is the use of affinity chromatography-attached immunoglobulin G (IgG) for rat ghrelin [1–11], which was effective in purification from a small amount of peptide.

2. PURIFICATION OF GHRELIN

The number of ghrelin-producing cell per unit area of stomach in nonmammalian vertebrates is approximately one-tenth of that in the rat stomach (H. Kaiya, unpublished observation), meaning that ghrelin content in nonmammalian vertebrates is significantly less than that in rat. As such, we have established an effective method using affinity chromatography to purify ghrelin from nonmammalian vertebrates.

2.1. Affinity gel for ghrelin

To effectively purify ghrelin with fatty acid modification, we developed an affinity column using IgG for rat ghrelin [1–11]. This was not used in the first ghrelin purification by Kojima et al. (1999).

2.1.1 *Purification of IgG for ghrelin*

1. Degas Protein A gel (Protein A sepharose 4 Fast Flow, Pharmacia) and store at 4 °C for 10 min.
2. Fill the gel (3.5 ml) in a disposable chromatography column and equilibrate with 15 ml of binding buffer (ImmunoPure Gentle Ag/Ab Buffer System, PIERCE).
3. Dilute 4 ml of anti-rat ghrelin [1–11] antisera with equal volume of binding buffer, load onto the washed column, and then collect the eluates (3 ml × 2 tubes and 2 ml × 1 tube).
4. Wash the loaded column with binding buffer (3 ml × 12 times).
5. Elute the adsorbed IgG with 6 ml of elution buffer (ImmunoPure Gentle Ag/Ab Buffer System, PIERCE) (1 ml × 2 tubes and 500 μl × 8 tubes).
6. Measure the absorbance of the eluates diluted with elution buffer to 1/100 at 280 nm.

2.1.2 *Conservation of Protein A gel*

1. Wash with 40 ml of 0.1 *M* glycine-HCl (pH 2.8).
2. Wash with 40 ml of 0.1 *M* CH_3COONa/0.5 *M* NaCl (pH 4.0).
3. Replace with 20 ml of MilliQ-grade water.
4. Fill up with 20 ml of 0.1% NaN_3.
5. Store at 4 °C.

2.1.3 *Desalting with Sephadex G-25*

1. Fractions containing IgG are desalted two times using a PD-10 column (Pharmacia Biotech).
2. Wash the PD-10 column with 20 ml of 0.1 *M* sodium phosphate buffer (Na_2HPO_4/NaH_2PO_4, pH 7.4).
3. Load samples from fraction 5 to 8, and wash with fractions 4 and 9. Collect the eluates from every loading.
4. Load 500 μl 0.1 *M* sodium phosphate buffer (pH 7.4) for nine times. Collect eluates from every loading.
5. Measure the absorbance of the eluates at 280 nm.
6. Wash the used column with 20 ml of 0.1 *M* sodium phosphate buffer (pH 7.4) again.
7. Load fractions that show high absorbance (i.e., fraction 8–10) to the washed PD-10 column, and wash with fractions 7 and 11. Collect the eluates from every loading.
8. Load 500 μl 0.1 *M* sodium phosphate buffer (pH 7.4) nine times. Collect the eluates from every loading.

9. Measure the absorbance of the eluates at 280 nm.
10. Calculate protein content (PC) in fraction(s) that showed high absorbance with the following formula: $PC(mg) = OD \times 51 \times 0.5 \times 0.71$ (OD is the sum of the optical density of the fractions).

2.1.4 Coupling ghrelin-specific IgG to Affi-gel 10

1. Affi-gel 10 (Bio-Rad) has a protein capacity of 15 mg/ml gel. Prepare two volumes of the gel according to the PC calculated above (Section 2.1.3, #10).
2. Fill a disposable chromatography column with an appropriate volume of Affi-gel 10.
3. Wash with one gel-volume of MilliQ-grade water.
4. Wash with one gel-volume of propanol.
5. Wash with two gel-volumes of chilled MilliQ-grade water.
6. Close the bottom of the column with a cap.
7. Load IgG fractions previously prepared by PD-10 into the column.
8. Close the top of the column with a cap.
9. Mix by a see-saw shaker at 4 °C overnight.
10. Handle carefully and restore the column at room temperature.
11. Remove the bottom cap and drain the solution in the column.
12. Add two gel-volumes of 0.1 *M* Tris–HCl/0.5 *M* NaCl (pH 8.0) and rinse the wall of the column and the top cap thoroughly.
13. Drain the former solution and replace with two gel-volumes of 0.1 *M* CH_3COONa/0.5 *M* NaCl (pH 4.0).
14. Repeat procedures #12 and #13 four times.
15. Wash two times with two gel-volumes of 0.1 *M* sodium phosphate buffer (pH 7.4).
16. Drain and replace with two gel-volumes of MilliQ-grade water.
17. Wash with two gel-volumes of 0.1% NaN_3.
18. Store with two gel-volumes of 0.1% NaN_3 at 4 °C.

2.2. Extraction of peptides from stomach

Since ghrelin is mainly produced in the stomach (Kojima et al., 1999), stomach is used as the primary site for peptide extraction.

2.2.1 Peptide extraction and preparation of peptide-enriched fraction

1. Mince fresh tissues in a beaker on ice or quickly pulverize the frozen tissues by a precooled pulverizer without melting the frozen tissues, and transfer the broken frozen tissues to a beaker on dry ice.

2. Put the tissues in a beaker with five tissue volumes of boiled MilliQ-grade water and boil for 10 min for inactivating intrinsic proteases.
3. Quickly chill the boiled sample on ice and adjust to 1 *M* acetic acid (AcOH) by adding glacial AcOH (60 μl/ml water).
4. Homogenize the boiled stomach tissues using a Polytron mixer for 1–2 min.
5. Centrifuge the crude acid extracts at 13,500 × *g* for 30 min.
6. Dilute the supernatant with an equal volume of MilliQ-grade water or 0.1% trifluoroacetic acid (TFA).
7. Load the sample (#6) into a Sep-Pak C18 cartridge (Waters, Milford, MA) which was washed with three gel-volumes of 60% acetonitrile/0.1% TFA and pre-equilibrated with three gel-volumes of 0.5 *M* AcOH or 0.1% TFA.
8. Wash the cartridge with three gel-volumes of 10% acetonitrile/0.1% TFA.
9. Elute the adsorbed peptides sequentially with three gel-volumes of 25% and 60% acetonitrile/0.1% TFA, respectively. Ghrelin is eluted in the latter fraction.
10. Evaporate the fraction that was eluted by 60% acetonitrile/0.1% TFA, and lyophilize.

2.2.2 Ion-exchange chromatography with SP-Sephadex C-25 (H^+ form)

1. Reconstitute the lyophilized sample with 1 *M* AcOH.
2. Put the sample into an SP-Sephadex C-25 column (H^+ form) (Amersham Pharmacia Biotech Inc., Buckinghamshire, UK). SP-Sephadex C-25 has a capacity of 100 mg/ml gel. Prepare three times greater than the amount of the gel to the lyophilized peptide weight.
3. Collect the flow-through as unbounded substances, namely SP-I.
4. Replace the gel with one gel-volume of MilliQ-water.
5. Elute successively with three gel-volumes of 2 *M* pyridine and 2 *M* pyridine–AcOH (pH 5.0), respectively, and collect the other two fractions, SP-II and SP-III, respectively. Ghrelin is eluted in the last strong basic peptide-enriched SP-III fraction.

2.3. Gel-permeation chromatography

The crude extract of peptides was first separated by molecular sieving chromatography to assemble the target molecules. It uses column chromatography if there is a large amount of peptides (e.g., eel ghrelin, Kaiya et al., 2003c). However, this gel-permeation chromatography process was not performed in most cases of nonmammalian ghrelin purification in order to prevent the loss of peptide. Instead, elution with 25% acetonitrile/0.1% TFA (Section 2.2.1, #9) was effective in excluding excessive peptides. Recently, we have used High-performance liquid chromatography (HPLC)-type gel-permeation chromatography when the amount of peptide was small.

2.3.1 Column using Sephadex G-50

1. Equilibrate a Sephadex G-50 column (column size: 2.9 × 144.5 cm, fine grade, Amersham Pharmacia Biotech.) with 1 *M* AcOH at a flow rate of 15 ml/h.
2. Dissolve the lyophilized sample in 5–10 ml (1/200 – 1/100 gel-bed volume) of 1 *M* AcOH.
3. Apply the sample and collect eluate in every 15-ml fractions.
4. Flow the column with 1 *M* AcOH for 3 days.
5. Measure optical density of collected samples at 280 nm.

2.3.2 HPLC using TSK-GEL G2000SW

1. Equilibrate a TSK-GEL G2000SW column (column size: 21.5 mm × 30 cm [order made]) with 35% acetonitrile/0.1% TFA at a flow rate of 2 ml/min.
2. Inject sample, monitor optical density of eluate at 280 nm at the same time, and collect the eluate in 2-ml fractions.

2.4. Ion-exchange HPLC

In almost all cases of nonmammalian ghrelin purification, the SP-III fraction is separated directly with ion-exchange HPLC because of the small amount of peptide and therefore to prevent peptide loss. Carboxymethyl (CM) column (TSK-gel CM-2SW column, 7.8 × 300 mm, Tosoh, Tokyo, Japan) is used for our ion-exchange HPLC. SP-III fraction is purified by a Sep-Pak C18 cartridge, and the eluate is lyophilized.

1. Dissolve lyophilized sample in 10 m*M* ammonium formate (pH 4.8) containing 10% acetonitrile (solvent A).
2. Inject the sample and elute with the following conditions: flow rate, 2 ml/min; two-step gradient profile, first from solvent A to 25% solvent B (1 *M* ammonium formate (pH 4.8) containing 10% acetonitrile) for 5 min, then to 55% solvent B until 65 min, and flow with an additional wash program (55–100% solvent B for 5 min and 100% solvent B for 10 min, total time 80 min).
3. Collect eluate in 1-ml fractions during 80 min. In almost all cases, ghrelin is eluted during the 65-min elution, but there were also some exceptions (Kaiya et al., 2009b; Kawakoshi et al., 2007).

2.5. Affinity chromatography

1. Active fractions in CM-ion-exchange HPLC were desalted by a Sep-Pak C18 column (Sep-Pak light C18 or Plus C18, Waters) before performing the affinity chromatography using antighrelin [1–11] IgG, and the eluate was then lyophilized.
2. Fill up 50 μl of Affi-gel attached antighrelin [1–11] IgG in a brief column made by a tip.
3. Wash the column with at least three gel-volumes of 60% acetonitrile/ 0.1% TFA and neutralize with at least six gel-volumes of 0.1 *M* sodium phosphate buffer (pH 7.4).
4. Dissolve the lyophilized sample in 500 μl of 0.1 *M* sodium phosphate buffer (pH 7.4).
5. Apply the sample to the column and collect the flow-through in a clean tube.
6. Apply the flow-through to the column again.
7. Wash the column with at least three gel-volumes of 0.1 *M* sodium phosphate buffer (pH 7.4).
8. Elute the adsorbed substances with three gel-volumes of 60% acetonitrile/0.1% TFA.
9. Evaporate the eluate until acetonitrile is removed from the sample and proceed to the next separation step (Section 2.6) using reverse-phase (RP)-HPLC.
10. Wash the used Affi-gel column with at least five gel-volumes of 0.1 *M* Tris–HCl/0.5 *M* NaCl (pH 8.0).
11. Wash with five gel-volumes of 0.1 *M* sodium phosphate buffer (pH 7.4).
12. Replace with five gel-volumes of 0.1% NaN_3.
13. Store the column at 4 °C.

2.6. Reverse-phase-HPLC

Before final purification, fractions are prepared by RP-HPLC, and the obtained ghrelin-active fraction is further purified by another RP-HPLC step with a different column or different elution condition until a single peptide peak is isolated.

2.6.1 Preparative RP-HPLC

1. Add 20 μl of 0.1% Triton X-100 to the evaporated sample to prevent the loss of peptide.

2. Inject the sample for RP-HPLC using a Symmetry C18 column (3.9 × 150 mm, Waters) at a flow rate of 1 ml/min under a linear gradient from 10% to 60% acetonitrile/0.1% TFA for 40 min.
3. Collect the eluates in 1-ml fractions (40 tubes) from immediately after injection to the end.

2.6.2 Final purification by RP-HPLC with a different column

1. Evaporate the ghrelin-active fraction until acetonitrile is removed from the sample.
2. Add 5 μl of 0.1% Triton X-100 to the evaporated sample to prevent loss of peptide.
3. Inject the sample for another RP-HPLC step using a diphenyl column (219TP5215, 2.1 × 150 mm, Vydac, Hesperia, CA) for 40 min under a linear gradient from 10% to 60% acetonitrile/0.1% TFA at a flow rate of 0.2 ml/min.
4. Collect at each absorbance peak.
5. Sometimes, some peaks may contain excessive peptides even in this stage. Hence, further purification is performed using another column, Chemcosorb 3-ODS-H column (2.1 × 75 mm, Chemco) for 80 min under a linear gradient from 10% to 60% acetonitrile/0.1% TFA at a flow rate of 0.2 ml/min.

2.7. Protein sequencing

Fundamentally, protein sequencing was done according to the manufacturer's protocol. In our laboratory, approximately 5–10 pmol of the finally purified peptide is analyzed by a protein sequencer (model 494 HT, Applied Biosystems, Foster City, CA). In this case, 10–15 amino acids can be determined. In the case of ghrelin, the third amino acid cannot be analyzed because of acyl modification.

3. MEASUREMENT OF GHRELIN ACTIVITY

During the purification process, ghrelin activity is monitored by measuring the changes in intracellular Ca^{2+} concentrations using a fluorescence-imaging plate reader (FLIPR) system (Molecular Devices, Sunnyvale, CA) in CHO or HEK293 cell-expressing rat GHS-R1a. In the case of Kojima et al. (1999), a cell line that stably expressed rat GHS-R1a (CHO-GHSR62) was prepared and used for the assay (Kojima et al., 1999).

3.1. Preparation of GHS-R1a-expressing cells

3.1.1 Transient expression of GHS-R1a protein

1. Prepare the open reading frame of rat GHS-R1a cDNA by RT-PCR using the rat brain cDNA library.
2. Clone the cDNA into pcDNA3.1/V5-His TOPO TA cloning vector (Invitrogen) with neomycin resistance.
3. Transfect the plasmid vector into CHO or HEK293 cell (1×10^6 cell/10-cm dish) using an appropriate transfection reagent (e.g., FuGENE 6 [Roche Diagnostic inc.]). At least 20 h after incubation, these transfected cells can be used for assessing ghrelin activity.

3.1.2 Selection of stably GHS-R1a-expressed cells

For the preparation of a stably GHS-R1a-expressed cell line, select the cells by treatment of neomycin (e.g., G418). It is better to use CHO because HEK293 cells express endogenous receptors of certain gut hormones.

1. Add 40 μl/10 ml culture medium of G418 solution (250 mg/ml in a buffered solution).
2. Culture the selected CHO cells in culture medium containing G418 approximately for 2 weeks with regular medium replacement.

3.2. FLIPR assay

1. Prepare the assay sample (100–200 mg tissue equivalent/fraction) containing 0.1% BSA/10 μl/tube.
2. Plate GHS-R1a-expressing CHO or HEK293 cells onto black-walled 96-well microplates (Costar, Corning Incorporated, Cornig, NY) at 5×10^4 cells/well for CHO, or 3×10^4 cells/well for HEK293. Note that HEK293 cells are plated on poly-D-lysin-coated microplates.
3. Culture the cell in a humidified environment of 95% air:5% CO_2 for at least 18 h prior to assay.
4. Aspirate the culture medium and incubate the cell for 1 h in a CO_2 incubator after addition of 100 μl mixture containing 4.4 μ*M* Fluo-4 AM (Invitrogen), 0.045% pluronic acid, and 1% fetal calf serum in an assay buffer (Hanks' balanced salt solution, 20 m*M* HEPES, 2.5 m*M* probenecid [Sigma]). Pluronic acid can be replaced by Power load 100 × concentrate (Invitrogen).
5. Wash the incubated cells four times by an automatic cell washer in the assay buffer.

6. Dissolve the lyophilized samples in 120 μl of the assay buffer containing 0.001% Triton X-100, and arrange the dissolved samples in their designated plate.
7. Treat automatically with 100 μl of each sample to the GHS-R1a-expressing cells by FLIPR.

4. CDNA CLONING AND DETERMINATION OF AMINO ACID SEQUENCE OF PREPRO-GHRELIN

Partial amino acid sequence of a mature ghrelin can be found by purification of the ghrelin peptide and subsequent protein sequencing. On the basis of this information, degenerated sense primers are designed and 3′-RACE PCR is performed. Based on the nucleotide sequence determined by 3′-RACE PCR, gene-specific antisense primers are designed, and 5′-RACE PCR is conducted. Based on the amino acid sequence of mature ghrelin, it is easy to design primers for the nested PCR, which is important for effective PCR amplification of the target. We have used GeneRacer Kit (Invitrogen) for making the template of the RACE PCR.

4.1. 3′-RACE PCR

1. Design the degenerate 24–27 bps sense primers for primary PCR based on the N-terminal seven-amino acid sequence of ghrelin (e.g., GSSFLSP) and for the nested PCR based on the amino acid sequence of the determined ghrelin in addition to the fifth, sixth, and seventh amino acids (Leu-Ser-Pro) of ghrelin.
2. Extract the total RNA from the stomach.
3. Synthesize the first-strand cDNAs from 1 μg of total RNA using a 3′-oligo-dT adaptor primer supplied by the kit.
4. Perform primary PCR using degenerate sense primers based on the N-terminal seven-amino acid sequence of ghrelin (GSSFLSP), a 3′-primer supplied with the kit, and DNA polymerase (e.g., EXtaq [TaKaRa, Shiga, Japan]).
5. Purify the amplified product by a Wizard PCR Preps DNA Purification Kit (Promega) to remove excess primers in the reaction.
6. Perform nested PCR using 1/10 of the purified primary PCR product as starting material. In this PCR, the negative control was set up by using either the sense or antisense primer alone and is run simultaneously to confirm specific amplification with combination of sense and antisense primers.

7. Subclone the putative ghrelin cDNA fragment into an appropriate TOPO TA cloning vector (e.g., pCR II-TOPO vector, Invitrogen).
8. Determine the nucleotide sequence by DNA sequencing (e.g., DNA sequencer model 3130, Applied Biosystems), according to the BigDye Terminator Cycle Sequencing Kit (Applied Biosystems) using the M13 forward or reverse primer.

4.2. 5′-RACE PCR

1. Design three gene-specific antisense primers that piled up the 3′-side mutually, based on the nucleotide sequence determined by the 3′-RACE PCR.
2. Treat total RNA (5 μg) with calf intestinal phosphatase and tobacco acid pyrophosphatase, and ligate with the GeneRacer RNA Oligo, which is supplied in the kit, according to the manufacturer's protocol.
3. Synthesize first-strand cDNA from half of the treated total RNA with either oligo-$dT_{12–18}$ primer or gene-specific antisense primer (primer 1) using an appropriate reverse transcriptase.
4. Perform primary PCR using a gene-specific antisense primer (primer 2), the 5′ primer supplied in the kit, and an appropriate DNA polymerase (e.g., HotSTAR, QIAGEN).
5. Conduct nested PCR using a gene-specific antisense primer (primer 3) and 1/10 of the primary PCR product purified from the Wizard PCR Preps DNA Purification Kit.
6. Subclone the putative ghrelin fragment into a TOPO TA cloning vector and determine the nucleotide sequence.

4.3. RT-PCR using a proofreading Taq

After determining the full length of ghrelin cDNA with 5′- and 3′-cDNA fragments, the gene-specific primer is designed at either 5′- or 3′-ends of the cDNA and is used to amplify the full-length cDNA using a proofreading Taq, so as to confirm the identity of the nucleotide sequence.

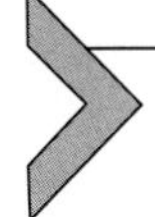

5. DETERMINATION OF SPECIES OF FATTY ACID MODIFICATION BY MASS SPECTROMETRIC ANALYSIS

Ghrelin has a unique acyl modification attributed to the various medium-chain fatty acids such as *n*-octanoic acid, *n*-decanoic acid, or other unsaturated fatty acids at the third serine (Ser^3) or threonine residue (Thr^3, Kaiya et al., 2001). The type of fatty acid can be determined on the basis of the molecular weight of the native purified peptide by measuring the *m/z* difference of the amino acid sequence between the measured and the theoretical masses obtained from calculations of peptide purification and cDNA cloning. Approximately 1 pmol is needed for the determination of molecular weight by mass spectrometer (e.g., MALDI-TOF) using α-cyano-4-hydroxycinnamic acid as the matrix.

6. SUMMARY

Accumulating evidence indicate that ghrelin is a multifunctional peptide that elicits various physiological actions other than orexigenic and GH-releasing activity in mammals (Baatar et al., 2011; Briggs and Andrews, 2011; Kojima and Kangawa, 2010; Steiger et al., 2011; Varela et al., 2011; Zhang et al., 2010) and nonmammals (Kaiya et al., 2007, 2008, 2009a 2011). To use a homologous endogenous peptide for experiment is important to look at the original effects that the peptide has because there are cases where a species-specific effect of homologous ghrelin has been demonstrated (Kitazawa et al., 2007). Here, we described the procedures of ghrelin determination. What is important for the success of ghrelin purification is how to evaluate the activity, meaning that a good receptor-expressing cell with sufficient sensitivity to ghrelin has to be prepared. In addition, facilities are also keys for the isolation of high-quality peptide. Entry of many new researchers would contribute to the understanding of biological actions of ghrelin in nonmammalian vertebrates.

ACKNOWLEDGMENTS

This work was supported in part by a Grant-in-Aid for Scientific Research from the Ministry of Education, Science, Sports, and Culture of Japan to H. K., M. M., and K. K., and in part by a SUNBOR GRANT to H. K.

REFERENCES

Baatar, D., et al., 2011. The effects of ghrelin on inflammation and the immune system. Mol. Cell. Endocrinol. 340, 44–58.

Briggs, D.I., Andrews, Z.B., 2011. Metabolic status regulates ghrelin function on energy homeostasis. Neuroendocrinology 93, 48–57.

Kaiya, H., et al., 2001. Bullfrog ghrelin is modified by n-octanoic acid at its third threonine residue. J. Biol. Chem. 276, 40441–40448.

Kaiya, H., et al., 2002. Chicken ghrelin: purification, cDNA cloning, and biological activity. Endocrinology 143, 3454–3463.

Kaiya, H., et al., 2003a. Peptide purification, complementary deoxyribonucleic acid (DNA) and genomic DNA cloning, and functional characterization of ghrelin in rainbow trout. Endocrinology 144, 5215–5226.

Kaiya, H., et al., 2003b. Identification of tilapia ghrelin and its effects on growth hormone and prolactin release in the tilapia, *Oreochromis mossambicus*. Comp. Biochem. Physiol. B 135, 421–429.
Kaiya, H., et al., 2003c. Amidated fish ghrelin: purification, cDNA cloning in the Japanese eel and its biological activity. J. Endocrinol. 176, 415–423.
Kaiya, H., et al., 2004. Structural determination and histochemical localization of ghrelin in the red-eared slider turtle, *Trachemys scripta elegans*. Gen. Comp. Endocrinol. 138, 50–57.
Kaiya, H., et al., 2005. Purification, cDNA cloning, and characterization of ghrelin in channel catfish, *Ictalurus punctatus*. Gen. Comp. Endocrinol. 143, 201–210.
Kaiya, H., et al., 2007. Ghrelin in birds: its structure, distribution and function. J. Poult. Sci. 44, 1–18.
Kaiya, H., et al., 2008. Ghrelin: a multifunctional hormone in non-mammalian vertebrates. Comp. Biochem. Physiol. A 149, 109–128.
Kaiya, H., et al., 2009a. Current knowledge of the roles of ghrelin in regulating food intake and energy balance in birds. Gen. Comp. Endocrinol. 163, 33–38.
Kaiya, H., et al., 2009b. Ghrelin-like peptide with fatty acid modification and O-glycosylation in the red stingray, *Dasyatis akajei*. BMC Biochem. 10, 30.
Kaiya, H., et al., 2011. Recent advances in the phylogenetic study of ghrelin. Peptides 32, 2155–2174.
Kawakoshi, A., et al., 2007. Identification of a ghrelin-like peptide in two species of shark, *Sphyrna lewini* and *Carcharhinus melanopterus*. Gen. Comp. Endocrinol. 151, 259–268.
Kitazawa, T., et al., 2007. Contractile effects of ghrelin-related peptides on the chicken gastrointestinal tract *in vitro*. Peptides 28, 617–624.
Kojima, M., Kangawa, K., 2010. Ghrelin: more than endogenous growth hormone secretagogue. Ann. N. Y. Acad. Sci. 1200, 140–148.
Kojima, M., et al., 1999. Ghrelin is growth-hormone-releasing acylated peptide from stomach. Nature 402, 656–660.
Miura, T., et al., 2009. Purification and properties of ghrelin from the intestine of the goldfish, *Carassius auratus*. Peptides 30, 758–765.
Steiger, A., et al., 2011. Ghrelin in mental health, sleep, memory. Mol. Cell. Endocrinol. 340, 88–96.
Varela, L., et al., 2011. Ghrelin and lipid metabolism: key partners in energy balance. J. Mol. Endocrinol. 46, R43–R63.
Zhang, G., et al., 2010. Ghrelin and cardiovascular diseases. Curr. Cardiol. Rev. 6, 62–70.

SECTION 2

Molecular Forms of Ghrelin and Measuring of the Concentrations

CHAPTER SIX

Morphological Analysis of Ghrelin Neurons in the Hypothalamus

Haruaki Kageyama*, Fumiko Takenoya*,†, Seiji Shioda*,1
*Department of Anatomy, Showa University School of Medicine, Tokyo, Japan
†Department of Physical Education, Hoshi University School of Pharmacy and Pharmaceutical Science, Tokyo, Japan
[1]Corresponding author: e-mail address: shioda@med.showa-u.ac.jp

Contents

Abstract

Ghrelin, which is mainly produced in the A/X-like cells of the oxyntic glands of the stomach, transduces an appetite-stimulatory signal from peripheral tissues to the central nervous system. Ghrelin is also localized in the hypothalamic arcuate nucleus of rodents. While ghrelin acts on the hypothalamus to promote feeding behavior and energy metabolism, it is important to clarify the neuronal circuits that involve ghrelin so as to elucidate the action of ghrelin in the brain. Immunoelectron microscopy reveals that ghrelin neurons send synaptic outputs to other feeding-regulating neurons (e.g., to neurons containing orexin, proopiomelanocortin, or neuropeptide Y) and receive synaptic inputs from other feeding-regulating neurons (proopiomelanocortin or neuropeptide Y). This chapter describes the immunohistochemical techniques employed to elucidate the neuronal interactions between ghrelin and other kinds of feeding-regulating peptide-containing neurons in the hypothalamus based on evidence at both light microscopic and ultrastructural levels.

Methods in Enzymology, Volume 514
ISSN 0076-6879
http://dx.doi.org/10.1016/B978-0-12-381272-8.00006-4

1. INTRODUCTION

Ghrelin is mainly produced in the A/X-like cells of the oxyntic glands of the stomach and is the only circulating hormone that acts on the hypothalamus to affect feeding behavior and energy metabolism (Kojima et al., 1999). The ghrelin-producing cells are round to ovoid and of a closed type that, although they have no contact with the lumen in the stomach, are positioned in close proximity to the capillary network (Date et al., 2000).

Ghrelin-containing neuronal cell bodies are localized in the hypothalamic arcuate nucleus (ARC) that integrates signals for energy homeostasis, and ghrelin-containing nerve fibers are widely distributed in the brain. Accumulated evidence shows that hypothalamic neuropeptides such as neuropeptide Y (NPY), orexin, and proopiomelanocortin (POMC) are involved in the regulation of feeding behavior and energy homeostasis via neuronal circuits in the hypothalamus. Ghrelin conveys information to orexin-containing neurons in the lateral hypothalamus or to NPY- or POMC-containing neurons in the ARC (Guan et al., 2003, 2008; Lu et al., 2002; Toshinai et al., 2003). Ghrelin–ghrelin neuronal interactions also exist in the ARC, where they form ultrashort circuits and stimulate each other via positive feedback system (Hori et al., 2008). Hence, ghrelin neurons also form feeding-regulating neuronal circuits with other feeding-regulating peptide-containing neurons within the hypothalamus. These protocols should provide valuable guidance for many investigators wishing to observe ghrelin-immunopositive neurons using both light and electron microscopy techniques.

2. OBSERVATION OF GHRELIN-CONTAINING CELLS AT THE LIGHT MICROSCOPE LEVEL

Ghrelin mRNA is present in the stomach and hypothalamus of rodents (Kojima et al., 1999). We previously detected ghrelin-immunopositive reactivity in the stomach of rats (Fig. 6.1A) and demonstrated using immunohistochemical analyses that ghrelin-immunoreactive neurons are localized in the hypothalamic ARC of colchicine-treated rats (Fig. 6.1B and C) (Hori et al., 2008; Kojima et al., 1999; Lu et al., 2002). Ghrelin-containing fibers in the brain can be found in close apposition with NPY-containing (Fig. 6.2A) or orexin-containing cell bodies and dendritic processes (Fig. 6.2B) (Toshinai et al., 2003). Thus, ghrelin

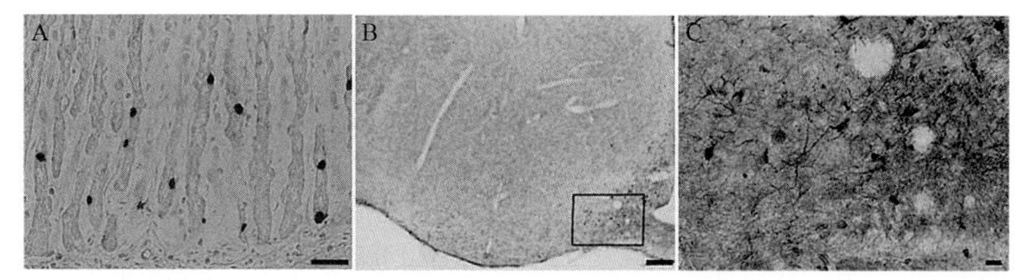

Figure 6.1 Ghrelin immunoreactivity in the hypothalamus and stomach of rat identified using the ABC–DAB method. Ghrelin-immunopositive cells are present in the gastric gland (A). Ghrelin-immunopositive neurons are present in the hypothalamic arcuate nucleus (B and C). C is an enlarged image of the boxed area in B. Scale bar is 50 μm (A), 200 μm (B), and 20 μm (C).

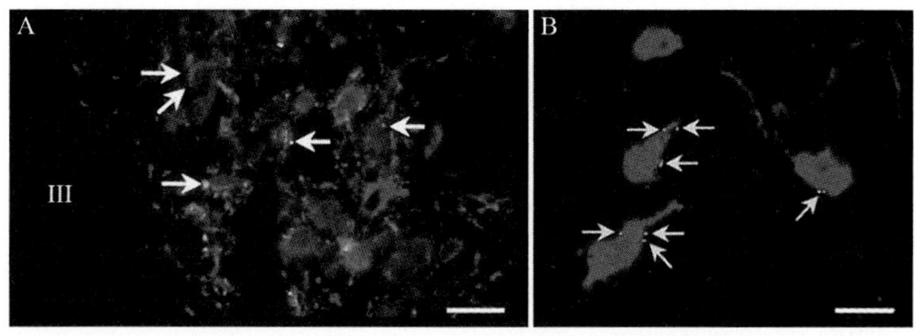

Figure 6.2 The innervation of ghrelin-immunoreactive axons to neuropeptide Y (NPY) (A) or orexin-immunoreactive cell bodies (B). (A) Ghrelin immunoreactivity is shown in red (Alexa Fluor 546). NPY immunoreactivity is shown in green (Alexa Fluor 488). Ghrelin-immunoreactive fibers can be seen in close proximity to NPY-immunopositive neurons. Arrows show that ghrelin-immunopositive fibers are in close apposition with NPY-immunopositive cell bodies. (B) ghrelin immunoreactivity is shown in red (Alexa Fluor 546) and orexin immunoreactivity in green (Alexa Fluor 488). Ghrelin-immunoreactive fibers are found in close proximity to orexin-immunoreactive neurons. Arrows indicate the apposition of ghrelin fibers to orexin neurons. III; The third ventricle. Scale bar is 10 μm in (A) and (B). (See Color Insert.)

neurons form feeding-regulating neuronal circuits. However, immunohistochemistry studies reflect anomalies in the use of different antisera against ghrelin (Cowley et al., 2003). Because ghrelin is thought to form a complicated 3D structure by acylation, it is very difficult to make a specific antibody against it.

2.1. Colchicine treatment

Colchicine treatment enhances the ghrelin immunoreaction. Colchicine is a tubulin polymerization inhibitor and is known to inhibit cytoplasmic transport and exocytosis (Thyberg and Moskalewski, 1985). Hence, treatment of animals with colchicine is commonly used to promote the accumulation of secretory proteins in the cytoplasm.

1. Deeply anesthetize rat with sodium pentobarbital (40 mg/kg, intraperitoneal injection) and inject colchicine (200 μg/5 μl saline) into the lateral ventricle (0.8 mm posterior to bregma; 1.3 mm lateral from the midline; and 3.5 mm below the outer surface of the skull, according to the Paxinos and Watson atlas of rat brain (Paxinos and Watson, 1986)).
2. Forty-eight hours later, confirm tetraplegia of the colchicine-treated rat.

2.2. Preparation of sections for ABC–DAB staining or immunofluorescence

1. Deeply anesthetize rat by overdose with an intraperitoneal injection of sodium pentobarbital (50 mg/kg, Dainippon Pharmaceutical, Osaka, Japan)
2. Perfuse animal with 50 ml of saline (preheated to 37 °C) followed by 250–300 ml of fixative solution (4% paraformaldehyde in 0.1 *M* phosphate buffer (PB) (pH 7.4)).
3. Remove the brain tissue, trim, and immerse in the same fixative for 12 h at 4 °C.
4. After washing with 0.1 *M* PB, transfer the tissue to a 0.1 *M* PB solution containing 20% sucrose for 2 days at 4 °C and then embed in O.C.T. compound (Sakura Finetechnical, Tokyo, Japan).
5. Immediately freeze in liquid nitrogen–cooled isopentane and store at –80 °C.
6. Prepare 7 μm-thick cryosections cut from frozen tissue using a cryostat (MICROM HM 500; MICROM, Heiderberg, Germany) at –20 to –30 °C.

2.3. Immunofluorescence

1. Dry the section at room temperature (RT).
2. Remove the extra O.C.T. compound around the tissue by wiping with water-moistened paper.
3. Use a DAKO pen (DAKO, Glostrup Denmark) to mark the area around the tissue to create a hydrophobic barrier.
4. Wash the section for 5 min in phosphate-buffered saline (PBS) at RT.
5. Block the sections for 1 h with 5% normal horse serum in PBS.
6. Incubate with rabbit antighrelin antibody (available from Dr. Kenji Kangawa, National Cerebral and Cardiovascular Center (Kojima

et al., 1999)) or goat antighrelin antibody (1:10000 dilution, Santa Cruz Biotechnology, Inc. properly diluted) for 16 h at 4 °C.

7. Wash sections three times for 5 min in PBS at RT.
8. Incubate with fluorochrome-labeled donkey anti-rabbit IgG (1:400 dilution, Invitrogen Corp., Carlsbad, CA) or fluorochrome-labeled donkey anti-goat IgG (1:2000 dilution, Rockland, Inc., Gilbertsville, PA) for 1.5 h at RT.
9. Wash the sections in PBS for 5 min at RT three times.
10. To demonstrate simultaneously the presence of ghrelin-like immunoreactivity on the same section as that in which another peptide-like immunoreactivity is also detected, a double immunofluorescence technique must be used. Incubate with the second antibody of interest for 16 h at 4 °C.
11. Wash the section three times for 5 min in PBS at RT.
12. Incubate with fluorochrome-labeled anti-IgG (1:400 dilution, Invitrogen Corp.) for 1.5 h at RT.
13. Double immunolabeling is detected with the aid of a fluorescence microscope or a confocal laser microscope.

2.4. ABC–DAB staining for viewing sections with a light microscope

This protocol enables conjugation of a biotinylated secondary antibody with a preformed avidin:biotinylated peroxidase complex and has been termed the "ABC" technique. Because avidin has such an extraordinarily high affinity for biotin (over 1 million times higher than antibody for most antigens), the binding of avidin to biotin is essentially irreversible. Biotinylated peroxidase reacts with the substrate 3,3′-diaminobenzidine (DAB) to produce a brown color.

1. Block the sections for 1 h in 5% normal horse serum at RT.
2. Incubate with rabbit antighrelin antibody for 16 h at 4 °C.
3. Wash the sections three times for 5 min in PBS.
4. Incubate for 2 h with biotinylated anti-rabbit IgG (1:200 dilution, Vector laboratories, Inc., Burlingame, CA).
5. Wash the section in PBS for 5 min three times.
6. Incubate with ABC (VECTASTAIN ABC kits, Vector laboratories, Inc.) for 1 h at RT.
7. Wash the section three times for 5 min in PBS.
8. Treat with DAB in 0.05 *M* Tris–HCl (pH 7.6) buffer containing 0.005% hydrogen peroxide (DAB Peroxidase Substrate Kit, Vector laboratory) for about 10 min in the dark.

9. Wash the section in water for 5 min.
10. After counterstaining, dehydrate in a graded ethanol series and then penetrate in xylene.

3. Observation of Ghrelin-Containing Cells at Electron Microscopic Level

Ghrelin-like immunoreactivity in immunoreactive cell bodies, processes, and axon terminals was detected mainly in dense granular vesicles about 110 nm in diameter (Lu et al., 2002). Immunohistochemical studies at the ultrastructural level have demonstrated that ghrelin-containing axon terminals make synapses with NPY- or POMC-positive dendritic processes in the ARC (Cowley et al., 2003; Guan et al., 2003, 2008). Since orexin neurons receive synaptic inputs from NPY neurons (Horvath et al., 1999), the possibility exists that ghrelin stimulates the activity of orexin neurons via NPY neurons. However, physiological studies using antiorexin immunoglobulin G and a Y1 NPY receptor antagonist have demonstrated that ghrelin individually activates both the NPY and orexin systems to induce feeding behavior (Toshinai et al., 2003). Furthermore, ghrelin-immunopositive dendritic processes receive synaptic inputs from NPY-, POMC-, and ghrelin-containing axon terminals in the ARC (Guan et al., 2003, 2008; Hori et al., 2008) (Fig. 6.3). To identify ultrastructural aspects of ghrelin-immunopositive neurons in the rat hypothalamus, the following protocol can be used:

3.1. Preparation of section for electron microscopy

1. Deeply anesthetize a colchicine-treated rat by overdose with an i.p. injection of sodium pentobarbital (50 mg/kg, Dainippon Pharmaceutical)
2. Perfuse animals with 50 ml of saline (preheated to 37 °C) followed by 250–300 ml of 4% paraformaldehyde in 0.1 *M* PB.
3. Remove the brain tissue, trim, and immerse in the same fixative for 12 h at 4 °C.
4. Prepare 40-μm-thick floating sections using a microslicer (e.g., an Oxford vibratome).

3.2. Immunostaining for subsequent observation by an electromicroscopy

1. Block the floating sections for 1 h in a 5% normal horse serum at RT.
2. Perform the same as described above in Section 2.4.

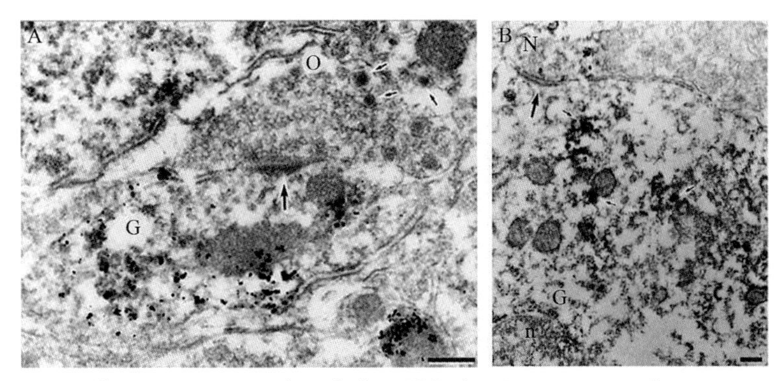

Figure 6.3 Electron micrographs of the ARC showing synapses between orexin-immunoreactive axon terminals and a ghrelin-immunoreactive neuron (A) and between NPY-immunoreactive axon terminals and ghrelin-immunopositive neurons (B). A, Orexin-immunoreactive axon terminals (O) make synapse with ghrelin-immunoreactive dendritic process (G). Large arrow shows asymmetric synapse. Small arrow shows a dense-cored vesicle. (B) NPY-immunoreactive axon terminals (N) make synapse with ghrelin-immunoreactive perikaryon (G). n, nucleus of ghrelin neuron. Large arrow shows symmetrical synapse. Small arrow shows a dense-cored vesicle. Scale bar is 500 nm in (A) and 200 nm in (B). Immunohistochemistry using antighrelin antibody complemented with SGI (black spot indicates reaction of SGI) was performed on this section, followed by immunohistochemistry with antiorexin or anti-NPY antibodies.

3.3. Silver–gold intensification

To enhance immunoreactivity or to perform double immunoelectromicroscopy, sections can be further treated with silver–gold intensification (SGI) to darken the brown-colored DAB reaction material to black (Guan et al., 2003, 2008; Teclemariam-Mesbah et al., 1997; van den Pol and Gorcs, 1986).

1. Following the DAB reaction (Section 2.4), incubate the floating sections with 10% thioglycolic acid for 2 h at RT.
2. Rinse the sections for 30 min in 2% sodium acetate.
3. Develop each section in ABC solution until the color highlighting immunoreactivity change from brown to black.

   3.1. Preparation of solutions A, B, and C.

   Solution A: 10% anhydrous sodium carbonate.

   Solution B: Add 2 g of ammonium nitrate, 2 g of silver nitrate, and 10 g of 12-tungsto (VI) silicic acid into 800 ml of distilled water and mix well. Adjust the volume of the solution to 1 l with distilled water.

   Solution C: Formalin (37% formaldehyde).

3.2. Add 20 μl of solution C to 5 ml of solution B.
3.3. Add 5 ml of the mixture of solution B and C to 5 ml of solution A.
3.4. Mix well. The ABC solution is clear without any precipitate.

4. Stop the development by adding1% acetic acid for 5 min.
5. Rinse the sections in 2% sodium acetate for 10 min.
6. Transfer the sections in 0.05% gold (III) chloride trihydrate and stand them for 10 min.
7. Rinse the sections in 2% sodium acetate for 10 min.
8. Fix the sections in 3% sodium thiosulfate for 10 min.
9. Rinse the sections twice in 2% sodium acetate for 10 min.
10. Wash the sections in PBS.
11. Perform second immunostaining with ABC–DAB method.

3.4. Postfixation

1. Following the DAB method (protocol 3.3.11), wash the sections in PBS.
2. Postfix twice for 15 min with 1% OsO_4 in 0.1 *M* PB (pH 7.4) at RT.
3. Dehydrate in a graded ethanol series (50–100% anhydrous ethanol)
4. Substitute a mixture of Quetol-812 (Nisshin EM Corporation, Tokyo, Japan)-QY1(Nisshin EM Corporation) (1:1) for ethanol
5. Embed the section in Quetol-812.
6. Cut ultrathin sections.
7. Observe the sections with the aid of an electron microscope.

4. SUMMARY

It has been demonstrated that a part of the peripheral ghrelin signal activates NPY in the ARC via the noradrenergic pathway ascending from the nucleus of the solitary tract to the ARC (Date et al., 2006). These results suggest that ghrelin in the hypothalamus stimulates food intake in harmony with appetite-stimulating signals from peripheral organs, particularly the stomach.

ACKNOWLEDGMENTS

This work was supported by two Grants-in-Aid for Scientific Research (C) from the Ministry of Education, Culture, Sports, Science, and Technology of Japan (to H. K. # 21590222 and to F. T. # 23500863) and in part by a Grant-in-Aid for Scientific Research on Innovative Areas from the Ministry of Education, Culture, Sports, Science, and Technology of Japan (to S. S. # 22126004).

REFERENCES

Cowley, M.A., et al., 2003. The distribution and mechanism of action of ghrelin in the CNS demonstrates a novel hypothalamic circuit regulating energy homeostasis. Neuron 37, 649–661.

Date, Y., et al., 2000. Ghrelin, a novel growth hormone-releasing acylated peptide, is synthesized in a distinct endocrine cell type in the gastrointestinal tracts of rats and humans. Endocrinology 141, 4255–4261.

Date, Y., et al., 2006. Peripheral ghrelin transmits orexigenic signals through the noradrenergic pathway from the hindbrain to the hypothalamus. Cell Metab. 4, 323–331.

Guan, J.L., et al., 2003. Synaptic interactions between ghrelin- and neuropeptide Y-containing neurons in the rat arcuate nucleus. Peptides 24, 1921–1928.

Guan, J.L., et al., 2008. Synaptic relationships between proopiomelanocortin- and ghrelin-containing neurons in the rat arcuate nucleus. Regul. Pept. 145, 128–132.

Hori, Y., et al., 2008. Synaptic interaction between ghrelin- and ghrelin-containing neurons in the rat hypothalamus. Regul. Pept. 145, 122–127.

Horvath, T.L., et al., 1999. Synaptic interaction between hypocretin (orexin) and neuropeptide Y cells in the rodent and primate hypothalamus: a novel circuit implicated in metabolic and endocrine regulations. J. Neurosci. 19, 1072–1087.

Kojima, M., et al., 1999. Ghrelin is a growth-hormone-releasing acylated peptide from stomach. Nature 402, 656–660.

Lu, S., et al., 2002. Immunocytochemical observation of ghrelin-containing neurons in the rat arcuate nucleus. Neurosci. Lett. 321, 157–160.

Paxinos, G., Watson, C., 1986. The Rat Brain in Stereotaxic Coordinates. Academic Press, Inc, San Diego, CA, USA.

Teclemariam-Mesbah, R., et al., 1997. A simple silver-gold intensification procedure for double DAB labeling studies in electron microscopy. J. Histochem. Cytochem. 45, 619–621.

Thyberg, J., Moskalewski, S., 1985. Microtubules and the organization of the Golgi complex. Exp. Cell Res. 159, 1–16.

Toshinai, K., et al., 2003. Ghrelin-induced food intake is mediated via the orexin pathway. Endocrinology 144, 1506–1512.

van den Pol, A.N., Gorcs, T., 1986. Synaptic relationships between neurons containing vasopressin, gastrin-releasing peptide, vasoactive intestinal polypeptide, and glutamate decarboxylase immunoreactivity in the suprachiasmatic nucleus: dual ultrastructural immunocytochemistry with gold-substituted silver peroxidase. J. Comp. Neurol. 252, 507–521.

CHAPTER SEVEN

High-Performance Liquid Chromatography Analysis of Hypothalamic Ghrelin

Takahiro Sato[1], Masayasu Kojima

Institute of Life Science, Kurume University, Kurume, Japan

[1]Corresponding author: e-mail address: satou_takahiro@kurume-u.ac.jp

Contents

Abstract

Ghrelin, first identified in the stomach, is a ligand of an orphan G-protein coupled receptor. Early studies indicated that the growth hormone secretagogue receptor (GHS-R; ghrelin receptor) is ubiquitously distributed in the brain. In addition, centrally administered ghrelin and ghrelin receptor agonist have effects on central neurons in many regions, including the hypothalamus, caudal brain stem, and spinal cord. These effects are due to ghrelin secreted from the brain, rather than from the stomach; ghrelin does not cross efficiently through the blood–brain barrier. Identification of ghrelin in the hypothalamus demonstrated that, as with stomach ghrelin, hypothalamic ghrelin also has two molecular forms, namely, octanoyl ghrelin and des-acyl ghrelin. Hypothalamic ghrelin plays diverse roles in processes including feeding regulation and thermoregulation. Thus, the analysis of hypothalamic ghrelin will provide new information about the action of ghrelin in the central nervous system. In this chapter, we outline high-performance liquid chromatography and real-time PCR analysis of hypothalamic ghrelin.

Methods in Enzymology, Volume 514
ISSN 0076-6879
http://dx.doi.org/10.1016/B978-0-12-381272-8.00007-6

1. INTRODUCTION

In early work, Bowers et al. (1980) observed that some opioid peptide derivatives had weak growth hormone (GH)-releasing activity. They referred to these compounds as growth hormone secretagogues (GHSs). Thereafter, many types of GHSs were identified, such as GHRP-6 and L-163,191 (MK-0677), and the action of the GHSs was gradually elucidated (Bowers et al., 1984; Cheng et al., 1993; Patchett et al., 1995). Growth hormone-releasing hormone (GHRH), which promotes GH secretion from GH-secreting cells in the anterior pituitary, acts on the GHRH receptor to increase intracellular cAMP, which serves as a second messenger (Akman et al., 1993; Blake and Smith, 1991; Cheng et al., 1989, 1991; Popovic et al., 1996). GHSs also act on a different receptor expressed by GH-secreting cells in the anterior pituitary, increasing intracellular Ca^{2+} concentration via an inositol 1,4,5-trisphosphate (IP_3) signal transduction pathway. These results led us to predict the existence of an endogenous ligand of GHS.

Growth hormone secretagogue receptor (GHS-R) was identified as a receptor with which GHSs stimulate phospholipase C, resulting in an increase in IP_3 and intracellular Ca^{2+} (Howard et al., 1996). GHS-R is expressed in the pituitary, hypothalamus, and hippocampus. Therefore, a search for its endogenous ligand was actively undertaken using the orphan receptor strategy, focusing especially on the hypothalamus. However, working with rats, Kojima et al. (1999) discovered the endogenous ligand, the 28-amino acid peptide they named ghrelinin, in an unexpected organ: the stomach.

Subsequently, through high-performance liquid chromatography (HPLC) analysis, we determined that *n*-octanoyl ghrelin is present in the hypothalamus, where it is synthesized (Sato et al., 2005). We also showed that, as in the stomach, *n*-octanoyl and des-acyl ghrelin are the two major molecular forms of ghrelin in the hypothalamus.

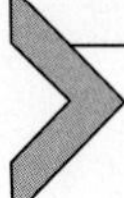

2. IDENTIFICATION OF HYPOTHALAMIC GHRELIN

2.1. Sample preparation

To identify hypothalamic ghrelin, it is essential to boil the samples in order to inactivate the intrinsic protease (Hosoda et al., 2000). In practice, we use a water bath over a strong flame (e.g., a stove burner) in order to prevent a decrease in water temperature after the addition of samples (Fig. 7.1). The temperature of water should remain above 95 °C during the boiling of samples. In order

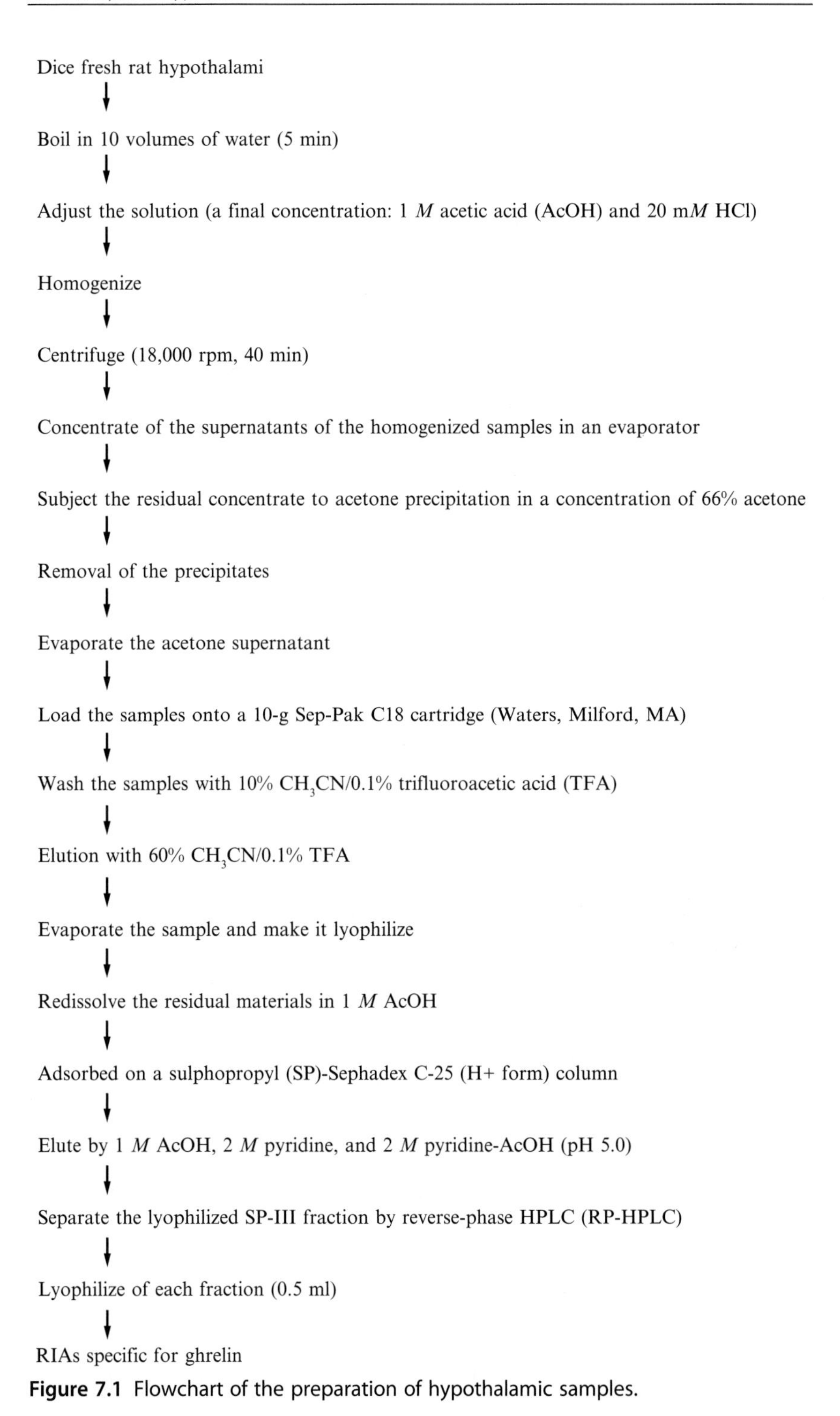

Figure 7.1 Flowchart of the preparation of hypothalamic samples.

to obtain sufficient quantities of the peptide sample, it is necessary to sacrifice ten or more rats. To date, we have been unable to successfully identify hypothalamic ghrelin in mice.

Procedure

1. Remove rat brains, dissect out the hypothalami, and mince them (not less than 20 g).
2. Boil for 5 min in 10 volumes of water to inactivate intrinsic proteases.
3. Adjust the solution to a final concentration of 1 *M* acetic acid (AcOH) and 20 m*M* HCl.
4. Homogenize the boiled hypothalami using a Polytron mixer.
5. Centrifuge the homogenized sample at 15,000 rpm ($18,000 \times g$) for 40 min.
6. Concentrate the supernatants from homogenized samples to ~20 ml in an evaporator.
7. Subject the residual concentrate to acetone precipitation in a concentration of 66% acetone.
8. After removal of the precipitates, evaporate the acetone supernatant.
9. Load the sample onto a 10-g Sep-Pak C18 cartridge (Waters, Milford, MA) and wash with 10% CH_3CN/0.1% trifluoroacetic acid (TFA).
10. Elute with 60% CH_3CN/0.1% TFA.
11. Lyophilize the sample.
12. Redissolve the residual materials in 1 *M* AcOH.
13. Adsorb on a sulfopropyl (SP)-Sephadex C-25 (H^+ form) column, pre-equilibrated in 1 *M* AcOH.
14. Elute successively with 1 *M* AcOH, 2 *M* pyridine, and 2 *M* pyridine–AcOH (pH 5.0) and designate these three fractions as SP-I, SP-II, and SP-III.
15. Separate the lyophilized SP-III fraction by reverse-phase HPLC (RP-HPLC) using a μBondasphere C18 column (3.9×150 mm; Waters). A linear gradient of CH_3CN from 10% to 60% in 0.1% TFA for 40 min serves as the solvent system, at a flow rate of 1 ml/min (Fig. 7.2A).
16. Lyophilize each fraction (0.5 ml) and subject to RIAs specific for ghrelin.

2.2. RIAs for rat ghrelin

To characterize the molecular forms of immunoreactive ghrelin, we recommend RIA analysis using two polyclonal antibodies: no. 6-6 for amino-terminal RIA (N-RIA) and no. 1-7 for carboxyl-terminal RIA (C-RIA), raised against the amino terminal (Gly^1-Lys^{11} with *O*-*n*-octanoylation at Ser^3) and C-terminal (Gln^{13}-Arg^{28}) fragments of rat ghrelin, respectively.

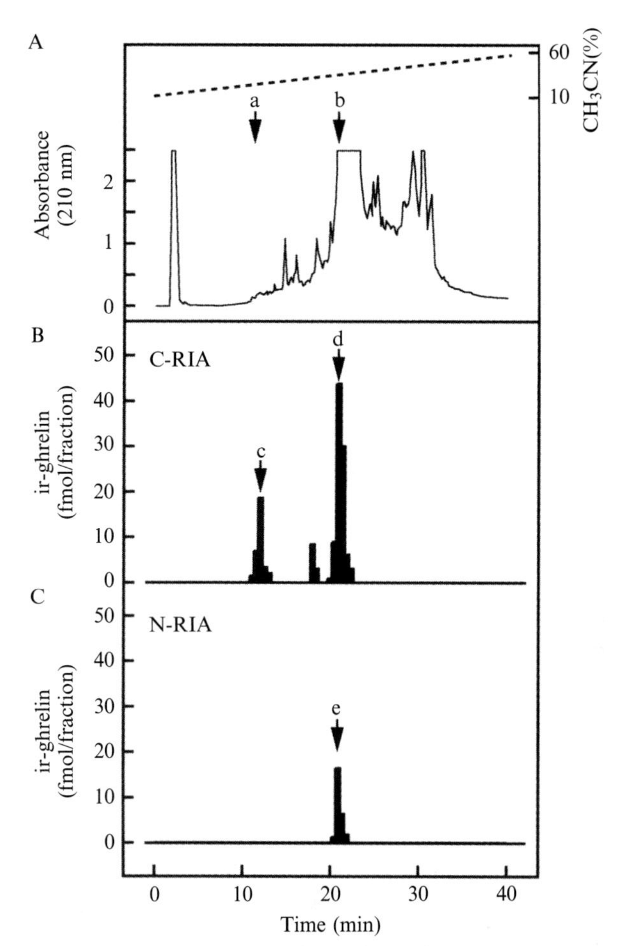

Figure 7.2 Representative RP-HPLC profiles of ghrelin immunoreactivity in the rat hypothalamus. A linear gradient of 10–60% CH_3CN containing 0.1% TFA was run for 40 min at 1.0 ml/min. (A) Chromatograph of rat hypothalamic extract. RP-HPLC of rat hypothalamus was monitored by C-RIA (B) and N-RIA (C) for ghrelin, using fraction volumes of 0.5 ml. The arrows indicate the elution points of des-acyl rat ghrelin-(1–28) (arrow a) and *n*-octanoylated rat ghrelin-(1–28) (arrow b). The two major peaks observed were consistent with the elution points of des-acyl rat ghrelin-(1–28) (arrow c) and *n*-octanoylated rat ghrelin-(1–28) (arrows d and e) [Sato et al., 2005].

2.2.1 Preparation for RIA

1. RIA buffer:

   To make the RIA buffer, mix the following reagents in distilled water: 50 m*M* PBS (pH 7.4), 80 m*M* NaCl, 25 m*M* EDTA-2Na, 0.05% NaN_3, and 0.5% Triton X-100. Add 4 N NaOH to adjust the solution to pH 7.4. To prevent nonspecific binding (NSB) of peptides to RIA tubes, bovine serum albumin (BSA) should be added to the RIA buffer. Add 29.23 ml of 8.55% BSA into 400 ml of the RIA buffer and then adjust the volume of the RIA buffer to 500 ml.

2. Standard preparation:

For ghrelin RIA, use rat or human ghrelin peptide with the *n*-octanoyl modification. Prepare 1 nmol/ml of peptide standard solution by RIA buffer. One nanomole of rat ghrelin (MW = 3,315) is 3.315 μg; 1 nmol of human ghrelin (MW = 3,371) is 3.371 μg. Dilute the standard peptide for duplicate assays by serial dilution as follows: 8196, 512, 256, 128, 64, 32, 16, 8, 4, 2, 1, 0.5, 0.25, and 0.125 fmol/100 μl.

2.2.2 RIA

1. Day 1
   - a. Add RIA buffer to the lyophilized samples. Vortex well.
   - b. Centrifuge the samples to remove insoluble materials (4 °C, 3,000 rpm, 10 min).
   - c. Pipet 100 μl of standard peptides and unknown samples into appropriately labeled tubes, in duplicate.
   - d. Pipet 300 μl of the RIA buffer into the NSB tubes. Add 10 μl per tube of NRS (1/10 dilution).
   - e. Pipet 200 μl antiserum solution into all tubes except the NSB tubes and record the total counts (TCs). The antiserum solution is constituted as follows: ghrelin serum, x μl; NRS (1/10 dilution), 10 μl; and RIA buffer, [200 (x + 10)] μl. For N-terminal RIA, dilute no. 6-6 serum 1:2,500,000 (final); for C-terminal RIA, dilute no. 1-7 serum 1:12,000 (final).
   - f. Vortex well.
   - g. Incubate at 4 °C for 12 h.
2. Day 2
   - a. Dilute ^{125}I-labeled ghrelin to 20,000 cpm/100 μl RIA buffer in each tube. Pipet 100 μl tracer into all tubes.
   - b. Incubate at 4 °C for 36 h.
3. Day 4
   - a. Dilute secondary antibody (goat anti-rabbit IgG) 1:35 with the RIA buffer. Pipet 100 μl of the diluted secondary antibody into all tubes.
   - b. Incubate at 4 °C for 24 h.
4. Day 5
   - a. Centrifuge the tubes at 4 °C for 30 min at 3,000 rpm.

b. After centrifugation, place the tubes on ice and then aspirate the supernatant.
c. Quantitate the radioactivity in the pellet with a γ counter using an apparatus such as the ARC-1000M (Aloka, Tokyo, Japan) (Fig. 7.2B and C).

3. QUANTIFICATION OF IMMUNOREACTIVE GHRELIN IN RATS

3.1. Preparation of tissue samples

Broadly, this protocol is similar to the one described in Section 2.1.

1. Sacrifice rats and quickly remove the whole hypothalamus from each brain.
2. Mince the hypothalami and boil for 5 min in 10 volumes of water to inactivate the intrinsic proteases.
3. After cooling, adjust the solutions to a final concentration of 1 *M* AcOH and 20 m*M* HCl.
4. Homogenize tissues with a Polytron mixer and centrifuge at 15,000 rpm for 10 min; preserve the supernatants.
5. Load the supernatants onto Sep-Pak C18 cartridges (Waters). Wash the cartridges in 0.9% NaCl and 10% CH_3CN/0.1% TFA. Elute the bound protein with 60% CH_3CN/0.1% TFA. Lyophilize and subject the eluate to ghrelin-specific ELISA.

3.2. ELISA for rat ghrelin

Hypothalamus ghrelin levels can be easily measured using an Active Ghrelin ELISA Kit (Mitsubishi Kagaku Iatron, Inc., Tokyo, Japan) to assess the *n*-octanoyl-modified ghrelin and a Desacyl-Ghrelin ELISA Kit (Mitsubishi Kagaku Iatron, Inc.) to measure des-acyl ghrelin. To measure ghrelin levels by ELISA, samples must be obtained from whole rat hypothalami. The protocols should follow the manufacturer's instructions. In the case of plasma samples, it is important to calculate a corrected volume because of the addition of 1 N HCl to plasma.

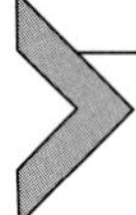

4. QUANTIFICATION OF RAT GHRELIN MRNA IN HYPOTHALAMIC GHRELIN IN RATS

To understand the process of ghrelin secretion, it is necessary to investigate the synthesis of ghrelin. Real-time PCR is an useful tool for the analysis of ghrelin mRNA levels. Because the level of hypothalamic ghrelin mRNA is very low, purification of poly(A)$^+$ RNA is necessary. In addition, in order to obtain accurate readings, it is essential to carefully compare samples with a "no template" control.

4.1. Synthesis of cDNA

1. Extract the total RNA from frozen hypothalami using TRIzol (Invitrogen, Tokyo, Japan).
2. Purify poly(A)$^+$ RNA from 75 μg or more total hypothalamic RNA using Oligotex-dT30 <Super> (Roche, Tokyo, Japan), according to the manufacturer's instructions.
3. Synthesize cDNA from the poly(A) $^+$ RNA (0.4 μg per animal or more).
4. Incubate the reaction mixtures at 37 °C for 60 min.
5. Stop the reaction by incubation at 70 °C for 15 min.

4.2. Real-time PCR

1. cDNA amplification is performed using SYBR Green PCR Core Reagents (PE Applied Biosystems) and uracil-*N*-glycosylase (Invitrogen), to prevent contamination by carried-over PCR products, as suggested by the manufacturer. Samples are amplified in a single MicroAmp Optical 96-well reaction plate (PE Applied Biosystems). The results reflect duplicate runs of at least two independent experiments. Primer pairs for ghrelin gene are designed using Primer3 software as follows:

   sense primer 5′-GAAGCCACCAGCTAAACTGC-3′;
   antisense primer 5′-GCTGCTGGTACTGAGCTCCT-3′.

   Researchers can also purchase the primer-pair-optimized real-time PCR as a regent product from a company such as Takara Bio, Inc.
2. Each standard well contains a pGEM-T Easy vector carrying the standard cDNA fragment. The concentrations of the standards cover at least six orders of magnitude. We also include no template controls on each plate.
3. PCR cycling conditions are initiated by a 2-min incubation at 50 °C to eliminate any deoxyuridine triphosphate-containing PCR products resulting from carry-over contamination. After a 15-min period at 95 °C to activate HotStarTaq DNA polymerase, PCR fragments are amplified by 40 cycles of 95 °C for 30 s, 60 °C for 30 s, and 1 min at 72 °C.
4. Experimental samples with a threshold cycle value within 2 SD of the mean threshold cycle value for the "no template" controls are considered to be below the limits of detection. The relative levels of mRNA are

standardized to a housekeeping gene, such as glyceraldehyde-3-phosphate dehydrogenase or ribosomal protein S18, to correct for any bias among the samples caused by RNA isolation, RNA degradation, or efficiencies of RT. After amplification, PCR products are analyzed by a melting curve to confirm amplification specificity. Amplicon size and reaction specificity are confirmed by agarose gel electrophoresis.

5. IDENTIFICATION OF GHRELIN-PRODUCING NEURONS IN RATS

Hypothalamus is central to the regulation of autonomic functions. Therefore, in order to investigate the projections of ghrelin-producing neurons, it is useful to obtain information regarding the neuronal circuit of autonomic function. Ghrelin immunohistochemical staining is performed using the avidin–biotin–peroxidase complex (ABC) system, for example, the VECTASTAIN ABC-PO kit (Vector Laboratories Inc., Burlingame, CA). In this section, we briefly introduce an immunohistochemical method for use in porcine hypothalamus (Fig. 7.3). Other researchers have demonstrated the presence of ghrelin-producing neurons in the hypothalami of rats treated with colchicine (Canpolat et al., 2006; Kojima et al., 1999; Mondal et al., 2005).

1. Immerse the porcine hypothalamus in 4% paraformaldehyde solution overnight.
2. Immerse the sample in a series of 10%, 20%, and 30% sucrose solutions with 10% alabia gum every 24 h.
3. Embed the tissues in OCT compound (Tissue-Tek Miles, Elkhart, IN).
4. Cut sections to a thickness of 20 μm using a cryostat (CM 3050S; Leica Microscopy and Scientific Instruments Group, Heerbrugg, Switzerland) and mount them on Matsunami adhesive-coated slides (Matsunami, Osaka, Japan).
5. Dry the sections at 37 °C for 30 min.
6. Wash the sections in 10 m*M* PBS (pH 7.4).
7. Pretreat the sections with 3% hydrogen peroxide in methanol for 10 min to block endogenous peroxidase activity.
8. Treat the sections with 0.01% saponin in PBS for 20 min.
9. After rinsing with PBS, treat the sections with 3% normal goat serum for 1 h.
10. Incubate the sections in polyclonal rabbit anti-ghrelin antibody (no. 6-6; dilute 1:80,000) for 16 h at 4 °C.
11. Rinse the sections with PBS.

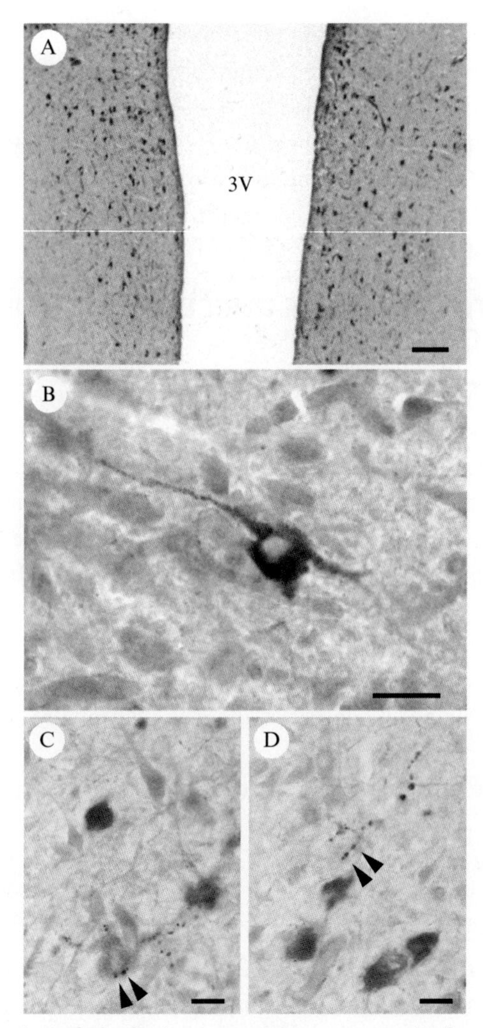

Figure 7.3 Localization of ghrelin-immunopositive neurons in the porcine hypothalamus. (A) Ghrelin neuron distribution in the paraventricular nucleus. (B) A ghrelin-producing neuron in paraventricular nucleus. A subset of ghrelin-positive neurons projected to cell bodies of either additional ghrelin-positive neurons (C, arrowheads) or ghrelin-negative neurons (D, arrowheads). 3V, Third ventricle. Bar, 200 μm (A), 20 μm (B–D) [Sato et al., 2005]. (See Color Insert.)

12. Incubate the sections with biotinylated anti-rabbit IgG for 40 min.
13. Rinse the sections with PBS.
14. Incubate the sections with VECTASTAIN ABC Reagent for 1 h.
15. Rinse the sections with PBS.

16. Develop the samples in 3,3′-diaminobenzidine using the Dako liquid diethylaminobenzidine substrate-chromogen system (Dako, Kyoto, Japan)..
17. Enclose the samples with cover glass, and image.

6. SUMMARY

Analysis of hypothalamic ghrelin secretion is important to our understanding of ghrelin function. The wide distribution of GHS-R led us to propose a range of functions for ghrelin in the central nervous system. HPLC analysis is useful for the detection of hypothalamic ghrelin. However, the analysis of hypothalamic ghrelin of rats requires meticulous attention to details, because the concentration of hypothalamic ghrelin is low. Therefore, we recommend the combined use of HPLC analysis and real-time PCR analysis to accurately measure hypothalamic ghrelin synthesis and secretion.

ACKNOWLEDGMENTS

This work was supported by a Grant-in-Aid for Young Scientists (B) from the Japan Society for the Promotion of Science (JSPS) and the Foundation for Growth Science.

REFERENCES

Akman, M.S., et al., 1993. Mechanisms of action of a second generation growth hormone-releasing peptide (Ala-His-D-Beta Nal-Ala-Trp-D-Phe-Lys-NH_2) in rat anterior pituitary cells. Endocrinology 132, 1286–1291.

Blake, A.D., Smith, R.G., 1991. Desensitization studies using perifused rat pituitary cells show that growth hormone-releasing hormone and His-D-Trp-Ala-Trp-D-Phe-Lys-NH_2 stimulate growth hormone release through distinct receptor sites. J. Endocrinol. 129, 11–19.

Bowers, C.Y., et al., 1980. Structure-activity relationships of a synthetic pentapeptide that specifically releases growth hormone in vitro. Endocrinology 106, 663–667.

Bowers, C.Y., et al., 1984. On the in vitro and in vivo activity of a new synthetic hexapeptide that acts on the pituitary to specifically release growth hormone. Endocrinology 114, 1537–1545.

Canpolat, S., et al., 2006. Effects of pinealectomy and exogenous melatonin on immunohistochemical ghrelin staining of arcuate nucleus and serum ghrelin leves in the rat. Neurosci. Lett. 410, 132–136.

Cheng, K., et al., 1989. The synergistic effects of His-D-Trp-Ala-Trp-D-Phe-Lys-NH_2 on growth hormone (GH)-releasing factor-stimulated GH release and intracellular adenosine 3′,5′-monophosphate accumulation in rat primary pituitary cell culture. Endocrinology 124, 2791–2798.

Cheng, K., et al., 1991. Evidence for a role of protein kinase-C in His-D-Trp-Ala-Trp-D-Phe-Lys-NH_2-induced growth hormone release from rat primary pituitary cells. Endocrinology 129, 3337–3342.

Cheng, K., et al., 1993. Stimulation of growth hormone release from rat primary pituitary cells by L-692,429, a novel non-peptidyl GH secretagogue. Endocrinology 132, 2729–2731.

Hosoda, H., et al., 2000. Ghrelin and des-acyl ghrelin: two major forms of rat ghrelin peptide in gastrointestinal tissue. Biochem. Biophys. Res. Commun. 279, 909–913.

Howard, A.D., et al., 1996. A receptor in pituitary and hypothalamus that functions in growth hormone release. Science 273, 974–977.

Kojima, M., et al., 1999. Ghrelin is a growth-hormone-releasing acylated peptide from stomach. Nature 402, 656–660.

Mondal, M.S., et al., 2005. Identification of ghrelin and its receptor in neurons of the rat arcuate nucleus. Regul. Pept. 126, 55–59.

Patchett, A.A., et al., 1995. Design and biological activities of L-163,191 (MK-0677): a potent, orally active growth hormone secretagogue. Proc. Natl. Acad. Sci. USA 92, 7001–7005.

Popovic, V., et al., 1996. Evaluation of pituitary GH reserve with GHRP-6. J. Pediatr. Endocrinol. Metab. 9 (Suppl. 3), 289–298.

Sato, T., et al., 2005. Molecular forms of hypothalamic ghrelin and its regulation by fasting and 2-deoxy-D-glucose administration. Endocrinology 146, 2510–2516.

CHAPTER EIGHT

Standard Sample Collections for Blood Ghrelin Measurements

Hiroshi Hosoda*,[1], Kenji Kangawa[†]

*Department of Regenerative Medicine and Tissue Engineering, National Cerebral and Cardiovascular Center Research Institute, Suita, Osaka, Japan

[†]National Cerebral and Cardiovascular Center Research Institute, Suita, Osaka, Japan

[1]Corresponding author: e-mail address: hosoda.hiroshi.ri@mail.ncvc.go.jp

Contents

Abstract

Background: Octanoyl modification of ghrelin is rapidly hydrolyzed to des-acyl ghrelin in blood samples. Owing to the increased interest in ghrelin measurement, a standardized method of sample collection is required.

Methods: This chapter investigates the effect of a variety of anticoagulants and storage conditions on ghrelin stability. Experiment 1 evaluates the effects of anticoagulants on ghrelin measurements. Experiment 2 evaluates the effect of plasma pH on ghrelin stability. Experiment 3 evaluates the mechanisms of degradation of the active form of ghrelin in plasma. Experiment 4 investigates the kinetics of ghrelin following intravenous injection of rat ghrelin.

Methods in Enzymology, Volume 514
ISSN 0076-6879
http://dx.doi.org/10.1016/B978-0-12-381272-8.00008-8

Results: In whole blood and plasma, octanoylated ghrelin is highly unstable. The collection of blood samples with EDTA–aprotinin under cooled conditions was appropriate to maintain ghrelin stability. Acidification of plasma to pH 3–4 and storage at 4 °C maintained ghrelin stability. The degradation of ghrelin was shown to be due to the hydrolysis to des-acyl ghrelin. After intravenous administration to rats, plasma ghrelin levels rapidly decreased with a half-life of 8 min.

Conclusion: The results showed that the ghrelin values measured in human blood samples were markedly affected by the conditions of collection and storage, the pH, and the RIA method in measurement. Measuring the ghrelin values of the active form is useful for studying plasma ghrelin changes over short time periods. As ghrelin is highly unstable, it is necessary to standardize the preparation of samples to ensure reliable ghrelin measurements.

1. INTRODUCTION

Ghrelin is an acylated peptide with growth-hormone-releasing activity (Kojima et al., 1999, 2001). It was first isolated from rat and human stomach during the search for an endogenous ligand to the orphan G-protein-coupled receptor, growth-hormone secretagogue receptor (Howard et al., 1996; McKee et al., 1997). The peptide contains 28 amino acids, and *n*-octanoylation of the serine 3 hydroxyl group is necessary for biological activity. Attempts to understand the posttranslational mechanism of the octanoyl modification opened a new field in protein chemistry, as octanoylation had not been previously observed as a regulatable peptide modification. Ghrelin *O*-acyltransferase (GOAT), which belongs to a family of hydrophobic membrane-bound acyltransferases, was identified as the specific ghrelin-octanoylating enzyme (Yang et al., 2008). The gene expression of GOAT is largely restricted to the stomach and intestine, which are the major ghrelin-secreting tissues. Most studies have focused on the somatotrophic and orexigenic role of ghrelin; therefore, little is known about the kinetics of ghrelin. As the ester bond is both chemically and enzymatically unstable, elimination of the octanoyl modification of ghrelin can occur during storage, handling, and/or dissolution in the culture medium (Kanamoto et al., 2001). Thus, "active" octanoylated ghrelin could be highly unstable in blood and plasma samples.

We have established two sensitive RIAs for ghrelin: one recognizes only the active, octanoyl-modified ghrelin, and the other, recognizing the C-terminal portion of ghrelin, can be used to measure the total ghrelin (Date et al., 2000; Hosoda et al., 2000). Differences in assay methodologies as well as sample handling, such as the method of storage, effect of anticoagulants, or previous freeze–thawing of the samples, could influence the reported ghrelin levels.

In this study, we focused on the active form of ghrelin and investigated the effects of anticoagulants and storage conditions on ghrelin stability. To clarify the mechanisms of degradation of ghrelin, we studied the metabolites and pharmacokinetics of ghrelin.

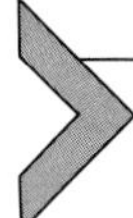

2. MATERIALS AND METHODS

2.1. Materials

Rat and human ghrelin were synthesized as previously described (Kojima et al., 1999) with the hydroxyl group of Ser^3 acylated with *n*-octanoic acid by the action of 1-ethyl-3-(3-dimethylaminopropyl) carbodiimide in the presence of 4-(dimethylamino) pyridine. The synthesized peptides were purified by reverse-phase (RP) HPLC.

2.2. RIA for ghrelin

Two RIAs measuring plasma ghrelin were performed as described (Hosoda et al., 2000). Briefly, ghrelin levels were measured by two RIAs using two polyclonal rabbit antibodies raised against the N-terminal [1–11] (Gly^1-Lys^{11}) or C-terminal [13–28] (Gln^{13}-Arg^{28}) fragments of rat ghrelin. Two tracer ligands were synthesized: C-terminally Tyr-extended rat ghrelin with octanoylated Ser^3 for anti-rat ghrelin [1–11] antiserum and N-terminally Tyr-extended rat ghrelin [13–28] for anti-rat ghrelin [13–28] antiserum. RIA incubation mixtures, containing 100 μl of either standard ghrelin or diluted plasma/serum sample with 200 μl of antiserum diluted in RIA buffer containing 0.5% normal rabbit serum, were initially incubated for 12 h. Next, 100 μl of ^{125}I-labeled tracer (15,000 cpm) was added and incubated for 36 h. Anti-rabbit IgG goat serum (100 μl) was added prior to an additional 24-h incubation. Free and bound tracers were then separated by centrifugation at $2000 \times g$ for 30 min. Following aspiration of the supernatant, radioactivity in the pellet was quantitated using a gamma counter (ARC-600, Aloka, Tokyo). All assays were performed in duplicate at 4 °C. The anti-rat ghrelin [1–11] antiserum (#G606) specifically recognizes rat *n*-octanoylated ghrelin and does not detect des-acyl ghrelin. The anti-rat ghrelin [13–28] antiserum (#G107) recognizes both the acylated and des-acyl rat ghrelin forms equally. Both antisera were equally cross-reactive with human ghrelin and did not recognize the other enteric peptides (Ariyasu et al., 2002; Date et al., 2000; Hosoda et al., 2000, 2003). In the following sections, the RIA procedures using G606 antiserum is termed N-RIA, while that utilizing the G107 antiserum is termed C-RIA. The minimal detectable quantities of N- and C-RIA were 5.0 and 50 fmol/ml, respectively. The respective intra- and interassay coefficients of variation were 3% and 6%, respectively, for N-RIA and 6% and 9% for C-RIA ($n=8$ assays).

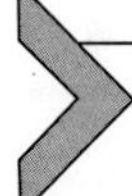

3. STUDY PROTOCOLS

3.1. Experiment 1

This study sought to compare ghrelin measurements in serum and plasma samples collected in different anticoagulants and after different storage conditions. All blood samples were taken from three healthy volunteers who gave written informed consent. Blood was taken from the forearm vein and immediately divided into tubes for serum and plasma preparation using (1) EDTA-2Na (1 mg/ml) with aprotinin (500 kIU/ml), (2) EDTA-2Na, (3) heparin sodium, or (4) no anticoagulant. Synthetic human ghrelin was added to each blood sample at a final concentration of 40 ng/ml and sequentially divided into two aliquots for incubation at either 4 °C or 37 °C. After 0-, 30-, and 60-min incubation, blood samples were centrifuged, diluted 1:200 in the RIA buffer, and subjected to ghrelin-specific RIAs.

3.2. Experiment 2

We examined the effect of plasma pH on ghrelin stability. The EDTA–aprotinin-treated plasma ($n=3$) was divided into five samples; the pH was then adjusted to 3, 4, 5, 6, or 7.4 with 1 mol/l HCl. Synthetic human ghrelin was then added to each sample aliquot at a final concentration of 75 ng/ml. Each of the five plasma aliquots was then subdivided into two, with one stored at 4 °C and the other at 37 °C. Measurements were repeated at 0, 1, 2, 4, and 6 h for samples stored at 37 °C or at 0, 3, and 6 h for samples incubated at 4 °C. We then evaluated the effects of repeated freezing and thawing on the stability of ghrelin. EDTA–aprotinin-treated plasma samples were divided into two pH groups, one of which was acidified to pH 4, while the other was not acidified (pH 7.4). Following the addition of synthetic human ghrelin (75 ng/ml), we subjected the samples to four freezing and thawing cycles. Test samples were obtained before freezing and after thawing; the recovery of human ghrelin was then assessed by ghrelin-specific RIAs.

3.3. Experiment 3

To investigate the degradation time course of ghrelin in plasma, synthetic human ghrelin was added to EDTA–aprotinin plasma aliquot described above ($n=3$) and incubated at 37 °C. After 0, 1, 3, 6, 12, and 24 h of incubation, each plasma sample was loaded onto Sep-Pak C18 cartridges (Waters, Milford, MA) (Hosoda et al., 2000). Each eluate was subjected to RP-HPLC on a μBondasphere C18 column (Waters). RP-HPLC was performed using a linear gradient of CH_3CN from 10% to 60% in 0.1% trifluoroacetic acid for 40 min. An aliquot of each fraction obtained by the RP-HPLC was evaporated and lyophilized, and then subjected to ghrelin C-RIA.

3.4. Experiment 4

We assessed the kinetics of rat ghrelin plasma concentrations after intravenous administration to anesthetized rats. Male Wistar rats weighing 300–350 g (Chales River Japan, Yokohama, Japan) were anesthetized with pentobarbital (50 mg/kg, intraperitoneally) and canulated in the jugular artery and vein for blood sampling and infusion, respectively. Synthetic rat ghrelin (100 μg per body) or saline was administrated intravenously ($n=3$). Blood (0.2 ml) was drawn prior to infusion and at 1, 3, 5, 7, 10, 15, 20, 30, 45, 60, 75, and 90 min after peptide administration. Each blood sample was collected in a polypropylene tube containing EDTA-2Na (2 mg/ml) with aprotinin (500 kIU/ml). Plasma was immediately separated by centrifugation, diluted, and subjected to ghrelin RIAs. The plasma half-life of ghrelin was calculated using the WinNonlin Standard program (Ver. 3.1) (2-compartment model, Scientific Consulting Inc.). All procedures were performed in accordance with the Japanese Physiological Society's guidelines for animal care.

3.5. Statistical analysis

Data were expressed as means ± SD except when otherwise indicated. Comparisons of the time course of ghrelin levels between subgroups were made by two-way ANOVA for repeated measures, followed by Scheffe's test. A value of $P < 0.05$ was considered statistically significant.

4. RESULTS

4.1. Ghrelin measurements in serum and plasma samples (experiment 1)

Figure 8.1 compares the effects of different anticoagulants on the detected ghrelin levels after storage of whole blood up to 60 min at 4 or 37 °C. Although the serum and three different plasma samples tested gave comparable results for total ghrelin levels by C-RIA, N-RIA demonstrated ghrelin levels that decreased significantly at 37 °C. When the ghrelin was measured by N-RIA, serum samples were highly affected by such treatment; samples stored for 60 min at 37 °C lost approximately 35% of the ghrelin in comparison with the basal levels at 0 min ($P < 0.05$). The levels of ghrelin in samples containing heparin as an anticoagulant were also significantly decreased ($P < 0.05$). The use of EDTA–aprotinin for plasma treatment resulted in lower decreases in ghrelin stability than other procedures. Storage at 4 °C also improved ghrelin stability, making this treatment appropriate for separation within 30 min of sample collection.

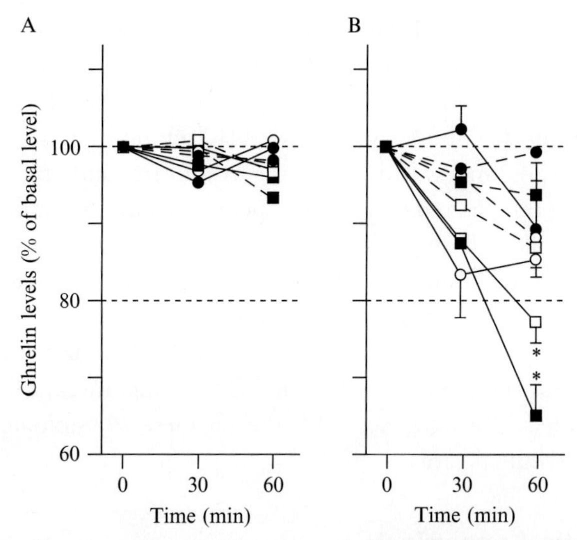

Figure 8.1 Ghrelin measurements in serum and different plasma samples. Synthetic ghrelin was incubated for 0, 30, and 60 min in whole blood collected with EDTA–aprotinin (●), EDTA (○), heparin sodium (□), or without anticoagulant (serum, ■). Samples were stored at 37 °C (solid line) or 4 °C (dotted line). Each point is expressed as mean percentage change from basal level ($n = 3$), assessed by C-RIA (A) and N-RIA (B). $^*P < 0.05$ compared with basal level.

4.2. Effect of plasma pH on ghrelin stability (experiment 2)

As ester bonds are often unstable in biological materials, we investigated the stability of ghrelin in human EDTA–aprotinin-treated plasma. Figure 8.2 summarizes the effect of acidification on ghrelin stability in plasma. When stored at 37 °C, ghrelin levels measured by N-RIA gradually decreased at all pH values tested. However, ghrelin was most stable in highly acidified plasma samples of pH 3–4. At plasma pH 3–5 and a storage temperature

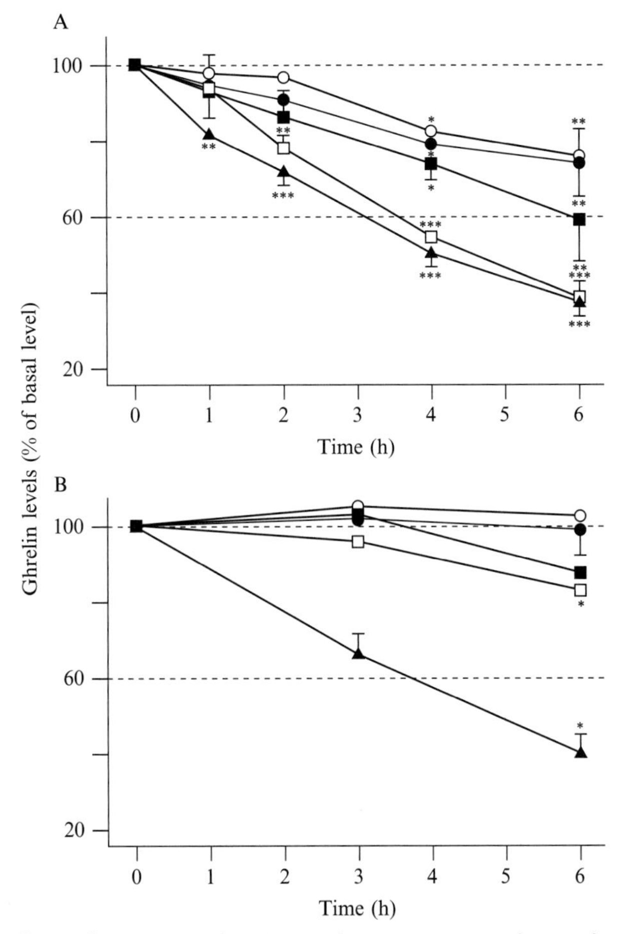

Figure 8.2 Effects of storage pH, duration, and temperature on plasma ghrelin stability. Synthetic ghrelin was incubated in EDTA–aprotinin-treated plasma at various pH values (pH 7.4, ▲; pH 6, □; pH 5, ■; pH 4, ●; pH 3, ○). Sample ghrelin levels were assayed by N-RIA. Each point is expressed as mean percentage change from basal level ($n=3$), incubated at 37 °C (A) and 4 °C (B). $^{*}P<0.05$, $^{**}P<0.01$, and $^{***}P<0.001$ compared with basal level.

of 4 °C, the stability of ghrelin did not change significantly over a 6-h period. By C-RIA, ghrelin levels remained stable across the different pH and storage temperature conditions.

Repeated freezing and thawing also influenced ghrelin stability. When measured by N-RIA, ghrelin levels in untreated plasma samples significantly decreased with each successive freeze–thaw cycle (Fig. 8.3A), whereas ghrelin stability remained relatively constant following acidification (Fig. 8.3B). Ghrelin levels by C-RIA were unchanged despite repeated freeze–thaw treatments in both acidified and untreated plasma samples.

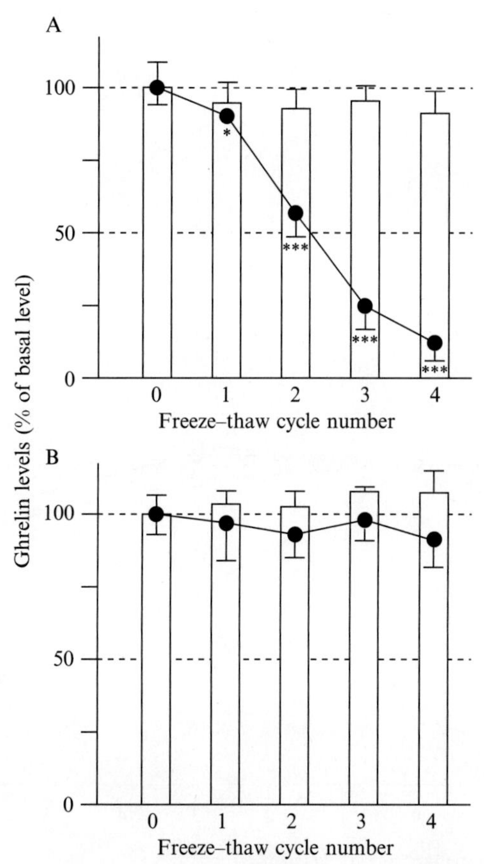

Figure 8.3 Effects of repeated freeze–thaw cycles on plasma ghrelin stability. Synthetic ghrelin in EDTA–aprotinin-treated plasma was subjected to four cycles of freezing and thawing. The plasma sample was not acidified (A) or acidified to pH 4 (B). Sample ghrelin levels were assayed by N-RIA (●) and C-RIA (bar). Each point is expressed as mean percentage change from basal level ($n=3$). *$P<0.05$ and ***$P<0.001$ compared with basal level.

4.3. Degradation time course of ghrelin in plasma (experiment 3)

To clarify the mechanism of degradation of ghrelin in plasma, RP-HPLC coupled with C-RIA was used to analyze ghrelin molecules. As shown in Fig. 8.4, the two peaks, A and B, of ir-ghrelin detected by C-RIA corresponded to des-acyl ghrelin and ghrelin, respectively. Ghrelin levels gradually decreased in the untreated plasma and completely disappeared after 24-h incubation at 37 °C. In contrast, the immunoreactivity of des-acyl ghrelin increased progressively.

4.4. Pharmacokinetics of ghrelin (experiment 4)

Mean plasma concentration versus time profiles of rat ghrelin after intravenous administration of 100 μg in male rats demonstrated that plasma ghrelin levels by C-RIA (865 ± 91 pmol/ml) were raised about 1.5-fold higher than those by N-RIA (580 ± 36 pmol/ml) at 1 min after ghrelin injection

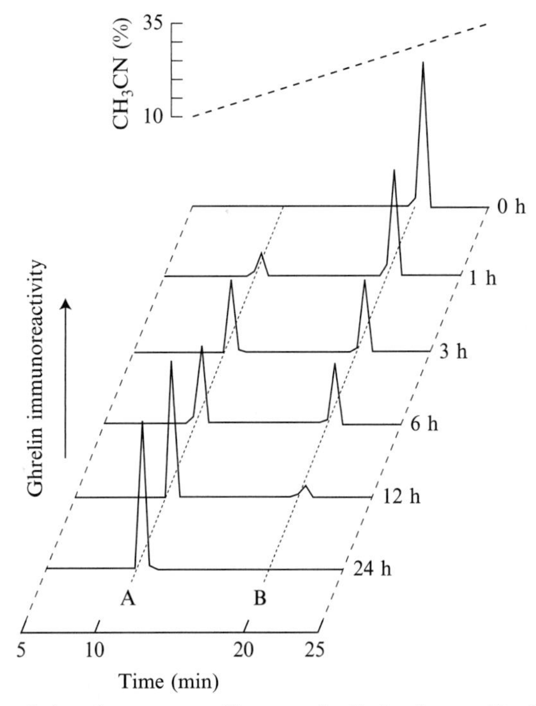

Figure 8.4 Degradation time course of human ghrelin in plasma. Ghrelin immunoreactivities in untreated plasma ($n = 3$) after incubation at 37 °C were characterized by C-RIA coupled with RP-HPLC. The two peaks indicate the elution positions of des-acyl ghrelin (A) and ghrelin (B).

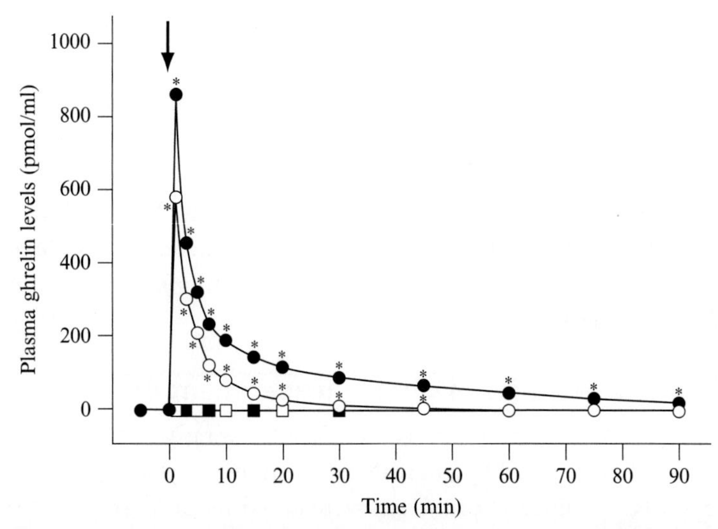

Figure 8.5 Plasma concentration after intravenous administration of rat ghrelin to anesthetized rats. The arrow indicates a bolus injection of ghrelin or saline. Each point is expressed as the mean of three individual samples of ghrelin (circle) and saline (square), assessed by C-RIA (closed symbol) and N-RIA (open symbol). $^{*}P < 0.05$ compared with saline group.

(Fig. 8.5). Levels assessed by C-RIA were sustained higher than those measured by N-RIA. The half-life of rat ghrelin after intravenous administration was 8 and 24 min by N-RIA and C-RIA, respectively.

5. DISCUSSION

Since the discovery of ghrelin, numerous studies have been reported regarding the pharmacological function of ghrelin as an orexigenic and adipogenic peptide (Asakawa et al., 2001; Kamegai et al., 2001; Shintani et al., 2001; Wren et al., 2001a,b). Ghrelin is synthesized principally in the stomach, and the concentration of circulating ghrelin increases under conditions of negative energy balance, such as starvation, cachexia, and anorexia nervosa. In contrast, circulating ghrelin decreases under conditions of positive energy balance, such as feeding, hyperglycemia, and obesity. In view of the increasing interest in the metabolic roles of ghrelin, it is important to be able to accurately measure ghrelin concentrations in different pathological metabolic syndromes. In addition to differences in assay methodologies, differences in sample handling, such as the method of storage, effects of anticoagulants, or previous freezing and thawing of the samples, could influence the reported levels (Boomsma et al., 1993; Evans et al., 2001; Flower et al., 2000; Miki and Sudo, 1998; Nelesen et al., 1992). Moreover, the instability of ghrelin has been a concern due to the ester bond within the structure of ghrelin. To distinguish the active form of ghrelin, we have established two ghrelin-specific radioimmunoassays: N-RIA recognizes the N-terminal, octanoyl-modified portion of the peptide, while C-RIA recognizes the C-terminal portion. Thus, the value determined by N-RIA specifically measures active ghrelin, while the value determined by C-RIA gives the total ghrelin concentration, including both active and des-acyl ghrelin. Both forms of the ghrelin peptide, that is, acylated ghrelin and des-acyl ghrelin, exist in human and rat plasma (Hosoda et al., 2000, 2003). The proportion of active ghrelin in plasma was 2–5% of total ghrelin levels in rodents (Ariyasu et al., 2002). In this study, the quantity of active ghrelin was approximately 10% of the total ghrelin levels in human plasma (data not shown). These findings imply that the inactive, des-acyl ghrelin circulates in the bloodstream at much higher levels than active ghrelin.

We therefore examined the pharmacokinetics of rat ghrelin by both N- and C-RIA after intravenous administration. Both plasma ghrelin forms had short half-lives; however, the signal disappeared earlier in N-RIA than in C-RIA, suggesting that there may be differential rates of metabolic turnover for acylated and des-acyl ghrelin. As shown in Fig. 8.4, octanoylated ghrelin was degraded into des-acylated ghrelin in normal plasma. Elimination of the octanoyl modification of ghrelin occurred during the storage of culture medium (Kanamoto et al., 2001). Thus, the "active" ghrelin can be rapidly hydrolyzed in circulating blood into "nonactive" des-acyl ghrelin through the cleavage of the ester bond.

The mechanism of posttranslational octanoyl modification, which is important in ghrelin biosynthesis, may be revealed through the use of N-RIA, possibly by measuring stomach ghrelin concentrations. Des-acyl ghrelin is relatively stable and its stability is not altered by the different storage conditions. An analogous situation has been reported for the activity of pancreatic beta cells, from which insulin and C-peptide are secreted in a 1:1 molar ratio. However, the half-life of C-peptide is much longer than that of insulin, leaving more C-peptide available in the circulation for quantification (Horwitz et al., 1975; Myrick et al., 1989). Measuring C-peptide levels provides an assessment of beta-cell secretory activity. Similarly, des-acyl ghrelin levels may serve as an indicator of ghrelin secretory function (Cummings et al., 2001).

Instability of peptides and proteins can be divided into two forms: chemical and physical (Reubsaet et al., 1998a,b). Chemical degradation of peptides is influenced by the pH of the aqueous solution; human parathyroid hormone and luteinizing hormone-releasing hormone derivative are examples (Hoitink et al., 1998; Nabuchi et al., 1997; Strickley et al., 1990). The stability of the octanoyl modification of ghrelin showed a marked decrease in strongly acidic (below pH 2), neutral, and alkaline solutions (data not shown). Similar to previous findings reporting ghrelin concentrations by C-RIA (Groschl et al., 2002), we demonstrated that, in whole blood and plasma, octanoylated ghrelin is unstable. When the results of these studies are taken together, the octanoyl modification may be degraded by at least two mechanisms: one involving enzymatic cleavage and the other involving spontaneous chemical hydrolysis of the ester bond. Acidification is a simple, reliable procedure protecting against the degradation of the acylated modification and resulted in highly improved stability at pH 4.

To acquire accurate data on ghrelin concentrations, this study recommends a standard procedure for the collection of blood samples:

- The collection of blood samples with EDTA–aprotinin is preferred.
- Blood samples should be chilled and centrifuged as soon as possible, at least within 30 min after collection.
- Because acidification is the best method for the plasma preservation of ghrelin, the plasma sample can be added to 1 mol/l HCl (10% of volume) for adjustment to pH 4.

ACKNOWLEDGMENTS

We thank H. Mondo and M. Miyazaki for their technical assistance. This work was supported by grants from the Ministry of Education, Science, Sports, and Culture of Japan; the Ministry of Health, Labor and Welfare of Japan; the Promotion of Fundamental Studies in Health Science from the Organization for Pharmaceutical Safety and Research (OPSR) of Japan; and the Takeda Science Foundation.

REFERENCES

Ariyasu, H., et al., 2002. Delayed short-term secretory regulation of ghrelin in obese animals: evidenced by a specific RIA for the active form of ghrelin. Endocrinology 143, 3341–3350.

Asakawa, A., et al., 2001. Ghrelin is an appetite-stimulatory signal from stomach with structural resemblance to motilin. Gastroenterology 120, 337–345.

Boomsma, F., et al., 1993. Optimal collection and storage conditions for catecholamine measurements in human plasma and urine. Clin. Chem. 39, 2503–2508.

Cummings, D.E., et al., 2001. A preprandial rise in plasma ghrelin levels suggests a role in meal initiation in humans. Diabetes 50, 1714–1719.

Date, Y., et al., 2000. Ghrelin, a novel growth hormone-releasing acylated peptide, is synthesized in a distinct endocrine cell type in the gastrointestinal tracts of rats and humans. Endocrinology 141, 4255–4261.

Evans, M.J., et al., 2001. Effect of anticoagulants and storage temperatures on stability of plasma and serum hormones. Clin. Biochem. 34, 107–112.

Flower, L., et al., 2000. Effects of sample handling on the stability of interleukin 6, tumour necrosis factor-alpha and leptin. Cytokine 12, 1712–1716.

Groschl, M., et al., 2002. Preanalytical influences on the measurement of ghrelin. Clin. Chem. 48, 1114–1116.

Hoitink, M.A., et al., 1998. Degradation kinetics of three gonadorelin analogues: developing a method for calculating epimerization parameters. Pharm. Res. 15, 1449–1455.

Horwitz, D.L., et al., 1975. Proinsulin, insulin, and C-peptide concentrations in human portal and peripheral blood. J. Clin. Invest. 55, 1278–1283.

Hosoda, H., et al., 2000. Ghrelin and des-acyl ghrelin: two major forms of rat ghrelin peptide in gastrointestinal tissue. Biochem. Biophys. Res. Commun. 279, 909–913.

Hosoda, H., et al., 2003. Structural divergence of human ghrelin. Identification of multiple ghrelin-derived molecules produced by post-translational processing. J. Biol. Chem. 278, 64–70.

Howard, A.D., et al., 1996. A receptor in pituitary and hypothalamus that functions in growth hormone release. Science 273, 974–977.

Kamegai, J., et al., 2001. Chronic central infusion of ghrelin increases hypothalamic neuropeptide Y and Agouti-related protein mRNA levels and body weight in rats. Diabetes 50, 2438–2443.

Kanamoto, N., et al., 2001. Substantial production of ghrelin by a human medullary thyroid carcinoma cell line. J. Clin. Endocrinol. Metab. 86, 4984–4990.

Kojima, M., et al., 1999. Ghrelin is a growth-hormone-releasing acylated peptide from stomach. Nature 402, 656–660.

Kojima, M., et al., 2001. Ghrelin: discovery of the natural endogenous ligand for the growth hormone secretagogue receptor. Trends Endocrinol. Metab. 12, 118–122.

McKee, K.K., et al., 1997. Molecular analysis of rat pituitary and hypothalamic growth hormone secretagogue receptors. Mol. Endocrinol. 11, 415–423.

Miki, K., Sudo, A., 1998. Effect of urine pH, storage time, and temperature on stability of catecholamines, cortisol, and creatinine. Clin. Chem. 44 (8 Pt 1), 1759–1762.

Myrick, J.E., et al., 1989. An improved radioimmunoassay of C-peptide and its application in a multiyear study. Clin. Chem. 35, 37–42.

Nabuchi, Y., et al., 1997. The stability and degradation pathway of recombinant human parathyroid hormone: deamidation of asparaginyl residue and peptide bond cleavage at aspartyl and asparaginyl residues. Pharm. Res. 14, 1685–1690.

Nelesen, R.A., et al., 1992. Plasma atrial natriuretic peptide is unstable under most storage conditions. Circulation 86, 463–466.

Reubsaet, J.L., et al., 1998a. Analytical techniques used to study the degradation of proteins and peptides: chemical instability. J. Pharm. Biomed. Anal. 17, 955–978.

Reubsaet, J.L., et al., 1998b. Analytical techniques used to study the degradation of proteins and peptides: physical instability. J. Pharm. Biomed. Anal. 17, 979–984.

Shintani, M., et al., 2001. Ghrelin, an endogenous growth hormone secretagogue, is a novel orexigenic peptide that antagonizes leptin action through the activation of hypothalamic neuropeptide Y/Y1 receptor pathway. Diabetes 50, 227–232.

Strickley, R.G., et al., 1990. High-performance liquid chromatographic (HPLC) and HPLC-mass spectrometric (MS) analysis of the degradation of the luteinizing hormone-releasing hormone (LH-RH) antagonist RS-26306 in aqueous solution. Pharm. Res. 7, 530–536.

Wren, A.M., et al., 2001a. Ghrelin enhances appetite and increases food intake in humans. J. Clin. Endocrinol. Metab. 86, 5992–5995.
Wren, A.M., et al., 2001b. Ghrelin causes hyperphagia and obesity in rats. Diabetes 50, 2540–2547.
Yang, J., et al., 2008. Identification of the acyltransferase that octanoylates ghrelin, an appetite-stimulating peptide hormone. Cell 132, 387–396.

SECTION 3

Enzymes for Processing of Ghrelin

CHAPTER NINE

From Ghrelin to Ghrelin's O-Acyl Transferase

Jesus A. Gutierrez*[,1], Jill A. Willency*, Michael D. Knierman*, Tamer Coskun[†], Patricia J. Solenberg[‡], Doug R. Perkins[§], Richard E. Higgs[¶], John E. Hale[#]

*Translational Science and Technologies, Eli Lilly and Company, Indianapolis, Indiana, USA
[†]Endocrinology, Eli Lilly and Company, Indianapolis, Indiana, USA
[‡]BioVenio LLC, Indianapolis, Indiana, USA
[§]Biotechnology Discovery Research, Eli Lilly and Company, Indianapolis, Indiana, USA
[¶]Global Statistics and Advanced Analytics, Eli Lilly and Company, Indianapolis, Indiana, USA
[#]Hale Biochemical Consulting LLC, Klamath Falls, Oregon, USA
[1]Corresponding author: e-mail address: Gutierrez_jesus_a@lilly.com

Contents

Methods in Enzymology, Volume 514
ISSN 0076-6879
http://dx.doi.org/10.1016/B978-0-12-381272-8.00009-X

Abstract

The hormone ghrelin is a unique signaling peptide with powerful metabolic effects, mediated by its acylated forms. The acyl modification of ghrelin is unique in that it takes place via a susceptible ester linkage in the conserved serine-3 of ghrelin and is composed principally of octanoyl and, to lesser extent, decanoyl fatty acids. The nature of this ester linkage makes it susceptible to esterases, which convert it to its des-acyl forms, and, if not adequately inhibited, the conversion to des-acyl ghrelin, particularly post sample collection, can lead to artifactual and misleading results. Here, we describe sample processing and mass spectrometric methodologies for the accurate and simultaneous quantification of acylated and des-acylated forms of ghrelin. We exploited these methodologies (1) to characterize circulating and tissue-specific forms of acyl and des-acyl ghrelin, (2) to optimize a cell system for acyl ghrelin production and search for the enzyme responsible for ghrelin's acylation, and (3) to demonstrate that GOAT is ghrelin's O-acyl transferase.

1. INTRODUCTION

Ghrelin is an extraordinary peptide hormone. By acting on its receptor, namely, the growth hormone secretagogue receptor 1a (GHSR1a), it stimulates growth hormone release from the pituitary gland, food intake, carbohydrate utilization, and adiposity and regulates insulin secretion and blood glucose (Kojima and Kangawa, 2005; Kojima et al., 1999; Tschop et al., 2000; van der Lely et al., 2004). Ghrelin is the only peptide hormone of peripheral tissue origin that increases food intake (Cummings and Overduin, 2007). To achieve these physiologically critical activities, it requires an unusual acyl modification by a labile ester linkage on its critical serine-3 residue. This modification, unique to ghrelin, involves the mid-chain fatty acids octanoate and decanoate (Kojima et al., 1999).

Several investigators have described highly sensitive radiologic or immunologic methods for acylated and des-acylated ghrelin in tissues and circulation (Akamizu et al., 2005; Groschl et al., 2002; Hosoda et al., 2000, 2004; Liu et al., 2008; Rauh et al., 2007). However, much debate has occurred in the ghrelin field as to the consistency of these observations because of the known instability of the acylated forms of ghrelin (Prudom et al., 2010). An understanding of this instability is increasingly important, as hypotheses have emerged involving the ratio of acyl and total ghrelin (acyl + des-acyl) as being physiologically relevant (van der Lely, 2009). Many investigators simply rely on measurements of total ghrelin, as these are reproducible across many studies. The core of this problem lies within the instability of the acyl ester linkage on the modified serine residue. This ester bond is particularly labile to tissue and circulating esterase activities that rapidly convert ghrelin to its inactive form, namely, des-acyl ghrelin (De Vriese et al., 2004). Much of this conversion takes place post tissue or sample collection and becomes a significant artifact of sample processing. As we embarked on the establishment of mass spectrometry methodologies to measure acyl and des-acyl ghrelin, we rapidly determined the need to establish sample collection and processing protocols to minimize this artifact. Observations made with these protocols enabled us to (1) measure reliably the specific acylated and des-acyl forms of ghrelin; (2) determine that ghrelin is modified, *in vivo*, by other short-chain fatty acids; (3) hypothesize as to genes that may be mediating the acylation of ghrelin; (4) establish a cell culture system to search for ghrelin's acyltransferase and describe its mechanism for acylation; and (5) demonstrate, *in vivo*, that GOAT is ghrelin's acyltransferase. In this chapter, we describe the specific methodologies, and their significance, that allowed us to make these relevant observations.

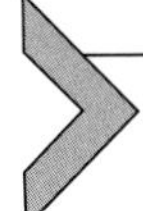

2. Quantitative MALDI-TOF MS Methodology for Ghrelin in Biological Matrices

2.1. Development of acyl and des-acyl ghrelin assays

In order to simplify the measurement of acyl and des-acyl ghrelin from biological samples, we chose to use matrix-assisted laser desorption/ionization time-of-flight mass spectrometry (MALDI-ToF MS). Our reasoning was that, with a single immunoprecipitation (IP), both species could be isolated, and with MALDI-ToF MS, we could detect both species in a single measurement. In addition, by including appropriate internal standards, the observations could be made quantitative, control for preanalytical variability, and measure artifactual des-acylation of ghrelin.

2.2. Sample preparation and processing

Blood samples in ghrelin preservative (see below) were thawed to room temperature, and 1-ml aliquots of blood were spiked with known amounts of stable-isotope-labeled (SIL; ^{13}C and ^{15}N, +24 mass units, Midwest Biotech, Indianapolis, IN) octanoylated and/or des-acyl ghrelin synthetic peptides. Samples were mixed and diluted with ethanol (80%, final concentration) for at least 2 h on ice to precipitate large proteins. After removal of the precipitate by centrifugation (2 kg for 10 min), nine volumes of an ether–methanol solution (85% diethyl ether and 15% methanol) was added to the supernatants to precipitate peptides for 1 h at room temperature. The precipitates were centrifuged (2 kg for 10 min), the supernatant was decanted, and the pellets were dried under a gentle stream of nitrogen gas. The precipitates were solubilized in 1 ml of Tris–Hepes buffer (50 m*M* Tris–HCl, 50 m*M* HEPES, 150 m*M* NaCl, and 0.1% *N*-octyl glucopyranoside detergent, pH 7.5). Insoluble particulates were removed by a brief centrifugation at 15,000 × *g* for 3 min and the clarified supernatants immunoprecipitated.

2.3. Immunoprecipitation reactions

Antibodies with specificity toward the carboxyl terminus of ghrelin peptide were covalently coupled to Invitrogen/Dynal Magnetic beads following the manufacturer's recommended protocol and used for IP reactions. Dynal magnetic beads, with approximately 1 μg of antighrelin antibody, were added to the clarified supernatants from the sample processing procedure and incubated for at least 2 h at 4 °C with gentle rotation. The antibody–antigen complexes were washed at room temperature with 500 μl of the following solutions: twice in 50 m*M* NH_4HCO_3 (pH 7.5), once in 0.1% SDS in NH_4HCO_3 (pH 7.5), once in 50 m*M* NH_4HCO_3 (pH 7.5), and once in distilled water. Ghrelin immunocomplexes were eluted from the antibody beads by acidification with 10 μl of a solution of 0.1% trifluoroacetic acid and concentrated on C18 ZipTips (Millipore, Billerica MA, catalogue No. ZTC1 8S) as recommended by the manufacturer. Ghrelin peptides were eluted from C18 ZipTip columns using 1.5 μl of 50% acetonitrile–0.1% TFA saturated with α-cyano-4-hydroxy cinammic acid matrix. The entire eluates (1.5 μl) were spotted on target plates coated with α-cyano-4-hydroxy cinammic acid matrix as described previously and analyzed using MALDI-ToF MS (Gutierrez et al., 2005).

2.4. MALDI-ToF MS

An Applied Biosciences 4700 or 4800 (Applied Biosystems, Foster City, CA) MALDI-ToF mass spectrometer was used for mass spectrometric analysis under optimized conditions for ghrelin peptide detection. The laser was operated at a fixed fluence just above the threshold value. Additional parameters were optimized for the detection of octanoylated and des-acyl ghrelin peptides. Spectra were automatically collected for each spot by a random, center-biased pattern. The acquired spectra were processed for

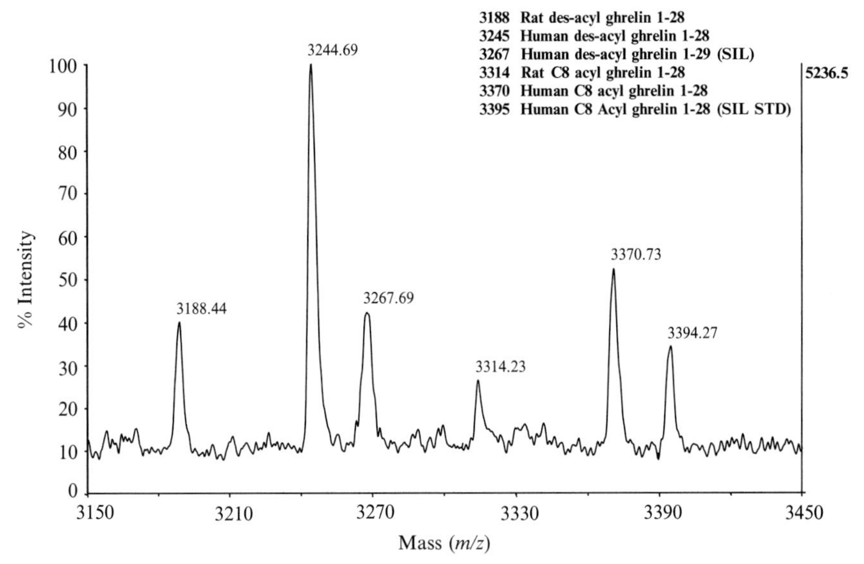

Figure 9.1 Representative mass spectrum generated from the ghrelin immunoprecipitation MALDI-ToF MS (IP-MS). Rat blood was exogenously supplemented with human ghrelin (both stable isotope labeled and unlabeled) peptides at 500 pg/ml for SIL octanoylated-ghrelin, 250 pg/ml for SIL des-acyl ghrelin, and 667 pg/ml for both unlabeled octanoylated and des-acyl ghrelin peptides, respectively. Peaks observed at *m/z* values 3188 and 3314 correspond to endogenous rat des-acyl and acyl ghrelin peptides, respectively.

quantification purposes using *maldi quant* software (Gutierrez et al., 2005). A representative spectrum of an IP of ghrelin from rat blood spiked with synthetic ghrelin standards is shown in Fig. 9.1.

2.5. Spiked-recovery analysis and assay validation

Prior to using the ghrelin IP-MS assay for quantitative purposes and in order to estimate a quantification working range, we implemented a spiked recovery experiment composed of a three-plate (each plate run on a different day), nine-point standard curve (run in triplicate). Validation samples consisted of eight concentrations between those used in the standard curve per plate (Lee et al., 2006). Whole rat blood collected in ghrelin preservative (see below) pooled and aliquoted into 1-ml fractions was spiked with SIL-labeled human octanoylated and des-acyl ghrelin peptides at 500 and 250 pg/ml, respectively. Samples were then spiked with the following concentrations of unlabeled human octanoylated and des-acyl ghrelin peptides: 5000, 3750, 2813, 2109, 1582, 1187, 890, 667, 501, 375, 282, 211, 158, 119, 89, 67, and 50 pg/ml. The blood samples spiked with 5000, 2813, 1582, 890, 501, 282, 158, 89, and 50 pg/ml were used as calibration standards, while the rest were used as validation points. Blood samples were processed as described above. The ratios of the SIL des-acyl ghrelin to unlabeled des-acyl ghrelin and SIL acyl ghrelin to unlabeled acyl ghrelin were calculated and used to quantify the peptides. Figure 9.2 illustrates the spiked recovery curves for acyl ghrelin (Fig. 9.2A) and des-acyl ghrelin (Fig. 9.2B). From the analysis of the measurement of error for the validation samples (denoted by Xs in the standard curves), the working range of the assay was calculated (Fig. 9.3A and B). For acyl ghrelin, the working range of the assay was 375–3750 pg/ml, and for des-acyl ghrelin, it was 211–2109 pg/ml.

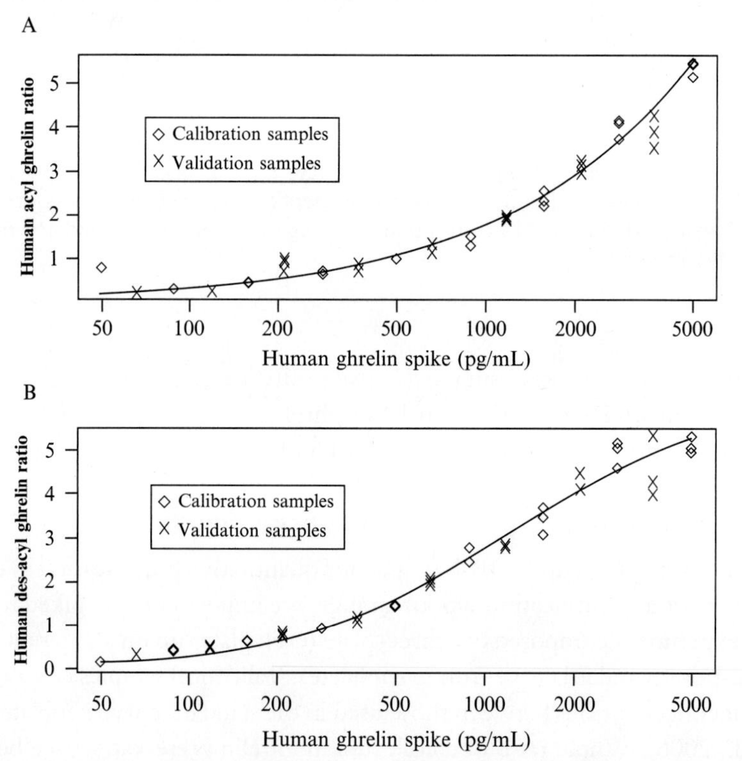

Figure 9.2 Representative results for the ratio of titrated human ghrelin to the corresponding internal standard, human acyl ghrelin (A) and human des-acyl ghrelin (B). Calibration dilutions are denoted by open diamonds and validation dilutions are denoted by Xs ($n=3$ replicates/dilution). A three-parameter logistic fit based on calibration dilutions is shown.

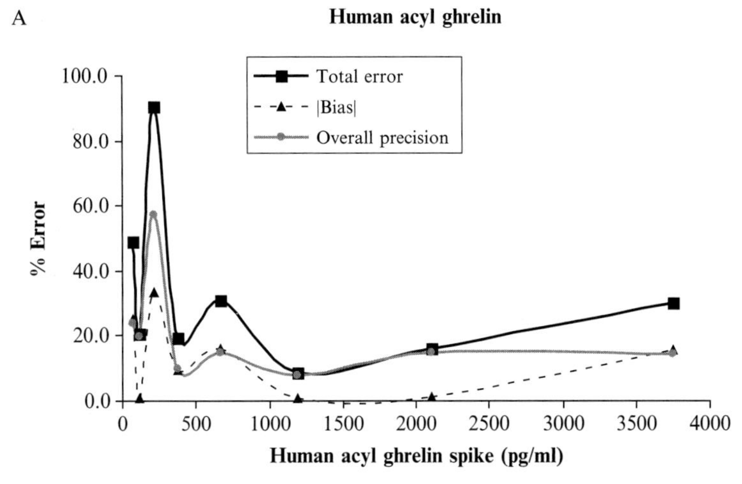

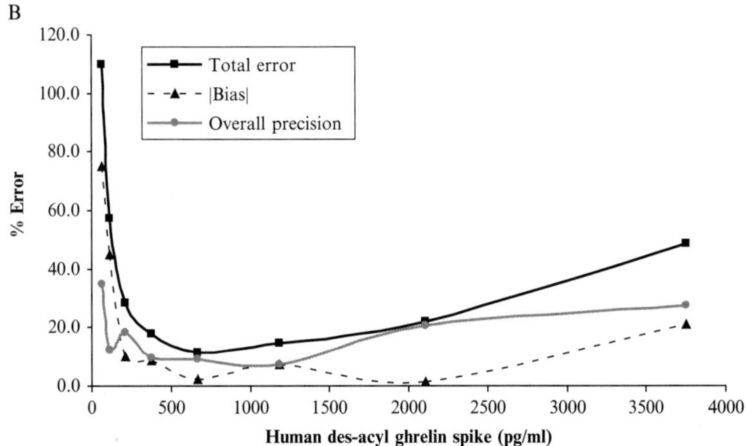

Figure 9.3 Bias, precision, and total error estimated from the between- and within-plate spiked recovery runs. Defining an assay working range by total error less than 30% yields a working range of approximately 10-fold for each form of ghrelin peptides at 375–3750 and 211–2109 pg/ml for (A) acyl and (B) des-acyl ghrelin peptides, respectively.

3. STABILIZING GHRELIN IN BIOLOGICAL MATRICES

One of the major problems with accurate measurements of ghrelin is the instability of the fatty acid ester bond on the modified serine residue. Biological fluids contain esterase activities that can rapidly convert acylated ghrelin into des-acyl ghrelin. This is a particular problem with blood samples, especially rodent blood. Acidification can quickly inactivate esterase activity, but this can also initiate protein aggregation. We evaluated a number of conditions for stabilization of ghrelin and incorporated additional steps for the inhibition of protein aggregation.

3.1. Stabilization of ghrelin in blood and media

A modification of the solution described by Reeve to stabilize cholecystokinin hormone was initially used to minimize ghrelin deacylation in blood (Reeve et al., 2003). One volume of 25 m*M* EDTA, 2 m*M* PMSF, 1 *M* NaCl, and 200 mN HCl was placed in a sample collection tube. In order to evaluate the effectiveness of the stabilization solution and to allow for quantification, SIL acyl ghrelin standards were included. At the time of blood draw, an equal volume of rat blood was added to the stabilization tube and placed on ice until centrifugation (within 30 min). Supernatants were removed and stored at −80 °C until further processing for mass spectral analysis.

3.2. Stabilization of ghrelin in stomach tissue

Stomach tissue from rats was immediately frozen upon collection by immersion into liquid nitrogen. The frozen tissue was pulverized with a mortar and pestle under liquid nitrogen until it became a fine powder. The pulverized tissue (0.1 g) was placed in a tube containing 1 ml of tissue extraction buffer (50 mN HCl, 150 m*M* NaCl, 12.5 m*M* EDTA, 1 mg/ml Roche Complete EDTA-Free Protease Inhibitor Cocktail, 0.1% *N*-octyl glucopyranoside, and 100 ng/ml each acyl and des-acyl SIL ghrelin standards). Samples were extracted with vortexing for 10 min at room temperature and then spun at 4000 × *g* for 10 min. The supernatant was removed and proteins were precipitated by adding 2.5 volumes of acetone on ice for 1 h. The precipitated proteins were centrifuged at 4000 × *g* for 10 min. Ghrelin remained in the supernatant, and it was precipitated by the addition of diethyl ether–methanol at a ratio of 1:8:1 supernatant:ether:methanol. The sample was mixed thoroughly and incubated at room temperature for 1 h. The precipitate was centrifuged at 4000 × *g* for 10 min, and the supernatant was discarded. The pellets were dried under a gentle stream of nitrogen gas for 30 min and solubilized in 0.5 ml of Tris–HEPES buffer (50 m*M* Tris–HCl, 50 m*M* HEPES, 150 m*M* NaCl, 0.1% *N*-octyl glucopyranoside, pH 7.5) prior to immunoprecipitation and MS analyses.

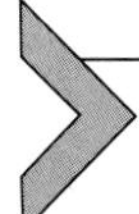

4. TT Cell Culture System for Ghrelin Production

Establishment of a cell culture system capable of reproducibly secreting acyl-modified forms of ghrelin was essential to initially implicate GOAT in the acylation of ghrelin. Kanamoto and collaborators demonstrated that the human medullary thyroid carcinoma cell line (TT cell line) produced des-acylated and acylated ghrelin, suggesting that these cells possessed ghrelin's acylation components (Kanamoto et al., 2001). We obtained TT cells from ATCC (Cat. No. CRL-1803) and cultured them in Ham's F12K medium (ATCC Cat. No. 30-2004) supplemented with 2 m*M* L-glutamine, 1.5 g/l sodium bicarbonate, 10% fetal bovine serum, 10 units/ml penicillin, and 10 mg/ml streptomycin. The cell medium was also supplemented with SIL ghrelin standard peptides for mass spectrometric analyses for octanoylated and des-acylated ghrelin at 0.4 ng/ml. The collected media were acidified to 50 nN HCl and stored at −80 °C until ready for IP-MS analysis.

4.1. Immunoprecipitation mass spectrometry method for TT cell culture produced ghrelin

Acidified media from cell culture studies were centrifuged at 2 kg for 5 min. The supernatants were extracted on equilibrated tC18 Sep Pak cartridges (Millipore Corp., Cat. No. WAT036805). The peptides were eluted from tC18 Sep Pack units using 60% acetonitrile in 0.1% trifluoroacetic acid and lyophilized to dryness. The dried pellets were suspended in 275 μl of Tris–HEPES buffer (50 m*M* Tris–HCl, 50 m*M* HEPES, 150 m*M* NaCl, and 0.1% *N*-octyl glucopyranoside, pH 7.5) and exposed to approximately 1 μg of antighrelin antibody bound to Dynal magnetic beads. The extracts and antibody beads were incubated overnight at 4 °C or 2 h at room temperature with gentle rotation. The antibody–antigen complexes were washed at room temperature with 500 μl of the following solutions: twice in Tris–HEPES buffer and twice in distilled water. Ghrelin immunocomplexes were separated from the antibody beads by acidification with 10 μl of a solution of 0.1% trifluoroacetic acid and further processed for MALDI-ToF MS analyses as described above (see Section 2.3).

4.2. Stabilizing ghrelin in cell culture conditions

Because acylated ghrelin was not stable in serum-containing media and it took at least 4 days to generate robust ghrelin signals, we reasoned that a feasible approach to prevent the des-acylation of ghrelin would be to include an antibody with specificity toward the octanoylated, modified N-terminus of the ghrelin peptide in the culture media. To explore this approach, we supplemented the TT cell culture media with a monoclonal antibody with specificity for the octanoylated N-terminus of ghrelin (Antibody C25a1). Results from these studies demonstrated that the supplementation of the TT cell medium with C25a1 antibody at 10 μg/ml protected acylated ghrelin for up to 10 days in the cell culture. These studies also revealed that, even though acylated ghrelin standard peptides could be protected and detected in the cell medium, TT cells were producing octanoylated ghrelin at levels below the detection limits for our ghrelin IP-MS assay (data not shown).

4.3. Stimulation of ghrelin production by octanoic acid in TT cells

Since des-acyl ghrelin could be readily detected in the TT cell media, we hypothesized that low mid-chain fatty acid levels in cell culture media may be limiting the acylation of ghrelin. To test this hypothesis, we supplemented TT cell culture media with varying levels of octanoic acid ranging from 0 to 1000 μg/ml. In cell media supplemented with octanoic acid at 62.5, 125, and 250 μg/ml, octanoylated forms of ghrelin 1–28 and 1–27 were readily observed. Peptide fragmentation and tandem MS/MS analyses confirmed that the octanoylation of TT-cell-mediated ghrelin occurred exclusively at serine-3, identical to tissue-extracted acylated ghrelin. These TT cell culture conditions provided the essential system to define the molecular components for the acylation of ghrelin. We selected 125 μg/ml as the optimal concentration of octanoic acid for the production of acylated ghrelin for subsequent studies (Gutierrez et al., 2008).

5. FUNCTIONAL SCREENING FOR GHRELIN'S O-ACYL TRANSFERASE

To select candidate sequences for transcript silencing in the TT cell ghrelin acylation functional assay system, we used the following criteria: (1) similarity to previously defined acyltransferases, (2) presence of a human homolog, and (3) genes of unknown function. With these criteria, we mainly identified members of the recently described membrane-bound O-acyl transferase (MBOAT) family of proteins (Hofmann, 2000). This led us to hypothesize that an orphan MBOAT protein mediated the acylation of ghrelin. At about the same time, and consistent with this hypothesis, Takada and collaborators demonstrated that protein WNT3a was palmitoylated at critical cysteine (C-77) and serine (S-209) residues by the MBOAT acyl transferase porcupine. Importantly, the palmitoylation at residue serine-209 was via an ester linkage like ghrelin's octanoylation (Takada et al., 2006).

5.1. Silencing hGOAT in TT cell culture ghrelin acylation system

Thyroid carcinoma cells for gene-silencing studies were cultured as described above. Transfection of RNA-silencing sequences into cells was achieved with an Amaxa Nucleofector II Device (Cologne, Germany) using 5×10^6 cells and 2 μg RNA-silencing sequences per transfection following the manufacturer's recommended protocol. The following five double-stranded RNA-silencing sequences were used to target GOAT transcript:

siRNA7-1, 5′-UGU UGCAGACAUUUGCCU UCU-3′ (siRNA 7-1);

siRNA7-3, 5′-AAU GCC UAA ACG UGG CAG UGA-3′ (siRNA 7-3);

stealth-1, 5′-CAG AUU CUU GGA CUA GAA UGC CUA A-3′ (siRNA 7-5);

stealth-2, 5′-CGG GAC UGA CUG AUU GCC AGC AAU U-3′ (siRNA 7-6);

and stealth-3, 5′-AGC UGA CUA CCU GAU UCA CUC CUU U-3′ (siRNA 7-7).

As controls, cells were treated with the nontargeting control siRNA (NTC-2; catalog no. D-001210-02-05 from Dharmacon). Double-stranded silencing RNAs stealth 1–3 were from Invitrogen, and siRNA7-1 and siRNA7-3 (catalog no. 1027020) were custom siRNAs from Qiagen. After transfection, cells in TT cell media were allowed to adhere overnight to T-25 tissue culture flasks. Cell media were replaced with TT cell culture media supplemented with 125 μg/ml octanoic acid, to stimulate ghrelin octanoylation; 0.4 ng/ml SIL human octanoylated and des acyl ghrelin forms; and 10 μg/ml antighrelin antibody (C2-5A1). The cells were allowed to incubate for 6 days. After this incubation period, cell media were collected, acidified to 50 mN HCl, and stored at −80 °C until ready for hGOAT transcript level determination and IP-MS analysis. TT cell treatment with reagents designed to specifically silence transcripts for GOAT robustly inhibited octanoyl ghrelin production (Fig. 9.4A and B).

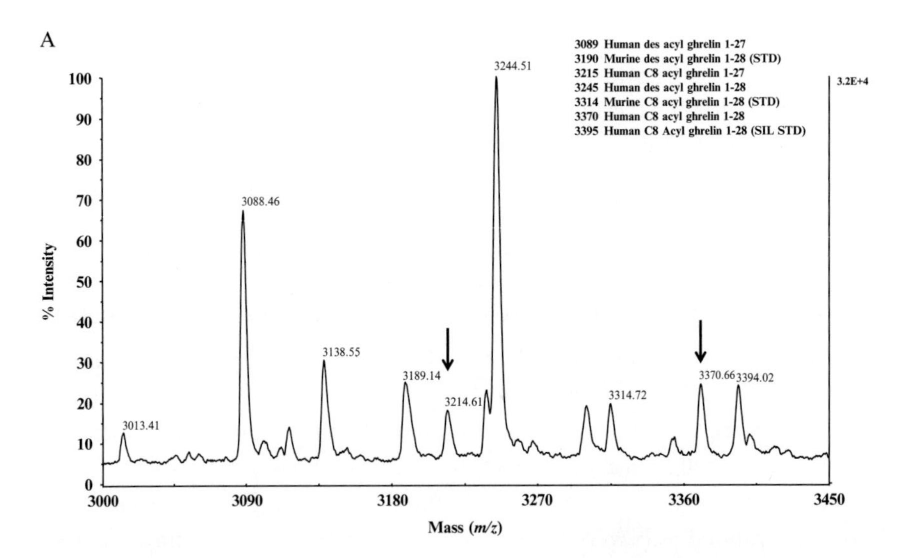

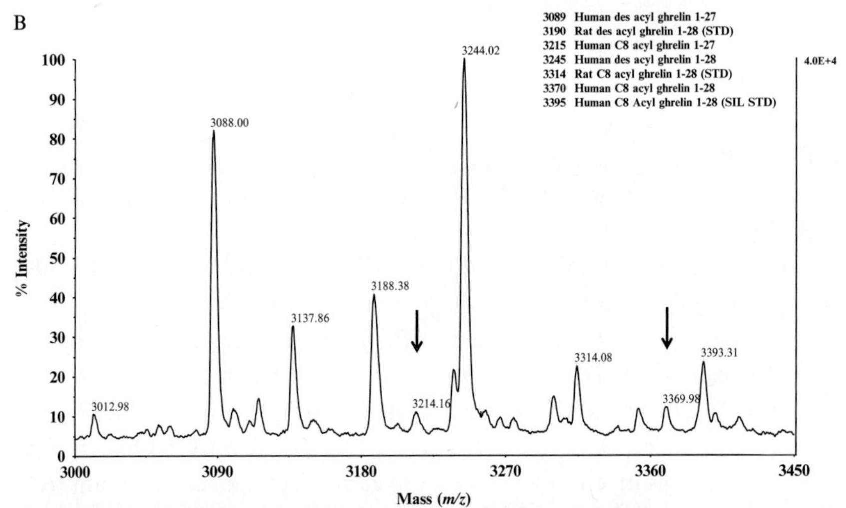

Figure 9.4 Silencing GOAT decreases ghrelin octanoylation in TT cells. TT cells were exposed to 2 μg of either nontargeting control (A; NTC-2) or GOAT-specific siRNAs (B; siRNA 7-3) and assayed for ghrelin octanoylation with the ghrelin IP-MS assay. Exposure of TT cells to GOAT-specific siRNAs diminished octanoylated ghrelin 1-28 (*m/z* 3370) and 1-27 (*m/z* 3214), denoted by downward arrows. Ghrelin peptide standards were included at the start of the culture period (*m/z* 3188, 3314, and 3393 for rat des acyl ghrelin, rat octanoylated ghrelin, and human SIL octanoylated ghrelin peptides).

These effects are dose dependent (data not shown) and can be achieved by targeting distinct regions of the GOAT gene. Consistent with the inhibition of ghrelin's acylation, similar decreases in GOAT transcripts were observed in TT cells (Gutierrez et al., 2008).

5.2. Exogenous cell expression studies: Ghrelin and GOAT are sufficient for ghrelin's acylation

While gene-silencing studies in TT cells clearly implicated GOAT in ghrelin's acylation, those studies do not demonstrate whether GOAT is sufficient to faithfully acylate ghrelin. It is possible that GOAT is part of a functional complex composed of other gene products also essential for this modification. To determine the sufficiency of GOAT in ghrelin's acyl modification, studies were carried out in HEK-293 cells, which do not normally express ghrelin or GOAT.

Transient expression of the ghrelin and GOAT cDNAs revealed the sufficiency and importance of GOAT in the acylation of ghrelin (Gutierrez et al., 2008). Expression of ghrelin alone shows the production of unmodified ghrelin forms. However, when both ghrelin and GOAT cDNAs are coexpressed in cells, the principal form of ghrelin observed corresponds to the octanoylated form. Tandem MS/MS fragmentation of the HEK293 cell culture produced ghrelin confirms that the octanoylation modification resides exclusively at the critical serine-3 residue, identical to the natural forms of ghrelin observed from stomach tissue, circulation, or TT cells. HEK 293 cells expressing both GOAT and ghrelin cultured under conditions lacking supplemental octanoic acid also reveal ghrelin forms modified with acetyl or butyryl groups at serine 3. These observations demonstrate that GOAT can acyl modify ghrelin with alternate fatty acids besides octanoate or decanoate and that sufficient levels of these fatty acids are present with HEK-293 cells to mediate this modification without media supplementation. These observations also led us to define the fatty acid selectivity by GOAT for ghrelin's acylation. GOAT acyl modifies ghrelin with fatty acids ranging from acetate (C2) to tetradecanoate (C14) (Gutierrez et al., 2008).

While our work was in revision for publication, Yang and collaborators described their work where they also hypothesized that an MBOAT may mediate the acylation of ghrelin, leading them to independently define MBOAT4 (GOAT) as ghrelin's acyl transferase (Yang et al., 2008).

6. GOAT IS GHRELIN'S ACYL TRANSFERASE

Our results from the TT cell culture system silencing GOAT and the HEK-293 cells recapitulating the production of acylated ghrelin clearly implicated GOAT in this modification and demonstrated the sufficiency of ghrelin and GOAT in the acyl modification of ghrelin. These data, however, did not demonstrate the essentiality *in vivo* of GOAT for ghrelin's octanoylation and its physiological functions. One approach to determine whether GOAT is the only gene capable of carrying out this important function is to generate genetically modified mice with a complete deletion of the GOAT gene. Ghrelin analyses in tissues or circulation from these animals would demonstrate whether another enzyme could also carry out ghrelin's modification.

6.1. Genetically deficient GOAT mice lack acylated ghrelin

GOAT-null mice were generated by Taconic/Artemis using C57BL/6N cells from C57/TacN mice, as previously described (Gutierrez et al., 2008; Kirchner et al., 2009). To determine the ghrelin profile in tissue and circulation for GOAT-null mice, blood and stomach tissues from homozygous GOAT-null or wild-type littermate animals were collected, stabilized for ghrelin analyses, and processed as described above (see Section 3). Ghrelin IP-MS results from these studies demonstrate the complete absence of octanoylated ghrelin in circulation in contrast to wild-type animals. Stomach-derived forms of ghrelin from GOAT-null animals also showed a complete absence of octanoylated and decanoylated ghrelin forms (Fig. 9.5). These data demonstrate that GOAT is ghrelin's acyl transferase and that there is no alternate compensatory mechanism for the acylation of ghrelin. Of interest, we noticed that GOAT-null mice display elevated levels of des-acyl ghrelin in circulation. This higher level of des-acyl ghrelin was substantially more than the expected levels for total ghrelin as defined by the sum of acylated and des-acylated ghrelin (Fig. 9.6). This observation was independently confirmed by Kang and collaborators with a different cohort of GOAT-deficient mice and perhaps points to an acyl ghrelin–GHSR1a feedback mechanism, where the absence of acylated ghrelin disrupts this mechanism (Kang et al., 2012). This disrupted signaling mechanism may be monitored by ghrelin-producing cells and lead to compensation in ghrelin production. The physiological significance of this feedback mechanism still needs to be discovered.

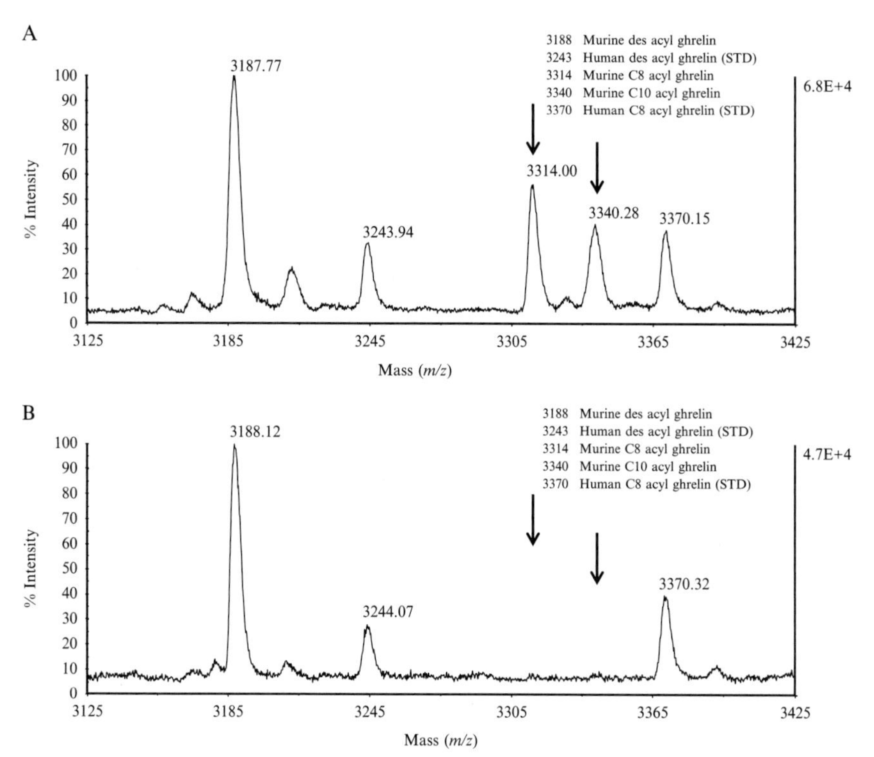

Figure 9.5 GOAT-gene-deficient mice lack acylated forms of ghrelin in stomach. Stomach tissue profiles for acylated and des-acyl ghrelin in either wild-type (A) or GOAT-gene-null mice (B) were profiled using the ghrelin IP-MS assay. Arrows denote the location for octanoylated and decanoylated ghrelin. Peaks at *m/z* 3188, 3314, and 3340 correspond to mouse des-acyl, octanoylated, and decanoylated ghrelin forms. Ghrelin peptide standards were human des-acyl ghrelin (*m/z* 3243) and human octanoylated ghrelin (*m/z* 3370).

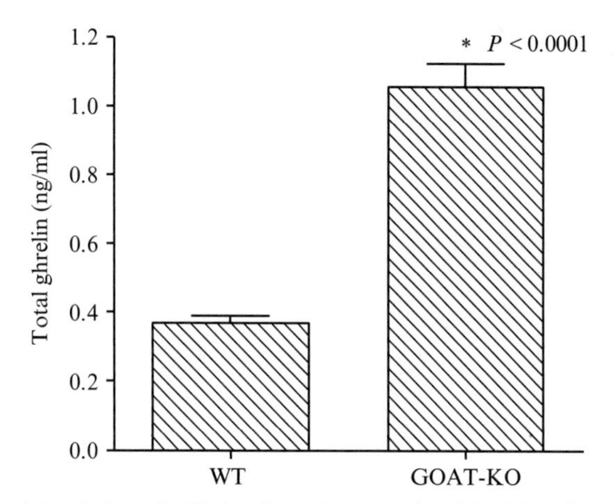

Figure 9.6 Total circulating ghrelin levels are increased in GOAT-knockout animals. Total ghrelin levels were determined by adding acylated and des-acyl ghrelin measurements as determined with the ghrelin IP-MS assay. The ghrelin levels in GOAT-knockout mice were exclusively due to des-acyl ghrelin. * denotes statistical significance ($P < 0.0001$).

7. SUMMARY

The metabolic hormone ghrelin is an extraordinary 28-amino acid peptide with a unique acyl ester modification on its serine 3 residue. This posttranslational modification is absolutely essential for its growth hormone, orexigenic, metabolic, and insulin secretion effects, yet the modification is highly susceptible to circulating esterases, which can convert the active form of ghrelin to the des-acylated form. This conversion of acyl to des-acyl ghrelin can occur within minutes after sample collection, and if left uncontrolled, the des-acylation reaction can lead to variability and distracting artifacts in acyl ghrelin measurements. As part of our efforts to establish mass spectrometric methodologies to simultaneously quantify acyl and des-acyl ghrelin forms, we observed this rapid artifactual conversion to des-acyl ghrelin and established protocols to minimize it. In this chapter, we described methods to stabilize acylated ghrelin in blood and tissue samples, applied these methods to quantify and profile ghrelin forms, and used these methodologies to describe the molecular basis for the acylation of ghrelin. Our studies defined a functional cell-based assay system to reliably octanoylate ghrelin and to search for ghrelin's acyl transferase, namely, GOAT. Further, we used these methodologies in studies with GOAT-null mice to demonstrate conclusively that GOAT is ghrelin's acyl transferase. Simply, these methodologies led us from the biomarker acyl ghrelin to the deorphanizing of an MBOAT to the new molecular target for therapeutic intervention, GOAT.

These methodologies are also likely useful for future studies in the fundamental physiology and regulation of GOAT, acyl ghrelin, and the ghrelin receptor, GHSR1a. The following fundamental questions in ghrelin GOAT biology remain to be addressed. First, what is the source of octanoic and decanoic acids in GOAT and ghrelin-positive cells? It is evident that dietary mid-chain fatty acids can be a source (Kirchner et al., 2009; Nishi et al., 2005), but it is also clear that it is not the only source. Do these cells produce octanoate and decanoate via beta oxidation or fatty acid synthesis mechanisms? Why is ghrelin principally acylated by octanoate and decanoate when GOAT is capable of acylating ghrelin with fatty acids ranging from acetate to tetradecanoate? Second, what is the mechanistic role that GOAT, acyl ghrelin, and GHSR1a play in the regulation of insulin secretion? Is pancreatic ghrelin acylated with octanoate and decanoate like stomach ghrelin? What is the source of fatty acid for acylation of ghrelin in the pancreas? Precisely, how does locally produced acylated ghrelin signal the insulin-producing beta cells to modulate insulin secretion? Is this a direct effect, or is it mediated by an intermediate signaling mechanism or cell type? Third, does GOAT have additional acylation substrates besides ghrelin? The acylation motif exploited by GOAT appears to be exclusively present in ghrelin at least by bioinformatic analyses. However, this is not conclusive evidence for the absence additional GOAT substrates. In addition, recent published studies propose the existence of cells in stomach that express GOAT but appear to lack ghrelin, suggesting that other proteins can be acyl modified by GOAT (Kang et al., 2012). Fourth, are there other physiologically relevant ester octanoylated proteins? To date, ghrelin is the only protein known to have this modification. Recent studies showed that WNT3a is palmitoylated via an ester linkage at serine 209. This modification is important for WNT3a's function, and it is likely that, if present, other serine octanoylated proteins will also play important physiological functions, just like octanoylated ghrelin.

ACKNOWLEDGMENTS

We wish to thank Drs. Mark L. Heiman and Jude E. Onyia for support, stimulating discussions, and input on the topic of ghrelin and the discovery of GOAT. We also thank Dr. Derrick R. Witcher for providing the antighrelin antibodies used in these studies.

REFERENCES

Akamizu, T., et al., 2005. Separate measurement of plasma levels of acylated and desacyl ghrelin in healthy subjects using a new direct ELISA assay. J. Clin. Endocrinol. Metab. 90, 6–9.

Cummings, D.E., Overduin, J., 2007. 'rointestinal regulation of food intake. J. Clin. Invest. 117, 13–23.

De Vriese, C., et al., 2004. Ghrelin degradation by serum and tissue homogenates: identification of cleavage sites. Endocrinology 145, 4997–5005.

Groschl, M., et al., 2002. Preanalytical influences on the measurement of ghrelin. Clin. Chem. 48, 1114–1116.

Gutierrez, J.A., et al., 2005. Quantitative determination of peptides using matrix-assisted laser desorption/ionization time of flight mass spectrometry. Biotechniques 38, S13–S17.

Gutierrez, J.A., et al., 2008. Ghrelin octanoylation mediated by an orphan lipid transferase. Proc. Natl. Acad. Sci. USA 105, 6320–6325.

Hofmann, K., 2000. A superfamily of membrane-bound O-acyltransferases with implications for Wnt signaling. Trends Biochem. Sci. 25, 111–112.

Hosoda, H., et al., 2000. Ghrelin and des-acyl ghrelin: two major forms of rat ghrelin peptide in gastrointestinal tissue. Biochem. Biophys. Res. Commun. 279, 909–913.

Hosoda, H., et al., 2004. Optimum collection and storage conditions for ghrelin measurement: octanoyl modification of ghrelin is rapidly hydrolyzed to desacyl ghrelin in blood samples. Clin. Chem. 50, 1077–1080.

Kanamoto, N., et al., 2001. Substantial production of ghrelin by a human medullary thyroid carcinoma cell line. J. Clin. Endocrinol. Metab. 86, 4984–4990.
Kang, K., et al., 2012. Mouse ghrelin-O-acyltransferase (GOAT) plays a critical role in bile acid reabsorption. FASEB J. 26, 1–13.
Kirchner, H., et al., 2009. GOAT links dietary lipids with the endocrine control of energy balance. Nat. Med. 15, 741–745.
Kojima, M., Kangawa, K., 2005. Ghrelin: structure and function. Physiol. Rev. 85, 495–522.
Kojima, M., et al., 1999. Ghrelin is a growth-hormone-releasing acylated peptide from stomach. Nature 402, 656–660.
Lee, J.W., et al., 2006. Fit-for-purpose method development and validation for successful biomarker measurement. Pharm. Res. 23, 312–328.
Liu, J., et al., 2008. Novel ghrelin assays provide evidence for independent regulation of ghrelin acylation and secretion in healthy young men. J. Clin. Endocrinol. Metab. 93, 1980–1987.
Nishi, Y., et al., 2005. Ingested medium-chain fatty acids are directly utilized for the acyl modification of ghrelin. Endocrinology 146, 2255–2264.
Prudom, C., et al., 2010. Comparison of competitive radioimmunoassays and two-site sandwich assays for the measurement and interpretation of plasma ghrelin levels. J. Clin. Endocrinol. Metab. 95, 2351–2358.
Rauh, M., et al., 2007. Simultaneous quantification of ghrelin and desacyl ghrelin by liquid chromatography- tandem mass spectrometry in plasma, serum, and cell supernatants. Clin. Chem. 53, 902–910.
Reeve, J.R., Green, G.M., Chew, P., Eysselein, V.E., Keire, D.A., 2003. CCK-58 is the only detectable endocrine form of cholecystokinin in the rat. Am. J. Physiol. Gastrointest. Liver Physiol. 285, 255–265.
Takada, R., et al., 2006. Monounsaturated fatty acid modification of Wnt protein: its role in Wnt secretion. Dev. Cell 11, 791–801.
Tschop, M., et al., 2000. Ghrelin induces adiposity in rodents. Nature 407, 908–913.
Van der Lely, A.J., 2009. Ghrelin and new metabolic frontiers. Horm. Res. 71, 129–133.
Van der Lely, A.J., et al., 2004. Biological, physiological, pathophysiological, and pharmacological aspects of ghrelin. Endocr. Rev. 25, 426–457.
Yang, J., et al., 2008. Identification of the acyltransferase that octanoylates ghrelin, an appetite-stimulating peptide hormone. Cell 132, 387–396.

CHAPTER TEN

Enzymatic Characterization of GOAT, ghrelin O-acyltransferase

Hideko Ohgusu, Tomoko Takahashi, Masayasu Kojima[1]

Molecular Genetics, Institute of Life Science, Kurume University, Kurume, Fukuoka, Japan

[1]Corresponding author: e-mail address: kojima_masayasu@kurume-u.ac.jp

Contents

Abstract

Ghrelin is a gastric peptide hormone in which serine 3 (threonine 3 in frogs) is modified primarily by an *n*-octanoic acid; this modification is essential for ghrelin's activity. The enzyme that transfers *n*-octanoic acid to the third serine residue of ghrelin peptide has been identified and named GOAT for ghrelin *O*-acyltransferase. GOAT is the only known enzyme that catalyzes the acyl modification of ghrelin and specifically modifies the third amino acid serine and does not modify other serine residues in ghrelin peptides. GOAT prefers *n*-hexanoyl-CoA over *n*-octanoyl-CoA as the acyl donor, although in the stomach *n*-octanoyl form is the main acyl-modified ghrelin and the concentration of *n*-hexanoyl form is very low. Moreover, a four-amino acid peptide derived from the N-terminal sequence of ghrelin can be modified by GOAT, indicating that these four amino acids constitute the core motif for substrate recognition by the enzyme.

Methods in Enzymology, Volume 514
ISSN 0076-6879
http://dx.doi.org/10.1016/B978-0-12-381272-8.00010-6

1. INTRODUCTION

Ghrelin is identified in almost all mammalian and nonmammalian vertebrate species, including frogs, birds, and fish (Kojima and Kangawa, 2005). A characteristic feature of ghrelin is an acyl modification at the third amino acid, which is typically serine or threonine. This acyl modification is necessary for ghrelin to bind to its receptor, namely, the GHS-R, and exert biological activity (Kojima et al., 1999). The structure of ghrelin, particularly that of the acyl-modification region, is highly conserved throughout all vertebrate species (Kojima et al., 2008). The primary fatty acid that acylates ghrelin is *n*-octanoic acid. However, the mechanism underlying this unique modification is still unknown. Thus, investigations characterizing the putative ghrelin ser *O*-acyltransferase are needed.

In 1998, Yang et al. reported that the membrane-bound acyltransferase MBOAT4 is a ghrelin *O*-acyltransferase (GOAT) that catalyzes *n*-octanoyl modifications of ghrelin in cultured cells (Yang et al., 2008). The observation that GOAT knockout mice lacked octanoylated ghrelin confirmed the role of GOAT in this process (Gutierrez et al., 2008). Using a combination of recombinant GOAT and ghrelin-specific immunoassays, we showed that GOAT catalyzes the *n*-octanoyl modification of ghrelin *in vitro* (Ohgusu et al., 2009). With this platform in hand, we analyzed the basic enzymatic characteristics such as the optimal temperature and pH, and profiled the peptide and acyl substrate specificities of GOAT.

2. GOAT ENZYMATIC ASSAY

2.1. Construction of stable GOAT-expressing cells and preparation of GOAT

To characterize GOAT enzymatic activity, we first established a stable GOAT-expressing cell line and prepared the enzyme from the cell cultures (Ohgusu et al., 2009).

1. Primer pairs were designed on the basis of the cDNA sequence of mouse GOAT (GenBank accesssion No. EU721729). The primer sequences were sense 5′-TCAAGCTTAGGATGGATTGGCTCCAGCTCTTTTTC TGCATCCTTTATC-3′ and antisense 5′-GACTCGAGTCAGTTAC GTTTGTCTTTTCTCTCCGCTAACAG-3′, containing a HindIII and a XhoI site, respectively.

2. Mouse GOAT cDNA was amplified by PCR from stomach cDNA using Pyrobest DNA polymerase (Takara Bio Inc., Ohtsu, Japan).
3. The amplified cDNA fragment was inserted into a pcDNA3.1 vector at the HindIII-XhoI site.
4. CHO cells were transfected with the GOAT-pcDNA3.1 vector using Lipofectamine 2000 reagent (Invitrogen, Carlsbad, CA) and cultured in α-MEM medium with G418 (1 mg/ml) for stable cell selection.
5. Cells that grew in the presence of G418 were tested for expression of GOAT mRNA by PCR. The primers used were the same as those employed for amplifying GOAT cDNA (Step 1). The cell with the highest GOAT expression level was used for the following experiments.
6. GOAT-expressing CHO cells were cultured to 80–90% confluence in 12 100-mm culture plates (IWAKI, Tokyo, Japan). Then, the cells were homogenized by a Teflon homogenizer in the extraction buffer: 100 m*M* Tris–HCl (pH 7.4) containing 1 m*M* PMSF (phenylmethylsulfonyl fluoride), 0.8 n*M* aprotinin, 15 μ*M* E-64, 20 μM leupeptin, 50 μ*M* bestatin, and 10 μ*M* pepstatin A.
7. First, the homogenate was centrifuged at 800 × *g* for 5 min. Then, the resultant supernatant was further centrifuged at 100,000 × *g* for 1 h. The pellet was dissolved in the same extraction buffer and stored at − 80 °C. The described subcellular fractionation procedure provided specific ghrelin *n*-octanoyl transferase activity in the 100,000 × *g* pellet but not the supernatant (Fig. 10.1). These results indicated that GOAT was a membrane-bound enzyme. In addition, control CHO cells demonstrated no GOAT activity (Fig. 10.1). We looked for GOAT activity in preparations of mouse stomach membrane using the same assay system described in this study (see Section 2.2). However, we could not detect any GOAT activity, perhaps due to low concentrations of the enzyme in the stomach (data not shown).

2.2. *In vitro* GOAT enzymatic assay

1. The standard *in vitro* assay for the ghrelin *n*-octanoyl modification contained the following: 200 μl of 50 m*M* Tris–HCl (pH 7.4), 0.5 μ*M* rat des-acyl ghrelin, 10 μ*M* *n*-octanoyl-CoA (Sigma-Aldrich Co., St. Louis, MO), 0.1% CHAPS, and 1–5 μl enzyme solution. Figure 10.2 shows that *n*-octanoyl ghrelin was produced from the acyl donor *n*-octanoyl-CoA but not from *n*-octanoic acid. Moreover, co-incubation of both

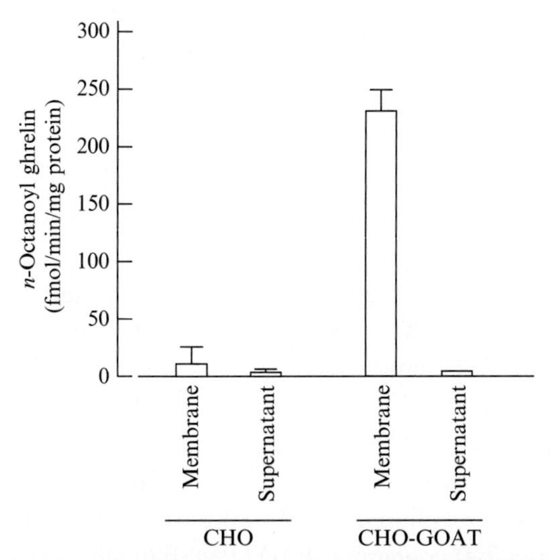

Figure 10.1 Subcellular localization of GOAT activity. Wild-type CHO and GOAT-expressing CHO (CHO-GOAT) cells were collected and separated by centrifugation at $100{,}000 \times g$ to obtain membrane fractions. A ghrelin *n*-octanoyl modification reaction was performed using the standard assay conditions. *n*-Octanoyl ghrelin concentrations were measured using the active ghrelin ELISA kit. The results are expressed as the means $\pm$ SD ($n = 3$).

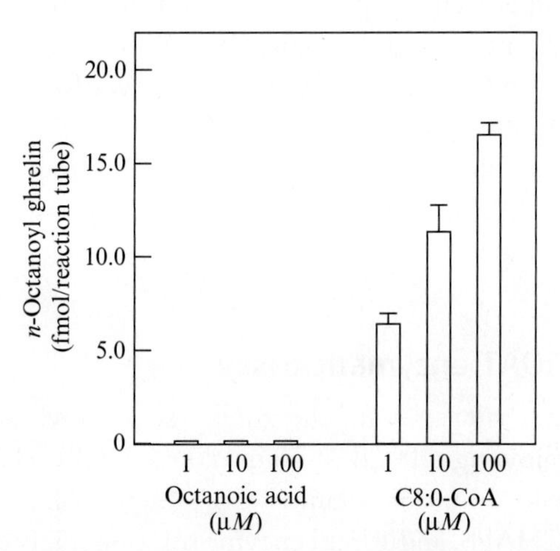

Figure 10.2 GOAT acyl donor selectivity. The production of *n*-octanoyl ghrelin was measured using the standard assay conditions with varying concentrations of *n*-octanoic acid or *n*-octanoyl-CoA as donor substrates. The results are expressed as the means $\pm$ SD ($n = 3$).

n-octanoic acid and CoA alone did not produce *n*-octanoyl ghrelin (data not shown). Thus, *n*-octanoyl-CoA was an acyl donor for ghrelin.

2. The reaction was initiated by adding the enzyme solution to the other components and incubating at 37 °C for 30 min.
3. The reaction was stopped by adding 20 μl 1 N HCl. The solution was stored at −30 °C until ghrelin concentrations were measured.

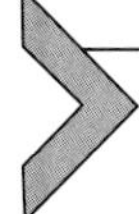

3. DETECTING THE MOLECULAR FORMS OF GHRELIN GENERATED BY *IN VITRO* GOAT ENZYMATIC ASSAYS

We used three approaches to identify the molecular forms of ghrelin produced during *in vitro* GOAT activity assays (Fig. 10.3).

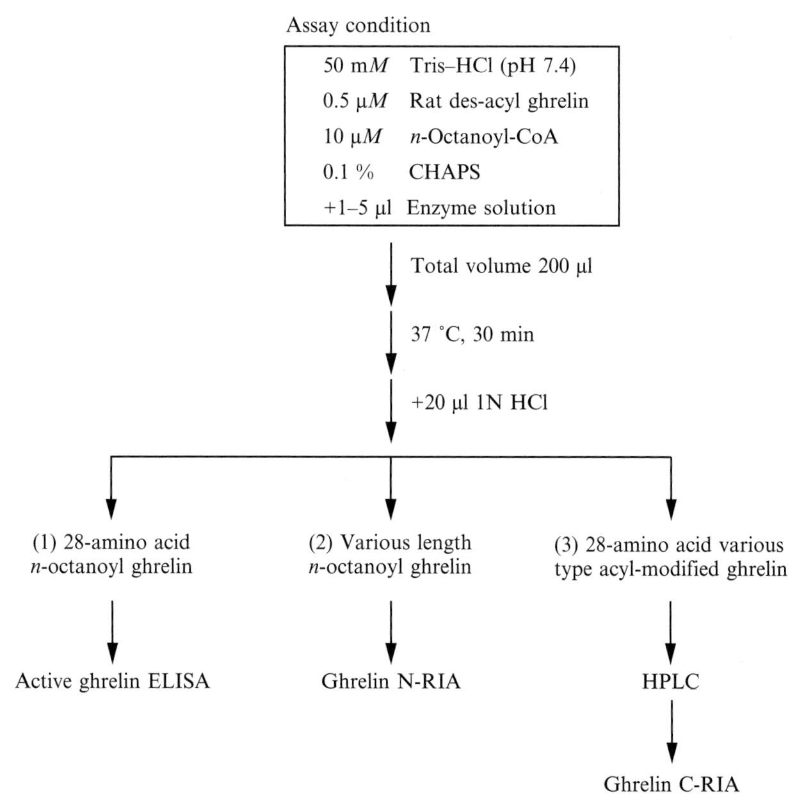

Figure 10.3 Assays to identify various molecular forms of ghrelin. Three assay systems, namely, active ghrelin ELISA, ghrelin N-RIA, and ghrelin C-RIA, were used to identify various lengths and acyl modifications of ghrelin peptides produced *in vitro*.

1. *n*-Octanoyl-modified ghrelin peptides less than 28 amino acids long were identified by a radioimmunoassay (RIA) specific to the N-terminal fragment of ghrelin (N-RIA), which detects the *n*-octanoyl moiety.
2. To analyze the various types of acyl modifications in ghrelin peptides, the assayed samples were fractionated by high-performance liquid chromatography (HPLC), and the ghrelin concentration in each fraction was detected by RIA specific to the C-terminal fragment of ghrelin (C-RIA). The various types of acyl modifications in ghrelin resulted in different retention times by HPLC; the modifications with longer acyl acids eluted earlier than those with shorter ones.
3. Full-length *n*-octanoyl ghrelin was detected by an active ghrelin ELISA kit (Mitsubishi Kagaku Iatron, Inc., Tokyo, Japan). This ELISA specifically detects *n*-octanoyl ghrelin of 28 amino acids in length.

3.1. RIA of ghrelin

1. Rabbit polyclonal antibodies were raised against N- and C-terminal rat ghrelin peptides: Gly^1-Lys^{11} with *O*-*n*-octanoylation at Ser^3 and Gln^{13}-Arg^{28}, respectively (Hosoda et al., 2000).
2. RIA incubation mixtures contained 100 μl of either standard ghrelin or an unknown sample with 200 μl of antiserum diluted in the RIA buffer: 50 m*M* sodium phosphate buffer, pH 7.4; 0.5% BSA; 0.5% Triton X-100; 80 m*M* NaCl; 25 m*M* EDTA-2Na; 0.05% NaN_3; and 0.5% normal rabbit serum.
3. Anti-N- and C-terminal rat ghrelin antisera were used at final dilutions of $1{:}1.5 \times 10^6$ and $1{:}1 \times 10^4$, respectively.
4. After a 12-h incubation at 4 °C, 100 μl ^{125}I-labeled ligand (20,000 cpm) was added, and the incubation was continued for an additional 36 h. Then, 100 μl of anti-rabbit goat antibody was added and the samples were incubated for 24 h at 4 °C.
5. Free and bound tracers were separated by centrifugation at 3000 rpm for 40 min. The radioactivity in the pellet was quantified in a γ-counter (ARC-600, Aloka, Tokyo, Japan). All assays were performed in duplicate.
6. Both antisera exhibited complete cross-reactivity with human, mouse, and rat ghrelins. The anti-N-terminal rat ghrelin antiserum, which specifically recognizes the *n*-octanoylated portion of ghrelin, exhibited 100% cross-reactivity with rat, mouse, and human *n*-octanoyl ghrelin but did not recognize des-acyl ghrelin.

7. The anti-C-terminal rat ghrelin antiserum equally recognized both des-acyl and all acylated forms of ghrelin peptides including *n*-hexanoyl, *n*-octanoyl, *n*-decanoyl, *n*-lauroyl, *n*-myristoyl, and *n*-palmitoyl ghrelins.
8. The ED_{50} values for N- and C-terminal ghrelin RIAs were approximately 8 and 32 fmol/tube, respectively. The minimal detection levels by the N- and C-terminal RIAs were 0.25 and 1.0 fmol/tube, respectively. All samples were diluted in the RIA buffer to fit the optimal detection range (between ED_{20} and ED_{80}) for each RIA.
9. Throughout the following sections, the RIA systems using the N- and C-terminal antisera are termed N- and C-RIAs, respectively.

3.2. ELISA of ghrelin

1. An active ghrelin ELISA Kit (Mitsubishi Kagaku Iatron, Inc.) was used to specifically measure *n*-octanoyl ghrelin, and a des-acyl ghrelin ELISA Kit (Mitsubishi Kagaku Iatron, Inc.) was used to specifically assess des-acyl ghrelin.
2. Samples produced by *in vitro* GOAT enzymatic assays were directly used for the ghrelin ELISAs.

3.3. HPLC analysis of *n*-octanoyl-modified ghrelin

We analyzed the molecular forms of ghrelin in the GOAT reaction to confirm that the product was a ghrelin peptide modified by *n*-octanoyl acid.

1. The reaction products were desalted with Sep-Pak C18 cartridges (Waters, Milford, MA) that had been pre-equilibrated in 10% acetonitrile (CH_3CN)/0.1% trifluoroacetic acid (TFA). The samples were loaded onto the cartridges, which were then washed with 10% CH_3CN/0.1% TFA. The peptides were eluted with 60% CH_3CN/0.1% TFA.
2. Next, the eluate was lyophilized. Then, the peptides were resuspended in 10% CH_3CN/0.1% TFA and separated by reverse-phase HPLC (RP-HPLC) using a μBondasphere C18 column (3.9 × 150 mm; Waters) and a linear 40-min gradient of 10–60% CH_3CN in 0.1% TFA at 1 ml/min. Fractions (500 μl) were collected throughout the gradient.
3. The fractions were lyophilized and subjected to ghrelin analysis by RIA or ELISA.
4. To determine the retention times of the standard ghrelin peptides, human *n*-octanoyl ghrelin and des-acyl ghrelin, the standards were mixed with the GOAT assay solution and (without performing a reaction) the mixture was immediately desalted with a Sep-Pak C18 cartridge, lyophilized, and resuspended in 10% CH_3CN/0.1% TFA.

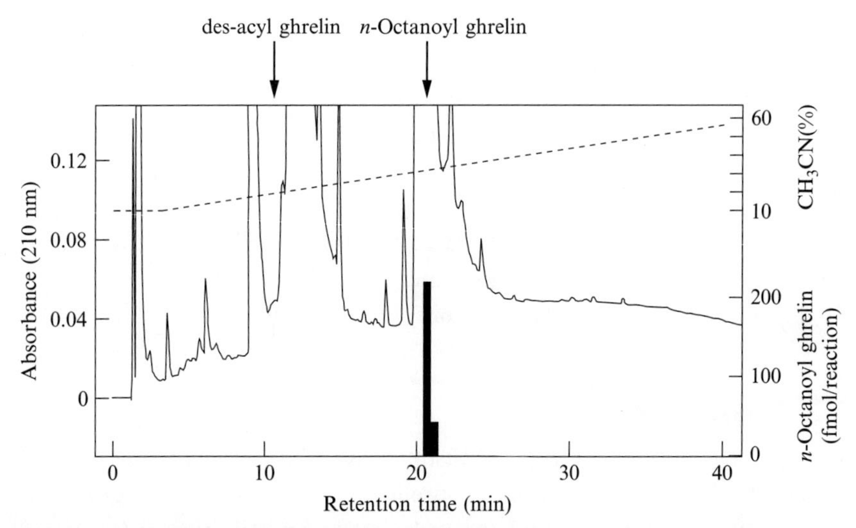

Figure 10.4 Characterization of ghrelin peptides produced by *in vitro* GOAT enzymatic reactions. RP-HPLC of GOAT reaction products was monitored by an ELISA specific for *n*-octanoyl ghrelin. The black bars indicate *n*-octanoyl ghrelin immunoreactivity. The gradient profile is indicated by the dotted line. The standard eluted positions of des-acyl ghrelin and *n*-octanoyl ghrelin are indicated by arrows.

5. Then, the peptide standards were separated by RP-HPLC and assayed by two ELISAs: the active ghrelin ELISA Kit (specific for the *n*-octanoyl peptide) and the des-acyl ghrelin ELISA Kit (specific for des-acyl peptide).
6. Standard *n*-octanoyl human ghrelin eluted in fraction 42 (at 20.5–21.0 min) and des-acyl ghrelin eluted in fraction 22 (at 10.5–11.0 min). Figure 10.4 shows the HPLC analysis of the reaction products from GOAT activity assays. Each fraction was measured by an active ghrelin ELISA. We detected immunoreactivity in fraction 42, the same elution position as the *n*-octanoyl ghrelin peptide standard. Thus, the ghrelin peptide produced by GOAT *in vitro* was identified as *n*-octanoyl ghrelin (1–28).

4. ENZYMATIC CHARACTERIZATION OF GOAT

4.1. Effect of detergents on GOAT activity

Subcellular fractionation revealed that *n*-octanoyltransferase activity partitioned with the cell membrane fraction, confirming that GOAT is a membrane-bound enzyme as suggested in the original GOAT papers (Gutierrez et al., 2008; Yang et al., 2008). We examined the effects of detergents on GOAT activity.

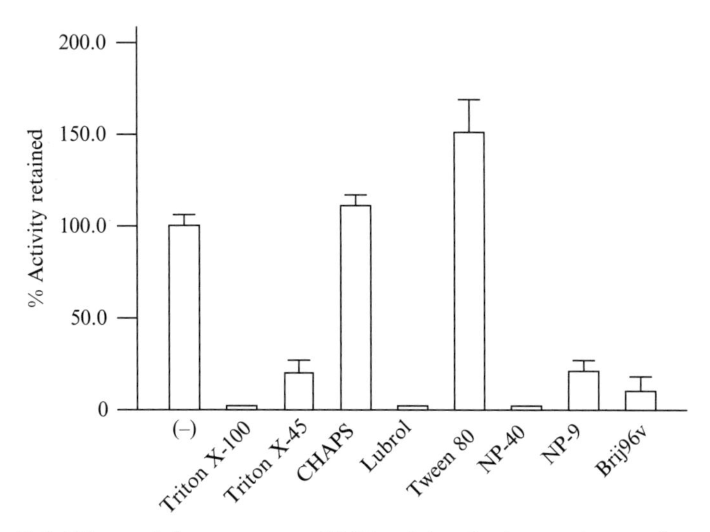

Figure 10.5 Effects of detergents on GOAT activity. Crude membranes from GOAT-expressing CHO cells were solubilized in 1% of the indicated detergent for 15 min at 37 °C. Then, the *n*-octanoyl transferase reaction was performed in the standard assay solution containing 0.1% of the indicated detergent.

1. Eight detergents were used: Triton X-100, Triton X-45, CHAPS, Lubrol, Tween 80, NP40, NP9, and Brij96v. These detergents were prepared in water to 5.0%.
2. The enzyme preparation from GOAT-expressing CHO cells was incubated in 0.1% detergent for 15 min at 37 °C. Then, the GOAT activity assay was performed according to the standard method (Section 2.2). We found that treatment of the membrane fraction with CHAPS or Tween 80 retained GOAT activity (Fig. 10.5). By contrast, treatment with the six other detergents attenuated or abolished GOAT activity. These results indicated that CHAPS and Tween 80 stabilized the conformation of GOAT and were useful for solubilization of the enzyme.

4.2. Exploration of GOAT acyl donors

To investigate whether GOAT utilizes only *n*-octanoyl-CoA as an acyl donor, we assessed the ability of recombinant GOAT to acylate ghrelin using a variety of *n*-acyl-CoAs, including *n*-hexanoyl-, *n*-decanoyl-, *n*-palmitoyl-, and *n*-myristoyl-CoA (Ohgusu et al., 2009). The reaction conditions were the same as for the standard reaction mixture except for the substitution of *n*-acyl-CoA substrates in place of *n*-octanoyl-CoA. The reaction products were subjected to HPLC analyses to confirm the molecular structure by comparison with synthetic standard peptides. To monitor the retention time of acyl-modified ghrelins, we used C-RIA, which recognizes the C-terminal portion of the ghrelin peptide and is not affected by acyl modifications.

1. Fatty acid-CoAs used for the acyl donor studies were purchased from Sigma-Aldrich Co.. These fatty acid-CoAs were used at the standard assay concentration of 10 μ*M*.
2. The reaction products were desalted with Sep-Pak C18 cartridges and subjected to RP-HPLC after evaporation of acetonitrile and resuspension in 10% CH_3CN/0.1% TFA. A 40-min linear gradient of 10–60% CH_3CN in 0.1% TFA was used at 1 ml/min. Fractions (500 μl) were collected throughout the gradient.
3. The fractions were lyophilized and analyzed by anti-ghrelin C-RIA. Ghrelin immunoreactive fractions were compared with the standard elution times of synthetic *n*-acyl-modified ghrelins. We found that, in addition to *n*-octanoyl CoA, GOAT modified the des-acyl ghrelin peptide with other medium-chain acyl acids, such as *n*-hexanoyl-CoA (Fig. 10.6A) and *n*-decanoyl-CoA (Fig. 10.6B). By contrast, long-chain fatty acids were not used as acyl donors. Next, we conducted kinetic studies using des-acyl ghrelin and three medium-chain acyl-CoAs as the donor substrates: *n*-hexanoyl-, *n*-octanoyl-, and *n*-decanoyl-CoA (Fig. 10.6C and D). Increasing the acyl-CoA concentrations resulted in increasing GOAT activity. The order of substrate preference as evaluated by V_{max}/K_m was *n*-hexanoyl-CoA > *n*-octanoyl-CoA > *n*-decanoyl-CoA. The K_m values of *n*-hexanoyl-CoA and *n*-octanoyl-CoA were 294 and 13.6 μ*M*, respectively. We could not calculate the K_m values of *n*-decanoyl-CoA because the concentration of the produced *n*-decanoyl ghrelin was very low.

4.3. Substrate specificity of GOAT for ghrelin peptides

We examined the peptide substrate specificity of GOAT (Ohgusu et al., 2009). Synthetic peptide substrates were derived from the N-terminal sequence of mammalian ghrelin. The length of these substrates (four to eight amino acids) was shorter than that of des-acyl ghrelin, and the C-terminus of the peptide substrates had an α-amide structure. N-RIA was used for the detection of *n*-octanoyl-modified peptides.

1. Short (four to eight amino acids) synthetic ghrelin peptides and *n*-octanoyl-CoA were used as the peptide and acyl donor substrates, respectively. The synthetic peptides were GSSF-NH_2, GSSFL-NH_2, GSSFLK-NH_2, GSSFLSP-NH_2, and GSSFLSPK-NH_2.

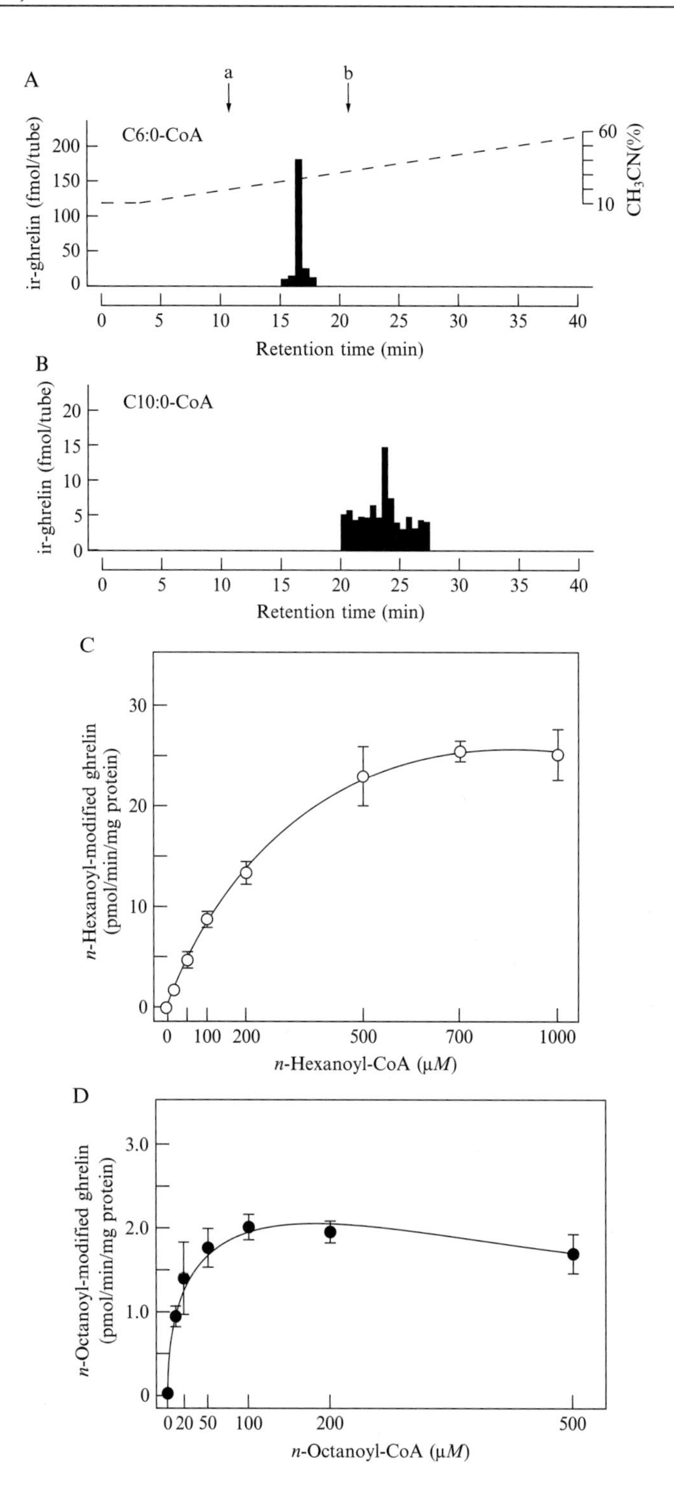
A
a
b
C6:0-CoA
ir-ghrelin (fmol/tube)
200
150
100
50
0
60
10
CH_3CN(%)
0 5 10 15 20 25 30 35 40
Retention time (min)
B
C10:0-CoA
ir-ghrelin (fmol/tube)
20
15
10
5
0
0 5 10 15 20 25 30 35 40
Retention time (min)
C
n-Hexanoyl-modified ghrelin (pmol/min/mg protein)
30
20
10
0
0 100 200 500 700 1000
n-Hexanoyl-CoA (μ*M*)
D
n-Octanoyl-modified ghrelin (pmol/min/mg protein)
3.0
2.0
1.0
0
0 20 50 100 200 500
n-Octanoyl-CoA (μ*M*)

2. The reaction products were directly subjected to RP-HPLC analysis using a 40-min linear gradient of 10–60% CH_3CN in 0.1% TFA at 1 ml/min. Fractions (500 μl) were collected throughout the gradient.
3. Fractions were lyophilized, reconstituted in RIA buffer, and analyzed by N-RIA. Figure 10.7 shows the HPLC retention times of the five synthetic ghrelin-derived substrates. The retention times of the reaction products were delayed for all of the peptide substrates assayed. The retention times were as follows: 23.5–24.5 min, GSSF-NH_2; 26.0–26.5 min, GSSFL-NH_2; 23.0–23.5 min, GSSFLK-NH_2; 25.0–25.5 min, GSSFLSP-NH_2; and 23.5–24.0 min, GSSFLSPK-NH_2. Moreover, the synthetic *n*-octanoyl-modified peptide GSS(C8:0)FLSPK-NH_2 had the same retention time as that of the GOAT-reacted GSSFLSPK-NH_2 product (data not shown). These results indicated that these five peptides served as GOAT substrates that were modified by *n*-octanoic acid. Furthermore, peptides as short as four amino acids were sufficient to serve as GOAT substrates.

4.4. Optimal temperature and pH for GOAT activity

The optimal reaction temperature was from 37 to 50 °C (Fig. 10.8A). GOAT activity was retained at 55 °C but was abolished above 60 °C. The specific activity of GOAT was determined over a range of pH values (Fig. 10.8B) using the following buffers: 50 m*M* MES (pH 5–7), Tris–HCl (pH 7.5–8.5), and $NaHCO_3$ (pH 9–10). The optimal pH for maximal specific activity was pH 7.0–7.5. The specific activity dropped off rapidly below pH 6.5 and above pH 8.5.

4.5. Effects of cations on GOAT activity

The effects of various cations on GOAT activity were compared (Fig. 10.9). Recombinant GOAT was activated by low concentrations but inhibited by high concentrations of Mg^{2+} and Ca^{2+}. By contrast, Fe^{3+} and Cu^{2+}

Figure 10.6 Acyl-CoA specificity of GOAT. The acyl-CoA specificity of GOAT toward (A) *n*-hexanoyl-CoA (C6:0) and (B) *n*-decanoyl-CoA (C10:0) was analyzed by incubating 0.5 μ*M* rat des-acyl ghrelin in the presence of 1 μl enzyme solution. Two reactions for each acyl-CoA were performed; these reactions were pooled for HPLC analyses. The reaction products were subjected to HPLC, and each fraction was assayed for immunoreactive ghrelin by ghrelin C-RIA. The arrows indicate the eluted positions of (a) des-acyl ghrelin and (b) *n*-octanoyl ghrelin. (C and D) Kinetic studies of recombinant GOAT. GOAT assays were performed by incubating increasing concentrations of (C) *n*-hexanoyl-CoA (C6:0), (D) *n*-octanoyl-CoA (C8:0), and (not shown) *n*-decanoyl-CoA (C10:0) under the same assay conditions. The concentrations of acylated ghrelin products were measured using ghrelin C-RIA after HPLC. The K_m values were calculated from these results.

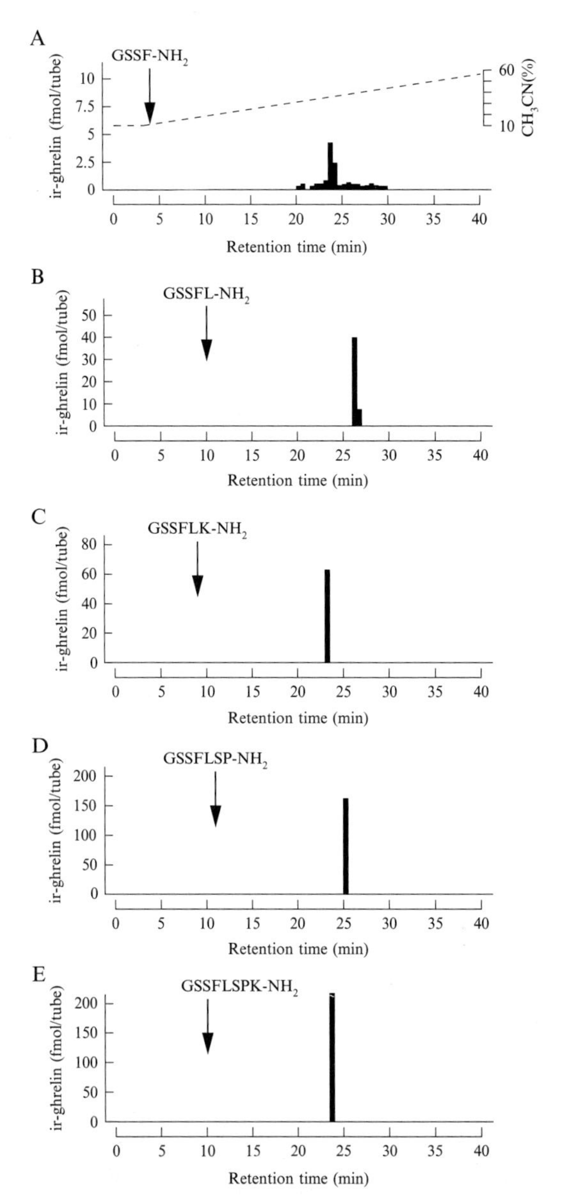

Figure 10.7 Analysis of GOAT substrate specificity. HPLC analyses of GOAT reaction products when (A) GSSF-NH_2, (B) GSSFL-NH_2, (C) GSSFLK-NH_2, (D) GSSFLSP-NH_2, and (E) GSSFLSPK-NH_2 were used as substrates. Reaction products were subjected to HPLC and each fraction was assayed for immunoreactive *n*-octanoyl ghrelin by N-RIA. The eluted positions of peptide substrates are indicated by arrows.

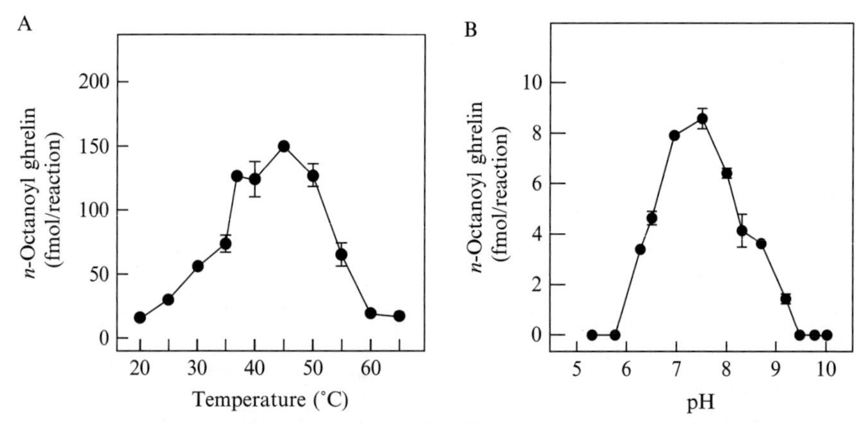

Figure 10.8 Optimal temperature and pH for GOAT activity. *n*-Octanoyl ghrelin concentrations were measured by the active ghrelin ELISA kit. Results are expressed as the mean $\pm$ SD ($n=3$). (A) Temperature dependence of GOAT activity. (B) pH dependence of GOAT activity.

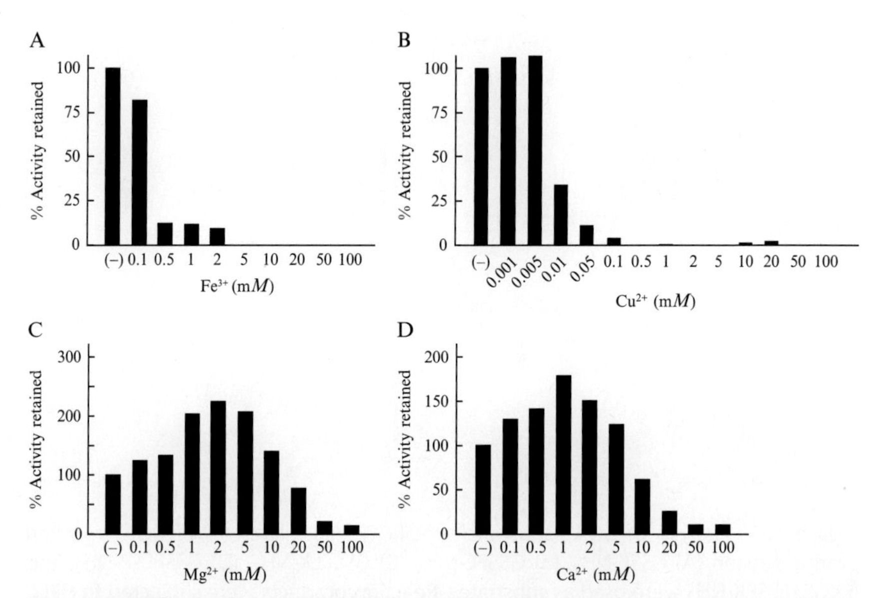

Figure 10.9 Effects of cations on GOAT activity. Enzymatic activity was measured using standard assay conditions with the addition of cations. The cations used were (A) $FeCl_3$, (B) $CuCl_2$, (C) $MgCl_2$, and (D) $CaCl_2$. Activity is expressed as the percent activity retained after cation treatment as compared to standard assay conditions. Data represent the average of two independent experiments.

potently inhibited GOAT activity. Specifically, GOAT activity was completely blocked over 5 m*M* Fe^{3+} and 0.5 m*M* Cu^{2+}. EDTA and EGTA did not affect GOAT activity, indicating that the enzyme had no absolute requirement for cations.

5. ALTERATIONS IN GOAT MRNA EXPRESSION IN THE STOMACH UNDER FASTING CONDITIONS

The most important factor regulating ghrelin expression in the stomach is how recently one has eaten. Ghrelin mRNA expression levels increased during fasting and decreased after refeeding. To examine the relationship of GOAT production to fasting/fed conditions, we investigated changes in the expression of GOAT mRNAs in the rat stomach after fasting and refeeding (Takahashi et al., 2009).

1. To measure GOAT mRNA expression levels in the rat stomach, real-time PCR was performed using a PRISM 7000 Sequence Detection system (PE Applied Biosystems, Foster City, CA, USA).
2. cDNA amplification was performed using SYBR Green PCR Core Reagents (PE Applied Biosystems). All samples were amplified on a single MicroAmp Optional 96-well reaction plate (PE Applied Biosystems). Results reflect duplicate values from at least two independent experiments.
3. Primer sequences used for PCR analysis were as follows: rat ghrelin, sense 5′-GAAGCCACCAGCTAAACTGC-3′ and antisense 5′-GCTGCTGGTACTGAGCTCCT-3′; rat GOAT, sense 5′-TTTGTATCCCAGTATCTCTTTCTGG-3′ and antisense 5′-CCAGTGGGAGTAGTAGGTGAGTTTA-3′.
4. After an initial 15 min at 95 °C to activate the HotStarTaq DNA polymerase, PCR fragments were amplified by 40 cycles as follows: 94 °C for 30 s, 60 °C for 30 s, and 72 °C for 30 s. Standard wells contained a TOPO vector (Invitrogen) bearing the standard cDNA fragment. The concentration of standards covered at least six orders of magnitude.
5. We also included no-template controls on each plate. Experimental samples with a threshold cycle value within 2 SD of the mean threshold cycle value of the no-template controls were considered to be below the limits of detection.
6. The relative mRNA levels were standardized to a housekeeping gene, namely, glyceraldehyde-3-phosphate dehydrogenase, to correct for any bias among the samples caused by RNA isolation, RNA degradation, or efficiencies of the reverse transcriptase.

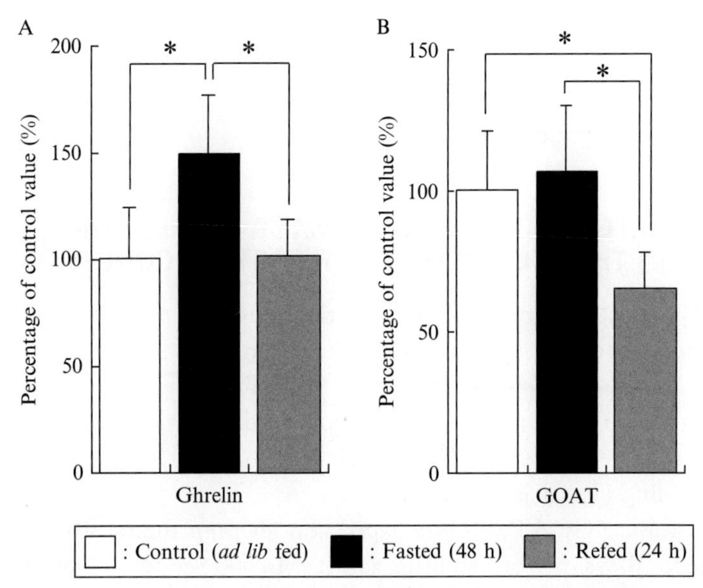

Figure 10.10 Real-time PCR analysis for mRNA levels in the stomach of rats fed *ad libitum* (control), 48-h fasted rats, or rats fasted for 48 h and refed. GAPDH was used as the internal control. Control values (*ad lib fed*) were normalized to 100%. (A) Ghrelin; (B) GOAT mRNA levels. Results are expressed as mean $\pm$ SD ($n=12$). Asterisks indicate the differences between each group ($P<0.05$).

7. After amplification, PCR products were analyzed according to their melting curve to confirm amplification specificity. Amplicon size and reaction specificity were confirmed by agarose gel electrophoresis. Ghrelin mRNA expression in the stomach after fasting for 48 h was significantly increased (by 49%) as compared to the *ad libitum* fed control (Fig. 10.10A). GOAT expression levels after fasting were not significantly changed as compared to the control (Fig. 10.10B). Although ghrelin mRNA expression returned to control levels after refeeding, the expression levels of GOAT were significantly decreased.

6. CONCLUSION

GOAT is a member of an acyltransferase family that comprises at least 16 enzymes (Hofmann, 2000). Among them, only GOAT shows the ability to acyl-modify ghrelin. Just as ghrelin is observed in all vertebrate species, GOAT is also found in mammals, birds, and fish (Yang et al., 2008). A thorough characterization of GOAT is an important step in understanding the molecular mechanisms underlying the acyl modification of ghrelin.

Recently, a GOAT-specific inhibitor was developed and demonstrated to improve glucose tolerance and suppress weight gain in wild-type mice but not in ghrelin-KO mice (Barnett et al., 2010). These results suggested that GOAT could be a clinical target for metabolic diseases.

REFERENCES

Barnett, B.P., et al., 2010. Glucose and weight control in mice with a designed ghrelin *O*-acyltransferase inhibitor. Science 330, 1689–1692.

Gutierrez, J.A., et al., 2008. Ghrelin octanoylation mediated by an orphan lipid transferase. Proc. Natl. Acad. Sci. USA 105, 6320–6325.

Hofmann, K., 2000. A superfamily of membrane-bound *O*-acyltransferases with implications for wnt signaling. Trends Biochem. Sci. 25, 111–112.

Hosoda, H., et al., 2000. Ghrelin and des-acyl ghrelin: two major forms of rat ghrelin peptide in gastrointestinal tissue. Biochem. Biophys. Res. Commun. 279, 909–913.

Kojima, M., Kangawa, K., 2005. Ghrelin: structure and function. Physiol. Rev. 85, 495–522.

Kojima, M., et al., 1999. Ghrelin is a growth-hormone-releasing acylated peptide from stomach. Nature 402, 656–660.

Kojima, M., et al., 2008. Structure of mammalian and nonmammalian ghrelins. Vitam. Horm. 77, 31–46.

Ohgusu, H., et al., 2009. Ghrelin *O*-acyltransferase (GOAT) has a preference for n-hexanoyl-CoA over n-octanoyl-CoA as an acyl donor. Biochem. Biophys. Res. Commun. 386, 153–158.

Takahashi, T., et al., 2009. Production of n-octanoyl-modified ghrelin in cultured cells requires prohormone processing protease and ghrelin *O*-acyltransferase, as well as n-octanoic acid. J. Biochem. 146, 675–682.

Yang, J., et al., 2008. Identification of the acyltransferase that octanoylates ghrelin, an appetite-stimulating peptide hormone. Cell 132, 387–396.

CHAPTER ELEVEN

The Study of Ghrelin Deacylation Enzymes

Motoyasu Satou, Hiroyuki Sugimoto[1]
Department of Biochemistry, Dokkyo Medical University School of Medicine, Mibu, Tochigi, Japan
[1]Corresponding author: e-mail address: h-sugi@dokkyomed.ac.jp

Contents

Abstract

Like other posttranslational modifications, fatty acid modification of amino acid residues in peptide chains is a critical determinant of their functional properties. A unique feature of ghrelin is the attachment of an acyl moiety at the third serine residue. Ghrelin is a hormone present in the circulation with roles in the release of growth hormone, control of behaviors related to appetite, and diverse cellular functions. Although lipid modification of ghrelin is essential for its binding to the ghrelin receptor, several lines of evidence suggest that deacylated ghrelin has physiological activity or activities similar to and distinct from the activities of the acylated form. Therefore, the understanding of deacylating process of ghrelin *in vivo* is key to accepting the physiological importance of ghrelin. In this review, we summarize results and methodology relevant to our recent efforts to determine the molecular mechanisms involved in ghrelin processing, including (1) immunological and mass spectrometry-based detection of ghrelin, (2) quantification of ghrelin deacylase activity, and (3) characterization of ghrelin deacylation enzymes isolated from biological fluids and using heterologous expression systems.

ABBREVIATIONS

A2M alpha-2-macroglobulin
APT acyl-protein thioesterase
DHHC-CRD Asp-His-His-Cys-cysteine-rich domain

Methods in Enzymology, Volume 514
ISSN 0076-6879
http://dx.doi.org/10.1016/B978-0-12-381272-8.00011-8

ELISA enzyme-linked immunosorbent assay
FBS fetal bovine serum
GH growth hormone
GHSR1a growth hormone secretagogue receptor 1a
GOAT ghrelin O-acyltransferase
GST glutathione *S*-transferase
HHAT hedgehog acyltransferase
MALDI-TOF MS matrix-assisted laser desorption/ionization mass spectrometry
MBOAT membrane-bound O-acyltransferase
MS mass spectrometry
PAFAH platelet activating factor acetylhydrolase
RIA radioimmunoassay
RP-HPLC reverse-phase high-performance liquid chromatography

1. INTRODUCTION

Processing of precursor hormones is generally thought to be a prerequisite for their physiological effects on the target tissues and organs. In the production of mature ghrelin (28 amino acids), proghrelin polypeptide (in humans, a 94-amino acid polypeptide) is cleaved by prohormone convertase 1/3, followed by addition of an acyl moiety, commonly octanoic acid, at the third serine by ghrelin O-acyl transferase (GOAT) (Yang et al., 2008; Zhu et al., 2006). Lipid modification of ghrelin is essential for binding to its specific receptor (GHS-R1a) and subsequent downstream responses, for example, GH release, appetite stimulation, and cell proliferation (Kojima and Kanagawa, 2005). The membrane-bound O-acyltransferase (MBOAT) family consists of more than 16 members including GOAT, Hedgehog acyltransferase (HHAT), and Porcupine (Chang and Magee, 2009; Table 11.1). Emerging *in vivo* and *in vitro* evidence regarding lipid modifications of polypeptides shows that two large families of acyltransferases, namely, MBOATs and the Asp-His-His-Cys cysteine-rich domain (DHHC-CRD) family of proteins (Resh, 2006), are important for the acylation of polypeptides. By contrast, the delipidation machinery for acylated polypeptides is yet to be elucidated, with the exception of acyl-protein thioesterase 1 (APT1) (Rocks et al., 2010; Satou et al., 2010).

Although intracellular maturation of ghrelin by proteolytic cleavage and acylation is indispensable for secretion of the active form of ghrelin, the dominant form of ghrelin in circulation is the deacylated form. The importance of acylation of ghrelin for physiological activity was at first commonly

Table 11.1 Properties and functions of lipid-modified secretory peptides and proteins

				Enzymes		
	Function	Modification	Position	Lipidation	Delipidation	References
Ghrelin	GH release, appetite, gastric movement	O-acylation	N-term	GOAT	APT1 and undefined plasma esterase(s)	Satou et al. (2010) Shanado et al. (2004) Yang et al. (2008)
Hedgehog	Neuronal differentiation, tumorigenesis	*S*-palmitoylation	N-term	HHAT	Undefined	Pepinsky et al. (1998) Buglino and Resh (2008)
Spitz	Embryonic development, as an EGFR ligand	*S*-palmitoylation	N-term	HHAT	Undefined	Miura et al. (2006)
Wnt	Cell polarity, tumorigenesis, mitochondrial biogenesis	*S*-palmitoylation	Internal	Porcupine	Undefined	Mikels and Nusse (2006) Schulte et al. (2005) Yoon et al. (2010)

accepted. Recently, however, the des-acyl form of ghrelin has been recognized as a physiologically important peptide that can modulate behavioral reactions (Asakawa et al., 2005; Inhoff et al., 2008) as well as cellular responses such as proliferation (Delhanty et al., 2006) and inflammation (Bulgarelli et al., 2009). Only the des-acyl form of ghrelin is detectable by immunocytochemistry in some of ghrelin-producing cells in the stomach. Moreover, the amount of des-acyl ghrelin secreted by the stomach is comparable to the amount of acyl ghrelin it secretes (Mizutani et al., 2009). Taken as a whole, the available evidence suggests that controlling the secretion of acyl versus des-acyl ghrelin, as well as deacylation of ghrelin already in circulation, is important. Both mechanisms may pleiotropically modulate food intake, energy homeostasis, and diverse cellular functions. In this chapter, we summarize the results of studies aimed at detection and characterization of ghrelin derivatives, as well as our recent findings relevant to enzymes that can generate the des-acyl form of ghrelin in circulation and cells.

2. PROCESSING OF GHRELIN

Ghrelin has been shown to be broken down both in circulation and in tissues, although through different mechanisms (De Vriese et al., 2004). The authors showed that exposure of acylated ghrelin to human serum can result in the hydrolysis of the acyl moiety of ghrelin and that the peptide bonds formed remain stable for at least several hours. However, tissue homogenates from rat stomach, kidney, and liver had high endopeptidase activity against ghrelin, especially for the N-terminal "active core" (Bednarek et al., 2000). In our laboratory, we obtained nearly equivalent results as these authors using animal sera (unpublished observations by M. Satou), but the peptide bonds formed by ghrelin were relatively unstable following treatment with bovine plasma (Satou et al., 2011). As mentioned above, processed forms of ghrelin appear to have different and relevant physiological effects. However, the interaction of these various forms with receptors and/or adaptors and subsequent signal transduction remain to be elucidated (Bulgarelli et al., 2009).

Previously, platelet activating factor (PAF) acetylhydrolase (De Vriese et al., 2004), carboxypeptidase (De Vriese et al., 2004), and cholinesterase (Dantas et al., 2011; De Vriese et al., 2007), which are the major components of esterases in serum, have been reported to exhibit ghrelin deacylase activity. More recently, we purified and characterized APT1 as a serum ghrelin deacylase (Satou et al., 2010), and Eubanks et al. reported that alpha-2-macroglobulin (A2M) has ghrelin deacylating activity (Eubanks et al., 2011). In the following sections, we discuss the various candidate deacylation enzymes present in circulation, including a discussion of biochemical approaches useful for studying them. Although much remains to be understood, recent advances in the field have shed light on the physiological significance of ghrelin-derived peptides.

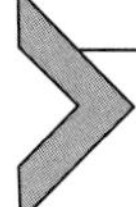

3. DETERMINATION OF GHRELIN DEACYLATION ACTIVITY

3.1. Detection of ghrelin

Immunoblotting is one of the most popular methods for detection of polypeptides. However, it is not easy to distinguish the precise structural differences between acylated and des-acylated ghrelin by separation with SDS-PAGE because of their low molecular weight (<4000 Da) (Fig. 11.1A). Tricine SDS-PAGE (Schägger, 2006) is recommended for the separation of small proteins and peptides including ghrelin (Fig. 11.1A–C). As shown in Fig. 11.1D, standard SDS-PAGE (acrylamide $>15\%$ w/v) can be used to detect ghrelin. When this is performed following mixing of plasma proteins with ghrelin and gel filtration, we could detect ghrelin in a higher molecular fraction (dotted line) as compared with the predicted mobility (solid line). These results suggest that ghrelin is simultaneously eluted with unknown ghrelin-associating macromolecules in plasma. Nevertheless, detection requires an immunoblotting system that can efficiently transfer small peptides to PVDF membranes. Although the use of denaturing conditions reduces its specificity, the antibody against ghrelin peptide is reactive with the ghrelin precursor in this context (Fig. 11.1B). None of the antibodies we have tried could distinguish the acylated from the des-acylated form of ghrelin (Fig. 11.1A), and the antibodies also failed to react with truncated forms (Satou et al., 2011; unpublished observation by M. Satou). Shanado et al. (2004) used a specific antibody for acylated ghrelin under denaturing conditions for purification of the ghrelin deacylating enzyme from stomach. Nishi and co-workers successfully obtained antibodies specific for the decanoyl-bound form of ghrelin for radioimmunoassay (RIA) (Hiejima et al., 2009; Nishi et al., 2011). RIA is very useful for the detection of small amounts of ghrelin in tissues (i.e., sub-fmol levels). The use of high-specificity antibodies against ghrelin would likely be a significant step forward. These reagents could presumably be useful not just for immunoblotting but also for RIA and enzyme-linked immunosorbent assay (Rauh et al., 2007).

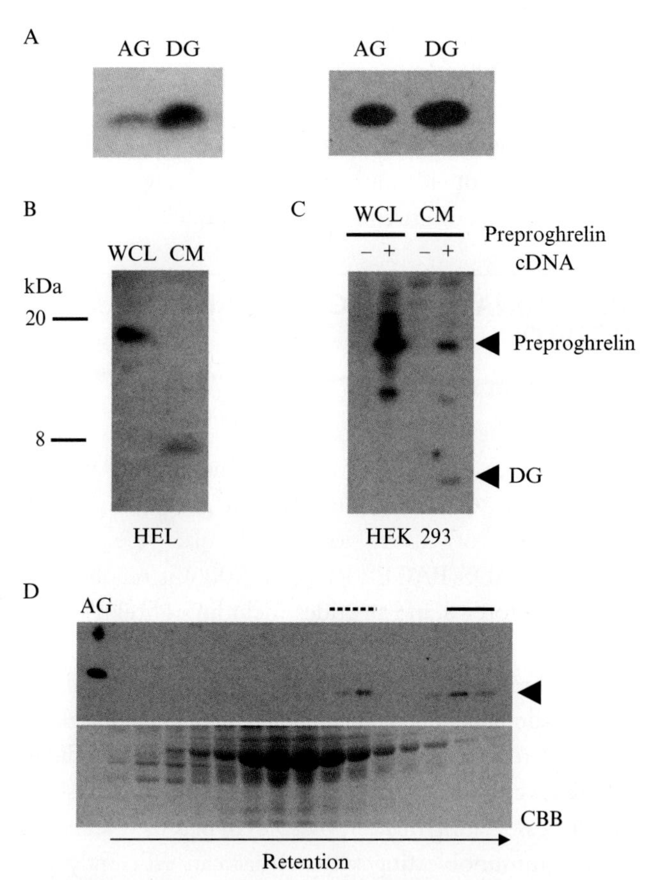

Figure 11.1 Detection of ghrelin by immunoblotting. Synthetic human acylated or des-acylated ghrelin is detected by commercially available antighrelin antibody from Alpha Diagnostic (left, 105 pmol/well) and Santa Cruz Biochemistry (right, 5 pmol/well). Note that acyl ghrelin (AG) reacts more weakly than the des-acyl form (DG) (A). Endogenous ghrelin can be detected in human erythroleukemia (HEL) cells, whole cell lysate (WCL), and conditioned medium (CM) (B). Human preproghrelin (NM_016362.3) can be heterologously overexpressed in HEK293 cells (C). These cells express endogenous ghrelin processing enzymes, that is, prohormone convertases but not the ghrelin acylating enzyme GOAT. Secreted ghrelin might be the des-acylated form. Bovine plasma mixed with synthetic human ghrelin was separated by Sephacryl S-300 (D). Immunoreacted ghrelin (arrowhead) was detected in molecular fractions higher than expected (dotted line), as compared with its predicted mobility (continuous line). AG indicates the control sample, acylated ghrelin without plasma. The lower panel shows Coomassie Brilliant Blue staining of separated plasma proteins. The blots were separated by Tricine-PAGE (A, B, C) or Tris-PAGE (D) prior to transfer to PVDF membranes.

Mass spectrometry (MS) is beneficial for analyzing ghrelin processing *in vitro*. The utility of MS for the determination of the molecular mass and levels of the degraded form of ghrelin was verified in our previous work (Satou et al., 2010). After synthetic octanoyl ghrelin was incubated with fetal bovine serum (FBS) for 30 min, ghrelin derivatives were adsorbed and concentrated with a Sep-Pak^{TM} C18 column and then eluted with acetonitrile. The eluates were analyzed using matrix-assisted laser desorption/ionization

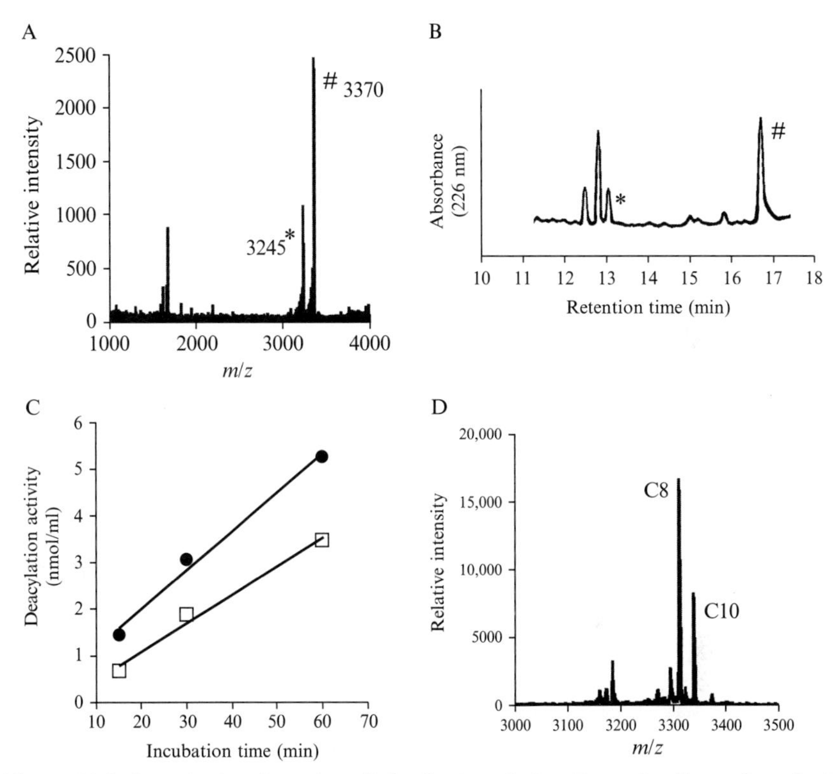

Figure 11.2 Quantitative detection of ghrelin deacylation. Two microliters of synthetic octanoyl ghrelin (136 pmol) and 1 μl of FBS were dissolved with 20 m*M* Tris–HCl (pH 7.5) containing 0.02% Triton X-100 (w/v) up to a final volume of 10 μl. The reaction mixture was incubated for 30 min at 37 °C and submitted to MADLI-TOF MS (A) or higher amount to RP-HPLC (B). The peaks obtained by MS were assigned theoretically to the corresponding values for ghrelin fragments: acyl (#*m/z*=3370) and des-acyl (**m/z*=3245) ghrelin. RIA followed by RP-HPLC (Nishi et al., 2005) revealed the presence of acyl (#) and des-acyl (*) ghrelin. A time course of ghrelin deacylation was evaluated by MS (closed circles) and RH-HPLC (white squares) (C). Equimolar amounts of mouse octanoyl (C8) and decanoyl (C10) ghrelin were mixed and incubated with mouse plasma. The reaction mixture was analyzed by MS as above (D). Fig. 11.2A–C was modified from Satou et al. (2010) with permission.

time-of-flight mass spectrometry (MALD-TOF MS). The molecular mass of human ghrelin is 3370, and following deoctanoylation, the size is reduced to 3245 (Fig. 11.2A). Deacylation activity can be calculated as the intensity 3245 (m/z) divided by the total (3370 + 3245) (m/z). Following this analysis, we confirmed that the MS data were comparable to those of reverse-phase high-performance liquid chromatography (RP-HPLC) followed by RIA (Fig. 11.2A–C; Date et al., 2000; Sato et al. 2005). Unless otherwise indicated, the ghrelin deacylation activities shown here were estimated using the MS-based method. A significant advantage of using MS to detect ghrelin is that it facilitates the detection and analysis of degraded ghrelin species derived from relatively higher levels of authentic ghrelin (i.e., sub-pmol amounts) following incubation with tissue samples. As it has a higher mass accuracy than SDS-PAGE separation and immunoblotting, MS allows the simultaneous analysis of degraded ghrelin species. As shown in Fig. 11.2D, the results indicate that, after incubation with mouse plasma, decanoyl ghrelin is relatively unstable as compared with its octanoyl counterpart. Nevertheless, the sensitivity of MS is not sufficient to determine the endogenous concentration of ghrelin in various tissues.

3.2. Ghrelin deacylation activity in biological samples

Because most circulating ghrelin is des-acylated, it seems reasonable that ghrelin is easily deacylated in circulation. We examined esterase activities against octanoyl ghrelin and butyrylthiocholine in animal sera and plasma (Fig. 11.3A). Cholinesterase activity was assayed following the method of De Vriese et al. (2004). Horse serum (stabilized for cell culture use) contains higher levels of cholinesterase activity than serum or plasma derived from bovines and has no obvious deacylation activity. Although it is not entirely clear what enzymes hydrolyze butyrylthiocholine, in this experiment there does not seem to be any correlation between ghrelin deacylase and cholinesterase activity. Serum esterases, such as platelet activating factor acetylhydrolase (PAFAH) and cholinesterases, have previously been described as the ghrelin deacylases present in blood (De Vriese et al., 2004, 2007; Tham et al., 2009). We performed a partial separation of bovine plasma by gel filtration and characterized the enzymatic properties of the resulting fractions (Fig. 11.3B). Ghrelin deacylation activity detected in fractions 7–10 correlates well with PAFAH activity. However, as we have reported previously, purified PAFAH did not exhibit deacylation activity against ghrelin (Satou et al., 2010).

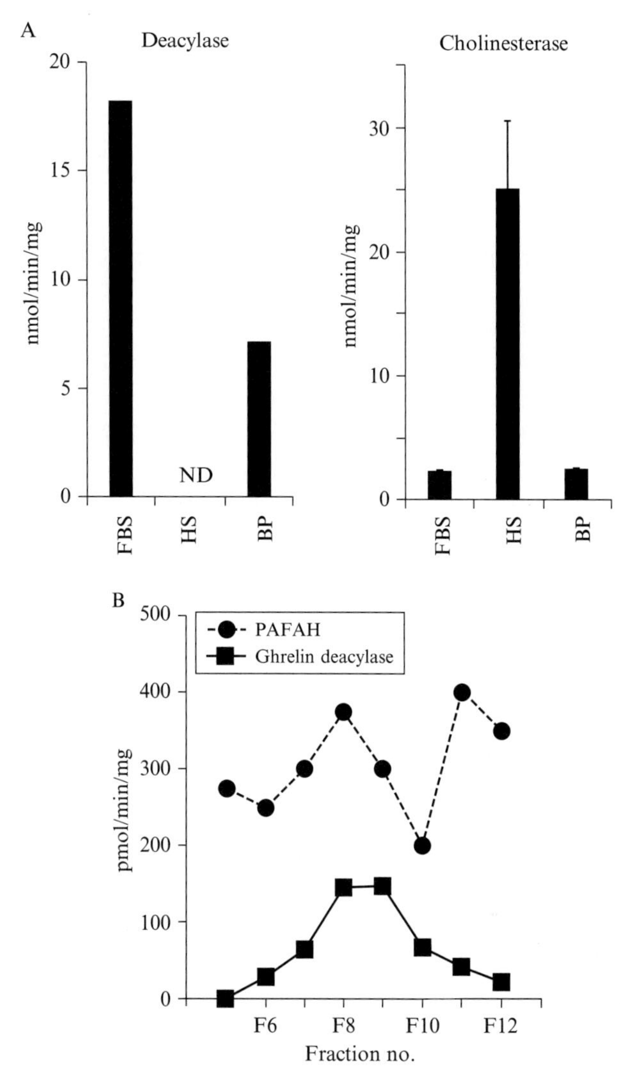

Figure 11.3 Esterase activities in sera and plasma. Enzymatic characterization of fetal bovine serum (FBS), horse serum (HS), and bovine plasma (BP) (A). The samples were submitted to a ghrelin deacylation assay by MS as described in Fig. 11.2. Cholinesterase activity was measured using colorimetric assay based on removal of thiocholine from butyrylthiocholine. Comparison with ghrelin deacylase (closed squares) and platelet activating factor acetylhydrolase (PAFAH) (closed circles) activities in fractions after gel filtration of bovine plasma (B). PAFAH activity was measured using a PAF acetylhydrolase assay kit (Cayman) according to the manufacturer's instructions.

3.3. Ghrelin processing enzymes in cultured cell models

As previously reported, endogenous APT1 can be released from mouse macrophage RAW264.7 cells in response to lipopolysaccharide treatment (Satou et al., 2010). In the case of HepG2 cells, a weak but detectable APT1 ghrelin deacylation activity could be purified from the conditioned medium following several sequential rounds of column chromatography. Heterologous overexpression of target genes is a standard method for the purification and characterization of enzymes. Indeed, GOAT was determined to be the ghrelin acylating enzyme using this approach (Yang et al., 2008). In that study, enzymatic activities for candidates as the ghrelin deacylation enzyme were determined by overexpression in some heterologous cells. Either cDNA encoding human butyrylcholinesterase (BchE, EC 3.1.1.8) or plasma PAFAH (EC 3.1.1) was transfected into Neuro2A cells, and the conditioned medium was collected from each sample. A significant increase in butyrylcholinesterase activity could be detected in BchE overexpressing cell-derived medium; indeed, this level exceeded that observed in animal sera (Fig. 11.4A). However, we could not detect ghrelin deacylase activity in either of these conditioned media (unpublished observations). Although BchE can hydrolyze a broad spectrum of substrates (Balasubramanian and Bhanumathy, 1993), it seems that ghrelin is excluded from its coverage. Very recently, A2M was identified as a ghrelin deacylation enzyme present in rat serum using a chemical crosslinking strategy, and stable trapping of ghrelin by the macromolecule was shown (Eubanks et al., 2011). In an attempt to replicate these results, we overexpressed A2M in Neuro2A cells (Fig. 11.4B). However, we were not able to detect ghrelin deacylation activity in the conditioned media, similar to what we found for conditioned media from cells overexpressing BchE (unpublished observations). The A2M protein is a protease inhibitor that acts via covalent binding to target proteins, which include trypsin and coagulant proteins found in circulation (de Boer et al., 1993; Swenson and Howard, 1979). Additional approaches should be tested in order to examine whether A2M and/or A2M-associated protein(s) exhibit esterase activity, especially for ghrelin deacylation.

Prokaryotic cells are an alternative platform for expression of recombinant proteins. Previously, we expressed glutathione *S*-transferase-fused APT1 in *E. coli* and were able to show that heterologously expressed APT1 can act as a serum ghrelin deacylase (Satou et al., 2010; Fig. 11.4C). This enzyme was originally purified as a cytosolic lysophospholipase from swine stomach (Sunaga et al., 1995) and rat liver (Sugimoto et al., 1996).

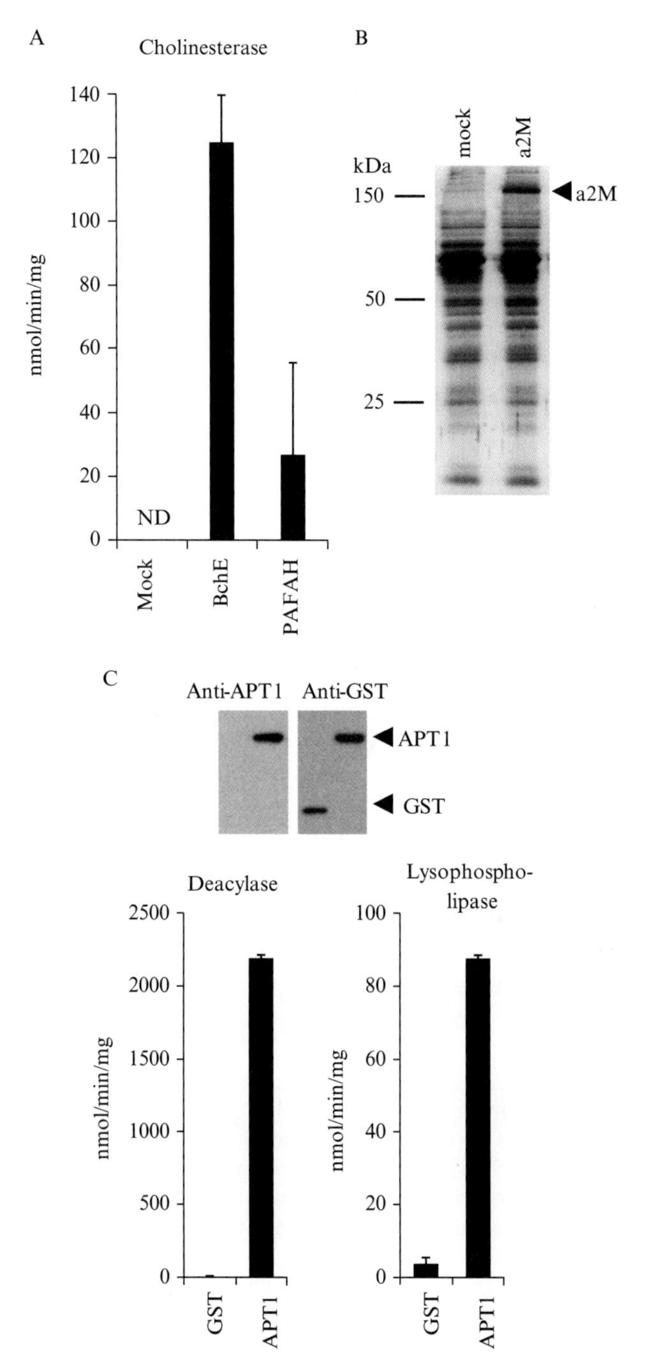

Figure 11.4 Heterologous expression of candidates for the ghrelin deacylation enzyme in Neuro2A cells. Human cDNA encoding cholinesterase (BchE, NM_000055.2), platelet

Consistent with this, we were able to confirm that recombinant APT1 hydrolyzes both lysophosphatidylcholine and acyl ghrelin (Fig. 11.4C). Recently, APT1 was reported to have phospholipase A_1-like activity for the production of *sn-1* lysophospholipids (Bolen et al., 2011). Thus, taken together, the evidence suggests that APT1 is a key deacylating enzyme that can act on a broad spectrum of substrates, including peptides, proteins, and phospholipids.

Conditioned media obtained from normally growing mammalian cell lines generally have a low level of deacylation activity which might be attributable to APT1. Recently, we found that conditioned medium from HepG2 cells contains a ghrelin deacylating activity that is distinct from the activity of APT1. The undetermined, partially purified ghrelin deacylating enzyme(s) from HepG2 conditioned medium shows a higher affinity to anion-exchange DEAE-Sepharose and hydroxyapatite as compared with APT1 (Fig. 11.5A). The presence of two active peaks in the isolated fractions suggests that one or more enzymes are responsible for ghrelin hydrolysis. As shown in Fig. 11.5B, our data suggest that the enzyme(s) can also be distinguished from BchE and A2M (Fig. 11.5B). This work is currently in progress.

4. FUTURE DIRECTIONS

As mentioned previously (Satou et al., 2011), processing of ghrelin, which includes digestion of the peptide chain and deacylation, appears to provide functional diversity. Thus, the molecular mechanisms underlying ghrelin metabolism are an essential subject of study in the field. Other than APT1, there has been no decisive evidence for the presence of a ghrelin deacylase. APT1 is thought to be a "universal" intracellular delipidation enzyme (Hirano et al., 2009; Rocks et al., 2010) that can also act

activating factor acetylhydrolase (PAFAH, NM_005084.3), or alpha-2-macroglobulin (A2M, NM_000014.4) was transfected into Neuro2A cells, and the conditioned medium was collected 48 h later. BchE overexpressing cell-derived conditioned media showed prominent cholinesterase activity (A). Silver staining of conditioned media derived from mock or A2M-transfected cells (B). Arrowhead indicates a band corresponding to the A2M monomer. Glutathione *S*-transferase (GST)-tagged APT1 (NM_013006.1) was expressed in *E. coli* and purified using a glutathione-sepharose column (C). Protein expression was confirmed by immunoblotting using anti-APT1 (Sugimoto et al., 1996; left) and anti-GST (Amersham Pharmacia, right) antibodies. Recombinant APT1 had both ghrelin deacylase and lysophospholipase activities. Figure 11.4C was modified from Satou et al. (2010) with permission.

extracellularly. What has been of particular interest to our group and others is to consider the presence of other enzyme(s) in plasma with ghrelin deacylating activity (Fig. 11.5).

Interestingly, more than half of the ghrelin released from the stomach is in the deacylated form (Mizutani et al., 2009). This finding brings into question whether the levels of the des-acyl form of ghrelin in circulation are increased by deacylation of ghrelin prior to secretion, or whether ghrelin is secreted in a form without cellular lipid modification. To help address this, we will have to determine whether there are alternative lipidation/delipidation enzymes, other than APT1, present in the secretion pathway.

Members of the MBOAT family of proteins are responsible for acylation of the molecules listed in Table 11.1. A unique means of lipidation, similar to what is observed for ubiquitin conjugation, is also a possibility to explore (Ichimura et al., 2000). The identification of delipidation enzymes that act on *S*-palmitylated proteins such as Hedgehog and Wnt *in vivo* has similarly remained elusive (Table 11.1), although these proteins can be delipidated by APT1 *in vitro* (Schulte et al., 2005). Our research has just begun to get us closer to the identification and characterization of delipidation enzymes.

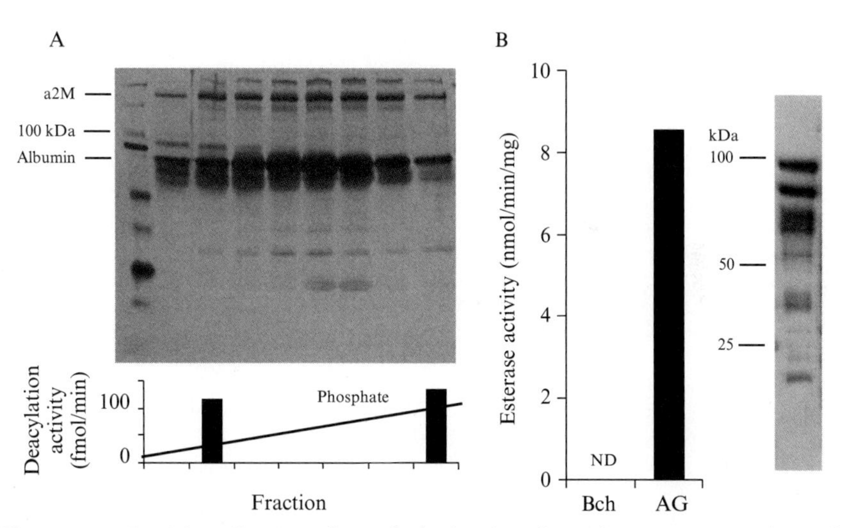

Figure 11.5 Partial purification of novel ghrelin deacylases from HepG2-conditioned medium. Hep2G cell-conditioned medium was collected and applied to Q-sepharose (GE Bioscience), and the active fractions were obtained via stepwise elution with NaCl. The sample was next applied to a hydroxyapatite column and eluted using a linear gradient of phosphate to 200 m*M*. These fractions were visualized by SDS-PAGE followed by silver staining (A). For a further purified fraction, we determined esterase activity against butyrylthiocholine (Bch) or ghrelin (AG) and analyzed the sample by SDS-PAGE (B).

ACKNOWLEDGMENTS

We are grateful to Dr. Yoshihiro Nishi for helpful discussion. We also thank the staff members at the Medical Research Center and acknowledge the Research Support Center, Dokkyo Medical University School of Medicine, for technical assistance. Ongoing research in this area in our laboratory was supported by a Grant-in-Aid for Young Scientists from the Ministry of Education, Culture, Sports, Science, and Technology of Japan.

REFERENCES

Asakawa, A., et al., 2005. Stomach regulates energy balance via acylated ghrelin and desacyl ghrelin. Gut 54, 18–24.

Balasubramanian, A.S., Bhanumathy, C.D., 1993. Noncholinergic functions of cholinesterases. FASEB J. 7, 1354–1358.

Bednarek, M.A., et al., 2000. Structure-function studies on the new growth hormone-releasing peptide, ghrelin: minimal sequence of ghrelin necessary for activation of growth hormone secretagogue receptor 1a. J. Med. Chem. 43, 4370–4376.

Bolen, A.L., et al., 2011. The phospholipase A1 activity of lysophospholipase A-I links platelet activation to LPA production during blood coagulation. J. Lipid Res. 52, 958–970.

Buglino, J.A., Resh, M.D., 2008. Hhat is a palmitoylacyltransferase with specificity for N-palmitoylation of sonic hedgehog. J. Biol. Chem. 283, 22076–22088.

Bulgarelli, I.L., et al., 2009. Desacyl-ghrelin and synthetic GH-secretagogues modulate the production of inflammatory cytokines in mouse microglia cells stimulated by beta-amyloid fibrils. J. Neurosci. Res. 87, 2718–2727.

Chang, S.C., Magee, A.I., 2009. Acyltransferases for secreted signalling proteins (review). Mol. Membr. Biol. 26, 104–113.

Dantas, V.G., et al., 2011. Obesity and variants of the GHRL (ghrelin) and BCHE (butyrylcholinesterase) genes. Genet. Mol. Biol. 34, 205–207.

Date, Y., et al., 2000. Ghrelin, a novel growth hormone-releasing acylated peptide, is synthesized in a distinct endocrine cell type in the gastrointestinal tracts of rats and humans. Endocrinology 141, 4255–4261.

de Boer, J.P., et al., 1993. Alpha-2-macroglobulin functions as an inhibitor of fibrinolytic, clotting, and neutrophilic proteinases in sepsis: studies using a baboon model. Infect. Immun. 61, 5035–5043.

De Vriese, C., et al., 2004. Ghrelin degradation by serum and tissue homogenates: identification of the cleavage sites. Endocrinology 145, 4997–5005.

De Vriese, C., et al., 2007. Ghrelin interacts with human plasma lipoproteins. Endocrinology 148, 2355–2362.

Delhanty, P.J., et al., 2006. Ghrelin and unacylated ghrelin stimulate human osteoblast growth via mitogen-activated protein kinase (MAPK)/phosphoinositide 3-kinase (PI3K) pathways in the absence of GHS-R1a. J. Endocrinol. 188, 37–47.

Eubanks, L.M., et al., 2011. Identification of $\alpha_2$macroglobulin as a major serum ghrelin esterase. Angew. Chem. Int. Ed Engl. 50, 10699–10720.

Hiejima, H., et al., 2009. Regional distribution and the dynamics of n-decanoyl ghrelin, another acyl-form of ghrelin, upon fasting in rodents. Regul. Pept. 156, 47–56.

Hirano, T., et al., 2009. Thioesterase activity and subcellular localization of acylprotein thioesterase 1/lysophospholipase 1. Biochim. Biophys. Acta 1791, 797–805.

Ichimura, Y., et al., 2000. A ubiquitin-like system mediates protein lipidation. Nature 408, 488–492.

Inhoff, T., et al., 2008. Desacyl ghrelin inhibits the orexigenic effect of peripherally injected ghrelin in rats. Peptides 29, 2159–2168.
Kojima, M., Kanagawa, K., 2005. Ghrelin: structure and function. Physiol. Rev. 85, 495–522.
Mikels, A.J., Nusse, R., 2006. Wnts as ligands: processing, secretion and reception. Oncogene 25, 7461–7468.
Miura, G.I., et al., 2006. Palmitoylation of the EGFR ligand spitz by rasp increases spitz activity by restricting its diffusion. Dev. Cell 10, 167–176.
Mizutani, M., et al., 2009. Localization of acyl ghrelin- and des-acyl ghrelin-immunoreactive cells in the rat stomach and their responses to intragastric pH. Am. J. Physiol. Gastrointest. Liver Physiol. 297, G974–G980.
Nishi, Y., et al., 2005. Ingested medium-chain fatty acids are directly utilized for the acyl modification of ghrelin. Endocrinology 146, 2255–2264.
Nishi, Y., et al., 2011. Structures and molecular forms of the ghrelin-family peptides. Peptides 32, 2175–2182.
Pepinsky, R.B., et al., 1998. Identification of a palmitic acid-modified form of human sonic hedgehog. J. Biol. Chem. 273, 14037–14045.
Rauh, M., et al., 2007. Simultaneous quantification of ghrelin and desacyl-ghrelin by liquid chromatography-tandem mass spectrometry in plasma, serum, and cell supernatants. Clin. Chem. 53, 902–910.
Resh, M.D., 2006. Palmitoylation of ligands, receptors, and intracellular signaling molecules. Sci. STKE 2006, re14.
Rocks, O., et al., 2010. The palmitoylation machinery is a spatially organizing system for peripheral membrane proteins. Cell 141, 458–471.
Sato, T., et al., 2005. Molecular forms of hypothalamic ghrelin and its regulation by fasting and 2-deoxy-d-glucose administration. Endocrinology 146, 2510–2516.
Satou, M., et al., 2010. Identification and characterization of acyl-protein thioesterase 1/lysophospholipase I as a ghrelin deacylation/lysophospholipid hydrolyzing enzyme in fetal bovine serum and conditioned medium. Endocrinology 151, 4765–4775.
Satou, M., et al., 2011. Understanding the functional significance of ghrelin processing and degradation. Peptides 32, 2183–2190.
Schägger, H., 2006. Tricine-SDS-PAGE. Nat. Protoc. 1, 16–22.
Schulte, G., et al., 2005. Purified wnt-5a increases differentiation of midbrain dopaminergic cells and dishevelled phosphorylation. J. Neurochem. 92, 1550–1553.
Shanado, Y., et al., 2004. Lysophospholipase I identified as a ghrelin deacylation enzyme in rat stomach. Biochem. Biophys. Res. Commun. 325, 1487–1494.
Sugimoto, H., et al., 1996. Purification, cDNA cloning, and regulation of lysophospholipase from rat liver. J. Biol. Chem. 271, 7705–7711.
Sunaga, H., et al., 1995. Purification and properties of lysophospholipase isoenzymes from pig gastric mucosa. Biochem. J. 308 (Pt. 2), 551–557.
Swenson, R.P., Howard, J.B., 1979. Structural characterization of human alpha2-macroglobulin subunits. J. Biol. Chem. 254, 4452–4456.
Tham, E., et al., 2009. Acylated ghrelin concentrations are markedly decreased during pregnancy in mothers with and without gestational diabetes: relationship with cholinesterase. Am. J. Physiol. Endocrinol. Metab. 296, E1093–E1100.
Yang, J., et al., 2008. Identification of the acyltransferase that octanoylates ghrelin, an appetite-stimulating peptide hormone. Cell 132, 387–396.
Yoon, J.C., et al., 2010. Wnt signaling regulates mitochondrial physiology and insulin sensitivity. Genes Dev. 24, 1507–1518.
Zhu, X., et al., 2006. On the processing of proghrelin to ghrelin. J. Biol. Chem. 281, 38867–38870.

SECTION 4

Synthesis of Ghrelin Agonists and Antagonists

CHAPTER TWELVE

Synthesis of Ghrelin: Chemical Synthesis and Semisynthesis for Large-Scale Preparation of Modified Peptides

Tomohiro Makino*, Masaru Matsumoto†, Yoshiharu Minamitake‡,1

*Faculty of Pharmacology II, Asubio Pharma Co. Ltd., Chuo-ku, Kobe, Japan
†R&D Group Vaccine Business Planning Department Business Intelligence Division, Daiichi Sankyo Co. Ltd., Edogawa-ku, Tokyo, Japan
‡Board Director, Asubio Pharma Co. Ltd., Chuo-ku, Kobe, Japan
[1]Corresponding author: e-mail address: minamitake.yoshiharu.d8@asubio.co.jp

Contents

Abstract

Most biologically active peptide hormones, including ghrelin, undergo numerous post-translational modifications and play many crucial roles in nature. Medium- or large-scale

Methods in Enzymology, Volume 514
ISSN 0076-6879
http://dx.doi.org/10.1016/B978-0-12-381272-8.00012-X

preparation methods are required to understand their biological functions and potential applications in life sciences and the biomedical fields.

Since ghrelin has an O-acyl modification in its Ser3, recombinant expression for its production has not solely been employed thus far. In this chapter, we provide two distinct protocols for the preparation of human ghrelin: a chemical synthesis method for medium-scale (up to hundreds of milligrams) and a semisynthesis method for large-scale (more than grams) preparation. Established Fmoc chemistry for solid-phase synthesis enables the highly efficient procedure for synthesizing ghrelin in the medium scale. Semisynthesis method, the coupling of chemically synthesized O-acylated ghrelin(1–7) with recombinantly expressed ghrelin(8–28), can be applied for larger scale preparation.

ABBREVIATIONS

AcOH acetic acid
ANP atrial natriuretic peptide
Boc *tert*-butoxycarbonyl
CHO chinese hamster ovary
DCM dichloromethane
DIPEA *N, N*-diisopropylethylamine
DMAP 4-dimethylaminopyridine
DMF *N,N*-dimethylformamide
E. coli *Escherichia coli*
EDC 1-ethyl-3-(3-dimethylaminopropyl)carbodiimide
ESI-MS electrospray ionization mass spectrometry
Fmoc 9-fluorenylmethoxycarbonyl
GHS-R growth hormone secretagogue receptor
GLP-1 glucagon-like peptide-1
HBTU 2-(1*H*-benzotriazole-1-yl)-1,1,3,3,-tetramethyluronium hexafluorophosphate
HOBt 1-hydroxybenzotriazole
IPTG isopropyl-β-D-thiogalactopyranoside
MeCN acetonitrile
NMP *N*-methyl-2-pyrrolidone
OD optical density
Pmc 2,2,5,7,8-pentamethylchroman-6-sulfonyl
PTH parathyroid hormone
RP-HPLC reversed-phase high-performance liquid chromatography
TBAF tetrabutylammonium fluoride
TBDMS *t*-butyldimethylsilyl
***t*Bu** *tert*-butyl
TFA trifluoracetic acid
TIPS triisopropylsilane
TRH thyrotropin-releasing hormone
Tris-HCl tris(hydroxymethyl)aminomethane-hydrochlorates
Trt trityl
UV ultraviolet

1. INTRODUCTION

Peptide hormones have widely been studied in the field of life sciences and used in therapeutic applications. There are currently more than 40 commercially available peptide-based drugs such as insulin, and ANP (atrial natriuretic peptide) and GLP-1 (glucagon-like peptide-1) analogs. In addition, over 100 new peptide therapeutics are currently being evaluated in clinical trials (Reichert, 2010). Their molecular weights differ dramatically from those of small molecules, such as the tripeptide TRH (thyrotropin-releasing hormone) analog, and of long polypeptides such as human insulin (51 residues) and PTH(1–84) (parathyroid hormone). Because of their range in size, many different methods are used to produce these peptides, such as chemical synthesis, recombinant expression, fermentation, and extraction from native tissues. Chemical synthesis or recombinant expression system is commonly utilized for medium- or large-scale preparation.

Most biologically active peptide hormones have a wide variety of posttranslational modifications such as C-terminal amidation, phosphorylation, and acetylation. These modifications are very important for their biological activity and stability in the blood stream (Matsubayashi, 2011; Reichert, 2010). Ghrelin is an acylpeptide consisting of 28 amino acids, with Ser3 esterified with octanoic acid (Kojima et al., 1999). This modification is essential for its biological activity (Kojima et al., 1999; Matsumoto et al., 2001a).

Here, we provide two different protocols for human ghrelin preparations, namely, a chemical synthesis method and a semisynthesis method on a laboratory scale.

For scales up to several hundreds of milligrams, a solid-phase chemical synthesis protocol can be utilized for the preparation of human ghrelin, as shown below. However, preparation at the gram scale is still challenging because (1) chemical synthesis imposes a limitation on the size of peptide that can be produced on a large scale and (2) even with recombinant expression, which is more suitable for the large-scale preparation, the majority of the commonly utilized expression hosts do not have endogenous posttranslational modification machinery for ghrelin.

Therefore, we also provide a semisynthesis method, combining chemical synthesis and recombinant expression system, that utilizes the advantages of both (Makino et al., 2005).

2. PROTOCOLS OF CHEMICAL SYNTHESIS OF HUMAN GHRELIN

2.1. Synthetic scheme of human ghrelin

The established 9-fluorenylmethoxycarbonyl (Fmoc)/Bu^t chemistry (Carpino and Han, 1972) for solid-phase synthesis is applicable to human ghrelin for small-scale synthesis (Matsumoto et al., 2001b). The introduction of an *n*-octanoyl moiety at the third Ser hydroxyl group is the key step in the synthetic procedure of this O-acylated peptide. Although the *n*-octanoyl group can be directly incorporated as Ser(octanoyl) in the course of chain elongation, O- to *N*-acyl rearrangement, β-elimination, and hydrolysis of the ester possibly take place after the following Fmoc removal step by base treatment. To avoid these side reactions, the *n*-octanoyl group is directly esterified onto the Ser3 hydroxyl group after completion of the peptide backbone construction on the resin.

The synthetic scheme is shown in Fig. 12.1. After construction of the peptide backbone on Wang resin (Wang, 1973), acylation of the target hydroxyl group is achieved by taking advantage of the preferential acid sensitivity of *O*-trityl (Trt) group for the Ser3 and *O*-*tert*-butyl group for the remaining three Ser residues. The protected resin-bound peptide is treated with dilute trifluoracetic acid (TFA) in the presence of triisopropylsilane (TIPS) to selectively remove the Trt at the Ser3, while leaving the Bu^t groups on the other Ser side chains intact. The deprotected Ser3 is reacted with *n*-octanoic acid by 1-ethyl-3-(3-dimethylaminopropyl)carbodiimide (EDC)/4-dimethylaminopyridine (DMAP) to give fully protected human ghrelin bound to the resin. After deprotection by TFA, the liberated crude

Fmoc-Arg(Pmc)^{28}-Wang resin
↓ -20% piperidine in NMP
↓ -HBTU/HOBT in NMP
Boc-Gly^{1}-Ser(*t*Bu)-Ser(Trt)-Phe-Leu^{5}-Ser(*t*Bu)----Arg(Pmc)^{28}-Wang resin
↓ -1%TFA w/5% TIPS in DCM
↓ -*n*-Otanoic acid, EDC, DMAP in NMP
Boc-Gly^{1}-Ser(*t*Bu)-Ser(octanoyl)-Phe-Leu^{5}-Ser(*t*Bu)--Arg(Pmc)^{28}-Wang resin
↓ -TFA/H_2O/phenol/TIPS=88/5/5/2 (V/V)
Crude human ghrelin
↓ -CM-column chromatography
↓ -RP-HPLC
Purified human ghrelin

Figure 12.1 Synthetic scheme for human ghrelin.

peptide is successively purified by ion-exchange column chromatography and reversed-phase high-performance liquid chromatography (RP-HPLC) to homogeneity.

2.2. Materials and instruments

2.2.1 *Fmoc-amino acid derivatives and reagents*

Amino acid derivatives used in this protocol are of the L-configuration. Fmoc-amino acid derivatives and reagents for solid-phase peptide synthesis can be purchased from Applied Biosystems.

2.2.2 *Peptide synthesizer*

Applied Biosystem Model 433A Peptide Synthesizer; User guide: http://www3.appliedbiosystems.com/cms/groups/psm_support/documents/generaldocuments/cms_041850.pdf.

2.2.3 *Analytical HPLC*

System: Shimadzu LC-10A System.

Columns: YMC-Pack PROTEIN-RP (ID 4.6 × 150 mm length; manufactured by YMC Co., Ltd., Kyoto, Japan).

Eluent condition: In 0.1% TFA, the acetonitrile (MeCN) concentration is linearly changed to a maximum of 50%. Flow rate: 1 ml/min.

Detection: Ultraviolet (UV) at 210 nm.

Column temperature: 40 °C.

2.2.4 *Mass spectrometry*

Finnigan MAT Corporation TSQ7000 instrument.

Ion source: ESI; detection ion mode: positive; spray voltage: 4.5 kV; capillary temperature: 250 °C; mobile phase: 0.2%; acetic acid (AcOH)–methanol (1:1) solution; flow rate: 0.2 ml/min; scan range: m/z from 300 to 1500.

2.2.5 *Amino acid composition analysis*

L-8500 amino acid analyzer (Hitachi, Ltd., Tokyo, Japan).

The sample is hydrolyzed with 6 *M* HCl containing 0.1% phenol at 110 °C for 24 h in a sealed tube.

2.2.6 *Preparative chromatogram for ion-exchange column chromatography and RP-HPLC*

System; Waters 600 Multisolvent Delivery System, Fraction collector; Pharmacia Fine Chemicals FRAC-100.

2.3. Synthesis of human ghrelin

2.3.1 Peptide chain elongation

A stepwise peptide chain elongation can be performed using an Applied Biosystems Model 433A peptide synthesizer by employing the manufacturer's standard protocol of Fmoc chemistry for 0.25-mmol scale. The N-terminal amino acid is exceptionally introduced as N^{α}-Boc derivative to avoid base treatment for Fmoc removal after introduction of the *n*-octanoyl ester. Starting with Fmoc-Arg(Pmc)-Wang resin (Applied Biosystems, 4-hydroxymethylphenoxy-methyl-copolystyrene resin, 472 mg of 0.53 mmol/g of resin, 0.25 mmol), the construction of 1770 mg of Boc-Gly-Ser(*t*Bu)-Ser(Trt)-Phe-Leu-Ser(*t*Bu)-Pro-Glu(OtBu)-His(Boc)-Gln(Trt)-Arg(Pmc)-Val-Gln(Trt)-Gln(Trt)-Arg(Pmc)-Lys(Boc)-Glu(OtBu)-Ser(*t*Bu)-Lys(Boc)-Lys(Boc)-Pro-Pro-Ala-Lys(Boc)-Leu-Gln(Trt)-Pro-Arg(Pmc)-Wang resin is accomplished by repeated introductions of Fmoc-amino acids (1 mmol, 4 equiv.) with 2-(1*H*-benzotriazole-1-yl)- 1,1,3,3,-tetramethyluronium hexa fluorophosphate (HBTU, 1 mmol, 4 equiv.) and 1-hydroxybenzotriazole (HOBt, 1 mmol, 4 equiv.) and removal of the Fmoc with 20% piperidine/*N*-methyl-2-pyrrolidone (NMP) . To ensure the coupling quality, the residues of Pro^{22}, Pro^{21}, Lys^{20}, Lys^{19}, Lys^{16}, Arg^{15}, Gln^{14}, and Gln^{13} are reacted twice in the process of peptide chain elongation based on the results of Kaiser test (Kaiser et al., 1970).

2.3.2 Octanoylation

The protected peptide resin (1770 mg, 0.25 mmol) is treated with 1% TFA in the presence of 5% TIPS/dichloromethane (DCM) (40 ml) at room temperature for 30 min, filtered, and washed successively with DCM, 1% *N, N*-diisopropylethylamine (DIPEA)/DCM, and DCM. The resulting peptide resin having liberated the hydroxyl group at Ser3 is reacted with *n*-octanoic acid (144 mg, 1.0 mmol, 4 equiv.) in NMP (8 ml) by EDC–HCl (192 mg, 1.0 mmol, 4 equiv.) in the presence of 31 mg (0.25 mmol, 1 equiv.) of DMAP. After gently stirring for 16 h, the resin is filtered and washed with NMP and DCM and dried under reduced pressure to obtain the protected and O-acylated peptide resin (ca. 1900 mg).

2.3.3 Deprotection and cleavage

Fifteen milliliters of deprotection cocktail consisting of TFA/H_2O/phenol/TIPS = 88/5/5/2 (v/v) is subsequently added to the resin and stirred for 2 h at room temperature to simultaneously remove the protecting groups and to cleave the peptide from the resin. After filtering the reaction mixture and concentrating it under reduced pressure, the peptide residue is precipitated and washed with diethylether. Finally, the powder is dried under reduced pressure to obtain 951 mg of the crude human ghrelin.

2.3.4 Purification

The crude human ghrelin (951 mg) is dissolved in 20 m*M* NH_4OAc buffer (pH 5.5, mobile phase) and filtered through a membrane filter to remove insoluble substances. The crude peptide solution is applied to ion-exchange chromatography (CM-Toyopearlpak 650M, TOSOH, ID 22 × 200 mm length), equilibrated with the mobile phase buffer. After a washing step using one column volume of the buffer followed by loading the crude peptide, human ghrelin is eluted using linear gradient flow from 33% to 100% for 60 min of the elution buffer (0.6 *M* NaCl in 20 m*M* NH_4OAc (pH 5.5)) at the flow rate of 7.5 ml/min, monitoring the optical density (OD) at 260 nm.

Fractions containing the main peak (200 ml) is pooled and directly applied onto the RP-HPLC column (YMC-PAK Protein-RP, YMC, ID 20 × 200 mm length) previously equilibrated with 5% AcOH. The peptide is eluted by a linear gradient at the flow rate of 10 ml/min from solvent A (5% AcOH) to solvent B (60% MeCN in 5% AcOH). The collected fraction is lyophilized to obtain 331 mg (based on amino acid analysis) of net human ghrelin (99.6% purity, 39% from the Fmoc-Arg(Pmc)-Wang resin).

The synthetic ghrelin prepared by this method is subjected to measure its amino acid composition, molecular mass spectrometry, and intracellular Ca^{2+} assay in growth hormone secretagogue receptor (GHS-R)-expressing Chinese hamster ovary (CHO) cells to verify the expected structure (for details, see Section 3.7).

2.3.5 Specification of synthetic human ghrelin

Analytical HPLC retention time: 16.8 min; purity: 99.6%; ESI-MS: found 3371 (calcd. 3370.9); amino acid composition analysis (theoretical): Ser 3.66 (4), Glx 5.94 (6), Gly 1.01 (1), Ala 1.01 (1), Val 0.98 (1), Leu 2, Phe 1.01 (1), Lys 4.01 (4), His 1.01 (1), Arg 2.99 (3), Pro 4.00 (4); EC_{50} in Ca^{2+} assay: 1.3 n*M*

3. PROTOCOLS OF SEMISYNTHESIS OF HUMAN GHRELIN

3.1. Semisynthesis scheme of human ghrelin

The strategy of producing semisynthetic ghrelin is outlined in Fig. 12.2. The N-terminal fragment, containing a C8 modification, can be synthesized chemically. Starting with a prolyl-2-chlorotrityl-resin, the peptide backbone is synthesized by Fmoc chemistry, giving Boc-Gly-Ser(*t*Bu)-Ser

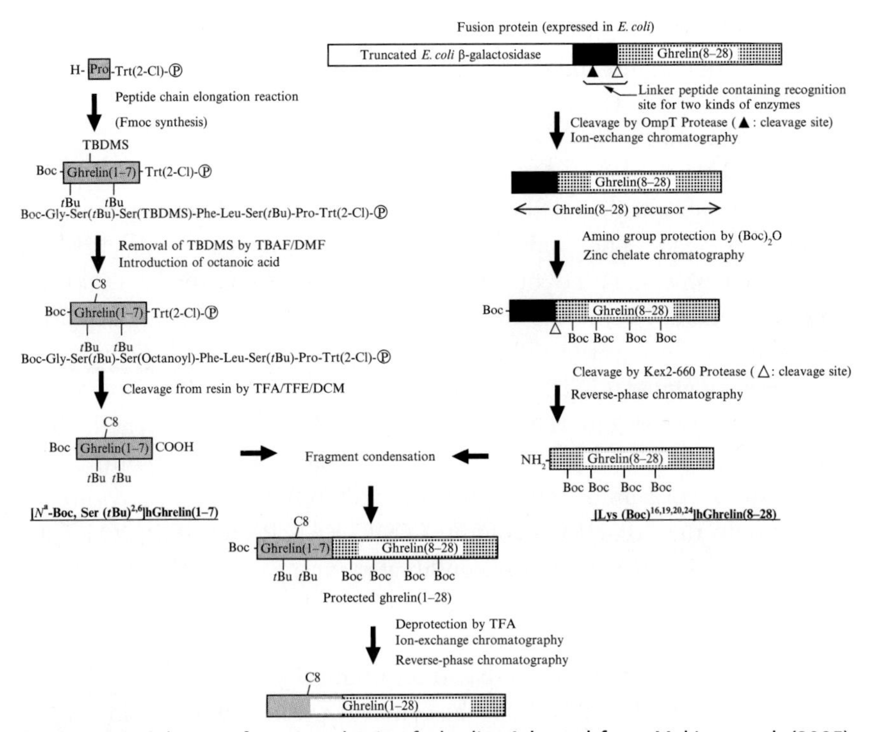

Figure 12.2 Scheme of semisynthesis of ghrelin. Adapted from Makino et al. (2005). (See Color Insert.)

(*t*-butyldimethylsilyl, TBDMS)-Phe-Leu-Ser(*t*Bu)-Pro-2-chlorotrityl-resin. The N-terminal fragment can be obtained by cleaving from the peptide resin after the resin is treated with tetrabutylammonium fluoride (TBAF), followed by acylation. For preparation of the C-terminal fragment (21 amino acids), its precursor is expressed recombinantly in *Escherichia coli* (*E. coli*). Successive enzymatic treatment of two distinct endoproteases (in this protocol, OmpT and Kex2-660 proteases are used) and Boc protection of amino groups enable the efficient preparation of a fragment that is suitable for coupling with the N-terminal fragment. After the two segments undergo fragment coupling, deprotection, and purification, highly purified human ghrelin can be produced on a large (gram) scale.

3.2. Materials and instruments

3.2.1 Preparative columns

Ion-exchange chromatography resin: SP Sepharose Big Beads.
Chelate chromatography resin: Chelating Sepharose Fast Flow.

Reversed-phase chromatography resin: Source 30RPC.
All the above can be purchased from GE Healthcare Bioscience (Piscataway, NJ, USA).

3.2.2 Analytical HPLC

System: A Shimadzu LC-10A System
Columns: YMC-Pack PROTEIN-RP, YMC-Pack ODS AP-302, and YMC-Pack PROTEIN-C8 (all ID 4.6 mm × 150 mm length; manufactured by YMC Co., Ltd.).
Eluent condition: In 0.1% TFA, the MeCN concentration is linearly changed to a maximum of 100%. Flow rate: 1 ml/min.
Detection: UV (210 or 214 nm).

3.2.3 Preparative chromatography system

AKTA explorer 10S and AKTA PILOT chromatography systems (GE Healthcare Bioscience).

3.2.4 Mass spectrometry

Described in Section 2.2.4.

3.2.5 Amino acid sequence analysis

Applied Biosystems 477A sequencer (Perkin–Elmer Yokohama, Japan).

3.2.6 Amino acid composition analysis

Described in Section 2.2.5.

3.3. Synthesis of the ([N^{α}-*t*-butyloxycarbonyl (Boc), Ser *t*-butyl (*t*Bu))^{2,6}]hGhrelin(1–7))N-terminal fragment

In an effort to avoid possible racemization upon coupling with the C-terminal fragment of ghrelin, Pro7 is selected as the coupling site. The solid-phase synthesis is performed using a prolyl-2-chlorotrityl-resin that is functionalized with the weak acid-labile linkage.

Starting with the prolyl-2-chlorotrityl-resin (46.52 g, 20.0 mmol, Novabiochem), the construction of Boc-Gly-Ser(*t*Bu)-Ser(TBDMS)-Phe-Leu-Ser(*t*Bu)-Pro-2-chlorotrityl-resin can be accomplished by repeated introductions of Fmoc-amino acids (2 equiv.) with HBTU (2 equiv.) and HOBt (2 equiv.) and removal of the Fmoc with 20% piperidine/NMP. Among a variety of protecting groups, the protection with TBDMS at Ser3 and with *t*Bu at Ser2 and Ser6 gives us the highest yield, purity, and efficacy throughout the process. The protected peptide resin is then treated with 0.1 *M* TBAF/*N,N*-dimethylformamide (DMF) (400 ml) at room temperature for 30 min, filtered, and washed successively with DMF and NMP. The resulting de-TBDMS peptide resin is reacted with *n*-octanoic acid (11.56 g, 4 equiv.) in NMP (163 ml), and EDC–HCl (16.90 g, 4.4 equiv.) is added in the presence of 7.33 g (1 equiv.) of DMAP. After stirring for 18 h, the resin is filtered and washed with NMP. To cleave the peptide from the resin, 300 ml of 0.1% TFA/20% trifluoroethanol/DCM is then added and stirred for 30 min at room temperature. After filtration and concentration, 200 ml of H_2O is added to precipitate them; the peptides are washed with H_2O and hexane, filtered again, and dried under reduced pressure to obtain 18.74 g (yield, 91%) of the N-terminal fragment ([N^{α}-Boc, Ser (*t*Bu)^{2,6}]hGhrelin(1–7)). A typical HPLC profile is shown in Fig. 12.5, from which the purity of the peptide is seen to be approximately 90%. This product can be confirmed to possess the same amino acid composition and molecular mass as the theoretical value (Table 12.1) by mass spectrometry and amino acid analysis.

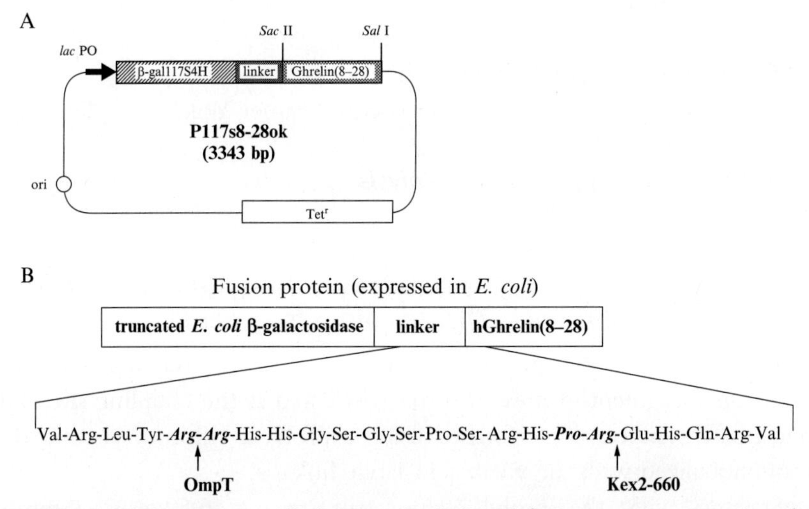

Figure 12.3 (A) Construction of a human ghrelin(8–28) fusion protein expression plasmid (p117s8-28ok). *lac*PO, lac promoter; Tetr, tetracycline resistant gene; ori, replication start point; Ghrelin(8–28), synthetic DNA oligo encoding human ghrelin(8–28); β-gal117S4H, DNA sequence of *E. coli* β-galactosidase derivative. (B) Schematic representation of human ghrelin(8–28) fusion protein with the amino acid sequences of the cleavage sites. The arrow indicates the cleavage site of OmpT and Kex2-660. Arg-Arg and Pro-Arg shown in italic bold are the recognition amino acid sequences for OmpT and Kex2, respectively.

Table 12.1 Amino acid composition and mass spectrometry of [N^{α}-Boc, Ser(*t*Bu)^{2,6}]hGhrelin(1–7) and [Lys(Boc)^{16,19,20,24}]hGhrelin(8–28)

	[N^{α}-Boc, Ser(*t*Bu)^{2,6}]hGhrelin (1–7)	**[Lys(Boc)^{16,19,20,24}]hGhrelin(8–28)**
ESI-MS	1032.6[a]	2696[b]
Amino acid	Gly: 1.01 (1); Leu: 1(1); Phe: 0.99 (1). Composition ratio, Pro: 1.00 (1); Ser, 2.70 (3)[c]	Ala: 1.01 (1); Arg: 2.84 (3); Glx: 5.75 (6); His: 0.92 (1): Leu: 1 (1); Lys: 3.92 (4); Pro: 2.92 (3); Ser: 0.90 (1); Val: 0.94 (1)

[a]Theoretical value: 1032.3.
[b]Theoretical value: 2969.4.
[c]Values in parentheses are theoretical.

3.4. Expression of the hGhrelin(8–28) derivative

For the expression of recombinant peptide using *E. coli*, a conventional fusion protein method is used for efficient preparation. In this method, the target peptide is fused to a "protection peptide" or "fusion partner" at its N-terminus, via an amino acid linker. Although an extra step is needed to cleave the target peptide from the fusion protein, this enables an increase in the levels of stability and expression in bacteria.

To separate the peptide from the fusion protein efficiently and precisely, several cleavage enzymes can be chosen depending on their specificity (Table 12.2).

For the preparation of the C-terminal portion of human ghrelin, the methods for efficient fragment coupling and suppressing side reactions are required to maximize the yield of product. Therefore, it is necessary to prepare a partially protected peptide fragment whose N-terminal α-amino group is free while all other side chain amino groups are blocked. For this purpose, the fusion protein is designed to consist of truncated *E. coli* β-galactosidase, followed by the target peptide connected with a linker designed to allow cleavage with two distinct proteases. The schematic representation of this protein is shown in Fig. 12.3. A cDNA encoding for the amino acid sequence of human ghrelin(8–28) is prepared using an annealing method with a synthetic DNA oligonucleotide (Sigma Aldrich Japan, Tokyo, Japan). The p117s8-28ok plasmid is a pBR322-based vector that expresses the amino acid sequence of ghrelin(8–28) as a fusion protein under the control of the *E. coli* lac promoter. The fusion protein consists of three parts: the N-terminal 117 amino acids of the *E. coli* β-galactosidase

Table 12.2 Site-specific endoprotease and its cleavage sequence

Protease	Amino acid sequence for cleavage
Lysyl endopeptidase	-Lys-↓
Trypsin	-Arg-↓, -Lys-↓
V8 protease	-Glu-↓
Kex2 protease	-Lys-Arg-↓, -Arg-Arg-↓, -Pro-Arg-↓
Enterokinase	-Asp-Asp-Aps-Lys-↓
Factor Xa	-Ile-Glu-Gly-Arg-↓
TEV protease	-Leu-Glu-Val-Leu-Phe-Gln-↓-Gly-
OmpT protease	-Arg-↓-Arg-, -Lys-↓-Arg-, -Lys-↓-Lys-

(Suzuki et al., 1998), an amino acid linker designed for facilitating protease digestion, and the amino acid sequence of ghrelin(8–28).

A schematic of the structure of p117s8-28ok and nomenclature of the ghrelin(8–28) fusion proteins used in this protocol are shown in Fig. 12.3. The expression plasmids are transformed into the *E. coli* strain W3110. The obtained transformants are stored in a medium containing 10% glycerol at –80 °C.

In this protocol, two proteases, namely, OmpT and Kex2-660, are employed to process the target peptide from the fusion protein. These two enzymes are well characterized and their cleavage sites can be relatively easily optimized. OmpT is an endogenous *E. coli* membrane-bound protease which cleaves a peptide bond at the center of a basic amino acid pair such as Arg-Arg, Lys-Lys, and Arg-Lys (Okuno et al., 2002; Sugimura and Higashi, 1988) (Table 12.2). Kex2-660 is a secretory type of Kex2 protease, belonging to the Kexin protease family, such as Furin and PC1/3 (Mizuno et al., 1987, 1989). Kex2-660 cleaves the C-terminal side of a Lys-Arg, Arg-Arg, or Pro-Arg pair (Bevan et al., 1998; Ledgerwood et al., 1995; Mizuno et al., 1987) (Table 12.2).

Fusion proteins are expressed in *E. coli* W3110 as inclusion bodies. Starting from the glycerol stock described above, the cells are grown at 37 °C in 100-ml Terrific broth (Sambrook and Maniatis, 1989) containing 100 μg/ml ampicillin. The seed culture is then used to inoculate 20 l of Terrific broth with containing 100 μg/ml ampicillin in a 30-l fermenter. The expression of the gene encoding for the fusion protein is induced by adding

1.0 mM of isopropyl-β-D-thiogalactopyranoside (IPTG) when the cell culture's OD at 660 nm is approximately 10.0.

After cultivation in a 20-l fermenter at 37 °C for 30 h, the cells are lysed by two passes through a high-pressure homogenizer. The cell debris and inclusion bodies are pelleted by centrifugation (6500 × g, 30 min) and resuspended in 820 ml of deionized water.

3.5. Preparation of the ([Lys(Boc)^{16,19,20,24}] hGhrelin(8–28)) C-terminal fragment

3.5.1 OmpT protease

The inclusion body suspension (820 ml) is incubated with OmpT protease in 9 l of 4 M urea, 20 mM Tris–HCl (pH 7.4), and 50 mM NaCl at 32° C. As the *E. coli* W3110 strain encodes this protease, endogenous OmpT cleaves the fusion protein under denaturing conditions, resulting in the generation of the C-terminal portion of ghrelin with a linker peptide at its N-terminus (Fig. 12.4a). The reaction can be monitored by HPLC and the cleavage efficiency is around 90% after 3 h of reaction time.

3.5.2 Cation-exchange chromatography

The residue is removed by centrifugation (9000 × g, 30 min) and the supernatant is purified by cation-exchange chromatography. The purpose of this step is to remove *E. coli* host cell proteins and the "partner peptide" that is derived from the fusion protein after the OmpT digestion. The supernatant is loaded on SP sepharose big beads column (ID 10 cm × 30 cm length), previously equilibrated with 1.5 M urea and 50 mM $NaHCO_3$ (pH 9) (equilibration buffer). After a washing step with five column volumes of the equilibration buffer, the hGhrelin(8–28) precursor (12.8 g) is eluted in 1.25 l of 1.5 M urea, 0.5 M NaCl, and 50 mM $NaHCO_3$ (pH 11).

3.5.3 Boc reaction

Purified hGhrelin(8–28) precursor is subjected to Boc protection by adding 3.57 l of 30% MeCN and 20 mM $(Boc)_2O$ (pH 10) and stirring for 2 h at room temperature. Through this process, the α-amino group at the N-terminal and the side chain Lys residues are protected by Boc groups (Fig. 12.4b). This reaction can be monitored by HPLC since the modified peptide increases in hydrophobicity during the reaction, leading to an increase in observed retention time (Fig. 12.4b). The reaction's efficiency is about 95% after 2 h.

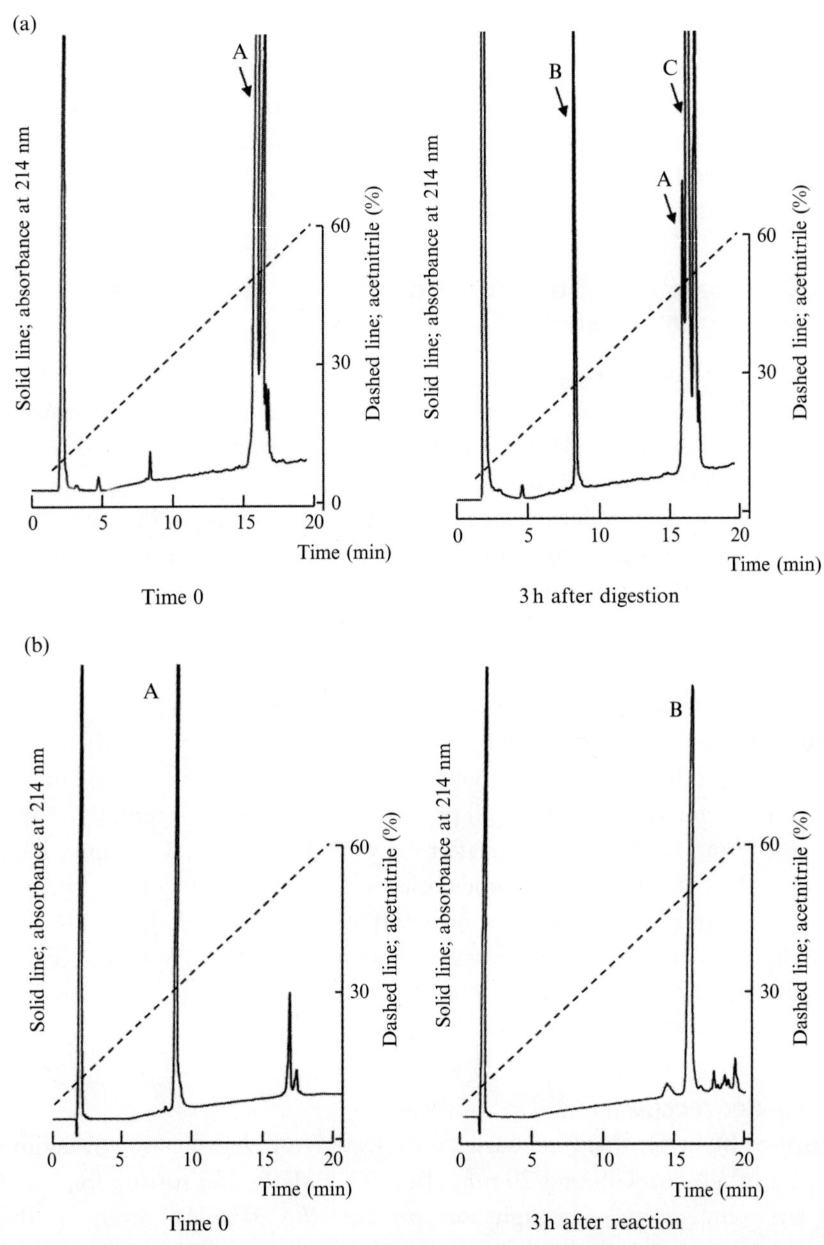

Figure 12.4 HPLC profile of (a) OmpT digestion before (left) and after 3-h (right) reaction. Compared with time 0, a new peak B appeared and was confirmed as the human ghrelin(8–28) precursor by mass spectrometry. Peak A: fusion protein; peak B: human ghrelin(8–28) precursor; peak C: truncated *E. coli* β-galactosidase. (b) Boc reaction before (left) and after 3-h (right) reaction. Peak A: Unprotected human ghrelin(8–28) precursor; peak B: [N^{α}-Boc, Lys(Boc)^{16,19,20,24}] hGhrelin(8–28) precursor.

continued

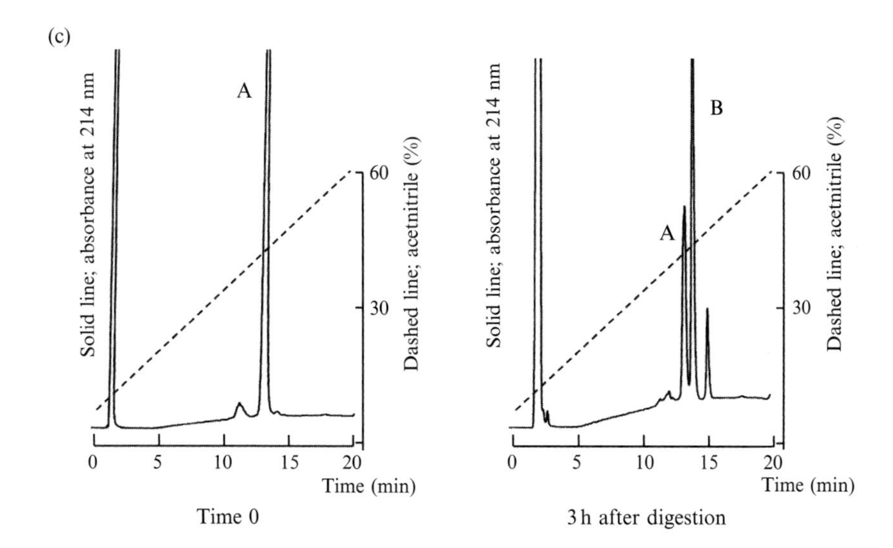

Figure 12.4—cont'd (c) Kex2 digestion before (left) and after 3-h (right) reaction. Peak A: [N^{α}-Boc,Lys(Boc)^{16,19,20,24}] hGhrelin(8–28) precursor; peak B: [Lys(Boc)^{16,19,20,24}] hGhrelin(8–28); peak C: [Lys(Boc)^{16,19,20,24}] hGhrelin(16–28). Analytical column YMC-PACK PROTEIN-C8 was used for all analysis.

3.5.4 Zn chelate chromatography

Zinc chelate chromatography can be applied by appending histidine tags at the amino acid linker sequence in the fusion protein for efficient purification. The [N^{α}-Boc, Lys(Boc)^{16,19,20,24}]hGhrelin(8–28) precursor is loaded on a chelate chromatography column (ID 5×30 cm length) previously equilibrated with 0.1 *M* $ZnSO_4$ and 20 m*M* imidazole, 20 m*M* Tris–HCl, pH 6.4, and eluted with 3.36 l of 0.1 *M* imidazole/20 m*M* Tris–HCl (pH 6.4). The eluent can be directly applied for the next step, that is, Kex2-660 protease treatments.

3.5.5 Kex2-660 protease

The purified precursor is digested with the second protease Kex2-660 (25,000 units/ml), which is a soluble form of Kex2, releasing [Lys (Boc)^{16,19,20,24}]hGhrelin(8–28) with a free N^{α}-amino group. The reaction is carried out in 3.57 l of 5 m*M* $CaCl_2$ (pH 8.3) at 37° C, giving roughly 90% of cleavage efficiency after 3 h, as monitored by HPLC (Fig. 12.4C).

3.5.6 Reversed-phase chromatography

The peptide is further purified by RP chromatography in order to obtain a highly purified product (Fig. 12.5a). The reaction solution is filtered through a 0.22-μm filter and loaded on to an RP chromatography column (Source

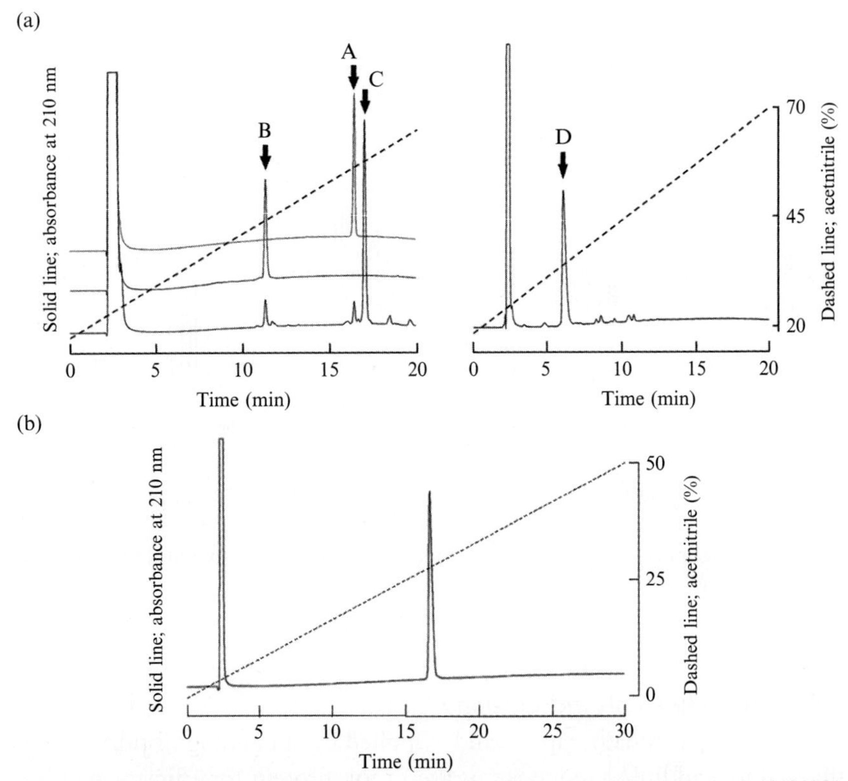

Figure 12.5 HPLC profile of [N^{α}-Boc, Ser(*t*Bu)^{2,6}]hGhrelin(1–7), [Lys(Boc)^{16,19,20,24}]hGhrelin(8–28), and human ghrelin. (a) peak A: purified [N^{α}-Boc, Ser(*t*Bu)^{2,6}]hGhrelin (1–7); peak B: purified [Lys(Boc)^{16,19,20,24}]hGhrelin(8–28); peak C: protected human ghrelin after coupling; peak D: human ghrelin after deprotection by TFA. (b) Purified human ghrelin after two-column chromatographic purifications. Analytical column YMC-PACK PROTEIN-RP was used for all analysis.

30RPC: ID 5 × 30 cm length). The peptide is eluted by a linear gradient from solvent A (10% MeCN, 0.1% TFA) to solvent B (50% MeCN, 0.1% TFA) for eight column volumes. The eluent (5.95 g, 730 ml) is lyophilized for storage. The exact amino acid composition and molecular mass can be determined (Table 12.1) by mass spectrometry and amino acid sequencing.

3.6. Fragment coupling and deprotection

The two fragments [N^{α}-Boc, Ser(*t*Bu)^{2,6}]hGhrelin(1–7) and [Lys(Boc)^{16,19,20,24}]hGhrelin(8–28) prepared in Sections 3.3 and 3.5 are then subjected to a coupling reaction. [N^{α}-Boc, Ser(*t*Bu)^{2,6}]hGhrelin(1–7)

(1.67 g, 1.6 mmol), HBTU (0.62 g, 1.6 mmol), and DIPEA (0.3 ml, 1.6 mmol) are dissolved in 8.3 ml of DMF and the solution is stirred for 30 min. [Lys(Boc)^{16,19,20,24}]hGhrelin(8–28) (5.53 g, 1.4 mmol) is dissolved in 27.7 ml of DMF, and the activated N-terminal fragment solution is added dropwise while stirring at room temperature. After 90 min, the reaction solvent is evaporated under reduced pressure. TFA (52.8 ml) is added to the resulting residue, and the mixture is slowly stirred at room temperature for 3 h. TFA is evaporated under reduced pressure, and 150 ml of isopropylether is added to precipitate the residue. The precipitate is then washed and dried to obtain 6.26 g of white, powdery crude peptide (purity of 77%). Figure 12.5a shows the HPLC profiles of the coupling reaction. Peak C (protected human ghrelin) appears in place of the disappearing peak A ([N^{α}-Boc, Ser(*t*Bu)^{2,6}]hGhrelin(1–7)) and peak B ([Lys(Boc)^{16,19,20,24}]hGhrelin(8–28)) during the reaction. Figure 12.5a shows the HPLC profile of the peptide after TFA deprotection.

3.7. Purification of human ghrelin

After synthesis, crude human ghrelin (3.32 g) is dissolved in 500 ml of 5% AcOH and loaded onto an ion-exchange chromatography column (SP Sepharose big beads: ID 2.6 × 30 cm length), pre-equilibrated with 50 m*M* sodium acetate (pH 5). After washing with 50 m*M* sodium acetate (pH 5), human ghrelin is eluted by 250 ml of 0.5 *M* NaCl/50 m*M* sodium acetate (pH 5). The eluent (3.29 g) is then further purified by an RP chromatography column (Source 30RPC: ID 2.6 × 30 cm length), previously equilibrated with 50 m*M* AcOH. The purified peptide is eluted using a linear gradient from solvent A (50 m*M* AcOH) to solvent B (60% MeCN and 50 m*M* AcOH) for eight column volumes. The final yield of purified ghrelin is 2.60 g with a purity of 98% (Fig. 12.5b). The total yield is about 40% from the OmpT reaction and 80% from the deprotection.

The amino acid composition, molecular mass, and amino acid sequence of the semisynthetic human ghrelin prepared by this method were confirmed to be identical to those expected from native human ghrelin.

The bioactivity of the semisynthetic ghrelin can also be checked by measuring calcium mobilization using GHS-R-expressing cells, as described below.

The assay is performed according to the method described by Hosoda et al. (2003) with minor modifications. CHO cells expressing GHS-R are prepared by conventional lipofection. These cells are plated on flat-bottom black-wall 96-well plates at 4×10^4 cells/well and incubated in Hanks' balanced salts solution in 20 m*M* Hepes and 2.5 m*M* probenecid containing 1% fetal calf serum at 37 °C under an atmosphere of 5% CO_2/air for 12–15 h before the assay. Prior to the measurement, the cells are incubated with 4 mM Fluo-4-AM fluorescent dye (Molecular Probes, Inc.) for 1 h. The cells are then washed four times with the solution. Samples are first dissolved in the same solution containing 0.1% BSA. The intracellular Ca^{2+} changes can be measured by a fluorometric imaging plate reader (FLEXstation; Molecular Devices Inc., Sunnyvale, CA, USA). The maximum change in fluorescence over baseline is used to determine agonist responses. These results are presented in the form of their EC_{50} values. Chemically synthesized human ghrelin is prepared by following Section 2 of this chapter.

The EC_{50} value of semisynthetic ghrelin for calcium uptake activity (1.2 n*M*) is almost identical to that of the chemically synthesized one (1.3 n*M*).

4. CONCLUSION

In this chapter, we presented protocols for both solid-phase chemical synthesis and semisynthesis of human ghrelin.

With respect to the chemical synthesis, the solid-phase peptide synthesis by Fmoc/Bu^t chemistry established by Carpino (Carpino and Han, 1972) was successfully applied to the synthesis of human ghrelin. In order to make sure the reactivity of each protected Fmoc-amino acid, Kaiser test (Kaiser et al., 1970) can be performed in the preliminary run. The residues, whose coupling is insufficient, are coupled twice in order to reduce the byproducts such as amino acid deletions or truncations and to maximize the yield. The *n*-octanoylation is a key step throughout the synthesis of this O-acylated peptide. The esterification should be performed after the construction of the peptide backbone in order to avoid the possible side reactions such as O- to *N*-acyl transfer in NH_2-Ser(octanoyl)^3 moiety on the resin, β-elimination, and hydrolysis of the ester when treating the base for Fmoc removal.

Site-specific *n*-octanoylation was successfully accomplished by utilizing the different acid sensitivities toward the triyl and *tert*-butyl groups by using 1% TFA with 5% TIPS, a Trt cation trapper, in DCM. Although the Trt group is commonly used for imidazole protection in His residue, the Boc group instead of Trt is preferable for protection of imidazole of the His9 to leave it protected during treatment of 1% TFA for the removal of Trt at Ser3. Octanoylation can be made by EDC/DMAP (4 equiv. of EDC in the presence of 1 equiv. of DMAP in NMP) on the resin without difficulty.

Deprotection of all the protecting groups and concurrent cleavage of the peptide from the Wang resin were performed by treating with the mixtures of TFA containing several cation scavengers such as water, phenol, and TIPS. Regarding the purification, two different column chromatography techniques, namely, cation-exchange chromatography and RP chromatography, were employed. Since the *n*-octanoyl ester in human ghrelin is unstable against either the low pH or neutral to high pH solution due to hydrolysis, or hydrolysis and β-elimination, respectively, these column purifications were conducted under weakly acidic buffers.

According to the protocol for the solid-phase peptide synthesis noted above, from 300 to 400 mg of purified human ghrelin was constantly obtainable from 0.25-mmol scale synthesis without special difficulty throughout the procedure. Our methodology and performance for chemical synthesis of human ghrelin are basically applicable and will be reproducible for both manual solid-phase synthesis and other automated peptide synthesizers.

In the semisynthesis part, the use of a 2-chlorotrityl-resin, known as a weak acid-cleavable resin, allows for the introduction of an octanoyl group to Ser3 selectively and quantitatively for the preparation of the N-terminal fragment of human ghrelin. Another advantage of using this type of resin is the suppression of the formation of diketopiperazine side reactions (Stavropoulos et al., 1991). The merit of this protocol is that (1) all reactions can take place on the solid-phase resin and (2) the desired product can be obtained in both high yield (91%) and high purity (90%) via a one-step precipitation with H_2O.

Several strategies can be employed for the preparation of C-terminal ghrelin suitable for coupling, such as chemical cleavage (Goeddel et al., 1979; Piszkiewicz et al., 1970), the formyl–deformyl method (Irino et al., 1994), the thioester method (Kawakami et al., 1998, 2000), and chemical ligation (Dawson et al., 1994). In this protocol, however, we have highlighted a method for the preparation of a partially protected peptide segment from the overexpressed peptide using two distinct proteases and the use of this peptide fragment for subsequent coupling with synthetic O-acylated segment. The major advantages of this protocol are as follows: (1) Since a recombinant expression system is employed, it is relatively easy to scale up the preparation. (2) Highly purified peptide can be obtained by engaging site-specific proteases, OmpT and Kex2-660. The highly purified peptide consequently allows the use of a strong coupling reagent such as HBTU, which can reduce side reaction/by-product formation dramatically for the latter coupling step.

This method can possibly be applied to the preparation of physiologically active peptides with various posttranslational modifications such as acylation, glycosylation, and phosphorylation.

ACKNOWLEDGMENTS

The authors would like to thank Dr. Masayasu Kojima and Dr. Kenji Kangawa for giving them an opportunity to write on ghrelin synthesis. They are also grateful to Dr. Yasuo Kitajima and Dr. Koji Magota for useful comments, and Dr. Georgios Skretas, Dr. Alexander Wong, and Brad Frideres for a critical reading of the chapter.

REFERENCES

Bevan, A., et al., 1998. Quantitative assessment of enzyme specificity in vivo: P2 recognition by Kex2 protease defined in a genetic system. Proc. Natl. Acad. Sci. USA 95, 10384–10389.

Carpino, L.A., Han, G.Y., 1972. The 9-fluorenylmethoxycarbonyl amino protecting group. J. Org. Chem. 37, 3404–3409.

Dawson, P.E., et al., 1994. Synthesis of proteins by native chemical ligation. Science 266, 776–779.

Goeddel, D.V., et al., 1979. Expression in Escherichia coli of chemically synthesized genes for human insulin. Proc. Natl. Acad. Sci. USA 76, 106–110.

Hosoda, H., et al., 2003. Structural divergence of human ghrelin. Identification of multiple ghrelin-derived molecules produced by post-translational processing. J. Biol. Chem. 278, 64–70.

Irino, S.O., *et al.* (1994). Eur Patent 1994, EP0664338.

Kaiser, E., et al., 1970. Color test for detection of free terminal amino groups in the solid-phase synthesis of peptides. Anal. Biochem. 34, 595–598.

Kawakami, T., et al., 1998. Synthesis of reaper, a cysteine-containing polypeptide, using a peptide thioester in the presence of silver chloride as an activator. Tetrahedron Lett. 39, 7901–7904.

Kawakami, T., et al., 2000. Polypeptide synthesis using an expressed peptide as a building block via the thioester method. Tetrahedron Lett. 41, 2625–2628.

Kojima, M., et al., 1999. Ghrelin is a growth-hormone-releasing acylated peptide from stomach. Nature 402, 656–660.

Ledgerwood, E.C., et al., 1995. Endoproteolytic processing of recombinant proalbumin variants by the yeast Kex2 protease. Biochem. J. 308 (Pt. 1), 321–325.

Makino, T., et al., 2005. Semisynthesis of human ghrelin: condensation of a Boc-protected recombinant peptide with a synthetic O-acylated fragment. Biopolymers 79, 238–247.

Matsubayashi, Y., 2011. Post-translational modifications in secreted peptide hormones in plants. Plant Cell Physiol. 52, 5–13.

Matsumoto, M., et al., 2001a. Structure-activity relationship of ghrelin: pharmacological study of ghrelin peptides. Biochem. Biophys. Res. Commun. 287, 142–146.

Matsumoto, M., et al., 2001b. Structural similarity of ghrelin derivatives to peptidyl growth hormone secretagogues. Biochem. Biophys. Res. Commun. 284, 655–659.

Mizuno, K., et al., 1987. Cloning and sequence of cDNA encoding a peptide C-terminal alpha-amidating enzyme from Xenopus laevis. Biochem. Biophys. Res. Commun. 148, 546–552.

Mizuno, K., et al., 1989. Characterization of KEX2-encoded endopeptidase from yeast Saccharomyces cerevisiae. Biochem. Biophys. Res. Commun. 159, 305–311.

Okuno, K., et al., 2002. An analysis of target preferences of Escherichia coli outer-membrane endoprotease OmpT for use in therapeutic peptide production: efficient cleavage of substrates with basic amino acids at the P4 and P6 positions. Biotechnol. Appl. Biochem. 36, 77–84.

Piszkiewicz, D., et al., 1970. Anomalous cleavage of aspartyl-proline peptide bonds during amino acid sequence determinations. Biochem. Biophys. Res. Commun. 40, 1173–1178.

Reichert, J. (2010). Development Trends for Peptide Therapeutics, *Peptide Therapeutics Foundation*. www.peptidetherapeutics.org

Sambrook, J.L., Maniatis, T., 1989. Molecular Cloning: A Laboratory Manual, 2nd ed. Cold Spring Harbor Laboratory, Cold Spring Harbor, NY.

Stavropoulos, G., et al., 1991. Solid-phase synthesis and spectroscopic studies of TRH analogues incorporating cis- and trans-4-hydroxy-L-proline. Acta Chem. Scand. 45, 1047–1054.

Sugimura, K., Higashi, N., 1988. A novel outer-membrane-associated protease in Escherichia coli. J. Bacteriol. 170, 3650–3654.

Suzuki, Y., et al., 1998. High-level production of recombinant human parathyroid hormone 1-34. Appl. Environ. Microbiol. 64, 526–529.

Wang, S.S., 1973. p-Alkoxybenzylalchol resin and and p-alkoxybenzyloxycarbonylhydrazine resin for solid phase synthesis. J. Am. Chem. Soc. 95, 1328–1333.

CHAPTER THIRTEEN

Ghrelin *O*-Acyltransferase Assays and Inhibition

Martin S. Taylor*,†, Yousang Hwang*, Po-Yuan Hsiao*, Jef D. Boeke†, Philip A. Cole*,[1]

*Department of Pharmacology & Molecular Sciences, The Johns Hopkins University School of Medicine, Baltimore, Maryland, USA

†Department of Molecular Biology & Genetics and High Throughput Biology Center, The Johns Hopkins University School of Medicine, Baltimore, Maryland, USA

[1]Correspondence author: e-mail address: pcole@jhmi.edu

Contents

Abstract

Ghrelin *O*-acyltransferase (GOAT) is responsible for catalyzing the attachment of the eight-carbon fatty acid octanoyl to the Ser3 side chain of the peptide ghrelin to generate the active form of this metabolic hormone. As such, GOAT is viewed as a potential therapeutic target for the treatment of obesity and diabetes mellitus. Here, we review recent progress in the development of cell and *in vitro* assays to measure GOAT action and the identification of several synthetic GOAT inhibitors. In particular, we discuss the design, synthesis, and characterization of the bisubstrate analog GO-CoA-Tat and its

Methods in Enzymology, Volume 514
ISSN 0076-6879
http://dx.doi.org/10.1016/B978-0-12-381272-8.00013-1

ability to modulate weight and blood glucose in mice. We also highlight current challenges and future research directions in our biomedical understanding of this fascinating ghrelin processing enzyme.

1. INTRODUCTION

Acyl ghrelin is a 28-amino acid peptide hormone that has been shown to modulate body weight and blood glucose. Discovered in 1999 by Kojima and colleagues, acyl ghrelin is produced primarily in the stomach, pancreas, and duodenum. Acyl ghrelin contains an unusual eight-carbon fatty acid posttranslational modification, which is essential for its biological function (Date et al., 2000; Heller et al., 2005; Kojima et al., 1999; Prado et al., 2004; Wierup et al., 2002).

The amino acid sequence of acyl ghrelin is highly conserved among mammals, differing only at two residues between humans and rodents. Acyl ghrelin homologs have also been identified in all vertebrates examined, including bullfrogs, chicken, and tilapia. Acyl ghrelin is the endogenous ligand of the growth hormone secretagogue receptor (GHS-R1a) (Kojima et al., 1999), a G-protein-coupled receptor that is present in the brain and other tissues. Upon acyl ghrelin binding to GHS-R1a, cellular phospholipase C is activated to generate inositol triphosphate (IP_3) and diacylglycerol, which in turn increases intracellular level of Ca^{2+}, resulting in growth hormone release (Korbonits et al., 1999). This pathway is distinct from that of the growth hormone-releasing hormone (GHRH), where binding to the GHRH receptor results in increase in cAMP levels.

GHS-R1a is predominantly expressed in the arcuate nucleus of the hypothalamus but is also found in the pituitary, the ventromedial nuclei, the hippocampus, and vagal afferent neurons, with lower levels of expression seen in nonneuronal cell types in the periphery, including the pancreas (Chen et al., 2004; Cowley et al., 2003; Guan et al., 1997; Howard et al., 1996). Acyl ghrelin may exert appetite stimulation as well as modulate metabolism via a variety of mechanisms (Chen et al., 2004; Kamegai et al., 2001; Morton and Schwartz, 2001; Willesen et al., 1999); its action is mediated at least in part by the uncoupling protein *UCP2* (Andrews et al., 2008).

In order to bind to GHS-R1a, acyl ghrelin requires a unique posttranslational modification in that the serine at position 3 is *n*-octanoylated (Kojima et al., 1999). Intriguingly, ghrelin acylation may be governed by the level of medium-chain triglycerides in the diet (Nishi et al., 2005). The biosynthesis of acyl ghrelin is outlined in Fig. 13.1. The 117-amino acid

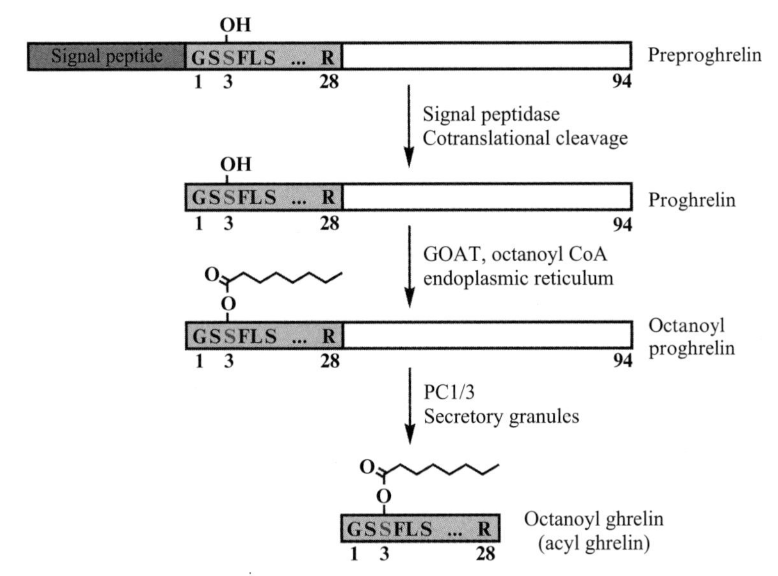

Figure 13.1 Ghrelin biosynthetic pathway. Ghrelin is synthesized as a 117-amino acid precursor, preproghrelin, containing a signal peptide, the 28-amino acid ghrelin sequence, and a 66-amino acid C-terminal peptide. The signal peptide is cotranslationally cleaved, releasing the 94-amino acid proghrelin into the lumen of the endoplasmic reticulum (ER). Attachment of the octanoate group to Ser3 of proghrelin occurs in the ER and is catalyzed by GOAT. In secretory granules, prohormone convertase 1/3 (PC1/3) then cleaves at the C-terminus of acyl proghrelin to give the mature acyl ghrelin. (See Color Insert.)

preprohormone contains a signal peptide and is cotranslationally cleaved, releasing the 94-amino acid proghrelin into the lumen of the endoplasmic reticulum (ER). Attachment of the octanoate group to Ser3 of proghrelin occurs in the ER and is catalyzed by ghrelin *O*-acyltransferase (GOAT). In secretory granules or perhaps as early as the trans-Golgi, prohormone convertase 1/3 (PC1/3) then cleaves at the C-terminus of acyl proghrelin to produce mature acyl ghrelin (Zhou et al., 1999; Zhu et al., 2006).

The ghrelin-octanoylating enzyme GOAT was molecularly identified by two groups and reported in 2008 (GOAT) (Gutierrez et al., 2008; Yang et al., 2008a). A 45-kDa multispanning transmembrane protein, GOAT is a member of the family of membrane-bound *O*-acyltransferases (MBOAT), which includes heghehog acyltransferase (Hhat) and porcupine (Porc), the substrates of which are the secretory proteins sonic hedgehog (Shh) and Wnt, respectively (Buglino and Resh, 2008; Chamoun et al., 2001; Hofmann, 2000; Kadowaki et al., 1996; Takada et al., 2006).

Ghrelin and Wnt3a are the only proteins known to contain acylated serine residues. MBOAT family members have diverse substrates including sterols, phospholipids, and proteins, but as integral membrane proteins the enzymatic properties of these proteins are poorly understood. To date, ghrelin is the only established substrate for GOAT, and is the only known octanoylated protein. In addition to the metabolic roles for acyl ghrelin, this hormone has also been implicated in many areas including but not limited to learning and memory, as well as food-related behavior (Carlini et al., 2002; Diano et al., 2006) and gastrointestinal motility (de la Cour Dornonville et al., 2004; Ogiso et al., 2011). These and other aspects of ghrelin physiology have been recently reviewed (Castaneda et al., 2010). Recent studies on GOAT knockout mice have suggested that this enzyme is important for survival under prolonged, severe starvation, although this characteristic may be context dependent (Sun et al., 2008; Yi et al., 2012; Zhao et al., 2010a). The potential therapeutic benefits of exploiting the ghrelin–GOAT system in managing obesity and diabetes are attractive but not yet fully explored. Much drug discovery work has been focused on GHS-R1a receptor modulators, but the discovery of GOAT offers new potential for the generation of GOAT-selective inhibitors; however, the potential scope of pharmacologic actions of such modulators is not yet known. This chapter discusses the practical issues involved in developing GOAT inhibitors, the initial progress in this area, as well as future challenges.

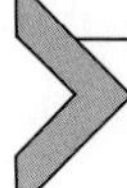

2. ANALYSIS OF ACYL GHRELIN LEVELS AND GOAT ACTIVITY

2.1. Assays measuring acyl and des-acyl ghrelin from blood and cells

In order to develop effective GOAT inhibitors, an important technical hurdle to overcome is reliable measurement of acyl and des-acyl ghrelin. A challenge in designing assays to measure ghrelin acylation is the instability of acyl ghrelin in biological systems, predominantly due to hydrolysis of the acyl group by esterases in plasma, cell culture medium, and cell extracts. Any assay must measure typically dilute concentrations of the hormone, preserve acylation through isolation, and then detect the acylation in the context of what is usually a larger amount of des-acyl ghrelin. A comparison of literature reports suggests wide ranges (10–100-fold or more) in concentrations of blood acyl ghrelin in studies on normal humans and rodents, underscoring the complexity of the measurements (Groschl et al., 2004; Liu et al., 2008a).

Early acyl ghrelin assays used the functionality of the hormone on the GHS-R1a receptor. While useful, these assays are rather complex and lack precision. Combinations of reversed-phase HPLC and immunoassays have proved increasingly reliable. Acyl ghrelin measurements relying on mass spectrometry have also been reported, although our lab has had limited success with this approach (Gutierrez et al., 2008; Satou et al., 2010). The current state of the art for ease of use and reproducibility seems to be sandwich ELISA assays.

Two-site immunoassays are generally more sensitive and specific than single-antibody assays and also do not cross-react with peptide fragments (Nussbaum et al., 1987). Some early publications used a two-site sandwich ELISA developed inhouse (Barkan et al., 2003), but this was not widely adopted. Liu et al. (2008a) developed two novel sandwich ELISAs specific for acyl and des-acyl full-length ghrelin. Capture was achieved using N-terminal acyl- and des-acyl-specific antibodies, and detection for both assays used an affinity-purified antibody to ghrelin's C-terminal amino acids 21–27. They also detailed an improved collection protocol in which blood is collected directly into chilled tubes preloaded with the protease inhibitor AEBSF (4-(2-aminoethyl)benzenesulfonyl fluoride), maintained on ice until prompt centrifugation, and then immediately acidified with 20% (v/v) 1 N HCl to protect the ghrelin ester from hydrolysis. These combined improvements represent the current state of the field, although the esterase inhibitor PHMB (*p*-hydroxymercuribenzoic acid) can be substituted for AEBSF and adding 1 *M* NaCl improves plasma separation.

Commercial two-site sandwich ELISA kits by Spi-Bio (now Bertin Pharma) are now available, sold through Cayman Chemical and Alpco Diagnostics. These kits have been used in recent studies (Barnett et al., 2010; Zhao et al., 2010b). The kits include wells coated with a C-terminal capture antibody and a modification-specific N-terminal antibody conjugated to acetylcholinesterase. The kits from the two companies are apparently identical except for the color of their packaging. We have validated their modification-specificity and sensitivity against both homemade standards and those supplied by the manufacturer. We also tested kits from Millipore with similar results. Other two-site kits are available from Mitsubishi Kagaku Iatron (Tokyo, Japan), using N-terminal-modification-specific antibody and C-terminal capture antibody, although we have not tested them.

2.2. Measuring acyl and des-acyl ghrelin levels in cell-based model systems

To establish a model system for ghrelin acylation, the field first turned to cell lines. The first cell line established was the TT cell, a medullary thyroid carinoma line (Kanamoto et al., 2001). Ghrelin production from these cells was similar to that in rat intestinal production, and approximately 20% of the ghrelin produced was found to be acylated. Ghrelin was secreted into the culture medium as well; the vast majority of secreted material was found to be des-acyl , and the different ratios between intracellular and secreted pools were attributed to degradation. This cell line was used to discover GOAT by Gutierrez et al. (2008) (see below). They demonstrated that the amount of acyl ghrelin in the medium could be increased by the addition of octanoic acid or protection of the acyl group with a modification-specific antibody and that octanoylation occurred only at Ser3.

The human erythroleukemia cell line (De Vriese et al., 2005) also produces acyl ghrelin, which was shown to be part of an autocrine loop leading to cell proliferation. Interestingly, they also demonstrate that the half-life of acyl ghrelin in culture medium is approximately 1 h. However, these cells are of limited utility as a model because the amount of acyl ghrelin produced is very low and ghrelin production is unstable (Takahashi et al., 2009).

Yang et al. (2008a) tested a number of cell lines for the ability to process proghrelin to ghrelin, measuring the retained intracellular ghrelin in cell lysates. HEK-293 and CHO-7 lysates contained only proghrelin, but the endocrine cell lines AtT-20, INS-1, and MIN-6 all contain mixtures of proghrelin and mature, processed ghrelin. Transfection of candidate acyltransferases into the INS-1 cell line was then used to independently discover GOAT, and all three cell lines were able to produce mature, octanoyl ghrelin when GOAT was transfected in.

Gutierrez et al. have transiently transfected plasmids expressing ghrelin and GOAT into GripTite HEK-293 MSR cells and, combined with addition of fatty acids in the medium, showed production of mature acyl ghrelin at much higher levels than seen in TT cells (Gutierrez et al., 2008). Interestingly, they detect mature, processed ghrelin in the medium of the precise expected mass. This demonstrates that HEK-293 cells can in fact process proghrelin to ghrelin, but the cleaved form may be promptly secreted. Enhanced production of acyl ghrelin was subsequently reported in three cell lines, namely, TT, AtT20, and COS-7, by cotransfecting a plasmid expressing ghrelin with either or both plasmids expressing GOAT and one of five proteases (Takahashi et al., 2009).

Three improved cell lines were recently isolated from ghrelinomas in ghrelin-promoter SV40-T-antigen transgenic mice: MGN3-1, PG-1, and SG-1 cells. These cell lines all express ghrelin, GOAT, and PC1/3, and have been shown to recapitulate physiologic ghrelin signaling to some extent, so they should be useful model systems going forward. MGN3-1 (Mouse Ghrelinoma 3-1) cells produce 5000 times more ghrelin than TT cells (Iwakura et al., 2010). Approximately 6–14% of the ghrelin produced in this cell line is acylated when octanoic acid is added to the medium, depending on the experiment, and acyl ghrelin is secreted into the culture medium in mature, cleaved form. siRNA knockdown of GOAT slightly depressed the ratio of acyl to des-acyl ghrelin produced.

The similar cell lines PG-1 and SG-1 were derived from pancreatic and stomach ghrelinomas, respectively (Zhao et al., 2010b). These lines produce comparably high levels of ghrelin to MGN3-1, although direct comparison is difficult because of different measurement techniques (single-site RIA vs. 2-site ELISA) and distinct normalizations (per cell vs. per microgram cellular protein). In PG-1 and SG-1 cells, up to 30% of secreted ghrelin is octanoylated when sodium octanoate-albumin is added to the culture medium.

We recently established the vector phPPG-mGOAT for stable expression of ghrelin and GOAT in the widely available 293T and HeLa cell lines (Barnett et al., 2010). This episomally maintained vector has a CMV promoter (Cyotomegalovirus immediate early promoter), ghrelin, internal ribosome entry site, and then GOAT, in that order, providing substantial production and a high substrate-to-enzyme ratio. By measuring intracellular ghrelin by ELISA, we then used this system to test structure–activity relationships for GOAT inhibitors. Note that this approach does not explicitly distinguish proghrelin from mature ghrelin forms and avoids the issues of ester hydrolysis that appear to accompany secretion into the medium. We have also found it most advantageous to consistently measure acyl and des-acyl ghrelin from the same samples and express the percentage of acyl ghrelin as a key parameter to circumvent the fluctuations of total ghrelin that may be related to cellular conditions or isolation procedures.

2.3. *In vitro* direct GOAT activity assays using microsomes

Yang et al. established an *in vitro* GOAT acyl transfer assay using membranes enriched for ER (microsomes) from insect cells infected with baculovirus encoding mouse GOAT to transfer ^{3}H-labeled octanoate onto proghrelin-

His8 (Yang et al., 2008b). Mouse GOAT microsomes were prepared as follows: GOAT was cloned into pFastBac HT-A, giving it an N-terminal His_{10}-TEV tag, and the baculovirus infected SF9 cells from which the microsomes were harvested. As a negative control, the MBOAT fingerprint residue H338 was mutated to alanine and control virus was also employed. Proghrelin-His_8 and mutant proghrelins were produced in bacteria using an N-terminal GST-TEV tag, such that TEV cleavage produced the authentic N-terminal sequence of proghrelin.

The *in vitro* octanoylation assay of Yang et al. was performed using 1 μM ^{3}H-octanoyl-CoA (high specific activity), 5 μg proghrelin-His_8 (8.6 μM), and 50 μg GOAT-containing microsomes. Reactions were quenched with buffer containing 0.1% SDS, labeled proghrelin was separated from the reaction mixture using nickel-affinity chromatography, and ^{3}H-octanoyl-proghrelin was quantified using liquid scintillation counting. Yang et al. found that addition of long-chain fatty coenzyme A (CoA) conjugates stimulate the reaction up to 3.5-fold by preventing hydrolysis of octanoyl-CoA to octanoate and therefore 50 μM palmitoyl-CoA was present in all further reactions.

The enzyme kinetics observed in these studies was complex and nonlinear, probably because of the presence of esterases, palmitoyl-CoA, and product inhibition. The apparent K_m values for octanoyl-CoA and proghrelin were found by Yang et al. to be 0.6 and 6 μM, respectively. As a control, it was shown that the S3A mutant ghrelin could not be octanoylated by GOAT. Octanoylation of ghrelin mutants G1S, G1A, and F4A was dramatically reduced, indicating the importance of residues G1, S3, and F4 in this recognition. L5A and S6A ghrelin mutations had smaller effects, indicating a lesser contribution of these side chains to recognition, and there was no effect of S2A or P7A mutations. Also, the addition of the two N-terminal residues Ser-Ala, which would be present were the signal peptide not cleaved, markedly reduced octanoylation. Together with the finding of acylated proghrelin in transfected INS-1 cells, this evidence suggests that the natural substrate for GOAT is proghrelin after signal peptide cleavage (Yang et al., 2008b). Yang et al. also demonstrated that truncated ghrelin pentapeptides could be acylated by microsomal GOAT, although they showed weaker apparent affinity for the enzyme.

Our group has developed a related assay for studying recombinant microsomal GOAT, but it has been prepared from human cells rather than insect cells (Barnett et al., 2010). HEK293T GnTI-cells were transfected with mouse GOAT containing a C-terminal 3xFlag tag cloned into a mammalian expression vector (CAG promoter). Microsomes were prepared in a manner similar to that reported by Yang et al., and GOAT assays were performed using a synthetic ghrelin tagged with a C-terminal biotin (Ghrelin27-Biotin), taking advantage of the robustness of streptavidin–biotin affinity. Assays were carried out with ^{3}H-octanoyl-CoA, with streptavidin beads used to isolate radiolabeled octanoyl ghrelin. Although signal was detected in this human cell expression system, there appears to be significantly greater signal using the insect cell expression system.

Garner and Janda (2010) have developed an elegant nonradioactive GOAT assay exploiting click chemistry. Replacing octanoyl-CoA with octynoyl-CoA containing a carbon–carbon triple bond at the 7–8 position, Garner and Janda reacted this with microsomal GOAT and bead-immobilized ghrelin pentapeptide, affording octynoyl ghrelin. This immobilized octynoyl ghrelin was then conjugated using copper-catalyzed cycloaddition to azido-HRP (horseradish peroxidase), and detection was achieved using the amplex red fluorogenic substrate. A strong signal-to-noise ratio was achieved, and the apparent K_m values measured for *n*-octynoyl-CoA and immobilized ghrelin (1–5) pentapeptide were 68 and 100 n*M*, respectively. The fact that the peptide is immobilized on a solid surface while the enzyme is still membrane-bound may explain the 1000-fold lower apparent K_m for the peptide and 10-fold lower apparent K_m for octynoyl-CoA in these conditions. The triple bond in the octynoyl-CoA may also affect its interaction with GOAT. This click assay appears to be potentially more amenable to high-throughput screening compared to the radioactive assay (Garner and Janda, 2010, 2011).

3. GOAT INHIBITOR DISCOVERY

Three classes of GOAT inhibitors have been described so far: product acyl-peptide analogs, a small molecule detected in a high-throughput screen, and a rationally designed bisubstrate analog.

When Ser3 in ghrelin (Fig. 13.2A, Compound 2) was replaced with DAP ((*S*)-2,3-diaminopropionic acid) creating an octanoyl-amide in place of ester (Fig. 13.2A, Compounds 3 and 4), both the 28-mer and 5-mer acyl ghrelins were potent GOAT inhibitors with IC_{50} values of 0.2 and 1 μ*M*, respectively. It is likely that these compounds correspond to the strong product inhibition of GOAT, but the lack of hydrolytic sensitivity of the amide linkage confers greater stability.

A

1 OH GSSFLSPEHQRVQ-QRKESKKPPAKLQPR

2 O O GSSFLSPEHQRVQ-QRKESKKPPAKLQPR

3 O NH GS(Dap)FLSPEHQRVQ-QRKESKKPPAKLQPR

4 O NH GS(Dap)FL-CONH$_2$

5 O R N O HO HN

R=*n*-hexyl, *n*-octyl

6 O SCoA NH GS(Dap)FLSPEHQ-Ahx-Tat

GO-CoA-Tat

7 O SCoA NH GS(Dap)FLSPEHQ-Ahx-Tat

8 O SCoA NH GS(Dap)FLSPEHQ-Ahx-Tat

9 O NH GS(Dap)FLSPEHQ-Ahx-Tat

10 O SCoA NH GS(Dap)FLSPEHQ-Ahx-Tat

11 O SCoA NH GS(Dap)FLSPEHQRVQ-QRKESKKPPAKLQPR

B

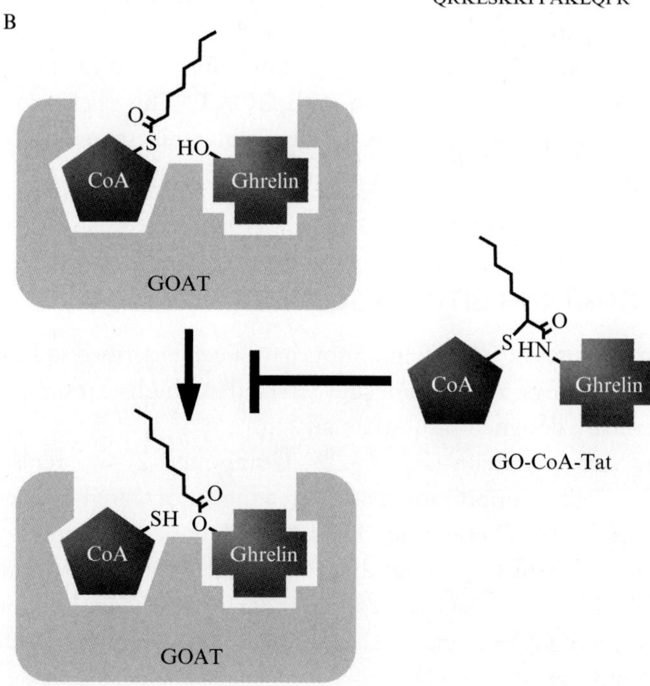

While showing high potency, product analogs have pharmacologic challenges for *in vivo* applications. As peptide compounds, their ability to penetrate cell membranes may be limited. Perhaps more importantly, they are likely potent agonists of GHS-R1a. Four residues of ghrelin functionally activate GHS-R1a about as efficiently as full-length ghrelin (Bednarek et al., 2000). We have also found that a Tat-conjugated 10mer-amide is also a potent GHS-R1a agonist (see below).

Garner and Janda (2011) carried out compound screening using their click assay. The assay's Z′ factor was determined to be 0.63, indicating high assay quality (Zhang et al., 1999). A small "credit card" library of drug-like small molecules was then screened for inhibition of GOAT, and two related small molecule inhibitors were discovered ($IC_{50}=7.5$ and 13 μ*M*, respectively, see Fig. 13.2, Compound 5). Interestingly, these compounds contain six- and eight-carbon alkyl chains, suggesting that they possibly compete for the octanoic acid binding site on GOAT. Although these compounds have not yet been explored in depth pharmacologically, they appear to represent attractive leads.

3.1. Bisubstrate analogs

It is now well established that mimics of the transition state of an enzyme-catalyzed reaction can serve as high-affinity inhibitors based on the premise that most enzymes have evolved to bind tightly to the transition state. For enzymes that use two substrates in a ternary complex mechanism, an attractive approach to rational inhibitor design involves covalent linkage of the two substrates to generate a bisubstrate analog, as shown schematically in Fig. 13.2B. Such compounds can show energetically favorable interactions with enzymes because dual occupancy of the substrate binding pockets is facilitated without the entropic penalty incurred with random collision of the individual substrate molecules. To be most effective, it is understood that a tether for the linkage

Figure 13.2 GOAT inhibitors. (A) Chemical structures of ghrelin and GOAT inhibitors. 1: Des-acyl ghrelin. 2: Acyl ghrelin. 3: Amide-linked octanoyl ghrelin. 4: Amide-linked 5-mer octanoyl ghrelin with C-terminus amidated. 5: Inhibitors discovered by Garner and Janda (2011). 6: GO-CoA-Tat. 7,8: Bisubstrate compounds with five and three amino acids of ghrelin. 9: GO-Tat: an octanoyl-amide Tat-tagged product analog. 10: Bisubstrate inhibitor with two-carbon acyl group. 11: Ghrelin28-Oct-CoA, a bisubstrate compound. (B) Mechanism-based design strategy of GO-CoA-Tat. The lipid–enzyme interaction is not shown but may be important. Also, the form of ghrelin acylated by GOAT is likely proghrelin; the smaller version is shown for clarity. (See Color Insert.)

must be able to approximate a mechanistically relevant orientation of the two substrates, ideally capturing elements of the transition state. In the best cases, bisubstrate analogs can show binding free energies to an enzyme that are equal to or greater than the sum of the binding energies of the individual substrate components to the same protein. Successful examples have been recorded of bisubstrate analogs for protein kinases and protein acetyltransferases inspired by enzyme mechanism considerations. By placing an acetyl bridge between ATP and peptide substrate sequences for kinases, compounds that show low micromolar to subnanomolar affinities have been achieved for the insulin receptor kinase, protein kinase A, Csk tyrosine kinase, cyclin-dependent kinase, Abl tyrosine kinase, and the epidermal growth factor tyrosine kinase (Bose et al., 2006; Cheng et al., 2006; Hines and Cole, 2004; Hines et al., 2005; Jencks, 1981; Levinson et al., 2006; Medzihradszky et al., 1994; Parang et al., 2001; Shen and Cole, 2003).

A related linker worked effectively for the histone acetyltransferase enzymes PCAF/GCN5 and p300/CBP which contain a CoA and peptide substrate fragments bridged by an acetyl spacer (Lau et al., 2000; Sagar et al., 2004). Several of these bisubstrate analogs have been useful in structural analysis of the enzyme reaction mechanism and substrate binding features (Liu et al., 2008b). On the other hand, these analogs have suffered from limited pharmacologic utility because of their large size, polarity, and the challenges of cell membrane penetration. However, the discovery of cell-penetrating peptide sequences derived from the HIV Tat protein have allowed for cell and *in vivo* applications for the bisubstrate analog HAT inhibitors (Bricambert et al., 2010; Cerchietti et al., 2010; Cleary et al., 2005; Guidez et al., 2005; Liu et al., 2008c; Marek et al., 2011; Oussaief et al., 2009; Spin et al., 2010; Wang et al., 2011; Zheng et al., 2005).

3.2. Development of GO-CoA-Tat, a potent and selective bisubstrate inhibitor of GOAT

Following the bisubsrate analog approach described above, we reported the development of GO-CoA-Tat (Barnett et al., 2010). GO-CoA-Tat (Fig. 13.2A, Compound 6) uses nonhydrolyzable amide and thioether linkages to combine octanoyl-CoA with the first 10 amino acids of ghrelin, which are 100% conserved in mammals. An HIV Tat-derived peptide sequence was attached to the C-terminus using a flexible linker to allow cell penetration. GO-CoA-Tat and a set of related analogs and control compounds (Fig. 13.2A, Compounds 6–11) were synthesized by a solid-phase strategy.

We then tested these compounds in HEK and HeLa cells expressing ghrelin and GOAT stably transfected with our phPPG-mGOAT vector. Cells were maintained in a medium supplemented with octanoic acid, preincubated with the compound for 24 h, and then lysed. Intracellular acyl and des-acyl proghrelin were measured by ELISA, with values validated using kits from two manufacturers and standards made inhouse. We first tested GO-CoA-Tat in these models, and the mean inhibitory concentration was ~5 μ*M*; control compound D4-Tat had no effect. Interestingly, maximum inhibition was achieved only after 24 h of incubation with GO-CoA-Tat. This could be due to the atypical behavior of the enzyme or inhibitor or due to preexisting intracellular stores of acyl ghrelin. To test this, we used the radioassay described in Section 2.3 and found substantial inhibition occurred within 5 min with 100 n*M* GO-CoA-Tat. This too suggests that there are intracellular stores of acyl ghrelin in these cells.

We examined structure–activity relationships required for inhibition with compounds used at 6 μ*M*. Consistent with what was seen in other assays, five residues of ghrelin were sufficient for inhibition but three residues were not. Inclusion of 10 residues increased potency, with a maximum of ~75% inhibition of acylation seen. CoA was also required for inhibition; this finding is discussed further in Section 4.3, and a version of the bisubstrate compound with a truncated two-carbon acyl group still showed some inhibition. Tat was required for inhibition, consistent with its role in entry into the cells and ruling out action on a surface receptor. None of the compounds was toxic to the cells in the low micromolar concentration range. GO-CoA-Tat's specificity is also reflected in its lack of inhibition of three acetyl-CoA-utilizing enzymes.

To further analyze GO-CoA-Tat's inhibition of GOAT, we developed a direct binding assay for GOAT, taking advantage of photocrosslinking technology. We first synthesized two chemically modified versions of our bisubstrate inhibitor, namely, GO-CoA-Tat-F4BP and GO-CoA-Tat-L5BP, in which Phe4 or Leu5, respectively, is replaced with a photoreactive amino acid benzoyl-phenylalanine and each is tagged with a biotin group (Barnett et al., 2010). We showed that this compound could covalently crosslink to recombinant solubilized or microsomal GOAT. This crosslinking could be blocked by an excess of unlabeled GO-CoA-Tat, providing evidence for specificity and demonstrating direct binding of GO-CoA-Tat to GOAT.

For these experiments, we used GOAT with a C-terminal 3xFlag tag, produced in SF9 cells using baculovirus. Microsomes were prepared as above, and the reaction was performed either in the microsome membranes or with GOAT purified to homogeneity using anti-Flag affinity chromatography and the Fos-Choline-16 detergent (Anatrace). This detergent was chosen because of its high ability to solubilize GOAT and because we reasoned that the long alkyl chain is less likely to interfere with the octanoic acid binding site on GOAT.

Photocrosslinking reactions were performed in a small water-jacketed quartz cuvette, custom made for this purpose by Quark Glass. The cuvette was connected to the water line and suspended above a magnetic stir plate. A small teflon stir bar was added, with medium agitation. A mercury UV lamp with a ~360-nm long-wave filter, such as UVP #B-100AP, was positioned with the center of the lamp approximately 2 cm from the cuvette, positioning the sample at the position of peak intensity. A time course experiment (not shown) demonstrated that the reaction had neared completion by 30 min. Crosslinked membranes were then solubilized and immunoprecipitated. Biotinylation was visualized using SDS-PAGE and streptavidin–HRP or, for more sensitivity, streptavidin followed by polyclonal anti-streptavidin.

3.3. Glucose and weight control in mice with GO-CoA-Tat

Treatment of C57BL6 mice on medium-chain triglyceride (MCT) diets (Kirchner et al., 2009) with GO-CoA-Tat at 40 mg/kg dose, but not with the control compound D4-Tat or vehicle, decreased plasma acyl ghrelin levels without changing the des-acyl ghrelin levels. Maximum inhibition was seen after 6 h, but some acyl ghrelin suppression was still detectable 24 h after GO-CoA-Tat treatment. Because of the daily fluctuations between animals and *ad lib* feeding, we found that the acyl to des-acyl ghrelin ratio was a more sensitive and specific measure of inhibition.

We explored the effect on weight gain over a 1-month period in mice placed on an MCT diet. Daily IP injections of GO-CoA-Tat as above reduced the weight gain seen in vehicle-treated mice. As measured by QMR spectroscopy, the difference in weight was due to significantly reduced fat mass in the GO-CoA-Tat-treated animals. In contrast, GO-CoA-Tat- versus vehicle-treated ghrelin-knockout mice showed no statistically significant difference in weight or body composition. To investigate the potential for GO-CoA-Tat toxicity, we examined the blood chemistries and cell counts in the mice after 1 month of treatment with the agent. There was no apparent untoward effect on normal blood chemistries or cell counts under these conditions. Interestingly, WT mice treated with GO-CoA-Tat showed reduced IGF-1 and lower blood glucose, consistent with suppression of ghrelin-mediated somatotroph signaling.

To investigate the role of acute pharmacologic inhibition of acyl ghrelin in insulin signaling and glucose homeostastis, we pretreated with GO-CoA-Tat and then measured the response to a glucose challenge, first in isolated pancreatic islets and then in mice. The insulin response was increased in islets and mice, where the response was accompanied by reduced blood glucose. In contrast, there was no effect when the studies were repeated in ghrelin-knockout animals, suggesting that GO-CoA-Tat's effects on insulin are due to the inhibition of ghrelin acylation. Finally, we showed by quantitative PCR that islets isolated from mice pretreated with GO-CoA-Tat had a 20-fold reduction in expression of uncoupling protein 2 mRNA (*UCP2*, which suppresses insulin secretion), but there was no change in *UCP2* expression in the gastric fundus. Together, these data show a tissue-specific role for GOAT inhibition in augmentation of insulin secretion. Regulation of *UCP2* also highlights the importance of ghrelin acylation in obesity and type 2 diabetes, underscoring the need for more drug-like GOAT inhibitors (Andrews et al., 2008; Dezaki et al., 2008; Joseph et al., 2002; Sun et al., 2006; Tong et al., 2010; Zhang et al., 2001).

4. CHALLENGES AND FUTURE DIRECTIONS

While there has been significant progress in GOAT enzymology and inhibition reported in the past few years, many challenges remain.

4.1. Targeting GOAT versus GHS-R1a

One hurdle in developing therapeutic agents to target GOAT is the extensive overlap in the pharmacophore recognized by GOAT and GHS-R1a. We have recently tested a number of compounds using a version of the GHS-R1a assay reported by Kojima et al. (Barnett et al., 2010; Kojima *et al.*, 1999) (Table 13.1). Dose–response traces from individual wells treated with acyl ghrelin are shown in Fig. 13.3A. The responses to selected compounds are shown in Fig. 13.3B. Acyl ghrelin and amide ghrelin (Ser3 Dap-Octanoyl-Amide, Fig. 13.2 compounds 2 and 3, respectively) are indistinguishable at the receptor. In contrast, GO-CoA-Tat does not activate GHS-R1a at the concentrations tested, which include concentrations higher than those used in mice. We also showed that the activity of 1 μ*M* or 100 n*M* ghrelin at GHS-R was not inhibited by GO-CoA-Tat at 60 nM, 600 nM, or 6 μM. Surprisingly, a 28-mer bisubstrate compound Ghrelin28-Oct-CoA (Compound 11) could activate the receptor with reduced affinity.

4.2. Moving toward studies of purified GOAT

Studying the activity and mechanism of purified GOAT is critical for improved inhibitor development. To date, all reported GOAT assays, with the exception of one example of our photocrosslinking-based binding assay, have been carried out in complex microsome mixtures containing thousands of other proteins. GOAT is only a small fraction of the total protein in these experiments, and reactions are usually carried out in the presence of relatively high concentrations of palmitoyl-CoA to inhibit esterases and other CoA-utilizing enzymes in these mixtures. Further, only very low conversion percentages are achievable. However, GOAT has not yet been solubilized in an active form. Progress in these areas will be critical to developing better inhibitors and for structural studies of GOAT.

4.3. Structural and mechanistic studies of GOAT and GOAT topology

Currently, we know very little about the structure and mechanism of GOAT. The specific and potent binding of the bisubstrate inhibitor GO-CoA-Tat argues for a ternary complex mechanism, but other mechanisms are still formally possible and further studies in this area are needed. We

Table 13.1 EC_{50} values in the GHS-R1a assay

Compound	**EC_{50}**
Acyl ghrelin	18 ± 6 nM
Des-acyl ghrelin	> 10 μM
Amide ghrelin	19 ± 8 nM
GO-Tat	23 ± 4 nM
GO-CoA-Tat	> 10 μM
Ghrelin28-Oct-CoA	270 ± 70 nM
D4-Tat	> 10 μM
GO-Tat S_2Oct	> 10 μM
GO-Tat S_6Oct	> 10 μM

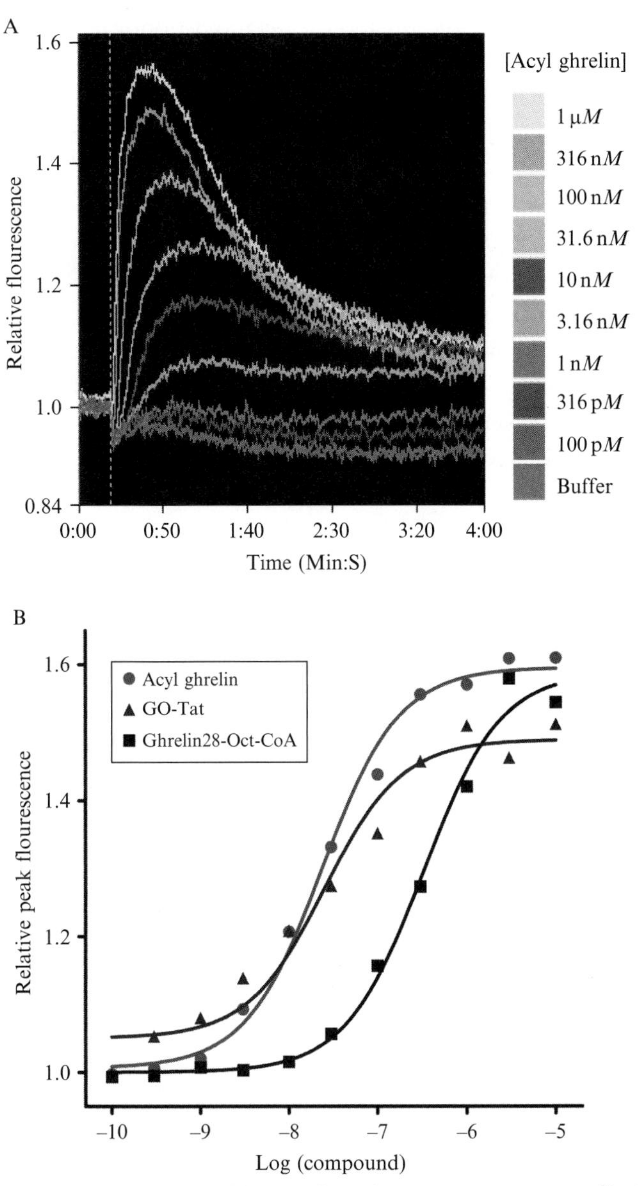

Figure 13.3 GHS-R1a assay in stably transfected HEK-293T-GHS-R1a cells. (A) Typical dose–response traces for acyl ghrelin, with concentrations on half-log scale from 1 μM to 100 pM, with buffer-only control. (B) Agonism for acyl ghrelin, GO-Tat (Fig. 13.2, Compound 9), and the bisubstrate compound Ghrelin28-Oct-CoA (Fig. 13.2, Compound 11). EC_{50} values are reported in Table 13.1. (See Color Insert.)

do not know where proghrelin and the octanoyl-donor bind, and the identity of the octanoyl donor has not been proven. Also, the identity of the active site has not been confirmed.

A map of the topology of GOAT will be helpful to answer some of these questions. Based on sliding-window Kyte–Doolittle hydropathy plots, mouse GOAT was predicted to contain eight transmembrane helices (TM) (Yang et al., 2008a); however, no experiments have yet been reported to further probe GOAT's topology. To acylate ghrelin, GOAT's active site should face the ER lumen where ghrelin is localized; this logic also applies to other MBOATs that acylate secreted and GPI-anchored proteins. The two most conserved residues in the MBOAT fingerprint are N307 and H338 in mouse GOAT, and only H338 is conserved throughout the entire MBOAT family. The homologous histidine in another MBOAT, the human cholesterol acyltransferase ACAT1, was mapped to the lumenal boundary of a TM or perhaps in a short-loop ER lumen (Guo et al., 2005). Like GOAT, mutation of this histidine in ACAT1 abolished catalytic activity. The topology of distantly related yeast MBOAT members Ale1p, Are1p, and Gup1p was recently studied in detail (Pagac et al., 2011); these enzymes acylate lipids, cholesterol, and GPI-anchored proteins, respectively, and Are1p is the yeast ortholog of ACAT1. The conserved histidine in all cases was shown to be in the ER lumen, but none of the active sites has yet been mapped.

These conserved residues have been called "catalytic residues"; however, there is a lack of mechanistic data that firmly establishes this point. Although GOATs with alanine mutants of these conserved His and Asn residues are inactive, the homologous histidine residue was recently shown to be dispensable for palmitoylation of Shh by Hhat. In this case, the H379A caused reduced binding affinity of Shh as reflected in an increased K_m without changing V_m (Buglino and Resh, 2010). Therefore, at this point, the identity of the active site of GOAT and other MBOATs is unclear.

With the active site of GOAT in the ER lumen, one question raised by the success of the microsomal GOAT assays is, how do the ghrelin peptides and octanoyl group reach the active site of GOAT? Microsomes are believed to be sealed bilayer vesicles, so in order to octanoylate ghrelin, some of the microsomes are presumably inside-out, with normally lumenal contents exposed to the assay buffer.

Another unanswered question is, if octanoyl-CoA is the correct acyl donor, how do acyl-CoAs, localized predominantly in the cytoplasm, gain access to the ER lumen? Although it has been hypothesized that GOAT might have a role in transport of the octanoyl group across the membrane, there is no specific evidence yet reported. GO-CoA-Tat appears to rely on Tat-mediated delivery of the agent into the cytoplasm (Potocky et al., 2003; Schwarze et al., 1999), but interestingly GO-Tat does not appear to block cellular GOAT even though it is homologous to potent product inhibitors that block GOAT *in vitro*. Perhaps, the CoA moiety in GO-CoA-Tat is crucial for ER entry through the proposed transport properties of GOAT. In addition to the possibility of GOAT participating in substrate transport, other enzyme mechanisms are also formally possible. Octanoate may first be transferred from octanoyl-CoA to an intermediate host, GOAT, another protein, or a lipid. It is even possible that an additional protein collaborates with GOAT to effect acyl transfer.

4.4. Exploring other MBOATs

The MBOATs porcupine and Hhat, acylating Wnt, and hedgehog proteins, respectively, share many features in common with GOAT. A detailed understanding of these pathways is critical for the progress in understanding how these signaling ligands modulate development, stem cell renewal and differentiation, and initiation and maintenance of cancer (Clevers, 2006; Pasca di Magliano and Hebrok, 2003; Reya and Clevers, 2005; Zhao et al., 2009). Techniques learned in studying the enzymology, structure, and function of the GOAT/ghrelin system should be readily translated to these related cases, leading to new inhibitors and insights into structure and function. Porcupine, in particular, is the only other protein known to acylate a serine and should therefore be most mechanistically like GOAT. A family of small molecules targeting porcupine was recently reported (Zhao et al., 2009); analogs of these compounds may also inhibit GOAT or other MBOATs and should be investigated.

4.5. Toward potent small-molecule inhibitors of GOAT

With the development of the first small-molecule inhibitor of GOAT and a high-throughput-ready screen (ELCCA), the prospect of a potent, specific small-molecule inhibitor of GOAT is exciting (Garner and Janda, 2011). Additionally, new *in vitro* and cell-based assay systems will surely lead to new mechanistic and structural insights and may also be amenable to screening approaches (Barnett et al., 2010; Iwakura et al., 2010; Yang et al., 2008b). With new model systems for GOAT inhibition now established in cells and mice, the efficacy of new compounds can now be evaluated. These compounds could be very promising leads in the treatment of obesity, diabetes, and other metabolic disorders and may provide much needed tools to map out a complete understanding of GOAT in biology.

ACKNOWLEDGMENTS

We thank Brad Barnett, Dan Leahy, Jun Liu, Paul Pfluger, Henriette Kirchner, Matthias Tschöp, and Mehboob Hussain for contributions to these studies and for a critical reading of this chapter. We also thank Don Steiner for helpful discussion. Further, we thank the NIH and Pfeiffer Foundation, Kaufman Foundation, and Keck Foundation for financial support.

REFERENCES

Andrews, Z.B., et al., 2008. UCP2 mediates ghrelin's action on NPY/AgRP neurons by lowering free radicals. Nature 454, 846–851.

Barkan, A.L., et al., 2003. Ghrelin secretion in humans is sexually dimorphic, suppressed by somatostatin, and not affected by the ambient growth hormone levels. J. Clin. Endocrinol. Metabol. 88, 2180–2184.

Barnett, B.P., et al., 2010. Glucose and weight control in mice with a designed ghrelin O-acyltransferase inhibitor. Science 330, 1689–1692.

Bednarek, M.A., et al., 2000. Structure-function studies on the new growth hormone-releasing peptide, ghrelin: minimal sequence of ghrelin necessary for activation of growth hormone secretagogue receptor 1a. J. Med. Chem. 43, 4370–4376.

Bose, R., et al., 2006. Protein tyrosine kinase-substrate interactions. Curr. Opin. Struct. Biol. 16, 668–675.

Bricambert, J., et al., 2010. Salt-inducible kinase 2 links transcriptional coactivator p300 phosphorylation to the prevention of ChREBP-dependent hepatic steatosis in mice. J. Clin. Invest. 120, 4316–4331.

Buglino, J.A., Resh, M.D., 2008. Hhat is a palmitoylacyltransferase with specificity for N-palmitoylation of Sonic Hedgehog. J. Biol. Chem. 283, 22076–22088.

Buglino, J.A., Resh, M.D., 2010. Identification of conserved regions and residues within Hedgehog acyltransferase critical for palmitoylation of Sonic Hedgehog. PLoS One 5, e11195.

Carlini, V.P., et al., 2002. Ghrelin increases anxiety-like behavior and memory retention in rats. Biochem. Biophys. Res. Commun. 299, 739–743.

Castaneda, T.R., et al., 2010. Ghrelin in the regulation of body weight and metabolism. Front. Neuroendocrinol. 31, 44–60.

Cerchietti, L.C., et al., 2010. BCL6 repression of EP300 in human diffuse large B cell lymphoma cells provides a basis for rational combinatorial therapy. J. Clin. Invest. 120, 4569–4582.

Chamoun, Z., et al., 2001. Skinny hedgehog, an acyltransferase required for palmitoylation and activity of the hedgehog signal. Science 293, 2080–2084.

Chen, H.Y., et al., 2004. Orexigenic action of peripheral ghrelin is mediated by neuropeptide Y and agouti-related protein. Endocrinology 145, 2607–2612.

Cheng, K.Y., et al., 2006. The role of the phospho-CDK2/cyclin A recruitment site in substrate recognition. J. Biol. Chem. 281, 23167–23179.

Cleary, J., et al., 2005. p300/CBP-associated factor drives DEK into interchromatin granule clusters. J. Biol. Chem. 280, 31760–31767.

Clevers, H., 2006. Wnt/beta-catenin signaling in development and disease. Cell 127, 469–480.
Cowley, M.A., et al., 2003. The distribution and mechanism of action of ghrelin in the CNS demonstrates a novel hypothalamic circuit regulating energy homeostasis. Neuron 37, 649–661.
Date, Y., et al., 2000. Ghrelin, a novel growth hormone-releasing acylated peptide, is synthesized in a distinct endocrine cell type in the gastrointestinal tracts of rats and humans. Endocrinology 141, 4255–4261.
de la Cour Dornonville, C., et al., 2004. Ghrelin stimulates gastric emptying but is without effect on acid secretion and gastric endocrine cells. Regul. Pept. 120, 23–32.
De Vriese, C., et al., 2005. Ghrelin is produced by the human erythroleukemic HEL cell line and involved in an autocrine pathway leading to cell proliferation. Endocrinology 146, 1514–1522.
Dezaki, K., et al., 2008. Ghrelin is a physiological regulator of insulin release in pancreatic islets and glucose homeostasis. Pharmacol. Ther. 118, 239–249.
Diano, S., et al., 2006. Ghrelin controls hippocampal spine synapse density and memory performance. Nat. Neurosci. 9, 381–388.
Garner, A.L., Janda, K.D., 2010. cat-ELCCA: a robust method to monitor the fatty acid acyltransferase activity of ghrelin O-acyltransferase (GOAT). Angew. Chem. Int. Ed Engl. 49, 9630–9634.
Garner, A.L., Janda, K.D., 2011. A small molecule antagonist of ghrelin O-acyltransferase (GOAT). Chem. Commun. (Camb.) 47, 7512–7514.
Groschl, M., et al., 2004. Evaluation of the comparability of commercial ghrelin assays. Clin. Chem. 50, 457–458.
Guan, X.M., et al., 1997. Distribution of mRNA encoding the growth hormone secretagogue receptor in brain and peripheral tissues. Brain Res. Mol. Brain Res. 48, 23–29.
Guidez, F., et al., 2005. Histone acetyltransferase activity of p300 is required for transcriptional repression by the promyelocytic leukemia zinc finger protein. Mol. Cell. Biol. 25, 5552–5566.
Guo, Z.Y., et al., 2005. The active site His-460 of human acyl-coenzyme A: cholesterol acyltransferase 1 resides in a hitherto undisclosed transmembrane domain. J. Biol. Chem. 280, 37814–37826.
Gutierrez, J.A., et al., 2008. Ghrelin octanoylation mediated by an orphan lipid transferase. Proc. Natl. Acad. Sci. USA 105, 6320–6325.
Heller, R.S., et al., 2005. Genetic determinants of pancreatic epsilon-cell development. Dev. Biol. 286, 217–224.
Hines, A.C., Cole, P.A., 2004. Design, synthesis, and characterization of an ATP-peptide conjugate inhibitor of protein kinase A. Bioorg. Med. Chem. Lett. 14, 2951–2954.
Hines, A.C., et al., 2005. Bisubstrate analog probes for the insulin receptor protein tyrosine kinase: molecular yardsticks for analyzing catalytic mechanism and inhibitor design. Bioorg. Chem. 33, 285–297.
Hofmann, K., 2000. A superfamily of membrane-bound O-acyltransferases with implications for wnt signaling. Trends Biochem. Sci. 25, 111–112.
Howard, A.D., et al., 1996. A receptor in pituitary and hypothalamus that functions in growth hormone release. Science 273, 974–977.
Iwakura, H., et al., 2010. Establishment of a novel ghrelin-producing cell line. Endocrinology 151, 2940–2945.
Jencks, W.P., 1981. On the attribution and additivity of binding energies. Proc. Natl. Acad. Sci. USA 78, 4046–4050.
Joseph, J.W., et al., 2002. Uncoupling protein 2 knockout mice have enhanced insulin secretory capacity after a high-fat diet. Diabetes 51, 3211–3219.

Kadowaki, T., et al., 1996. The segment polarity gene porcupine encodes a putative multitransmembrane protein involved in Wingless processing. Genes Dev. 10, 3116–3128.

Kamegai, J., et al., 2001. Chronic central infusion of ghrelin increases hypothalamic neuropeptide Y and Agouti-related protein mRNA levels and body weight in rats. Diabetes 50, 2438–2443.

Kanamoto, N., et al., 2001. Substantial production of ghrelin by a human medullary thyroid carcinoma cell line. J. Clin. Endocrinol. Metabol. 86, 4984–4990.

Kirchner, H., et al., 2009. GOAT links dietary lipids with the endocrine control of energy balance. Nat. Med. 15, 741–745.

Kojima, M., et al., 1999. Ghrelin is a growth-hormone-releasing acylated peptide from stomach. Nature 402, 656–660.

Korbonits, M., et al., 1999. The growth hormone secretagogue receptor. Growth Horm. IGF Res. 9, 93–99.

Lau, O.D., et al., 2000. HATs off: selective synthetic inhibitors of the histone acetyltransferases p300 and PCAF. Mol. Cell 5, 589–595.

Levinson, N.M., et al., 2006. A Src-like inactive conformation in the abl tyrosine kinase domain. PLoS Biol. 4, e144.

Liu, J., et al., 2008a. Novel ghrelin assays provide evidence for independent regulation of ghrelin acylation and secretion in healthy young men. J. Clin. Endocrinol. Metabol. 93, 1980–1987.

Liu, X., et al., 2008b. The structural basis of protein acetylation by the p300/CBP transcriptional coactivator. Nature 451, 846–850.

Liu, Y., et al., 2008c. A fasting inducible switch modulates gluconeogenesis via activator/coactivator exchange. Nature 456, 269–273.

Marek, R., et al., 2011. Paradoxical enhancement of fear extinction memory and synaptic plasticity by inhibition of the histone acetyltransferase p300. J. Neurosci. 31, 7486–7491.

Medzihradszky, D., et al., 1994. Solid-phase synthesis of adenosine phosphopeptides as potential bisubstrate inhibitors of protein-kinases. J. Am. Chem. Soc. 116, 9413–9419.

Morton, G.J., Schwartz, M.W., 2001. The NPY/AgRP neuron and energy homeostasis. Int. J. Obes. Relat. Metab. Disord. 25, S56–S62.

Nishi, Y., et al., 2005. Ingested medium-chain fatty acids are directly utilized for the acyl modification of ghrelin. Endocrinology 146, 2255–2264.

Nussbaum, S.R., et al., 1987. Highly sensitive two-site immunoradiometric assay of parathyrin, and its clinical utility in evaluating patients with hypercalcemia. Clin. Chem. 33, 1364–1367.

Ogiso, K., et al., 2011. Ghrelin: a gut hormonal basis of motility regulation and functional dyspepsia. J. Gastroenterol. Hepatol. 26, 67–72.

Oussaief, L., et al., 2009. Phosphatidylinositol 3-kinase/Akt pathway targets acetylation of Smad3 through Smad3/CREB-binding protein interaction: contribution to transforming growth factor beta1-induced Epstein-Barr virus reactivation. J. Biol. Chem. 284, 23912–23924.

Pagac, M., et al., 2011. Topology of 1-acyl-sn-glycerol-3-phosphate acyltransferases SLC1 and ALE1 and related membrane-bound O-acyltransferases (MBOATs) of Saccharomyces cerevisiae. J. Biol. Chem. 286, 36438–36447.

Parang, K., et al., 2001. Mechanism-based design of a protein kinase inhibitor. Nat. Struct. Biol. 8, 37–41.

Pasca di Magliano, M., Hebrok, M., 2003. Hedgehog signalling in cancer formation and maintenance. Nat. Rev. Cancer 3, 903–911.

Potocky, T.B., et al., 2003. Cytoplasmic and nuclear delivery of a TAT-derived peptide and a beta-peptide after endocytic uptake into HeLa cells. J. Biol. Chem. 278, 50188–50194.

Prado, C.L., et al., 2004. Ghrelin cells replace insulin-producing beta cells in two mouse models of pancreas development. Proc. Natl. Acad. Sci. USA 101, 2924–2929.
Reya, T., Clevers, H., 2005. Wnt signalling in stem cells and cancer. Nature 434, 843–850.
Sagar, V., et al., 2004. Bisubstrate analogue structure-activity relationships for p300 histone acetyltransferase inhibitors. Bioorg. Med. Chem. 12, 3383–3390.
Satou, M., et al., 2010. Identification and characterization of acyl-protein thioesterase 1/lysophospholipase I as a ghrelin deacylation/lysophospholipid hydrolyzing enzyme in fetal bovine serum and conditioned medium. Endocrinology 151, 4765–4775.
Schwarze, S.R., et al., 1999. In vivo protein transduction: delivery of a biologically active protein into the mouse. Science 285, 1569–1572.
Shen, K., Cole, P.A., 2003. Conversion of a tyrosine kinase protein substrate to a high affinity ligand by ATP linkage. J. Am. Chem. Soc. 125, 16172–16173.
Spin, J.M., et al., 2010. Chromatin remodeling pathways in smooth muscle cell differentiation, and evidence for an integral role for p300. PLoS One 5, e14301.
Sun, Y., et al., 2006. Ablation of ghrelin improves the diabetic but not obese phenotype of ob/ob mice. Cell Metab. 3, 379–386.
Sun, Y., et al., 2008. Characterization of adult ghrelin and ghrelin receptor knockout mice under positive and negative energy balance. Endocrinology 149, 843–850.
Takada, R., et al., 2006. Monounsaturated fatty acid modification of Wnt protein: its role in Wnt secretion. Dev. Cell 11, 791–801.
Takahashi, T., et al., 2009. Production of n-octanoyl-modified ghrelin in cultured cells requires prohormone processing protease and ghrelin O-acyltransferase, as well as n-octanoic acid. J. Biochem. 146, 675–682.
Tong, J., et al., 2010. Ghrelin suppresses glucose-stimulated insulin secretion and deteriorates glucose tolerance in healthy humans. Diabetes 59, 2145–2151.
Wang, L., et al., 2011. The leukemogenicity of AML1-ETO is dependent on site-specific lysine acetylation. Science 333, 765–769.
Wierup, N., et al., 2002. The ghrelin cell: a novel developmentally regulated islet cell in the human pancreas. Regul. Pept. 107, 63–69.
Willesen, M.G., et al., 1999. Co-localization of growth hormone secretagogue receptor and NPY mRNA in the arcuate nucleus of the rat. Neuroendocrinology 70, 306–316.
Yang, J., et al., 2008a. Identification of the acyltransferase that octanoylates ghrelin, an appetite-stimulating peptide hormone. Cell 132, 387–396.
Yang, J., et al., 2008b. Inhibition of ghrelin O-acyltransferase (GOAT) by octanoylated pentapeptides. Proc. Natl. Acad. Sci. USA 105, 10750–10755.
Yi, C.X., et al., 2012. The GOAT-ghrelin system is not essential for hypoglycemia prevention during prolonged calorie restriction. PLoS One 7, e32100.
Zhang, J.H., et al., 1999. A simple statistical parameter for use in evaluation and validation of high throughput screening assays. J. Biomol. Screen. 4, 67–73.
Zhang, C.Y., et al., 2001. Uncoupling protein-2 negatively regulates insulin secretion and is a major link between obesity, beta cell dysfunction, and type 2 diabetes. Cell 105, 745–755.
Zhao, C., et al., 2009. Hedgehog signalling is essential for maintenance of cancer stem cells in myeloid leukaemia. Nature 458, 776–779.
Zhao, T.J., et al., 2010a. Ghrelin O-acyltransferase (GOAT) is essential for growth hormone-mediated survival of calorie-restricted mice. Proc. Natl. Acad. Sci. USA 107, 7467–7472.
Zhao, T.J., et al., 2010b. Ghrelin secretion stimulated by {beta}1-adrenergic receptors in cultured ghrelinoma cells and in fasted mice. Proc. Natl. Acad. Sci. USA 107, 15868–15873.
Zheng, Y., et al., 2005. Synthesis and evaluation of a potent and selective cell-permeable p300 histone acetyltransferase inhibitor. J. Am. Chem. Soc. 127, 17182–17183.

Zhou, A., et al., 1999. Proteolytic processing in the secretory pathway. J. Biol. Chem. 274, 20745–20748.
Zhu, X., et al., 2006. On the processing of proghrelin to ghrelin. J. Biol. Chem. 281, 38867–38870.

SECTION 5

Action of Ghrelin

CHAPTER FOURTEEN

Model-Based Evaluation of Growth Hormone Secretion

Johannes D. Veldhuis*,†,1, Daniel M. Keenan‡
*Department of Medicine, Endocrine Research Unit and Biophysics Section, Mayo School of Graduate Medical Education, Clinical Translational Science Center, Mayo Clinic, Rochester, Minnesota, USA
†Department of Physiology, Endocrine Research Unit and Biophysics Section, Mayo School of Graduate Medical Education, Clinical Translational Science Center, Mayo Clinic, Rochester, Minnesota, USA
‡Department of Statistics, University of Virginia, Charlottesville, Virginia, USA
[1]Corresponding author: e-mail address: veldhuis.johannes@mayo.edu

Contents

Abstract

A minimal-model framework is that growth hormone (GH) secretion is controlled by an ensemble of interlinked peptides, namely, GH-releasing hormone (GHRH), somatostatin (SS), and ghrelin. Clinical studies, laboratory experiments, rare sporadic mutations, targeted gene silencing, and biomathematical models establish that at least three

Methods in Enzymology, Volume 514
ISSN 0076-6879
http://dx.doi.org/10.1016/B978-0-12-381272-8.00014-3

signals regulate GH secretion. A clarion implication of the concept of integrative control is that no one peptidic effector operates alone or can be adequately studied alone. A major unanswered question is how pathophysiology disrupts the core regulatory ensemble, thereby forcing relative GH and IGF-1 deficiency or excess. However, salient technical hurdles exist, namely, the lack of reliable experimental strategies and the paucity of validated analytical tools to distinguish the interlinked roles of GHRH, SS, and ghrelin. To address these significant obstacles requires administering peptide secretagogues in distinct combinations akin to the classical insulin/glucose clamp and implementing an analytical formalism to parse the interactive roles of GHRH, SS, and ghrelin objectively.

1. INTRODUCTION

Three principal peptides control pulsatile growth hormone (GH) secretion, namely, somatostatin (SS), GH-releasing hormone (GHRH), and ghrelin, as highlighted in Fig. 14.1A. SS controls the size and the timing of GH pulses via (1) direct pituitary effects and (2) intrahypothalamic actions (Giustina and Veldhuis, 1998; Veldhuis et al., 2006). At the pituitary, SS blocks the exocytosis, but not the synthesis of GH molecules, and resensitizes somatotrope GHRH receptors via phosphatase activation. GHRH stimulates not only GH release but also GH biosynthesis, thereby augmenting GH stores even when SS is present (Barinaga et al., 1985). In contrast, ghrelin (a native GHRP) does not drive GH gene expression in the adult (Veldhuis et al., 2006). Both GHRH and ghrelin induce GH exocytosis when noncompetitive inhibition by SS is relieved (Stachura et al., 1988). Therefore, model structure could allow intermittent exposure of somatotropes to SS and GHRH to expand GH stores, which are released under subsequent drive by GHRH and ghrelin (Kraicer et al., 1986; Richardson and Twente, 1991).

2. ENSEMBLE GH-REGULATION MODEL

In the hypothalamus, SS inhibits GHRH neurons directly via synaptic connections (Fairhall et al., 1995). Transsynaptic inhibition blocks both the secretion of GHRH into pituitary portal blood and the firing of GHRH neurons. Therefore, cyclic SS outflow within the brain could govern not only the size but also the timing of GHRH and thereby GH secretory bursts (Fig. 14.1B). GHRH itself has no consistent neuronal actions. However, ghrelin induces GHRH secretion by exciting arcuate nucleus (ArC) neurons and antagonizing the neuronal effects of SS,

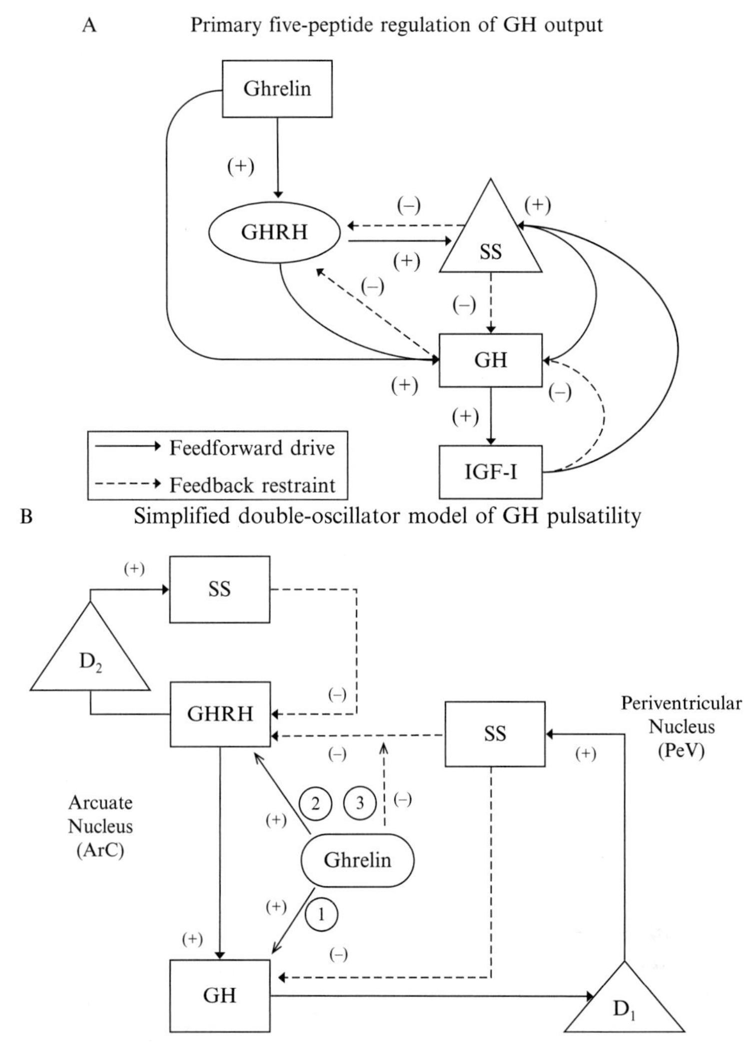

Figure 14.1 (A) Ensemble GH axis viewed in relation to five key peptide signals and their minimal interactions. (B) Formalized analytical construct of GH feedback-induced SS-directed GHRH oscillations. D1 and D2 are time delays. Note the topographic and functional separation of PeV and ArC SS neurons and integration of ghrelin in this model at three levels labeled ①, ②, and ③.

including SS's repression of GHRH release. Ghrelin does not diminish the secretion of SS into portal blood (Fletcher et al., 1996; Guillaume et al., 1994). Accordingly, in the brain SS is the prime inhibitor and ghrelin the major stimulator of the amount GHRH secreted into portal blood.

2.1. Timing role of SS in GH-pulse modeling

The role of SS in the generation of successive GH pulses can be viewed in a simplified model as follows: When a GH pulse occurs, rising GH concentrations act via negative feedback on CNS GH receptors to trigger periventricular-nuclear (PeV) release of SS. PeV SS outflow represses ArC GHRH neurons via direct synaptic contacts and inhibits pituitary GH release via the portal system (Bluet-Pajot et al., 1998; Veldhuis et al., 2006). The subsequent fall in systemic GH concentrations removes the drive to PeV SS neurons. Withdrawal of SS outflow disinhibits GHRH neurons and GH exocytosis, thus evoking a coordinated pair of GHRH and GH pulses. Available ghrelin amplifies the size of GHRH and GH pulses via facilitative actions on both ArC and the pituitary gland and putative opposition to SS at the same sites. In this construction, SS regulates the outflow of hypothalamic GHRH to, and the actions of GHRH and ghrelin on, somatotropes (Farhy and Veldhuis, 2003, 2004, 2005).

Cessation of a peripheral SS infusion elicits a rebound-like pulse of GHRH and GH. The infusion lowers systemic GH concentrations, which withdraws GH's feedback drive of SS outflow, thereby promoting GHRH release (Farhy and Veldhuis, 2005). Although secreted GHRH cannot stimulate GH exocytosis during the SS infusion, hypothalamic GHRH continues to induce the synthesis of GH stores (Giustina and Veldhuis, 1998; Veldhuis et al., 2006). Abruptly stopping the peripheral SS infusion allows secreted GHRH to evoke GH release (Epelbaum et al., 1989; Farhy and Veldhuis, 2003, 2004, 2005; Horvath et al., 1989). The GH pulse emerges because brain SS outflow to the pituitary gland is also reduced by the low GH feedback state. Accordingly, a peripheral SS clamp can be modeled as blocking pituitary GH exocytosis directly, decreasing systemic GH concentrations and triggering CNS feedback-dependent adaptations in hypothalamic GHRH and SS outflow. Modeling the magnitude and timing of GH responses to feedback withdrawal should confer a unique window into hypothalamic SS/GHRH regulation (Veldhuis et al., 2006).

2.2. Basic stimulatory effects of GHRH

The importance of GHRH is evident in studies in patients with mutations of the GHRH receptor (Roelfsema et al., 2001). The role of intermittent SS outflow is affirmed by the capability of continuous i.v. infusion of GHRH and GHRP-2 to evoke pulsatile GH secretion (Bowers et al., 2004; Evans et al., 2001; Veldhuis et al., 2002). Collective observations thus frame the following working model of GH pulsatility: (1) intermittent pulses of GHRH and SS set the timing and minimal size of GH secretory bursts (Farhy and Veldhuis, 2004; Farhy et al., 2001, 2002) and (2) ghrelin determines the maximal size of GH pulses without altering their number (Chihara et al., 1981; Clark et al., 1988; Farhy and Veldhuis, 2005; Guillaume et al., 1994; Robinson, 1991).

2.3. GH burst-amplifying model of GH-releasing peptides

GH-releasing peptides (GHRPs) were synthesized in 1977, the receptor was cloned in 1996, and the endogenous ligand, ghrelin, was identified in 1999 (Bowers et al., 1977; Howard et al., 1996; Kojima et al., 1999). Ghrelin is the most potent known GHRH *in vivo*. Laboratory data indicate that ghrelin exerts at least three major effects beyond direct stimulation of somatotrope GH release: (i) stimulation of GHRH secretion from ArC, (ii) antagonism of SS's inhibition of the pituitary gland and ArC, and (iii) potentiation of GHRH action *in vivo* (Di Vito et al., 2002; Fletcher et al., 1996; Guillaume et al., 1994; Hataya et al., 2001). On the other hand, ghrelin-receptor agonists do not (a) repress SS release into hypothalamo–pituitary portal blood, (b) synergize with GHRH *in vitro*, or (c) elicit maximal GH secretion *in vivo* when the pituitary stalk is interrupted.

Peripherally infused ghrelin can be modeled as acting upon both the hypothalamus and pituitary gland (Bowers, 1999; Fairhall et al., 1995; Guillaume et al., 1994; Tannenbaum et al., 2003). In the hypothalamus, ghrelin stimulates GHRH secretion, antagonizes neuronal inhibition by SS, and induces hunger (Guillaume et al., 1994; Veldhuis et al., 2006; Wren et al., 2001). Hypothalamic effects of ghrelin mediate synergy with GHRH *in vivo* because (i) synergy vanishes in patients with hypothalamo–pituitary disconnection (Popovic et al., 1995), (ii) there is no direct synergy on pituitary cells *in vitro* (Hataya et al., 2001), and (iii) GHRP-stimulated GH secretion is reduced by $>90\%$ in normal adults, given a GHRH-receptor antagonist (Pandya et al., 1998), and by $>98\%$ in rare patients with inactivating mutations of the GHRH receptor (Roelfsema et al., 2001). GHRP does not regulate the basic interneuronal SS–GHRH pulse-generating mechanism, in as much as continuous systemic infusion of GHRP augments the size but not the frequency of GH pulses (Bowers et al., 2004) (Fig. 14.1B).

2.4. Modeling GH-pulse frequency

Continuous i.v. infusion or tumoral secretion of GHRH markedly amplifies the size but not the frequency of GH pulses (Evans et al., 2001; Giustina and Veldhuis, 1998; Vance et al., 1985). The same is true for GHRP infusions. Thus, a basic model would be that frequency encoding of GH pulses arises from reciprocal synaptic interactions between hypothalamic SS and GHRH neurons (Farhy and Veldhuis, 2003, 2004, 2005; Veldhuis et al., 2002). The role of SS is supported by the fact that patients with inactivating mutations of the GHRH receptor maintain a normal frequency of GH pulses, albeit at 30-fold lower amplitude (Roelfsema et al., 2001).

2.5. Species and model selectivity

Laboratory and clinical studies indicate that coordinate interactions among systemic, hypothalamic, and pituitary effectors, such as peptides, sex steroids, thyroxine, cortisol, or free fatty acids, modulate pulsatile GH secretion (Bluet-Pajot et al., 1998; Farhy and Veldhuis, 2003, 2004; Farhy et al., 2001, 2002; Giustina and Veldhuis, 1998; Muller et al., 1999; Pellegrini et al., 1996; Robinson, 1991; Veldhuis et al., 2006). Nonetheless, the species specificity of sex-steroid action makes inferences gained in the rat, mouse, pig, and sheep illustrative of, rather than definitive to, the human (Giustina and Veldhuis, 1998; Veldhuis et al., 2006). For example, estrogens repress and nonaromatizable androgens increase pulsatile GH secretion in the rat but exert opposite effects in the human. Thus, models of GH pulsatility, including autonomous GHRH oscillations (Wagner et al., 1998) and feedback-driven SS oscillations (Farhy and Veldhuis, 2004), may not be identical among species.

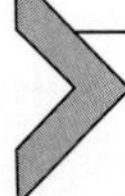

3. TECHNICAL ASSESSMENT OF MULTIPATHWAY GH REGULATION

3.1. Primary interactions

Regulation of the amount of GH secreted in pulses is significant for several key reasons: (a) Aging primarily reduces whereas neonatal life and puberty augment pulsatile (burst-like) GH production. (b) In humans, pulsatile (rather than basal, time-invariant) GH secretion constitutes the majority (88–94%) of total daily GH output. (c) The size of GH pulses is determined by a well-articulated ensemble of key peptides, namely, GHRH, SS, and ghrelin/GHRP, as well as feedback by GH and IGF-1. Because of the reciprocal, time-delayed nature of multipeptide interactions, formally validated methods are needed to reconstruct, test, and quantify ensemble control of GH pulsatility (Farhy and Veldhuis, 2003, 2004, 2005; Farhy et al., 2001, 2002; Straume et al., 1995a,b; Wagner et al., 1998, 2009).

3.2. Illustrative problem

A consequence of interactive (multipathway) control is that modulation of any given regulatory locus will perforce influence the behavior of all interconnected signals (Fig. 14.1). For example, the finding that ghrelin amplifies GH secretory-burst mass following a maximal GHRH stimulus could signify that ghrelin (a) upregulates somatotrope GHRH receptors, (b) increases GHRH-releasable GH stores, (c) decreases and/or attenuates central SSergic inhibition, and (d) mutes endogenous negative feedback by systemic GH or IGF-1 (Farhy and Veldhuis, 2005; Hataya et al., 2001; Turner and Tannenbaum, 1995). To overcome this experimental impasse, we have formulated a combined investigative platform comprising (i) studies in a sex-steroid-clamped milieu, (ii) simultaneous delivery of complementary secretagogues as "peptide clamps," and (iii) ensemble-model-assisted analyses.

3.3. Technical issues

Reliable analytical tools are needed to quantify complex physiological adaptations. For example, the first step in estimating GH secretion accurately requires quantifying the interrelated processes that collectively govern serially measured GH concentrations: the unknown timing and mass (size) underlying of GH secretory bursts; concomitant basal GH release; hormone-specific biexponential elimination kinetics; and random variability associated with sample collection, processing, and assay (Farhy and Veldhuis, 2003, 2004; Farhy et al., 2001, 2002; Keenan et al., 2003, 2004).

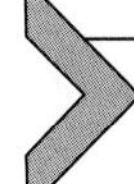

4. SPECIFIC METHODOLOGICAL STRATEGIES

4.1. Experimentally fixed testosterone (T) or estradiol (E_2) concentrations

A paradigm to explore sex-steroid action on the human GH axis could entail delivering placebo (Pl) or T/E_2 transdermally to men/women in amounts to mimic physiological concentrations during GnRH agonist-induced downregulation of the pituitary–gonadal axis (Erickson et al., 2004) (Fig. 14.2). To this end, depot leuprolide (3.75 mg i.m.) may be given twice 3 weeks

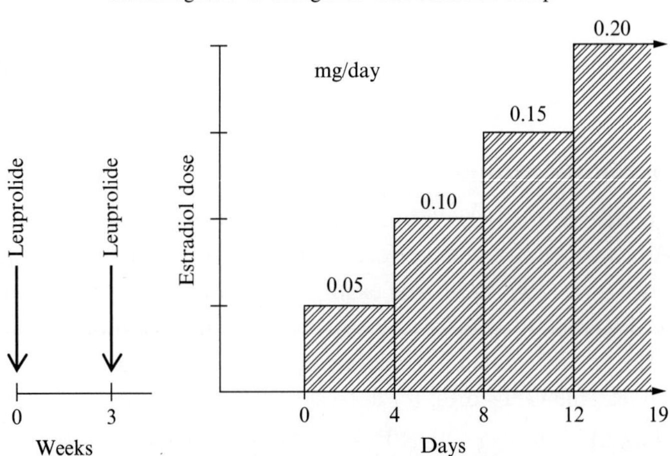

Figure 14.2 Illustrative graded estradiol (E_2) clamp paradigm. Alternative clamps comprise placebo or testosterone (T) addback after GnRH agonists or antagonists.

apart. Then transdermal Pl or T/E_2 can be administered at replacement doses over a designated interval, such as 7 days. Placebo addback provides a hypogonadal state for comparison. Ideally, dose–response and time-course studies would be performed first.

4.2. Deconvolution analysis

GH secretion can be estimated from GH concentration–time series using a flexible-waveform deconvolution model (Veldhuis et al., 2008) (Fig. 14.3). The approach suggested yields a maximum-likelihood estimate (MLE), which is statistically conditioned on biexponential kinetics and *a priori* estimates of decreasing sets of candidate pulse-onset times (Fig. 14.3A). The rapid and slow phase of half-lives of GH can be estimated or assumed (e.g., 3.5 and 20.8 min with fractional contributions of 37% and 63% (Faria et al., 1989)). Potential pulse times are identified first by image boundary-resolution methods, as described. The distribution of final interburst intervals is represented algebraically as a Weibull probability density defined by a pulse rate (number of events per 24 h, lambda) and interpulse-interval regularity(gamma) (Keenan et al., 2003). A value of $\gamma > 1.0$ signifies greater regularity than that of the classic Poisson distribution of random event times. The waveform of secretory bursts (time-plot of instantaneous secretion rates) is represented by a three-parameter generalized Gamma function, thus allowing for either symmetric or variably asymmetric

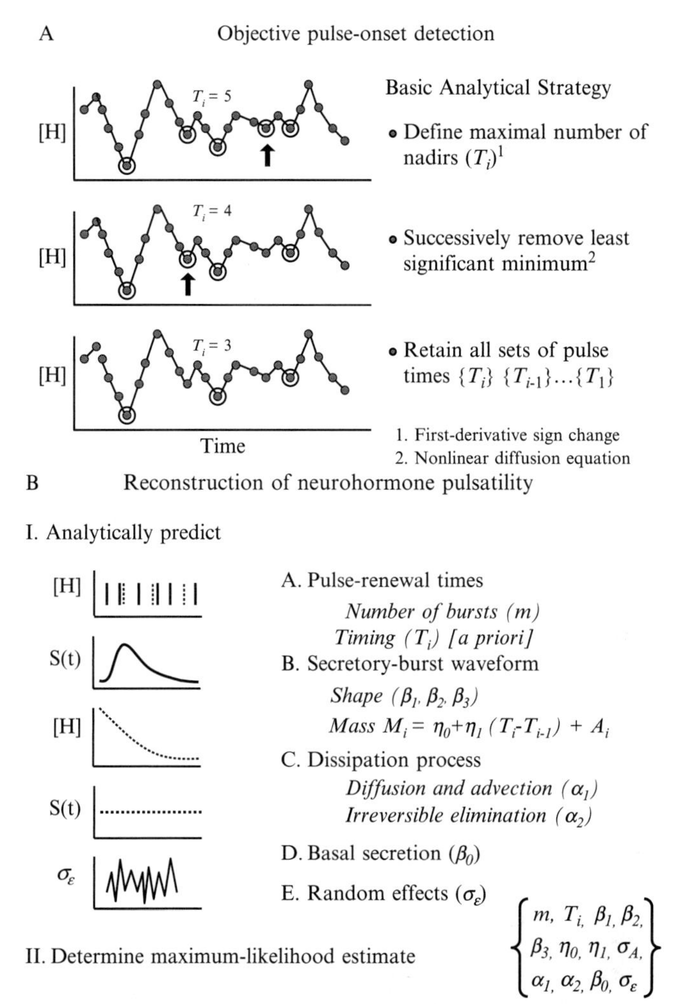

Figure 14.3 Variable-waveform deconvolution method comprising recursive alterations between probabilistic selection of pulse times (A) and nonlinear estimation of all secretion and elimination parameters simultaneously (B). The algorithm was validated experimentally and verified mathematically.

bursts (Keenan et al., 2003) (Fig. 14.3B). One measure of shape is the modal time (min) required to reach maximal secretion within bursts. The burst mode has been shown to be controlled by estrogen and peptidyl secretagogues.

The formalism overcomes a fundamental statistical impasse described in 1995 (Veldhuis et al., 1995), namely, the need for an MLE-based deconvolution methodology that is mathematically verifiable (Keenan et al., 2003, 2004). The solution is conditioned mathematically on selecting probabilistically among *a priori* candidate sets of objectively estimated pulse-onset times (Keenan et al., 2003) (Fig. 14.3A). The construct was validated empirically by frequent (5 min) and extended (4–12 h) sampling of hormone release into hypothalamo–pituitary portal and jugular-venous blood in the conscious sheep and horse (Keenan et al., 2004). The statistical framework was verified by direct mathematical proof of an unbiased asymptotic MLE solution for the full (simultaneous) set of parameters (PhD theses of W. Sun (1995), R. Yang (1997), and S. Chattopadhyay (2001) (Keenan et al., 2003, 2004)). The parameters estimated by the deconvolution procedure are schematized in Fig. 14.3B. According to this methodology, ghrelin stimulates pulsatile GH secretion by augmenting the amount (mass) of GH released per burst by twofold and by accelerating initial release of GH within individual secretory bursts by 1.8-fold. The latter denotes prompt exocytosis of somatotrope stores.

All GH concentration time series in a given sex-steroid milieu and infusion type can also be analyzed together under sex-steroid and agonist-specific conditions. GH secretory-burst mass and waveform (shape) in each subject are represented algebraically by the cohort mean plus an individual random effect. Standard errors of all parameters are derived analytically and verified by Monte Carlo resampling methods, as summarized in the appendices of Erickson et al. (2004).

4.3. Ensemble-model-based analyses

One may also formalize models of reversible, time-delayed, dose-responsive interactions among ghrelin, GHRH, SS, and GH/IGF-1, akin to those developed for other neuroendocrine networks (Farhy and Veldhuis, 2003, 2004, 2005; Farhy et al., 2001, 2002; Keenan et al., 2004). The ensemble formulation allows one to (a) examine clinical and experimental intuition based upon objective multipeptide linkages, (b) test integrative implications of the model, and (c) explore the regulatory hypotheses in follow-up experiments.

One ensemble model incorporates the published capabilities of ghrelin/GHRP to (i) stimulate pituitary GH secretion directly by two- to fourfold, (ii) induce hypothalamic GHRH release, and (iii) oppose the actions of SS in the ArC and on pituitary gland (Farhy and Veldhuis, 2005). Other constructs are needed, including models with stochastic allowance that permit analytical parameter estimation.

4.4. Simulations

Model-based simulations can be utilized to test whether *a priori* postulated pathways and mechanisms could in principle account for observed GH secretion in any given setting. Alternative inferences are tested in the same manner. This is a necessary but not sufficient condition for validity of inference.

4.5. Rationale of peptide-clamped GH feedback analysis

An indirect experimental paradigm may be suggested to estimate the stimulation of GH pulses by endogenous GHRH and ghrelin and inhibition by endogenous SS. The concept exploits the time-dependent feedback effects of a single pulse of GH (Veldhuis et al., 2006). The rationale is that a pulse of GH induces hypothalamic SS outflow, which (a) at the pituitary inhibits GH exocytosis (Kraicer et al., 1986) and (b) in the hypothalamus represses GHRH release from the ArC and antagonizes the ArC effects of ghrelin (Fig. 14.1). Specifically, the rise in GH concentrations would be modeled as triggering the release of SS from PeV into pituitary portal blood and to ArC neurons. The gradual metabolic removal of injected GH then causes GH concentrations to fall, thus removing feedback-induced SS secretion. The decrease in SS restraint evokes the next pulse of hypothalamic GHRH and consequently of pituitary GH (Farhy and Veldhuis, 2005) (Fig. 14.4A). Reemergence of pulsatile GH secretion would definitionally reflect the combined actions of GHRH and ghrelin in the face of low SS outflow.

To distinguish the roles of endogenous GHRH and ghrelin in driving pulsatile GH secretion after GH feedback, one could infuse a maximally effective dose of each separately during feedback recovery (Fig. 14.4B). The rationale is that constant exposure to a maximally effective GHRH stimulus should supplant any influence of endogenous GHRH pulses while potentiating the actions of endogenous ghrelin (and vice versa for an infusion of ghrelin). Potentiation between exogenous and endogenous (heterotypic) peptides would occur because GHRH and ghrelin are synergistic *in vivo* (Hataya et al., 2001). Thereby, one may estimate the effect of endogenous ghrelin from GH secretion during the fixed GHRH clamp, and that of endogenous GHRH during the exogenous ghrelin clamp. In contradistinction, according to a minimal three-peptide model (Fig. 14.1), when GHRH and ghrelin are infused together, pulsatile GH secretion would depend solely upon the timing and degree of recovery of

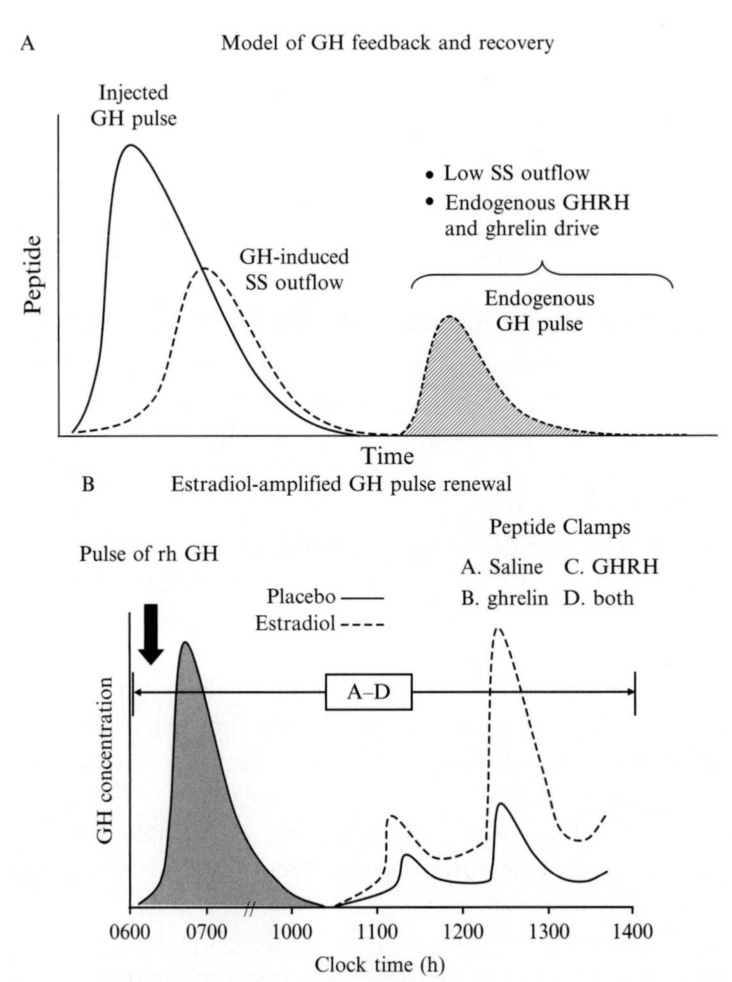

Figure 14.4 (A) Model of exogenous GH feedback-induced initial outflow and subsequent withdrawal of SS (interrupted line), allowing endogenous GHRH and ghrelin to act unopposed and evoke an amplified endogenous GH pulse (shaded). (B) Experimental GH feedback to induce reversible hypothalamic SS outflow during constant i.v. infusion of saline, GHRH, and/or ghrelin. The paradigm could be performed after placebo or sex-steroid (e.g., E_2 or T) treatment.

hypothalamic SS outflow (Farhy and Veldhuis, 2005). The protocol design should allow one to appraise how factors such as sex steroids, age, and body composition determine endogenous outflow (release and actions) of each of GHRH, ghrelin, and SS, so long as somatotropes *per se* are fully responsive.

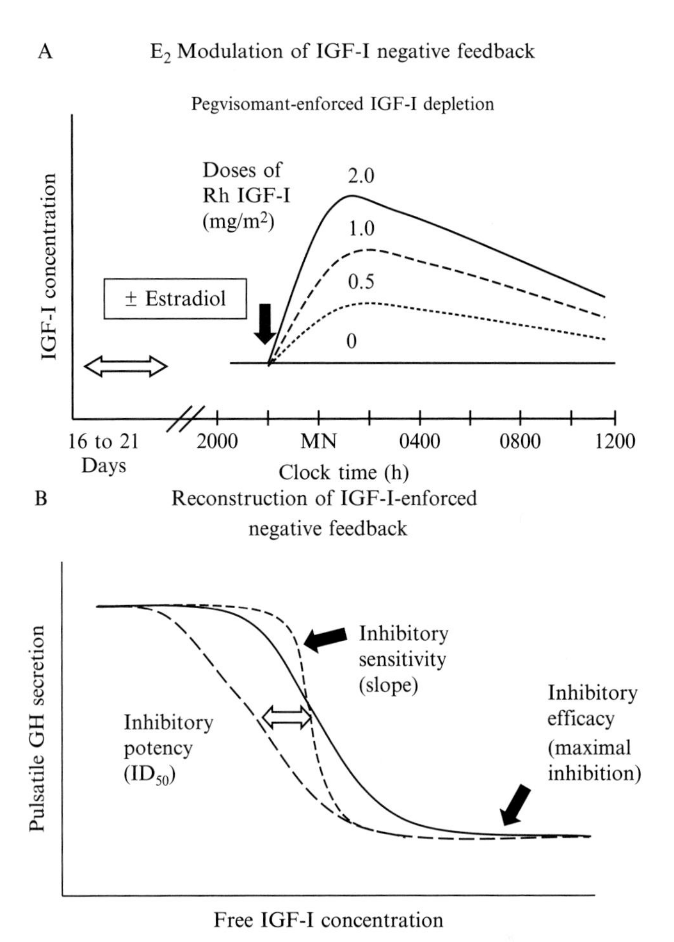

Figure 14.5 (A) Protocol to test E_2 effects on dose-varying IGF-I negative-feedback imposition under pegvisomant (GH-receptor) blockade to deplete endogenous IGF-I, with injecting recombinant human (Rh) IGF-I in several doses. (B) Inhibitory dose–response functions relating free IGF-I concentration to pulsatile GH secretion. The experiment and analytical estimates could be performed under low-E_2 (placebo, Pl) or high-E_2 ($+E_2$) repletion.

4.6. Approximate-entropy analyses of GH feedback

The approximate entropy (ApEn) statistic may be used as an independent, specific (> 90%), sensitive (> 90%), and model-free measure of feedback control in interlinked systems (Pincus, 2000). ApEn quantifies subpattern regularity or orderliness (Veldhuis et al., 2001a,b), which is maintained by negative feedback. Feedback-dependent orderliness of GH release is reduced in acromegaly, puberty, aging, and hypopituitarism (Pincus et al., 1999). An important point is that ApEn is calculated without pulse analysis, thus allowing independent statistical assessment of a feedback change (Veldhuis et al., 2001a,b).

4.7. Dose-response reconstruction

To test, for example, free IGF-1's negative feedback on GH secretion would require experimentally infusing distinct doses of IGF-1 and analytically constructing a four-parameter logistic function defined by baseline (saline-infused) GH secretion, inhibitory potency (IC_{50}), sensitivity (slope), and efficacy (asymptotically maximal suppression) (Keenan et al., 2004). The independent variable would be the measured free IGF-1 concentrations and the dependent variable the deconvolved pulsatile GH secretion rate (Fig. 14.5). To obtain a group estimate,

Step 1 would entail statistical estimation of the nominal negative-feedback lag time (latency). This could be done by cross-correlation analysis, wherein free IGF-1 concentrations are related to discretized GH secretory rates in each paired series, estimated by deconvolution analysis (Veldhuis et al., 2008). The approach was suggested recently for pairwise LH–testosterone and testosterone–LH interactions (Veldhuis et al., 2004b). Significant lag times are those for which r is negative at protected $P < 0.01$ (Veldhuis et al., 1994).

Step 2 then requires analytical construction of the group-specific inhibitory dose–response function (Keenan and Veldhuis, 2003; Keenan et al., 2004). One such (nonlinear mixed-effect) feedback model allows for possible random effects on inhibitory potency, sensitivity, or efficacy,to accommodate biological nonuniformity of feedback-interface properties (Keenan et al., 2004; Veldhuis et al., 2004a). The time-varying free IGF-1 concentration constitutes the negative-feedback signal $F_{\mathrm{I}}(t)$, which inhibits GH secretion $Z_{\mathrm{G}}(t)$ after a subject-specific time delay (estimated in Step 1). The time-lagged free IGF-1 concentration is propagated analytically through the simultaneously estimated inhibitory dose–response function to yield the GH secretion rates $Z_{\mathrm{G}}(t)$, as follows:

$$
Z_{\mathrm{G}}(t) = \begin{cases}
\beta_{\mathrm{G}} + \dfrac{\eta_{2,\mathrm{G}} + A_{\mathrm{G}}^{j}}{1 + \exp\{-(\eta_{0,\mathrm{G}} + \eta_{1,\mathrm{G}} \times F_{\mathrm{I}}(t))\}}, & T_{\mathrm{G}}^{j} \le t \le T_{\mathrm{G}}^{j+1},\ j = 1, \ldots, m, \\
\quad \text{[allowable variation } (A_{\mathrm{G}}'s) \text{ in lower asymptote:efficacy]} & T_{\mathrm{G}}^{j}{}'s: \textit{GH pulse times} \\
\beta_{\mathrm{G}} + \dfrac{\eta_{2,\mathrm{G}}}{1 + \exp\{-(\eta_{0,\mathrm{G}} + (\eta_{1,\mathrm{G}} + A_{\mathrm{G}}^{j}) \times F_{\mathrm{I}}(t))\}}, & T_{\mathrm{G}}^{j} \le t \le T_{\mathrm{G}}^{j+1},\ j = 1, \ldots, m, \\
\quad \text{[allowable variation } (A_{\mathrm{G}}'s) \text{ in slope:sensitivity]} & \eta_{1,\mathrm{G}} < 0 \\
\beta_{\mathrm{G}} + \dfrac{\eta_{2,\mathrm{G}}}{1 + \exp\{-((\eta_{0,\mathrm{G}} + A_{\mathrm{G}}^{j}) + \eta_{1,\mathrm{G}} \times F_{\mathrm{I}}(t))\}}, & T_{\mathrm{G}}^{j} \le t \le T_{\mathrm{G}}^{j+1},\ j = 1, \ldots, m \\
\quad \text{[allowable variation } (A_{\mathrm{G}}'s) \text{ in } EC_{50}\text{:potency]} &
\end{cases}
$$

The parameters under a frugal model will comprise (1) multiple individual (subject-specific) baseline GH secretion rates (reflecting a zero dose IGF-I); (2) three global (group-defined) feedback parameters, namely, inhibitory potency, sensitivity, and efficacy; (3) allowable random effects on efficacy; and (4) simultaneously estimated feedback time lags. The MLE of the combined parameter set in each cohort is based therefore upon all paired samples of GH and IGF-1 (no. of samples/h × h × no. of sessions per cohort) (Keenan and Veldhuis, 2003; Keenan et al., 2004). Standard errors are computed analytically for each of the primary feedback parameters in each study group (Keenan et al., 2004).

5. SUMMARY

Cycles of inhibition and disinhibition within an ensemble of interlinked nodes are necessary and sufficient to drive recurrent GH secretory bursts in simplified model systems (Farhy and Veldhuis, 2003, 2004, 2005; Farhy et al., 2001, 2002). Combined interventional experiments (selective sex steroids and peptide clamps), deconvolution and ApEn analyses, model formulation, and dose–response constructs are needed to probe GH regulation noninvasively and generate new hypotheses for further testing.

ACKNOWLEDGMENTS

We thank Jill Smith for the preparation of the manuscript and Ash Bryant for graphics. This work was supported in part via the Center for Translational Science Activities (CTSA) Grant Number 1 UL 1 RR024150 from the National Center for Research Resources (Rockville, MD), and AG019695 and AG029362 from the National Institutes of Health (Bethesda, MD). The content is solely the responsibility of the authors and does not necessarily represent the official views of the National Institute on Aging or the National Institutes of Health. MATLAB versions of ApEn and deconvolution analyses are available from veldhuis.johannes@mayo.edu.

REFERENCES

Barinaga, M., et al., 1985. Independent effects of growth hormone releasing factor on growth hormone release and gene transcription. Nature 314, 279–281.

Bluet-Pajot, M.T., et al., 1998. Hypothalamic and hypophyseal regulation of growth hormone secretion. Cell. Mol. Neurobiol. 18, 101–123.

Bowers, C.Y., 1999. GH releasing peptides (GHRPs). In: Kostyo, J., Goodman, H. (Eds.), Handbook of Physiology. Oxford University Press, New York, pp. 267–297.

Bowers, C.Y., et al., 1977. Effect of the enkephalins and enkephalin analogs on release of pituitary hormones *in vitro*. In: MacIntyre, I., Szelke, M. (Eds.), Molecular Endocrinology. Elsevier/North Holland, Amsterdam, pp. 287–292.

Bowers, C.Y., et al., 2004. Sustained elevation of pulsatile growth hormone (GH) secretion and insulin-like growth factor I (IGF-I), IGF-binding protein-3 (IGFBP-3), and IGFBP-5 concentrations during 30-day continuous subcutaneous infusion of GH-releasing peptide-2 in older men and women. J. Clin. Endocrinol. Metab. 89, 2290–2300.

Chihara, K., et al., 1981. Intraventrically injected growth hormone stimulates somatostatin release into rat hypophyseal portal blood. Endocrinology 109, 2279–2281.

Clark, R.G., et al., 1988. The rebound release of growth hormone (GH) following somatostatin infusion in rats involves hypothalamic GH-releasing factor release. J. Endocrinol. 119, 397–404.

Di Vito, L., et al., 2002. The GH-releasing effect of ghrelin, a natural GH secretagogue, is only blunted by the infusion of exogenous somatostatin in humans. Clin. Endocrinol. (Oxf) 56, 643–648.

Epelbaum, J., et al., 1989. Combined autoradiographic and immunohistochemical evidence for an association of somatostatin binding sites with growth hormone-releasing factor-containing nerve cell bodies in the rat arcuate nucleus. J. Neuroendocrinol. 1, 109–115.

Erickson, D., et al., 2004. Dual secretagogue drive of burst-like growth hormone secretion in postmenopausal compared with premenopausal women studied under an experimental estradiol clamp. J. Clin. Endocrinol. Metab. 89, 4746–4754.

Evans, W.S., et al., 2001. Continuous 24-hour intravenous infusion of recombinant human growth hormone (GH)-releasing hormone-(1,44)-amide augments pulsatile, entropic, and daily rhythmic GH secretion in postmenopausal women equally in the estrogen-withdrawn and estrogen-supplemented states. J. Clin. Endocrinol. Metab. 86, 700–712.

Fairhall, K.M., et al., 1995. Central effects of growth hormone-releasing hexapeptide (GHRP-6) on growth hormone release are inhibited by central somatostatin action. J. Endocrinol. 144, 555–560.

Farhy, L.S., Veldhuis, J.D., 2003. Joint pituitary-hypothalamic and intrahypothalamic autofeedback construct of pulsatile growth hormone secretion. Am. J. Physiol. Regul. Integr. Comp. Physiol. 285, R1240–R1249.

Farhy, L.S., Veldhuis, J.D., 2004. Putative GH pulse renewal: periventricular somatostatinergic control of an arcuate-nuclear somatostatin and GH-releasing hormone oscillator. Am. J. Physiol. 286, R1030–R1042.

Farhy, L.S., Veldhuis, J.D., 2005. Deterministic construct of amplifying actions of ghrelin on pulsatile GH secretion. Am. J. Physiol. Regul. Integr. Comp. 288, R1649–R1663.

Farhy, L.S., et al., 2001. A construct of interactive feedback control of the GH axis in the male. Am. J. Physiol. 281, R38–R51.

Farhy, L.S., et al., 2002. Unequal autonegative feedback by GH models the sexual dimorphism in GH secretory dynamics. Am. J. Physiol. 282, R753–R764.

Faria, A.C.S., et al., 1989. Half-time of endogenous growth hormone (GH) disappearance in normal man after stimulation of GH secretion by GH-releasing hormone and suppression with somatostatin. J. Clin. Endocrinol. Metab. 68, 535–541.

Fletcher, T.P., et al., 1996. Growth hormone-releasing and somatostatin concentrations in the hypophysial portal blood of conscious sheep during the infusion of growth hormone-releasing peptide-6. Domest. Anim. Endocrinol. 13, 251–258.

Giustina, A., Veldhuis, J.D., 1998. Pathophysiology of the neuroregulation of growth hormone secretion in experimental animals and the human. Endocr. Rev. 19, 717–797.

Guillaume, V., et al., 1994. Growth hormone (GH)-releasing hormone secretion is stimulated by a new GH-releasing hexapeptide in sheep. Endocrinology 135, 1073–1076.

Hataya, Y., et al., 2001. A low dose of ghrelin stimulates growth hormone (GH) release synergistically with GH-releasing hormone in humans. J. Clin. Endocrinol. Metab. 86, 4552.

Horvath, S., et al., 1989. Electron microscopic immunocytochemical evidence for the existence of bidirectional synaptic connections between growth hormone- and somatostatin-containing neurons in the hypothalamus of the rat. Brain Res. 481, 8–15.

Howard, A.D., et al., 1996. A receptor in pituitary and hypothalamus that functions in growth hormone release. Science 273, 974–977.

Keenan, D.M., Veldhuis, J.D., 2003. Mathematical modeling of receptor-mediated interlinked systems. In: Henry, H., Norman, A. (Eds.), Encyclopedia of Hormones. Academic Press, San Diego, CA, pp. 286–294.

Keenan, D.M., et al., 2003. Physiological control of pituitary hormone secretory-burst mass, frequency and waveform: a statistical formulation and analysis. Am. J. Physiol. 285, R664–R673.

Keenan, D.M., et al., 2004. Reconstruction of *in vivo* time-evolving neuroendocrine dose-response properties unveils admixed deterministic and stochastic elements. Proc. Natl. Acad. Sci. USA 101, 6740–6745.

Kojima, M., et al., 1999. Ghrelin is a growth-hormone-releasing acylated peptide from stomach. Nature 402, 656–660.

Kraicer, J., et al., 1986. Effect of somatostatin withdrawal and growth hormone (GH)-releasing factor on GH release *in vitro*: amount available for release after disinhibition. Endocrinology 119, 2047–2051.

Muller, E.E., et al., 1999. Neuroendocrine control of growth hormone secretion. Physiol. Rev. 79, 511–607.

Pandya, N., et al., 1998. Growth hormone (GH)-releasing peptide-6 requires endogenous hypothalamic GH-releasing hormone for maximal GH stimulation. J. Clin. Endocrinol. Metab. 83, 1186–1189.

Pellegrini, E., et al., 1996. Central administration of a growth hormone (GH) receptor mRNA antisense increases GH pulsatility and decreases hypothalamic somatostatin expression in rats. J. Neurosci. 16, 8140–8148.

Pincus, S.M., 2000. Irregularity and asynchrony in biologic network signals. Methods Enzymol. 321, 149–182.

Pincus, S.M., et al., 1999. Hormone pulsatility discrimination via coarse and short time sampling. Am. J. Physiol. 277, E948–E957.

Popovic, V., et al., 1995. Blocked growth hormone-releasing peptide (GHRP-6)-induced GH secretion and absence of the synergistic action of GHRP-6 plus GH-releasing hormone in patients with hypothalamopituitary disconnection: evidence that GHRP-6 main action is exerted at the hypothalamic level. J. Clin. Endocrinol. Metab. 80, 942–947.

Richardson, S.B., Twente, S., 1991. Pre-exposure of rat anterior pituitary cells to somatostatin enhances subsequent growth hormone secretion. J. Endocrinol. 128, 91–95.

Robinson, I.C.A.F., 1991. The growth hormone secretory pattern: a response to neuroendocrine signals. Acta Paediatr. Scand. Suppl. 372, 70–78.

Roelfsema, F., et al., 2001. Growth hormone (GH) secretion in patients with an inactivating defect of the GH-releasing hormone (GHRH) receptor is pulsatile: evidence for a role for non-GHRH inputs into the generation of GH pulses. J. Clin. Endocrinol. Metab. 86, 2459–2464.

Stachura, M.E., et al., 1988. Combined effects of human growth hormone (GH)-releasing factor-44 (GRF) and somatostatin (SRIF) on post-SRIF rebound release of GH and prolactin: a model for GRF-SRIF modulation of secretion. Endocrinology 123, 1476–1482.

Straume, M., et al., 1995a. Systems-level analysis of physiological regulation interactions controlling complex secretory dynamics of growth hormone axis: a connectionist network model. Methods Neurosci. 28, 270–310.

Straume, M., et al., 1995b. Realistic emulation of highly irregular temporal patterns of hormone release: a computer-based pulse simulator. Methods Neurosci. 28, 220–243.

Tannenbaum, G.S., et al., 2003. Interrelationship between the novel peptide ghrelin, somatostatin and growth hormone-releasing hormone in regulation of pulsatile growth hormone secretion. Endocrinology 144, 967–974.

Turner, J.P., Tannenbaum, G.S., 1995. *In vivo* evidence of a positive role for somatostatin to optimize pulsatile growth hormone secretion. Am. J. Physiol. 269, E683–E690.

Vance, M.L., et al., 1985. Pulsatile growth hormone secretion in normal man during a continuous 24-hour infusion of human growth hormone releasing factor (1-40). J. Clin. Invest. 75, 1584–1590.
Veldhuis, J.D., et al., 1994. Assessing temporal coupling between two, or among three or more, neuroendocrine pulse trains: cross-correlation analysis, simulation methods, and conditional probability testing. Methods Neurosci. 20, 336–376.
Veldhuis, J.D., et al., 1995. Complicating effects of highly correlated model variables on nonlinear least-squares estimates of unique parameter values and their statistical confidence intervals: estimating basal secretion and neurohormone half-life by deconvolution analysis. Methods Neurosci. 28, 130–138.
Veldhuis, J.D., et al., 2001a. Impact of pulsatility on the ensemble orderliness (approximate entropy) of neurohormone secretion. Am. J. Physiol. 281, R1975–R1985.
Veldhuis, J.D., et al., 2001b. Secretory process regularity monitors neuroendocrine feedback and feedforward signaling strength in humans. Am. J. Physiol. 280, R721–R729.
Veldhuis, J.D., et al., 2002. Impact of estradiol supplementation on dual peptidyl drive of growth-hormone secretion in postmenopausal women. J. Clin. Endocrinol. Metab. 87, 859–866.
Veldhuis, J.D., et al., 2004a. Estradiol supplementation modulates growth hormone (GH) secretory-burst waveform and recombinant human insulin-like growth factor-I-enforced suppression of endogenously driven GH release in postmenopausal women. J. Clin. Endocrinol. Metab. 89, 1312–1318.
Veldhuis, J.D., et al., 2004b. Erosion of endogenous testosterone-driven negative feedback on pulsatile LH secretion in healthy aging men. J. Clin. Endocrinol. Metab. 89, 5753–5761.
Veldhuis, J.D., et al., 2006. Somatotropic and gonadotropic axes linkages in infancy, childhood, and the puberty-adult transition. Endocr. Rev. 27, 101–140.
Veldhuis, J.D., et al., 2008. Motivations and methods for analyzing pulsatile hormone secretion. Endocr. Rev. 29, 823–864.
Wagner, C., et al., 1998. Genesis of the ultradian rhythm of GH secretion: a new model unifying experimental observations in rats. Am. J. Physiol. 275, E1046–E1054.
Wagner, C., et al., 2009. Interactions of ghrelin signaling pathways with the growth hormone neuroendocrine axis: a new and experimentally tested model. J. Mol. Endocrinol. 43, 105–119.
Wren, A.M., et al., 2001. Ghrelin enhances appetite and increases food intake in humans. J. Clin. Endocrinol. Metab. 86, 5992.

CHAPTER FIFTEEN

Ghrelin in the Control of Energy, Lipid, and Glucose Metabolism

Kristy M. Heppner*, Timo D. Müller†, Jenny Tong*, Matthias H. Tschöp*,†,1

*Metabolic Diseases Institute, Department of Medicine, University of Cincinnati, Cincinnati, Ohio, USA
†Institute for Diabetes and Obesity, Helmholtz Center and Technical University Munich, Munich, Germany
[1]Corresponding author: e-mail address: tschoep@helmholtz-muenchen.de

Contents

Abstract

The discovery of ghrelin as the endogenous ligand for the growth hormone secretagogue receptor (GHS-R) led to subsequent studies characterizing the endogenous action of this gastrointestinal hormone. Accordingly, exogenous administration of ghrelin was found to increase food intake and adiposity in a variety of species, including rodents, nonhuman primates, and humans. Later work supported these findings and confirmed that ghrelin acts through hypothalamic neurons to mediate its effects on energy metabolism. Ghrelin acts specifically through GHS-R to promote a positive energy balance as demonstrated by loss of ghrelin action after pharmacological blockade or genetic deletion of GHS-R. More recently, ghrelin was found to be a mediator of glucose metabolism and acts to inhibit insulin secretion from pancreatic β-cells. Together, the literature highlights a predominant role of ghrelin in regulating energy and glucose metabolism.

Methods in Enzymology, Volume 514
ISSN 0076-6879
http://dx.doi.org/10.1016/B978-0-12-381272-8.00015-5

1. INTRODUCTION

Before the discovery of ghrelin, work by Willesen et al. demonstrated that the growth hormone secretagogue receptor (GHS-R) was coexpressed on hypothalamic neuropeptide Y (NPY) neurons, suggesting that the ligand for this receptor is likely to be implicated in the regulation of system metabolism (Willesen et al., 1999). Shortly after this, ghrelin was found to be the endogenous ligand of GHS-R (Kojima et al., 1999), and the field exploded with publications characterizing the role of ghrelin as a mediator of energy metabolism. The work of Tschöp et al. revealed that exogenously administered ghrelin promotes a positive energy balance and induces adiposity in rodents (Tschop et al., 2000). Wren et al. examined acute peripheral and central action of ghrelin and demonstrated that ghrelin increases acute food intake in rats (Wren et al., 2000). Nakazato et al. confirmed the central orexigenic action of ghrelin by demonstrating that ghrelin activates hypothalamic NPY and Agouti-related peptide (AgRP) neurons and that coadministration of ghrelin and antagonists against NPY or AgRP abolishes ghrelin induction of food intake (Nakazato et al., 2001). Interestingly, the effects of ghrelin on adipogenesis are independent of ghrelin-induced hyperphagia (Theander-Carrillo et al., 2006).

In addition to regulating energy balance, ghrelin has also been shown to regulate glucose metabolism. Accordingly, ghrelin and its receptor are both expressed in pancreatic islet cells (Kageyama et al., 2005), and ghrelin inhibits glucose-stimulated insulin release in isolated islet and β-cells (Dezaki et al., 2004, 2007; Qader et al., 2008). Furthermore, in both rodents (Cui et al., 2008; Dezaki et al., 2004; Reimer et al., 2003) and humans (Broglio et al., 2001, 2004; Tong et al., 2010), acute peripheral administration of ghrelin inhibited fasting or glucose-stimulated insulin secretion leading to impairment in glucose tolerance. Together, all of these data suggest that ghrelin antagonism could be a promising therapy for treating obesity and type II diabetes.

2. CENTRAL ADMINISTRATION OF GHRELIN INCREASES ACUTE FOOD INTAKE IN RATS

Ghrelin is predominantly synthesized and secreted by X/A-like cells in the oxyntic glands of the mucosa of the gastric fundus (Sakata et al., 2002). Ghrelin's effect on food intake is, however, centrally mediated and includes the activation of hypothalamic NPY and AgRP neurons (Nakazato et al., 2001). The most consistent results to show that ghrelin-induced hyperphagia are after acute central administration (Wang et al., 2002; Wren et al., 2000). Ghrelin-induced hyperphagia in rodents has been reported after peripheral administration, although very high doses of ghrelin are required (Date et al., 2006; Wang et al., 2002; Wren et al., 2000, 2001). Chronic ghrelin administration reliably increases fat mass; however, in mice, a significant increase in food intake is not observed (Pfluger et al., 2011; Tschop et al., 2000), whereas chronic infusion in rats consistently induces both adiposity and hyperphagia (Perez-Tilve et al., 2011; Theander-Carrillo et al., 2006; Tschop et al., 2000). Another important note is that diet-induced obese (DIO) mice are resistant to the orexigenic effect of ghrelin (Gardiner et al., 2010; Perreault et al., 2004). This ghrelin resistance can be attributed to the impaired ability of ghrelin to activate NPY/AgRP neurons in the arcuate nucleus of the hypothalamus, which is likely due to decreased expression of hypothalamic *GHS-R* and *Npy/Agrp* in DIO animals (Briggs et al., 2010). Furthermore, acute exposure to high-fat diet (HFD) ablates ghrelin's ability to induce hyperphagia in rats (Perez-Tilve et al., 2011). Together, data in the literature indicate that administration of ghrelin centrally to rodents on a standard chow diet is the most reliable way to elicit an induction of acute food intake (Nakazato et al., 2001; Perez-Tilve et al., 2011; Sivertsen et al., 2011). Therefore, this section describes the methods for third-ventricle cannulation (i3vt) in rats and subsequent injection of ghrelin to induce acute food intake.

1. Rats should weigh approximately 250–300 g.
2. Anesthetize animals with ketamine/xylazine (1.0 ml/kg in 10:6.5 ratio) or equivalent anesthetic.
3. Secure animal into a stereotaxic device (Kopf Instruments, Tujunga, CA) with lambda and bregma at the same vertical coordinate.
4. Make a small midline incision on the top of the head to provide access to the cranium.
5. Make a small window in the cranium with a fine dremel bit and remove the piece of skull with forceps.
6. Carefully displace the superior sagittal sinus laterally with a metal probe.
7. Implant a 22-gauge stainless steel guide cannula (Plastics One, Roanoke, VA) into the third ventricle of the brain according to Paxinos atlas of the rat brain (Paxinos and Watson, 1998) using the stereotaxic coordinates: –2.2 mm posterior to bregma and to a depth of –7.5 mm from the surface of the brain (Fig. 15.1).

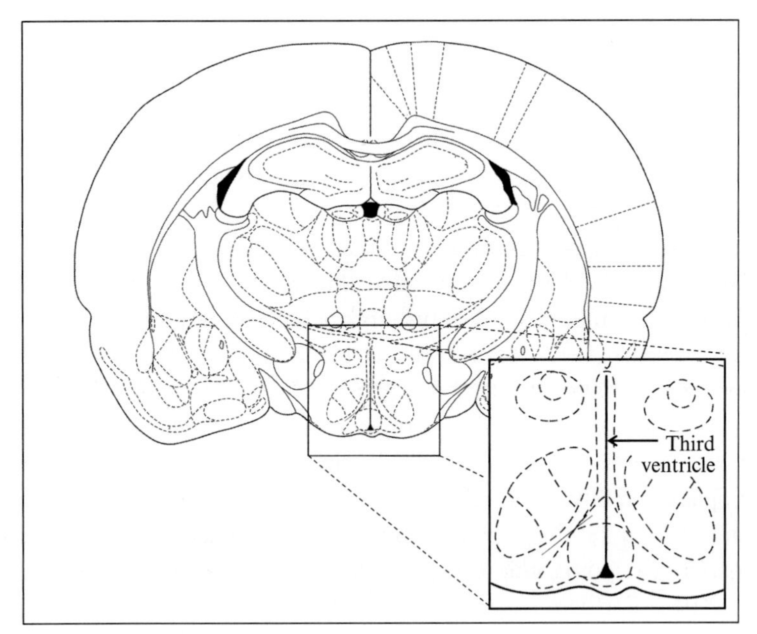

Figure 15.1 Coronal section illustrating the third cerebral ventricle as represented in Paxinos atlas of the rat brain (Paxinos and Watson, 1998). Coordinates for proper i3vt placement relative to bregma are −2.2 mm posterior and −7.5 mm.

8. Use anchor screws and dental acrylic to secure the cannula to the skull and insert an obturator extending 1 mm below the guide cannula.
9. Following surgery, inject animals subcutaneously with buprenorphine (0.28 mg/kg) or equivalent analgesic.
10. After 1 week of recovery, confirm the correct cannula placement by intracerebroventricular (icv) administration of 1 μg of angiotensin II in 1 μl of 0.9% saline. Rats that fail to drink a minimum of 5 ml of water within 60 min after the administration of angiotensin II should be removed from the studies.
    a. Cannula placement can also be verified postmortem using methylene blue injection into the cannula.
11. Allow a 5-day washout period of angiotensin II.
12. For acute injections, dissolve ghrelin in saline (0.9% sodium chloride) to a concentration of 10 μg/μl.
13. For the best results, perform injections at the beginning of the light phase when rats have had *ad libitum* access to food the night before.
14. Connect a 26-gauge internal cannula (Plastics One) to polyethylene-50 tubing and connect the tubing to a 25-μl Hamilton syringe (Hamilton, Reno, NV).

15. Draw up compound in syringe and inject 1 μl of ghrelin (10 μg).
16. Weigh food 2, 4, 8, and 24 h after injection.

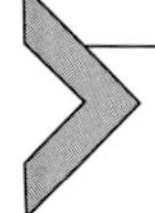

3. CHRONIC CENTRAL INFUSION OF GHRELIN GENERATES A POSITIVE ENERGY BALANCE IN RODENTS

In addition to the orexigenic action of ghrelin, this peptide potently increases body weight and adiposity when administered chronically (Tschop et al., 2000). Ghrelin induces adiposity and body weight gain through multiple mechanisms including modulation of food intake, energy expenditure, nutrient partitioning, adipocyte metabolism, and locomotor activity. Chronic infusion in rats consistently induces hyperphagia (Perez-Tilve et al., 2011; Theander-Carrillo et al., 2006; Tschop et al., 2000); however, chronic ghrelin treatment can induce adiposity in the absence of hyperphagia (Theander-Carrillo et al., 2006), suggesting that ghrelin's regulation of food intake and adiposity are through independent pathways. Furthermore, HFD exposure impairs the orexigenic action of ghrelin; however, ghrelin induction of adiposity remains intact (Perez-Tilve et al., 2011). This provides further evidence that separate hypothalamic circuits regulate the orexigenic and adipogenic effects of ghrelin.

Another mechanism through which ghrelin promotes a positive energy balance is by decreasing energy expenditure (Asakawa et al., 2001) in part through suppression of sympathetic nerve activity in brown adipose tissue (BAT) (Yasuda et al., 2003). In terms of nutrient partitioning, ghrelin causes a shift in macronutrient utilization to favor carbohydrates rather than fats as a major energy source (Pfluger et al., 2008; Tschop et al., 2000). Additionally, ghrelin acts through central mechanisms to alter adipocyte metabolism to cause inhibition of lipid oxidation and an increase in lipogenesis and triglyceride uptake in white adipose tissue (WAT) (Theander-Carrillo et al., 2006). Lastly, ghrelin has also been shown to further promote a positive energy balance by decreasing spontaneous locomotor activity (Pfluger et al., 2011). Collectively, ghrelin acts through multiple mechanisms to promote a positive energy balance. This section explains the methods used to induce and analyze ghrelin's chronic effects on energy metabolism in rodents.

3.1. Ghrelin induces adiposity in rodents

Here, we describe the most reliable way to induce adiposity in rodents through chronic central infusion of ghrelin. Furthermore, we suggest the most accurate method to analyze adiposity (Tschop et al., 2011).

3.1.1 Chronic icv infusion in rats

1. Dissolve ghrelin in saline (0.9% sodium chloride) at a final dose of 2.5 nmol/rat/day.
2. Prepare icv mini-pumps the day before surgeries according to manufacturer's instructions (Brain Infusion Kit 1 (3–5 mm); Osmotic mini-pump 2001; Alzet Durect, Cupertino, CA).
3. Anesthetize rats with 2.5% isoflurane in oxygen or equivalent anesthetic.
4. Place the animal in stereotaxic device (Kopf Instruments) with lambda and bregma at the same vertical coordinate.
5. Make a small midline incision on the top of the head to provide access to the cranium.
6. Stereotaxically implant cannula in the lateral cerebral ventricle of the brain (left or right ventricle can be used) using the coordinates –0.8 mm posterior to bregma and +/−1.4 mm lateral to the midsagittal suture and to a depth of –3.5 mm from the surface of the brain (Paxinos and Watson, 1998) (Fig. 15.2).

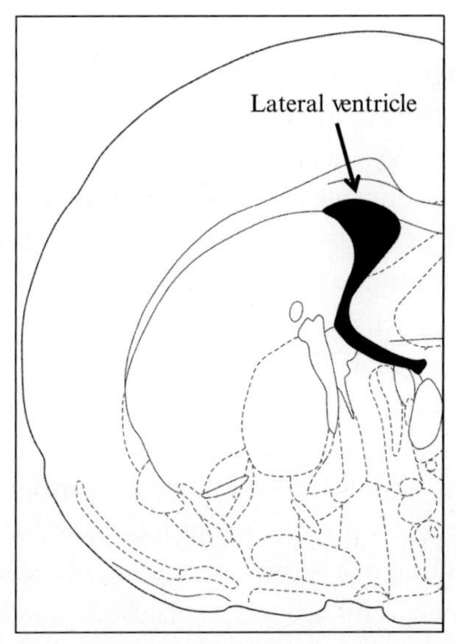

Figure 15.2 Coronal section illustrating the lateral ventricle as represented in Paxinos atlas of the rat brain (Paxinos and Watson, 1998). Coordinates for proper lateral ventricle placement relative to bregma are –0.8 mm posterior, +/−1.4 mm lateral, and –3.5 mm. For cannula placement in the lateral ventricle of mice, use the coordinates –0.7 mm posterior, +/−1.2 mm lateral, and –2.5 mm according to Paxinos atlas of the mouse (Franklin and Paxinos, 2008).

7. The cannula should be connected to the mini-pump with a polyethylene catheter, and the osmotic mini-pump should be subcutaneously implanted in the interscapular region.
8. Fix the cannula in place using screw anchors and dental acrylic.
9. Following surgery, inject animals subcutaneously with buprenorphine (0.28 mg/kg) or an equivalent analgesic.
10. Allow 7 days of infusion taking note of daily body weight and food intake.
11. Analyze body composition on day 0 and day 7 using one or more of the following methods:
    a. Nuclear magnetic resonance (NMR) technique (echoMRI, Houston, TX, USA).
    b. Dual-energy X-ray absorptiometry (DEXA).
    c. If specific fat pad analysis is desired, rodents can be sacrificed and fat pads can be excised and weighed with a balance.

3.1.2 Chronic icv infusion in mice

1. Dissolve ghrelin in saline (0.9% sodium chloride) at a final dose of 2.5 nmol/mouse/day.
2. Prepare icv-osmotic mini-pumps the day before surgeries according to manufacturer's instructions (Brain Infusion Kit 3 (1–3 mm), Osmotic mini-pump 1007D; Alzet, Cupertino, CA).
3. Anesthetize animals with 2.5% isoflurane in oxygen or equivalent anesthetic.
4. Place animal in stereotaxic device (Kopf Instruments) with lambda and bregma at the same vertical coordinate.
5. Make a small midline incision on the top of the head to provide access to the cranium.
6. Implant cannula in the lateral cerebral ventricle of the brain (left or right ventricle can be used) using the coordinates −0.7 mm posterior to bregma, +/−1.2 mm lateral to the midsagittal suture, and to a depth of −2.5 mm from the surface of the brain (Franklin and Paxinos, 2008).
7. The cannula should be connected to the mini-pump with a polyethylene catheter and the osmotic mini-pump should be subcutaneously implanted in the interscapular region.
8. Fix the cannula to the skull using tissue adhesive (VetBond St Paul, MN).
9. Following surgery, inject animals subcutaneously with buprenorphine (0.28 mg/kg) or equivalent analgesic.

10. Allow 7 days of infusion taking note of daily body weight and food intake.
11. Analyze body composition on day 0 and day 7 using one or more of following methods:
    a. NMR (echoMRI).
    b. DEXA.
    c. If specific fat pad analysis is desired, rodents can be sacrificed and fat pads can be excised and weighed with a balance.

3.2. Analysis of adipose tissue metabolism following ghrelin treatment

3.2.1 Analysis of respiratory quotient value in rodents

Ghrelin has been shown to increase the respiratory quotient (RQ) in rodents (Tschop et al., 2000). RQ, which is the ratio of the volume of CO_2 produced (V_{CO_2}) to the volume of O_2 consumed (V_{O_2}), indicates which macronutrient (carbohydrates or lipids) was preferentially oxidized. An RQ value of 1 indicates that carbohydrates were the main source of energy utilization, whereas an RQ of 0.7 indicates that lipids were the main source of energy utilization. The increase in RQ levels following ghrelin administration indicates that decreased lipid utilization contributes to ghrelin's adipogenic properties. To analyze which specific fuel sources are being utilized during chronic ghrelin infusion, indirect calorimetry is the preferred method to use (Nogueiras et al., 2007; Tschop et al., 2011).

1. Following icv mini-pump implantation, place animals in an indirect calorimetry system (Labmaster; TSE Systems, Bad Homburg v.d.H., Germany) which is an open-circuit system that determines O_2 consumption, CO_2 production, and RQ ($= V_{CO_2}/V_{O_2}$).
2. Allow at least a 48-h adaptation period to the system.
3. Extract data from system to determine the RQ value.

3.2.2 Analysis of adipocyte gene expression

Central ghrelin treatment influences both white and brown adipocyte metabolism (Theander-Carrillo et al., 2006). After chronic icv infusion of ghrelin, gene expression of WAT storage enzymes are upregulated, whereas expression of enzymes involved in fat oxidation are downregulated. The key WAT enzymes influenced by chronic ghrelin treatment are lipoprotein lipase (LPL), acetyl-CoA carboxylase α (ACC), fatty acid synthase (FAS), stearoyl-CoA desaturase-1 (SCD-1), and carnitine palmitoyl transferase-1α (CPT-1α). Furthermore, BAT enzymes uncoupling protein 1 and 3 (UCP1 and UCP3), which promote a negative energy balance by dissipating heat in response to sympathetic nerve activity, are downregulated following chronic ghrelin treatment. Interestingly, all of these changes in gene expression are independent of changes in food intake. Here, we describe the method for measuring the effect of central ghrelin treatment on adipocyte metabolism.

1. Implant icv mini-pump as outlined in Section 3.1.
2. After a week of infusion, decapitate animals and excise adipose tissue.
3. Store at –80 °C until WAT tissue is analyzed for LPL, ACC, FAS, SCD-1, and CPT-1α, and BAT is analyzed for UCP1 and UCP2 expression.
4. Analyze mRNA expression using methods described previously.

3.3. Ghrelin decreases locomotor activity

In addition to increasing adiposity, ghrelin also promotes a positive energy balance by decreasing locomotor activity (Pfluger et al., 2011; Tang-Christensen et al., 2004). Here, we describe the methods to analyze spontaneous locomotor activity in rodents, given chronic central treatment of ghrelin.

1. Implant icv mini-pump as outlined in Section 3.1.
2. Single-house animals in LabMaster chambers (TSE Systems).
3. Allow animals to adapt to the system for a minimum of 48 h.
4. Monitor home-cage locomotor activity by using the system's multidimensional infrared light beams on both the bottom and top of the cage.
5. Assess total locomotor activity by analyzing the number of beam breaks and dissect activity into fidgeting, ambulatory movement and rearing.
6. Assess fidgeting which is defined as consecutive breaks of one single light beam.
7. Assess ambulatory movement which is defined as breaks of any two different light beams.
8. Assess rearing which is defined as simultaneous break of beams on both cage bottom and top level.

4. GHRELIN REGULATES GLUCOSE HOMEOSTASIS

A large number of studies have focused on investigating the role of ghrelin in regulating glucose homeostasis. The action that ghrelin has on glucose metabolism is less clear than its effects on energy metabolism. Currently, the consensus in the field is that peripherally administered ghrelin inhibits glucose-stimulated insulin secretion, resulting in impairment in glucose tolerance. This section describes the methods for the acute effects of ghrelin on glucose tolerance and glucose-stimulated insulin secretion in mice.

4.1. Acute peripheral ghrelin administration suppresses glucose-stimulated insulin secretion and impairs glucose tolerance in mice

1. Fast the mice for 16 h before glucose injection.
2. Inject ghrelin (60 nmol) intraperitoneally (ip) 30 min prior to glucose injection.
3. Inject a bolus of glucose ip (2 g/kg body weight of 25% D-glucose)
4. Measure blood glucose via tail vein blood at 0, 15, 30, 60, and 120 min after glucose injection using a hand-held glucometer.
5. Collect blood samples (40 μl per time point) at 0, 15, and 60 min after glucose injection by tail vein bleeding into EDTA-coated capillary tubes.
6. Separate plasma and analyze samples for insulin using commercially available kit (Ultra Sensitive Mouse Insulin ELISA Kit; Crystal Chem).

5. SUMMARY

Ghrelin is a potent orexigenic hormone that acts through GHS-R expressed on hypothalamic neurons to mediate its effects on feeding. Being a centrally acting peptide, the most prominent effects of ghrelin on food intake are observed after acute icv administration. In addition to the short-term effects of ghrelin on feeding, chronic administration causes an increase in adiposity that is independent of hyperphagia. The potent orexigenic and adipogenic properties of ghrelin have made it a popular candidate for treating disorders such as cancer cachaxia and anorexia nervosa, which are associated with a negative energy balance. Unfortunately, ghrelin has not proven to be effective in alleviating these disorders (Miljic et al., 2006). Furthermore, ghrelin and ghrelin receptor antagonists have not been able to successfully reduce obesity in humans.

The exact physiological role that ghrelin plays in regulating glucose metabolism is still not well defined. The most consistent effects have been ghrelin's acute action to impair glucose tolerance by inhibiting insulin secretion from pancreatic β-cells. The development of an inhibitor for the ghrelin acylating enzyme (GOAT) provides further evidence that ghrelin plays a role in mediating glucose homeostasis (Barnett et al., 2010). Mice pretreated with the GOAT inhibitor experienced a significant increase in insulin response, which was accompanied by a reduction in blood glucose when mice were subjected to an ip glucose challenge. This effect was not apparent in ghrelin knockout mice, confirming that the improvement in glucose homeostasis was mediated by acyl ghrelin inhibition. An unanswered question in the field is whether ghrelin regulates glucose metabolism solely by the direct action on peripheral tissues or it also acts centrally to regulate glucose metabolism. This warrants further investigation, and a combination of pharmacological and genetic techniques could provide further insight into this question. Understanding these site-specific actions could help lead to the development of more effective therapeutics for treating both obesity and type II diabetes.

REFERENCES

Asakawa, A., et al., 2001. Ghrelin is an appetite-stimulatory signal from stomach with structural resemblance to motilin. Gastroenterology 120, 337–345.

Barnett, B.P., et al., 2010. Glucose and weight control in mice with a designed ghrelin O-acyltransferase inhibitor. Science 330, 1689–1692.

Briggs, D.I., et al., 2010. Diet-induced obesity causes ghrelin resistance in arcuate NPY/AgRP neurons. Endocrinology 151, 4745–4755.

Broglio, F., et al., 2001. Ghrelin, a natural GH secretagogue produced by the stomach, induces hyperglycemia and reduces insulin secretion in humans. J. Clin. Endocrinol. Metab. 86, 5083–5086.

Broglio, F., et al., 2004. Non-acylated ghrelin counteracts the metabolic but not the neuroendocrine response to acylated ghrelin in humans. J. Clin. Endocrinol. Metab. 89, 3062–3065.

Cui, C., et al., 2008. Ghrelin infused into the portal vein inhibits glucose-stimulated insulin secretion in Wistar rats. Peptides 29, 1241–1246.

Date, Y., et al., 2006. Peripheral ghrelin transmits orexigenic signals through the noradrenergic pathway from the hindbrain to the hypothalamus. Cell Metab. 4, 323–331.

Dezaki, K., et al., 2004. Endogenous ghrelin in pancreatic islets restricts insulin release by attenuating Ca^{2+} signaling in beta-cells: implication in the glycemic control in rodents. Diabetes 53, 3142–3151.

Dezaki, K., et al., 2007. Ghrelin uses Galphai2 and activates voltage-dependent $K+$ channels to attenuate glucose-induced Ca^{2+} signaling and insulin release in islet beta-cells: novel signal transduction of ghrelin. Diabetes 56, 2319–2327.

Franklin, K.B.J., Paxinos, G., 2008. The Mouse Brain in Stereotaxic Coordinates. Elsevier Academic Press, San Diego, CA.

Gardiner, J.V., et al., 2010. The hyperphagic effect of ghrelin is inhibited in mice by a diet high in fat. Gastroenterology 138, 2468–2476.

Kageyama, H., et al., 2005. Morphological analysis of ghrelin and its receptor distribution in the rat pancreas. Regul. Pept. 126, 67–71.

Kojima, M., et al., 1999. Ghrelin is a growth-hormone-releasing acylated peptide from stomach. Nature 402, 656–660.

Miljic, D., et al., 2006. Ghrelin has partial or no effect on appetite, growth hormone, prolactin, and cortisol release in patients with anorexia nervosa. J. Clin. Endocrinol. Metab. 91, 1491–1495.

Nakazato, M., et al., 2001. A role for ghrelin in the central regulation of feeding. Nature 409, 194–198.

Nogueiras, R., et al., 2007. The central melanocortin system directly controls peripheral lipid metabolism. J. Clin. Invest. 117, 3475–3488.

Paxinos, G., Watson, C., 1998. The Rat Brain in Stereotaxic Coordinates. Elsevier Academic Press, San Diego, CA.

Perez-Tilve, D., et al., 2011. Ghrelin-induced adiposity is independent of orexigenic effects. FASEB J. 25, 2814–2822.

Perreault, M., et al., 2004. Resistance to the orexigenic effect of ghrelin in dietary-induced obesity in mice: reversal upon weight loss. Int. J. Obes. Relat. Metab. Disord. 28, 879–885.

Pfluger, P.T., et al., 2008. Simultaneous deletion of ghrelin and its receptor increases motor activity and energy expenditure. Am. J. Physiol. Gastrointest. Liver Physiol. 294, G610–G618.

Pfluger, P.T., et al., 2011. Ghrelin, peptide YY and their hypothalamic targets differentially regulate spontaneous physical activity. Physiol. Behav. 105, 52–61.

Qader, S.S., et al., 2008. Proghrelin-derived peptides influence the secretion of insulin, glucagon, pancreatic polypeptide and somatostatin: a study on isolated islets from mouse and rat pancreas. Regul. Pept. 146, 230–237.

Reimer, M.K., et al., 2003. Dose-dependent inhibition by ghrelin of insulin secretion in the mouse. Endocrinology 144, 916–921.

Sakata, I., et al., 2002. Ghrelin-producing cells exist as two types of cells, closed- and opened-type cells, in the rat gastrointestinal tract. Peptides 23, 531–536.

Sivertsen, B., et al., 2011. Unique interaction pattern of a functionally biased ghrelin receptor agonist. J. Biol. Chem. 286, 20845–20860.

Tang-Christensen, M., et al., 2004. Central administration of ghrelin and agouti-related protein (83–132) increases food intake and decreases spontaneous locomotor activity in rats. Endocrinology 145, 4645–4652.

Theander-Carrillo, C., et al., 2006. Ghrelin action in the brain controls adipocyte metabolism. J. Clin. Invest. 116, 1983–1993.

Tong, J., et al., 2010. Ghrelin suppresses glucose-stimulated insulin secretion and deteriorates glucose tolerance in healthy humans. Diabetes 59, 2145–2151.

Tschop, M., et al., 2000. Ghrelin induces adiposity in rodents. Nature 407, 908–913.

Tschop, M.H., et al., 2011. A guide to analysis of mouse energy metabolism. Nat. Methods 9, 57–63.

Wang, L., et al., 2002. Peripheral ghrelin selectively increases Fos expression in neuropeptide Y-synthesizing neurons in mouse hypothalamic arcuate nucleus. Neurosci. Lett. 325, 47–51.

Willesen, M.G., et al., 1999. Co-localization of growth hormone secretagogue receptor and NPY mRNA in the arcuate nucleus of the rat. Neuroendocrinology 70, 306–316.

Wren, A.M., et al., 2000. The novel hypothalamic peptide ghrelin stimulates food intake and growth hormone secretion. Endocrinology 141, 4325–4328.

Wren, A.M., et al., 2001. Ghrelin causes hyperphagia and obesity in rats. Diabetes 50, 2540–2547.

Yasuda, T., et al., 2003. Centrally administered ghrelin suppresses sympathetic nerve activity in brown adipose tissue of rats. Neurosci. Lett. 349, 75–78.

CHAPTER SIXTEEN

Ghrelin and the Vagus Nerve

Yukari Date[1]
Frontier Science Research Center, University of Miyazaki, Kiyotake, Miyazaki, Japan
[1]Corresponding author: e-mail address: dateyuka@med.miyazaki-u.ac.jp

Contents

Abstract

Ghrelin, a gastrointestinal hormone, stimulates feeding and secretion of growth hormone (GH). Ghrelin is thought to directly affect neurons involved in feeding or GH secretion through growth hormone secretagogue receptor (GHS-R; ghrelin receptor); however, it is still unclear whether ghrelin crosses through the blood–brain barrier. Recently, several gastrointestinal hormones have been shown to transmit signals involved in feeding to the brain at least in part via the vagal afferent system. In fact, ghrelin's action on feeding or GH secretion is abolished or attenuated in rats that have undergone vagotomy or treatment with capsaicin, a specific afferent neurotoxin. GHS-R is present in the vagal afferent neurons as well as the brain and is transported to the afferent terminals. In addition, the firing rate of vagal afferent fibers significantly decreases after ghrelin administration. Taken together, these data show that the vagal afferent system is the major pathway conveying ghrelin's signals for feeding and GH secretion to the brain.

Methods in Enzymology, Volume 514
ISSN 0076-6879
http://dx.doi.org/10.1016/B978-0-12-381272-8.00016-7

1. INTRODUCTION

The afferent fibers of the vagus nerve are the major neuroanatomic linkage between the digestive tract and the nucleus of the solitary tract in the hindbrain (Ritter et al., 1994; Sawchenko, 1983; Schwartz et al., 2000; van der Kooy et al., 1984). Some meal-related metabolites, monoamines, and peptides, as well as mechanical and chemical stimuli, transmit their messages to the nucleus of the solitary tract via the vagal afferent system or to the hypothalamus via the bloodstream. Ghrelin, a peptide primarily produced in the stomach, functions in feeding control and growth hormone (GH) secretion by binding to the growth hormone secretagogue receptor (GHS-R) (Kojima et al., 1999; Nakazato et al., 2001; Tschöp et al., 2001; Wren et al., 2000). Our research group has demonstrated the importance of the vagal afferent system in conveying the ghrelin signals related to feeding or GH secretion to the brain. Here, we present the methodology indispensable for elucidating the role of the vagus nerve in ghrelin's action. These procedures may well be applicable for investigations into the relationship between the vagus nerve and other gastrointestinal hormones.

2. ANATOMY OF THE VAGUS NERVE

2.1. Vagal afferent neurons within the nodose ganglion

Vagal afferent neurons are present within the body and caudal pole of the nodose ganglion from which vagal afferent fibers innervating the abdominal viscera originate (Dockray and Sharkey, 1986; Green and Dockray, 1987, 1988; Sharkey and Williams, 1983). The rat nodose ganglion is composed of about 6000 neurons (Cooper, 1984).

1. To histologically identify vagal afferent neurons in the rat, perform transcardial perfusion with a fixative containing 4% paraformaldehyde.
2. Identify the nodose ganglion, which is a prominent swelling of the vagus nerve immediately before the entrance into the cranial cavity.
3. Remove the nodose ganglion and cut into 12-μm-thick slices at −20 °C using a cryostat. Sections that have been mounted on glass slides can be kept at −30 °C until staining.
4. To confirm the presence of vagal afferent neurons, immunohistochemical staining is useful. Substance P and calcitonin gene-related peptide are recognized as suitable markers for vagal afferent neurons and terminals.

2.2. Vagal afferent innervation

The digestive system is innervated by sensory fibers running through the vagi and the sympathetic nerves (Andrew, 1986; Mei, 1983). Although the sensory fibers innervating the digestive tract are intermingled with efferent fibers, the proportion of sensory fibers exceeds that of efferent fibers. Approximately 90% of the vagus nerves in the subdiaphragm are afferent and are unmyelinated or only thinly myelinated (Agostoni et al., 1957). Some afferent fibers terminate within the gastrointestinal mucosa and submucosa. These afferent fibers are optimally positioned to monitor the composition of gastrointestinal lumen or the concentration of bioactive substances released from enteroendocrine cells (Grundy and Scratcherd, 1989). It is well known that several gastrointestinal hormones, including cholecystokinin (CCK), peptide YY (PYY), glucagon-like peptide-1 (GLP-1), and ghrelin, transmit signals of satiety and starvation to the brain at least in part via the vagal afferent system (Abbott et al., 2005; Date et al., 2002; Koda et al., 2005; Smith et al., 1981). CCK, PYY, and GLP-1 function as satiety signals, whereas ghrelin is the first gastrointestinal peptide shown to act as a starvation signal in the periphery (Tschöp et al., 2001). This is noteworthy because the orexigenic peptide-based system in the periphery was unknown until ghrelin was discovered.

3. GHS-R IN THE VAGAL AFFERENT NEURONS

3.1. Immunoreactivity of GHS-R

1. After fixation with 4% paraformaldehyde (as shown in Section 2.1), remove the nodose ganglion and cut into 12-μm-thick slices at −20 °C using a cryostat.
2. After washing the sections with 0.01 mol/L phosphate-buffered saline (PBS, pH 7.4), incubate them overnight at 4 °C with rabbit anti-GHS-R antiserum (dilution: 1/1000) (Date et al., 2005).
3. Wash the sections with 0.01 mol/L PBS (pH 7.4) and then incubate them for 2 h with Alexa Fluor 594-conjugated chicken anti-rabbit IgG (Molecular Probes, Inc., Eugene, OR) to detect GHS-R immunoreactivity.
4. Assess the immunoreactivity by using fluorescence microscopy (BH2-RFC; Olympus, Tokyo, Japan).

3.2. Ghrelin binding to GHS-R

1. Make a crushing ligation to the vagus nerve with suture thread 20 mm caudal to the nodose ganglion.
2. Excise the portion of the vagus nerve containing the ligation site with the rostral and caudal fibers, embed it in Tissue-Tek O.C.T. Compound (Sakura Finetechnical Co., Ltd., Tokyo, Japan), and then freeze.
3. Using a cryostat, cut serial sections (10 μm thick) along the longitudinal axis of the nerve and then mount them onto gelatin-coated glass slides.
4. Incubate the sections at 37 °C with binding buffer (20 mmol/L HEPES, 150 mmol/L NaCl, 5 mmol/L $MgCl_2$, 1 mmol/L ethylene glycol-bis (β-aminoethyl ether)-*N,N,N′,N′*-tetraacetic acid, and 0.1% bovine serum albumin) for 60 min.
5. Put 3 nmol/L [^{125}I-Tyr^{29}]-rat ghrelin onto the nerve sections and incubate for 30 min.
6. Expose the sections to an imaging plate (Fuji Film, Tokyo, Japan) for 24 h and analyze using a BAS-2000 scanner (Fuji Film).
7. Nonspecific binding can be determined in the presence of excess (3 μmol/L) unlabeled ghrelin.

4. ACTION OF GHRELIN VIA THE VAGUS NERVE

4.1. Effect of vagotomy or capsaicin treatment on the action of ghrelin

4.1.1 Vagotomy

1. Anesthetize rats by intraperitoneal injection of sodium pentobarbital (40 mg/kg) (Abbot Laboratories, Chicago, IL).
2. Make a midline incision to sufficiently expose of the upper abdominal organs.
3. Expose the bilateral subdiaphragmatic trunks of the vagal nerve along the esophagus.
4. Split the trunks and cut them to produce vagotomized rats.
5. Close the abdominal incision.
6. In the sham operation, expose and split but not cut the bilateral trunks.

4.1.2 Capsaicin treatment

1. Anesthetize rats by intraperitoneal injection of sodium pentobarbital (40 mg/kg) (Abbot Laboratories).
2. Make a midline incision to sufficiently expose the upper abdominal organs.

3. Expose the bilateral subdiaphragmatic trunks of the vagal nerve along the esophagus.
4. Prepare cotton strings soaked in capsaicin (Sigma Chemical Co., St. Louis, MO), a specific afferent neurotoxin, dissolved in olive oil (5% w/v).
5. Loosely tie the bilateral subdiaphragmatic trunks of the vagal nerve with the capsaicin-soaked cotton strings.
6. Remove the cotton string 30 min later and close the abdominal incision.

4.1.3 The effect of vagotomy or capsaicin treatment on ghrelin-induced feeding and GH secretion

Experiments were performed 1 week after vagotomy or perivagal capsaicin application. Only rats that showed progressive weight gain and food intake were used in the feeding experiments. Rat ghrelin (1.5 nmol/rat) (Peptide Institute, Inc., Osaka, Japan) was intravenously administered at 9 a.m., when the endogenous orexigenic system is expected to be inactive, to rats that had undergone bilateral subdiaphragmatic vagotomy or capsaicin treatment, or to sham-operated rats after light anesthesia with ether. A single administration of ghrelin to sham-operated rats significantly increased 2-h food intake but did not increase food intake in rats that had undergone subdiaphragmatic vagotomy or capsaicin treatment (Fig. 16.1) (Date et al., 2002).

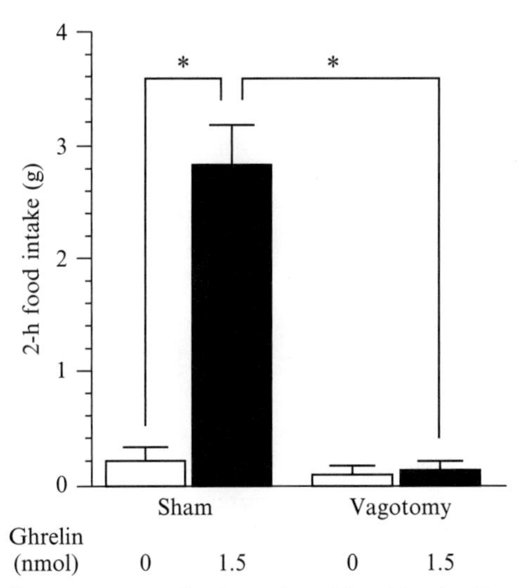

Figure 16.1 Effect of vagotomy on ghrelin-induced food intake. Food intake of rats that had undergone bilateral subdiaphragmatic vagotomy after single intravenous administration of ghrelin (1.5 nmol) is shown. $^{*}P < 0.001$ versus sham-operated rats.

To investigate the role of the vagus nerve system in GH secretion, rat ghrelin (1.5 nmol/100 μl) was intravenously administered at 9 a.m. to rats that had undergone bilateral subdiaphragmatic vagotomy or capsaicin treatment, or to sham-operated rats. Blood samples (20 μl) were obtained from the tail vein at 0, 5, 10, 15, 20, 30, and 60 min after administration. The plasma concentration of GH was measured with a Biotrak Rat GH RIA kit (Amersham, Buckinghamshire, England).

The GH response to ghrelin was profoundly attenuated by both vagotomy and capsaicin treatment (Date et al., 2002).

4.2. Ghrelin-induced alteration of the firing rate of vagal afferent fibers

Intravenous administration of ghrelin (1.5 nmol/rat) to rats significantly decreases the firing rate of gastric vagal afferent fibers, whereas des-acyl ghrelin, which lacks the *n*-octanoylation at Ser3 that is essential for ghrelin's binding activity to GHS-R, does not (Fig. 16.2). Des-acyl ghrelin also does not affect food intake or GH secretion. In this study, standard methods of extracellular recording from vagal nerve fibers were used, as described in detail elsewhere (Niijima and Yamamoto, 1994). In brief,

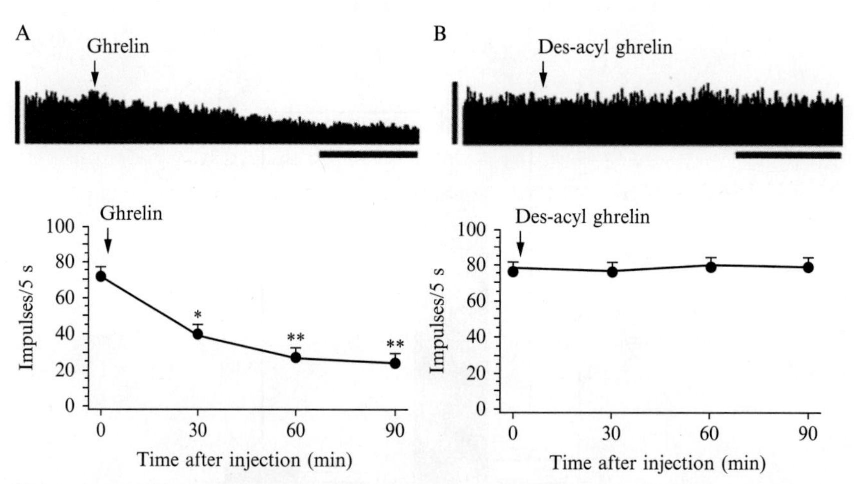

Figure 16.2 Effects of intravenous administration of ghrelin and des-acyl ghrelin on the firing rate of gastric vagal afferent fibers. Ghrelin (A), but not des-acyl ghrelin (B), suppresses the firing rate of gastric vagal afferent fibers. $^{*}P<0.001$; $^{**}P<0.0001$ versus value at 0 min.

1. Anesthetize rats by intraperitoneal injection of urethane (1 g/kg; Sigma). This study is performed under anesthetization throughout.
2. Intubate the rat trachea and record the electrocardiogram, maintaining the body temperature at 37 °C.
3. After laparotomy, insert a small catheter (Intramedic PE-10; Clay Adams, Parsippany, NJ) into the inferior vena cava and fix it at the rostral position 2 cm from the inserted site.
4. Visualize the gastric branch of the vagus nerve and attach the fibers isolated from the peripheral cut end of the gastric ventral branch to a pair of silver wire electrodes.
5. For display and storage of the firing rate, connect the silver wire electrodes to an oscilloscope and magnetic tape recorder through an AC-coupled differential amplifier (DAP-10E, Dia Medical System Co., Tokyo, Japan).

A window discriminator converts spikes to constant-amplitude pulses for analysis of spike frequency with a rate meter that resets at 5-s intervals. Output from the rate meter is recorded on chart paper (8 K20 recoder; NEC-SAN E1 Co., Tokyo, Japan).

5. SUMMARY

Ghrelin had been thought to enter the brain across the blood–brain barrier, where it induces food intake and GH secretion. However, considering that visceral sensory information, including humoral signals related to feeding, is transmitted to the brain through the vagal afferent pathway, it is plausible that this pathway may instead be the route via which peripheral ghrelin signals are conveyed to the brain. In fact, blockade of the vagal afferent pathway abolished ghrelin-induced feeding or attenuated ghrelin-induced GH secretion. To confirm whether ghrelin directly affects the vagal afferent pathway, investigations into the presence of GHS-R in the vagal afferent neurons located in the nodose ganglion, the binding of ghrelin to GHS-R, and the transportation of GHS-R to the vagal terminals are very important. In addition, alteration of the firing rate of vagal afferent fibers after ghrelin administration is strong evidence of a direct relationship between ghrelin and the vagal afferent pathway. The materials and methods shown here will be useful to elucidate the functional linkage of different gastrointestinal hormones with the vagal afferent pathway.

ACKNOWLEDGMENTS

I thank A. Niijima, N. Murakami, M. Nakazato, and K. Kangawa for technical advice and helpful discussions. This work was supported in part by the Grants-in-Aid from the Ministry of Education, Culture, Sports, Science, and Technology of Japan, and the Program for Promotion of Basic Research Activities for Innovative Bioscience (PROBRAIN).

REFERENCES

Abbott, C.R., et al., 2005. The inhibitory effects of peripheral administration of peptide YY (3-36) and glucagon-like peptide-1 on food intake are attenuated by ablation of the vagal-brainstem-hypothalamic pathway. Brain Res. 1044, 127–131.

Agostoni, E., et al., 1957. Functional and histological studies of the vagus nerve and its branches to the heart, lungs and abdominal viscera in the cat. J. Physiol. 135, 182–205.

Andrew, P.L.R., 1986. Vagal afferent innervation of the gastrointestinal tract. In: Cervero, F., Morrison, J.F.B. (Eds.), Progress in Brain Research. Viseral Sensation, vol. 67. Elsevier, New York, pp. 65–68.

Cooper, E., 1984. Synapse formation among developing sensory neurons from rat nodose ganglia grown in tissue culture. J. Physiol. 351, 263–274.

Date, Y., et al., 2002. The role of the gastric afferent vagal nerve in ghrelin-induced feeding and growth hormone secretion in rats. Gastroenterology 123, 1120–1128.

Date, Y., et al., 2005. Peripheral interaction of ghrelin with cholecystokinin on feeding regulation. Endocrinology 146, 3518–3525.

Dockray, G.J., Sharkey, K.A., 1986. Neurochemistry of visceral afferent neurons. Prog. Brain Res. 67, 133–148.

Green, T., Dockray, G.J., 1987. Calcitonin gene-related peptide and substance P in afferents to the upper gastrointestinal tract in the rat. Neurosci. Lett. 76, 151–156.

Green, T., Dockray, G.J., 1988. Characterization of the peptidergic afferent innervation of the stomach in the rat, mouse and guinea-pig. Neuroscience 25, 181–193.

Grundy, D., Scratcherd, T., 1989. Sensory afferents from the gastrointestinal tract. In: Schultz, S.G. (Ed.), Handbook of Physiology: The Gastrointestinal System. Motility and Circulation, vol. 1. Oxford University Press, New York, pp. 593–620.

Koda, S., et al., 2005. The role of the vagal nerve in peripheral PYY_{3-36}-induced feeding reduction in rats. Endocrinology 146, 2369–2375.

Kojima, M., et al., 1999. Ghrelin is a novel growth hormone releasing acylated peptide from stomach. Nature 402, 656–660.

Mei, N., 1983. Sensory structures in the viscera. In: Autrum, H., Ottoson, D., Perl, E.R., Schmidt, R.F., Shimazu, H., Willis, W.D. (Eds.), Progress in Sensory Physiology 4. Springer-Verlag, New York, pp. 2–42.

Nakazato, M., et al., 2001. A role for ghrelin in the central regulation of feeding. Nature 409, 194–198.

Niijima, A., Yamamoto, T., 1994. The effects of lithium chloride on the activity of the afferent nerve fibers from the abdominal visceral organs in the rat. Brain Res. Bull. 35, 141–145.

Ritter, S., et al., 1994. Induction of Fos-like immunoreactivity (Fos-li) and stimulation of feeding by 2,5-anhydro-D-mannitol (2,5-AM) require the vagus nerve. Brain Res. 646, 53–64.

Sawchenko, P.E., 1983. Central connections of the sensory and motor nuclei of the vagus nerve. J. Auton. Nerv. Syst. 9, 13–26.

Schwartz, M.W., et al., 2000. Central nervous system control of food intake. Nature 404, 661–671.

Sharkey, K.A., Williams, R.G., 1983. Extrinsic innervation of the rat pancreas: demonstration of vagal sensory neurones in the rat by retrograde tracing. Neurosci. Lett. 42, 131–135.
Smith, G.P., et al., 1981. Abdominal vagotomy blocks the satiety effect of cholecystokinin in the rat. Science 213, 1036–1037.
Tschöp, M., et al., 2001. Circulating ghrelin levels are decreased in human obesity. Diabetes 50, 707–709.
van der Kooy, D., et al., 1984. The organization of projections from the cortex, amygdala, and hypothalamus to the nucleus of the solitary tract in rat. J. Comp. Neurol. 224, 1–24.
Wren, A.M., et al., 2000. The novel hypothalamic peptide ghrelin stimulates food intake and growth hormone secretion. Endocrinology 141, 4325–4328.

CHAPTER SEVENTEEN

Measurement of AMP-Activated Protein Kinase Activity and Expression in Response to Ghrelin

Chung Thong Lim, Francesca Lolli, Julia D. Thomas, Blerina Kola, Márta Korbonits[1]

Centre for Endocrinology, William Harvey Research Institute, Barts and The London School of Medicine and Dentistry, Queen Mary University of London, London, United Kingdom

[1]Corresponding author: e-mail address: m.korbonits@qmul.ac.uk

Contents

Abstract

Ghrelin is a circulating brain–gut peptide that is known to exert several metabolic effects such as stimulating appetite, inducing adiposity, increasing bone formation, and influencing the cardiovascular functions. AMP-activated protein kinase (AMPK), a highly conserved heterotrimeric protein that plays a key role in energy homeostasis, has been shown to mediate many of these metabolic effects of ghrelin. Ghrelin is shown to stimulate hypothalamic AMPK activity and inhibit liver and adipose tissue AMPK activity. The effects of ghrelin on AMPK activity can be studied using an elegant kinase assay, which involves immunoprecipitating AMPK protein from the tissue of interest followed by quantifying

Methods in Enzymology, Volume 514
ISSN 0076-6879
http://dx.doi.org/10.1016/B978-0-12-381272-8.00017-9

its enzymatic activity using radiolabeled adenosine triphosphate (ATP) in the presence of a suitable substrate. As a surrogate marker of AMPK activity, AMPK Thr(172) phosphorylation can be measured by Western blotting. Information about the AMPK pathway can also be gained by studying the mRNA expression of various AMPK subunits and by Western blotting for phosphorylated acetyl-CoA carboxylase, a key AMPK target. These methods have been widely used and published for investigating the effects of ghrelin on AMPK activity. In this chapter, we look into these experiments' methodology in detail.

1. INTRODUCTION

Ghrelin is a circulating brain–gut peptide that is known to exert several physiological effects and regulate energy balance (Anderson et al., 2008; Andrews et al., 2008; Cao et al., 2011; Chen et al., 2011; Healy et al., 2011; Kola et al., 2005; Korbonits et al., 2004; Lage et al., 2011; Lim et al., 2011; Lopez et al., 2008; Sangiao-Alvarellos et al., 2009, 2011; Stevanovic et al., 2012; van Thuijl et al., 2008; Wang et al., 2010; Xu et al., 2008). These include releasing the growth hormone (Kojima et al., 1999), stimulating appetite (Wren et al., 2000, 2001), inducing adiposity (Tschop et al., 2000), increasing bone formation (Fukushima et al., 2005), and influencing cardiovascular functions (Leite-Moreira and Soares, 2007; van der Lely et al., 2004). Many of these metabolic effects are shown to be mediated by AMP-activated protein kinase (AMPK) (Andersson et al., 2004; Kohno et al., 2008; Kola et al., 2005, 2008a, 2008b), a highly conserved enzyme that plays a key role in energy homeostasis (Hardie, 2011; Minokoshi et al., 2004). AMPK is a serine/threonine kinase with three subunits (the α-, β-, and γ-subunits) forming a heterotrimeric protein (Fig. 17.1). The kinase activity is regulated by change in the AMP/ATP and ADP/ATP ratios (Hardie et al., 1999, 2011), leading to ultimate effects that influence the energy metabolism of the cell (Hardie, 2011; Hardie et al. 1999; Kola et al., 2006). Ghrelin has been shown to exert its orexigenic effect by stimulating hypothalamic AMPK activity (Anderson et al., 2008; Andrews et al., 2008; Kola et al., 2005, 2008a). However, intracerebroventricular injection of ghrelin in chickens has been shown to inhibit AMPKα1 and AMPKα2 mRNA expression (Xu et al., 2011). Peripherally, ghrelin inhibits AMPK activity in the adipose tissues, leading to increased adipogenesis and reduced fat consumption for metabolic fuel (Kola et al., 2005, 2008a). Ghrelin also inhibits AMPK activity in the liver, thereby promoting gluconeogenesis and decreasing fatty acid oxidation (Kola et al., 2005, 2008a) (Fig. 17.1). *In vivo* ghrelin treatment in rats has shown increased AMPK activity as well as

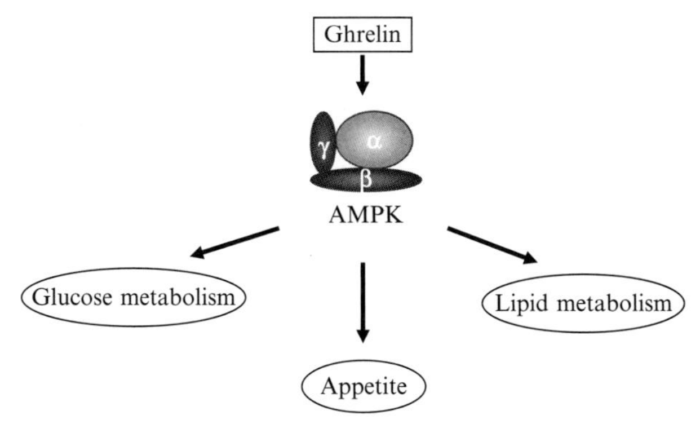

Figure 17.1 The heterotrimeric structure of AMPK, consisting of the α-, β-, and γ-subunits. AMPK has been shown to mediate several metabolic effects of ghrelin, which include stimulating appetite as well as affecting glucose and lipid metabolism. (See Color Insert.)

phosphorylated AMPK (pAMPK) and pACC protein expression in the myocardium (Kola et al., 2005), and *in vitro* ghrelin treatment of rat cardiomyocytes increased AMPK gene expression (Cao et al., 2011).

The effects of ghrelin on AMPK activity in tissues can be studied with a kinase assay that has been widely used and accepted for scientific purposes (Davies et al., 1989; Kola, 2008; Kola et al., 2005, 2008a). The assay involves immunoprecipitating the kinase protein from the tissue of interest that has been treated appropriately with ghrelin. This will then allow its enzymatic activity to be quantified using radiolabeled adenosine triphosphate (ATP) in the presence of a suitable substrate.

AMPK is activated by phosphorylation of its threonine-172 (Thr(172)). Therefore, quantification of AMPK Thr(172) phosphorylation by immunoblotting is a recognized surrogate for AMPK activity. pAMPK levels need to be corrected for total AMPK levels. The AMPK activity assay and the pAMPK immunoblotting method should provide similar, but not necessarily overlapping, results. AMPK has been shown to be phosphorylated on another site, serine-385, which leads to the inhibition of AMPK activity (Ning et al., 2011).

2. THE AMPK ASSAY METHODOLOGY

The following assay is based on the method first developed by Professor Grahame Hardie's laboratory (Carling et al., 1987, 1989; Davies et al., 1989). The four main steps of this assay are tissue homogenization and protein extraction, protein quantification, immunoprecipitation (IP) of AMPK, and AMPK activity determination (Fig. 17.2).

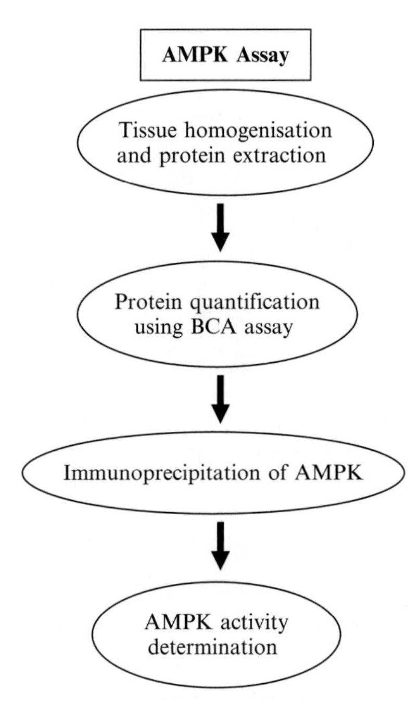

Figure 17.2 The four main stages of AMPK assay: tissue homogenization and protein extraction, protein quantification, immunoprecipitation of AMPK, and AMPK activity determination.

2.1. Preparation of buffers and working solutions

The AMPK kinase assay involves four main buffer solutions.

N.B. buffer solutions should be kept at 4 °C unless stated otherwise.

1. Tissue lysis buffer: For all tissues except hypothalamus: 50 m*M* Tris–hydrochloric acid (Tris–HCl, pH 7.4, 50 m*M* sodium fluoride (NaF), 5 m*M* sodium pyrophosphate, 1 m*M* EDTA, 10% (v/v) glycerol, 1% (v/v) Triton X-100, 1 m*M* DTT, 1 m*M* benzamidine, 1 m*M* phenylmethane sulfonyl fluoride (PMSF), 5 μg/ml soybean trypsin inhibitor (SBTI), distilled water to make up for the remaining volume to 1 l; For hypothalamic tissues: 50 m*M* Tris–HCl (pH 7.4), 50 m*M* NaF, 5 m*M* Na pyrophosphate, 1 m*M* EDTA, 250 m*M* sucrose, 1% (v/v) Triton X-100, 1 m*M* DTT, 1 m*M* benzamidine, 0.1 m*M* PMSF, 5 μg/ml SBTI, distilled water to make up for the remaining volume to 1 l.
2. IP buffer: 50 m*M* Tris–HCl (pH 7.4), 150 m*M* NaCl, 50 m*M* NaF, 5 m*M* sodium pyrophosphate, 1 m*M* EDTA, 1 m*M* EGTA, 1 m*M*

DTT, 0.1 m*M* benzamidine, 0.1 m*M* PMSF, 5 μg/ml SBTI, distilled water to make up for the remaining volume to 1 l.

3. High sodium immunoprecipitation (NaCl-IP) buffer: 50 m*M* Tris–HCl (pH 7.4), 1 *M* NaCl, 50 m*M* NaF, 5 m*M* Na pyrophosphate, 1 m*M* EDTA, 1 m*M* EGTA, 1 m*M* DTT, 0.1 m*M* benzamidine, 0.1 m*M* PMSF, 5 μg/ml SBTI, distilled water to make up for the remaining volume to 1 l.
4. HEPES-Brij buffer: 50 m*M* Na HEPES, 1 m*M* DTT, 0.2 g Brij 35 0.02%, distilled water to make up for the remaining volume to 1 l.

2.2. Tissue homogenization and protein extraction

There are different established techniques available for protein extraction and homogenization of tissue samples, the choice of which depends on several factors. Freeze-clamping has been used in many studies. Ultimately, the final outcome and quality should not differ significantly, provided that the technique of choice has been carried out appropriately and accurately.

Method

1. Tissues should be kept on dry ice at all times. In a sterile work area, incise a small (approximately 100 g) section from the first tissue sample using a surgical blade.
2. Using the first sample as a size guide, cut pieces from the remaining tissue samples. Use a fresh specimen dish for each sample and clean the blade with ethanol between samples to eliminate cross-contamination.
3. Homogenization can be carried out using a prechilled pestle and mortar or a homogenization machine. In our laboratory, a Precellys 24 homogenizer and CK14 tubes containing ceramic beads (Stretton Scientific, Stretton, United Kingdom) were used. Samples were placed in the tubes with 800 μl of protease inhibitor-treated tissue lysis buffer (400 μl hypothalamic lysis buffer for hypothalamic samples of 50 mg). Tubes were placed in the homogenizer and processed at 5000–6000 × *g* for 2 min with interval pauses. This step was repeated if visible fragments of tissue remained in the tube.
4. Allow the homogenized samples to settle on ice for 5 min; then centrifuge for 30 min at 13,000 × *g* at 4 °C.
5. Transfer the supernatant to 1.5-ml Eppendorf tubes. Avoid disturbing the beads and pellet at the bottom of the homogenate.

2.3. Protein quantification using BCA assay

The bicinchoninic acid (BCA) assay is a protein quantification technique that was first demonstrated by Paul K. Smith at the Pierce Chemical Company in 1985 (Smith et al., 1985). The principle of this technique is quite similar to the Lowry protein assay, which is one of the alternative protein quantification methods. BCA is a weak acid composed of two carboxylated quinoline rings and serves the purpose of the Folin reagent in the Lowry assay, namely, to react with the product (Cu^{1+} ions) of complexes between copper ions and peptide bonds to produce a purple end product that strongly absorbs at 562 nm (Smith et al., 1985). The amount of protein present in a solution can be quantified by measuring the absorption spectra and comparing them with those of protein solutions of known concentration. The advantage of BCA is that the reagent is fairly stable under alkaline conditions and can be included in the copper solution to allow a one-step procedure (Smith et al., 1985).

This standard BCA assay can be used to determine the protein concentrations of each homogenized sample.

2.4. Immunoprecipitation of AMPK

IP is a technique that involves precipitating a protein antigen out of a solution using an antibody that specifically binds to it. There are generally two methods of IP: (i) direct precipitation, which involves mixing the protein solution with its specific antibody that has been bound to a solid-phase substrate such as the superparamagnetic microbeads and (ii) indirect precipitation, which first involves mixing the protein solution with its specific antibody, followed by adding the solid-phase substrate after the protein antigen–antibody complexes have been formed. The presence of the solid-phase substrate allows the protein–antibody complex to be immobilized and precipitated.

IP of AMPK uses the direct precipitation principle. The specific AMPK antibodies ($\alpha1/\alpha2$ antibodies obtained from sheep) are bound to protein G sepharose beads first before mixing them with the protein solutions. The AMPK $\alpha1/\alpha2$ antibody binds to the α-subunit of AMPK molecule specifically, thus allowing this kinase protein to be immunoprecipitated. The $\alpha1$ and $\alpha2$ AMPK activity can be measured separately. In this case, the beads need to be prepared separately with the $\alpha1$ and $\alpha2$ antibodies and a sample can be incubated consecutively with $\alpha1$ and $\alpha2$ antibody-bound beads and then processed further separately. The precipitated protein can then be repeatedly washed and sedimented via centrifugation so that the end product is as pure as possible.

Method

A. Preparation of antibody–bead mixture

(*Note*: All steps are carried out at 4 °C unless stated otherwise.)

1. Place the required volume of Protein G sepharose (10 μl/sample taking into account the 20% ethanol) into a labeled 15-ml falcon tube. Make sure the beads are well suspended; 100% ethanol or IP buffer can be used to dilute the beads.
2. Wash the beads with 5 ml of IP buffer. This involves mixing them well with a vortex machine and centrifuge it for 1 min at 3000 × *g*. The supernatant is discarded, leaving the pellet of the sediment in the falcon tube.
3. Repeat step 2 for a further of four times. Suspend the final pellet of sediment in 5 ml of IP buffer.
4. Add the required amount of α1/α2 antibodies (2.5 μg α1 and 2.5 μg α2 antibody per sample) into the tube and mix it well on a roller for 45 min. If α1 and α2 activities are studied separately, then separate sets of beads need to be prepared with 5 μg α1 antibody and 5 μg α2 antibody, respectively.
5. Centrifuge the mixture for 2 min at 3000 × *g*. Remove the supernatant and wash the antibody–bead mixture five times with 5 ml of IP buffer (as in step 2).
6. Suspend the final pellet of sediment in IP buffer. The volume of IP buffer needed can be calculated on the basis of the number of samples (1 ml for every 20 samples). Note that the same number of samples is also used to determine the amount of antibodies and beads initially.

B. IP process

(*Note*: All steps are carried out at 4 °C unless stated otherwise.)

1. Add 50 μl of the antibody-coated beads to each protein sample (100–300 μg protein, depending on the BCA assay). Top up the volume by adding 300 μl of IP buffer to each sample.
2. Mix well on a roller for 2 h.
3. Once the samples have mixed, centrifuge for 5 min at 3000 × *g*. Remove the supernatant carefully, making sure that the sediment is not disturbed.
4. Wash the mixture once with 3 ml of NaCl-IP buffer. This involves mixing them well with a vortex machine. Centrifuge for 2 min at 3000 × *g*. The supernatant is discarded, leaving the pellet of sediment in the falcon tube.

5. Wash the mixture once with 2 ml of HEPES-Brij buffer (as in step 4).
6. Suspend the final pellet in 300 μl of HEPES-Brij buffer.
7. Mix the mixture well using a pipette and aliquot it into three new labeled 1.5-ml Eppendorf tubes for that sample (100 μl of mixture per tube). This allows at least duplicate readings per sample and thus more accurate and reliable results for the assay. Repeat this step for other protein samples.
8. Centrifuge all the tubes for 2 min at 13,000 × *g*. Remove 80 μl of the supernatant and proceed to AMPK assay immediately.

2.5. AMPK assay

When investigating an enzyme activity, there are several important factors that have to be taken into account, as they can have an influence on the assay. These include the enzyme and substrate concentrations, temperature and pH of the medium, presence or absence of enzyme inhibitors, and activators as well as co-factors. The same applies to AMPK assay, which is an enzyme study looking at the activity of the kinase protein. In this AMPK assay, the optimum temperature and pH of the buffer are used and remain constant throughout the study. A high concentration of substrate is used to ensure that it does not act as a limiting factor in the study. Cofactors can be included in the study but have to be equally present for every sample studied in the assay.

AMPK activity is commonly determined by radioactive labeling of artificial peptide substrates such as the SAMS peptide. SAMS is a synthetic peptide with a modified version of the sequence around the AMPK target site in rat acetyl-CoA carboxylase (ACC). SAMS is widely used as a substrate for AMPK because it is phosphorylated rapidly by AMPK and is more specific for the kinase than ACC enzyme itself, thus providing a convenient and sensitive assay for AMPK. In this study, the radioactive label can be initially attached onto the ATP molecule, forming γ^{32} P-ATP. In the presence of AMPK, the interaction between SAMS and γ^{32} P-ATP will lead to the phosphorylation of SAMS (γ^{32} P-SAMS) onto a charged membrane which can be detected by scintillation counting (Davies et al., 1989; Suter et al., 2006). The counts will then allow the rate of AMPK activity to be calculated.

Method

(*Note*: All steps are carried out at 4 °C unless stated otherwise. When dealing with radioactive compounds, ensure that the work is done behind protective screen and according to radioactive safety protocols.)

A. Preparation of Master Mixtures

1. First, prepare a working solution that contains the γ^{32} P-ATP:10 μl γ^{32} P-ATP 10 mCi/ml (specific activity 250–500 cpm/pmol), 10 μl ATP 100 m*M*, 25 μl magnesium chloride ($MgCl_2$) 1 *M*, 955 μl HEPES-Brij buffer to make a final volume of 1000 μl.
2. Prepare the positive and negative Master Mix solutions. The former contains the SAMS substrate, while the latter solution acts as a master mix for controls (the volume shown below (30 μl) is for one sample only).
   - Positive Master Mix: 10 μl working solution, 10 μl 1 m*M* AMP, 10 μl SAMS peptide 100 μ*M*.
   - Negative Master Mix: 10 μl working solution, 10 μl 1 m*M* AMP, 10 μl HEPES-Brij buffer.

B. AMPK activity

1. Mix the Master Mixture solutions well with a vortex machine.
2. Add 30 μl of the Master Mix solutions to the immunoprecipitated AMPK samples (every sample has triplicates, with one of the three acting as a control and thus should be mixed with the negative Master Mix solution. The remaining two precipitants will be mixed with the positive Master Mix solution). Repeat this step until all the samples have been mixed with the correct Master Mix solutions.
3. Prepare AMPK-positive and -negative controls to check that the positive Master Mix is working. Activated AMPK can be obtained from Upstates (stock solution of 100 mU/μl). Pipette 50 mU of active AMPK into two tubes (dilution can be carried out using HEPES-Brij buffer). Continue with these controls like the rest of the samples.
4. Transfer all the samples to a 30 °C shaker and incubate them for 20 min at 350 rpm.
5. After the incubation, mix the sample well with a pipette and then take out 35 μl and place it on a numbered P81 phosphocellulose paper square (see Fig. 17.3). After 2 s on the phosphocellulose paper square, transfer the square into a beaker containing 1% (v/v) phosphoric acid. This will stop any ongoing reaction.
6. Repeat this step for every sample. Ensure that a record of what number corresponds to which sample is kept.
7. Gently stir the paper squares in the phosphoric acid with a magnetic stirrer for 5 min. Drain the acid into a designated radioactive disposal sink.
8. Perform a second wash with 1% (v/v) phosphoric acid, again stirring for 5 min before draining the acid.

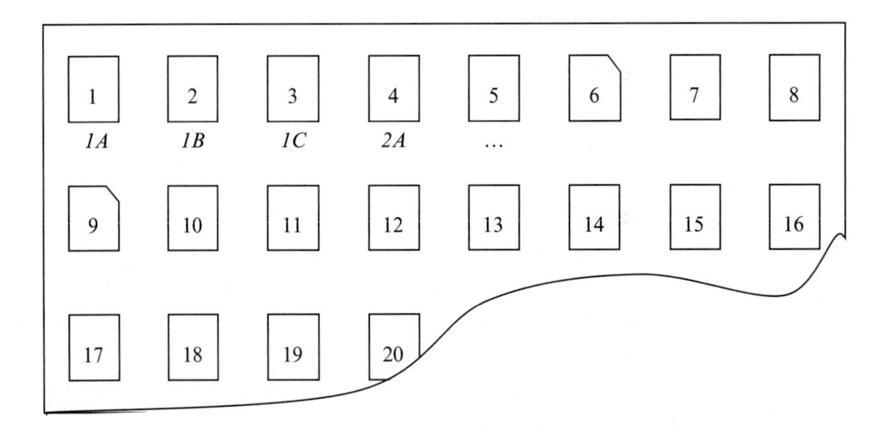

Figure 17.3 The arrangement of the P81 phosphocellulose paper square that has been numbered. The reacted samples (35 μl) are pipetted onto each phosphocellulose paper square, and at the same, a record of the corresponding number for the sample is kept to avoid error.

9. Rinse the squares once with distilled water, and carry out a final wash in distilled water (as in step 7 but without the phosphoric acid).
10. Transfer the squares onto a tray and place them in an oven set at 70 °C. Dry them for 15–20 min.
11. Prepare two extra phosphocellulose squares: one to be added with 5 μl of positive Master Mix solution and the other with 5 μl of negative Master Mix solution.
12. Transfer all the squares into an appropriate and separate medium that allows the scintillation counter to pick up readings.

C. Calculation of AMPK activity
1. AMPK activity is expressed in units of nanomoles of phosphate incorporated into the substrate peptide per minute.
2. To determine the specific radioactivity of the ATP, look at the count for 5 μl of positive Master Mix. This count corresponds to 1.66 nmol of ATP (and radioactive decay is corrected for automatically).
3. Calculate the nanomoles of ATP incorporated per minute (knowing the CPM value obtained by 1.66 μl of ATP) = (CPM × 1.66)/CPM positive Master Mix. Use CPM negative Master Mix for samples added with negative Master Mix.
4. Calculate the actual nanomoles of ATP incorporated per min by subtracting the step 3 value of negative Master Mix from the positive Master Mix value.

5. Calculate the quantity of protein that gives that CPM value = quantity of protein used at the initial step of IP divided by three (since the sample was divided into three aliquots), for example, 300 μg/3 (=100) multiplied by 35/50 (since we only used 35 μl from the 50 μl of total reaction volume onto the P81 square = 100 × 35/50 = 70 μg).
6. Calculate the nanomoles of ATP incorporated per minute per milligram of protein = nanomoles of ATP incorporated per minute × 1000/quantity of protein in micrograms.
7. Then calculate the average of AMPK activity obtained from the two positive aliquots of each sample.

3. IMMUNOBLOTTING pAMPK

This method measures the presence of Thr(172) pAMPK compared to the total AMPK.

3.1. Preparation of buffers

(*Note*: The buffer solutions should always be kept at 4 °C unless stated otherwise.)

1. SDS loading buffer: 1.514 g Tris–base (125 m*M*), 4 g SDS (sodium dodecyl sulfate) (4% w/v), 16 ml glycerol (20% w/v), 0.02 g bromophenol blue (0.002% w/v). Dissolve Tris–base in 100 ml of distilled water and adjust the pH to 6.8. Add the remaining reagents.
2. Running buffer: 30.3 g Tris–base, 14.14 g glycine, 10 g SDS. Make up to 1 l with distilled water, and dilute 1 in 10 for use.
3. Transfer buffer: 30.4 g Tris–base, 14.14 g glycine. Dissolve in 800 ml of distilled water and add 200 ml methanol. Mix well with a magnetic stirrer.
4. Washing buffer: Make up 10 × Tris-buffered saline (10 × TBS) with 24.2 g Tris–base, 80 g NaCl, and make up to 1 l with distilled water, adjusting the pH to 7.6 with hydrochloric acid. To make 1 l of the washing buffer, mix 100 ml of 10 × TBS with 900 ml distilled water and 1 ml Tween20.
5. Blocking buffer: 5% milk (2.5 g of nonfat milk powder in 50 ml 1 × TBS-Tween)
6. Stripping buffer: Add 100 ml of distilled water to 0.985 g Tris–HCl and 2 g SDS. Then, add 781 μl 2-mercaptoethanol to the mixture.

3.2. Immunoblotting protocol

1. Prepare the protein samples and determine their concentrations with BCA assay as previously described. Calculate the volumes required for each sample (10–20 μg protein) and aliquot it into 0.5-ml Eppendorf tubes. Add an equal volume of SDS loading buffer into the samples. Always keep the samples at 4 °C while working.
2. Denature the protein samples at 95 °C for 3–5 min, and then put them on ice. Centrifuge for 10 s.
3. Prepare an acrylamide gel (10% Tris–HCl ready gel) and insert it into the gel chamber and tank. Fill the tank center until it overflows to half-way up the outer tank with running buffer. Remove the white strip and well-protector comb.
4. Samples and marker (Kaleidoscope Prestained Standards, Bio-Rad) are then loaded into the wells (maximum loading volume is 50 μl).
5. Run at 100 V until the marker begins to separate (usually for about 15 min); then run at 150 V until the protein crosses the bottom of gel (usually for about 1.2 h).
6. Open the gel with a spatula. Trim the gel top and bottom.
7. Soak blotting paper and PVDF membrane (Immobilon-P, pore size 0.45 μm; Millipore UK Limited) in transfer buffer and place them on a rolling plate.
8. Dampen fingers and pick up gel from bottom. Place it briefly in the transfer buffer before putting it on the PVDF membrane. Cover with second dampened blotting paper, forming a "sandwich." Continue on rolling plate.
9. Transfer at 18 V for 1 h (0.8 A).
10. Wash twice with washing buffer.
11. Block with 40 ml of 5% milk for 90 min at room temperature on a roller.
12. Wash four times with washing buffer, twice fast and twice for 10 min.
13. Incubate the membranes with primary anti-pAMPK antibody (which recognizes the AMPK pan-α-subunit phosphorylated at Thr(172)) at a concentration of 1:1000 or with anti-phospho ACC antibody (1:1000, Upstate) in 5% milk in Tris-buffered saline-Tween overnight at 4 °C. AMPK antibodies are available from Prof. Hardie's laboratory, Upstate, Cell Signaling, or other suppliers.
14. Wash four times with washing buffer, twice fast and twice for 10 min.
15. To achieve cleaner bands, preincubate the membranes for pAMPK with unphosphorylated PT172 protein for 60 min before incubation

with the pAMPK antibody. Use donkey anti-sheep IgG HRP, 1:2000 (Santa Cruz, California, USA) as secondary antibody. Roll for 90 min at room temperature. Wash four times with washing buffer, twice fast and twice 10 min.

16. Strip the membrane total AMPK (mixture of primary antibody α1- and α2-AMPK 0.2 μg/ml for total AMPK) or total ACC antibody. Use GAPDH (primary antibody 1:2000 (Santa Cruz), secondary goat anti-rabbit antibody 1:10000 (DAKO, Denmark)), or β-actin (primary monoclonal antibody 1:5000 (Abcam, Cambridge, UK), and secondary goat anti-mouse antibody 1:10,000 (DAKO)) antibodies to normalize for equal loading and densitometric readings of the resulting band. Read on a Li-COR system.

4. AMPK Gene Expression

AMPK subunit mRNA expression can be studied with RT-PCR. To detect small changes, the use of real-time PCR is highly recommended. We used the method described below in a study not involving ghrelin (Christ-Crain et al., 2008), but others have studied AMPK mRNA expression in response to ghrelin treatment (Hardie et al., 2011; Xu et al., 2011).

4.1. Preparation of total RNA

1. Homogenize the tissue using the appropriate lysis buffer. For this purpose, homogenize 100 mg of fat adipose tissue in 1 ml of QIAzol reagent (Qiagen, Crawley, UK 79306) and purify the total RNA according to the manufacturer's instructions (Qiagen 74104). Homogenize 30 mg of liver and heart tissues in 175 μl of RNA lysis buffer (Promega, Z3105) using the Promega SV isolation kit (Promega, Southampton, UK), which includes a DNase step to eliminate the contaminating genomic DNA.
2. Assess the quality and concentration of RNA at 260/280 nm.
3. Assess the RNA integrity by gel electrophoresis: put in each tube 2 μl of RNA + 8 μl DEPC H_2O + 2 μl dye and warm samples before running at 65 °C for 5–10 min and then immediately on ice. Load 10 μl of each tube in 1% agarose gel electrophoresis (0.5 g agarose in 50 ml 1× TAE) and run at 150 V for 30 min.

4.2. Preparation of cDNA

1. Synthesize cDNA from 1 μg of RNA in a total volume of 25 μl, using random primer (Roche, Burgess Hill, UK, 11034731001) and reverse transcriptase (Invitrogen, Paisley, UK, 28025-013).
2. Prepare the RT Master Mix: 10 μl first strand buffer 5 × (Gibco) + 2 μl DTT (Gibco) + 2.5 μl dNTP (20 m*M*) + 0.25 μl random primers + 0.1 μl RNase inhibitor (Promega) + 1 μl MMLV reverse transcriptase (Invitrogen).
3. Heat the RNA mix with H_2O (max volume: 34.15 μl) on a PCR block at 65 °C for 5–10 min to denature RNA, and place it immediately on ice to cool.
4. Add 15.85 μl RT Master Mix to each tube and run the RT Program (stage 1: 26 °C for 10 min; Stage 2: 37 °C for 60 min; Stage 3: 92 °C for 10 min). Spin the samples afterward and keep at –20 °C.

4.3. Real-time PCR

1. Use real-time PCR to quantify the levels of AMPKα1 and AMPKα2 gene expression using specific assay-by-design primer and probe sets by Applied Biosystems (ABI, Warrington, UK) and the ABI PRISM 7900 Sequence Detector System. All gene expression assays have FAM TM reporter dye at the 5′ end of the TaqMan® MGB probe and a nonfluorescent quencher at the 3′ end of the probe. The TaqMan® MGB probes and primers have been premixed (20 ×) to a concentration of 18 μ*M* for each primer and 5 μ*M* for the probe (ABI). The endogenous control assay has a VIC® MGB Probe.
2. Perform PCR in a final volume of 10 μl. Run all the reactions in a duplex PCR reaction with β-actin (β-ACTB) as endogenous control (VIC® MGB Probe, Applied Biosystems) at these conditions: 5 μl TaqMan® Universal Master Mix (Applied Biosystems), 0.5 μl 20 × assay mix primer, 0.35 μl 20 × assay mix ACTB, 3.5 μl TE or H_2O, and 1 μl sample cDNA.
3. Run together with samples the control reactions for RT (containing RNA but no RT enzyme) and PCR (containing PCR mixture but no cDNA).
4. Run the reaction at 40 cycles at 50 °C for 2 min (stage 1), 95 °C for 10 min (stage 2), 95 °C for 15 s, and 60 °C for 1 min (stage 3).
5. Analyze the data using the standard curve method. Calculate the relative quantities of target transcripts from triplicate samples after normalization of the data against the housekeeping gene β-actin.

5. SUMMARY

We have previously shown that ghrelin exerts its orexigenic effect by stimulating the hypothalamic AMPK activity (Kola et al., 2005, 2008a). Ghrelin also stimulates cardiac AMPK activity and inhibits liver as well as adipose tissue AMPK activity peripherally (Kola et al., 2005, 2008a). These will influence the cardiovascular function, increase adipogenesis and gluconeogenesis, and decrease fatty acid oxidation (Kola et al., 2005, 2008a). These effects have also been confirmed by other studies (Anderson et al., 2008; Andersson et al., 2004; Andrews et al., 2008, 2010; Castaneda et al., 2010; Chen et al., 2004; Kohno et al., 2008; Rigault et al., 2007; Sangiao-Alvarellos et al., 2009; Sleeman and Latres, 2008). Both the AMPK assay and immunoblotting have been widely used for investigations of the effects of several hormones such as ghrelin, leptin, glucocorticoid, and adiponectin on the AMPK activity. These two simple yet elegant experiments will thus always be applicable for studying the effects of ghrelin on AMPK activity.

ACKNOWLEDGMENTS

We are very grateful to Professor Grahame Hardie and Dr. Simon Hawley (University of Dundee, United Kingdom) for their invaluable help and advice on setting up the AMPK assay in our laboratory. We would also like to thank all the staff at the William Harvey Research Institute, Centre for Endocrinology of Barts and The London School of Medicine and Dentistry, who have contributed to the development of the AMPK assay and immunoblotting protocols used in the laboratory.

REFERENCES

Anderson, K.A., et al., 2008. Hypothalamic CaMKK2 contributes to the regulation of energy balance. Cell Metab. 7, 377–388.

Andersson, U., et al., 2004. AMP-activated protein kinase plays a role in the control of food intake. J. Biol. Chem. 279, 12005–12008.

Andrews, Z.B., et al., 2008. UCP2 mediates ghrelin's action on NPY/AgRP neurons by lowering free radicals. Nature 454, 846–851.

Andrews, Z.B., et al., 2010. Uncoupling protein-2 decreases the lipogenic actions of ghrelin. Endocrinology 151, 2078–2086.

Cao, C., et al., 2011. Ghrelin inhibits insulin resistance induced by glucotoxicity and lipotoxicity in cardiomyocyte. Peptides 32, 209–215.

Carling, D., et al., 1987. A common bicyclic protein kinase cascade inactivates the regulatory enzymes of fatty acid and cholesterol biosynthesis. FEBS Lett. 223, 217–222.

Carling, D., et al., 1989. Purification and characterization of the AMP-activated protein kinase. Copurification of acetyl-CoA carboxylase kinase and 3-hydroxy-3-methylglutaryl-CoA reductase kinase activities. Eur. J. Biochem. 186, 129–136.

Castaneda, T.R., et al., 2010. Ghrelin in the regulation of body weight and metabolism. Front. Neuroendocrinol. 31, 44–60.
Chen, H.Y., et al., 2004. Orexigenic action of peripheral ghrelin is mediated by neuropeptide Y and agouti-related protein. Endocrinology 145, 2607–2612.
Chen, J.H., et al., 2011. Ghrelin induces cell migration through GHS-R, CaMKII, AMPK, and NF-kappaB signaling pathway in glioma cells. J. Cell. Biochem. 112, 2931–2941.
Christ-Crain, M., et al., 2008. AMP-activated protein kinase mediates glucocorticoid-induced metabolic changes: a novel mechanism in Cushing's syndrome. FASEB J. 22, 1672–1683.
Davies, S.P., et al., 1989. Tissue distribution of the AMP-activated protein kinase, and lack of activation by cyclic-AMP-dependent protein kinase, studied using a specific and sensitive peptide assay. Eur. J. Biochem. 186, 123–128.
Fukushima, N., et al., 2005. Ghrelin directly regulates bone formation. J. Bone Miner. Res. 20, 790–798.
Hardie, D.G., 2011. AMP-activated protein kinase—an energy sensor that regulates all aspects of cell function. Genes Dev. 25, 1895–1908.
Hardie, D.G., et al., 1999. AMP-activated protein kinase: an ultrasensitive system for monitoring cellular energy charge. Biochem. J. 338 (Pt. 3), 717–722.
Hardie, D.G., et al., 2011. AMP-activated protein kinase: also regulated by ADP? Trends Biochem. Sci. 36, 470–477.
Healy, J.E., et al., 2011. Peripheral ghrelin stimulates feeding behavior and positive energy balance in a sciurid hibernator. Horm. Behav. 59, 512–519.
Kohno, D., et al., 2008. Ghrelin raises $[Ca^{2+}]_i$ via AMPK in hypothalamic arcuate nucleus NPY neurons. Biochem. Biophys. Res. Commun. 366, 388–392.
Kojima, M., et al., 1999. Ghrelin is a growth-hormone-releasing acylated peptide from stomach. Nature 402, 656–660.
Kola, B., 2008. Role of AMP-activated protein kinase in the control of appetite. J. Neuroendocrinol. 20, 942–951.
Kola, B., et al., 2005. Cannabinoids and ghrelin have both central and peripheral metabolic and cardiac effects via AMP-activated protein kinase. J. Biol. Chem. 280, 25196–25201.
Kola, B., et al., 2006. Expanding role of AMPK in endocrinology. Trends Endocrinol. Metab. 17, 205–215.
Kola, B., et al., 2008a. The orexigenic effect of ghrelin is mediated through central activation of the endogenous cannabinoid system. PLoS One 3, e1797.
Kola, B., et al., 2008b. The role of AMP-activated protein kinase in obesity. Front. Horm. Res. 36, 198–211.
Korbonits, M., et al., 2004. Ghrelin—a hormone with multiple functions. Front. Neuroendocrinol. 25, 27–68.
Lage, R., et al., 2011. Ghrelin effects on neuropeptides in the rat hypothalamus depend on fatty acid metabolism actions on BSX but not on gender. FASEB J. 24, 2670–2679.
Leite-Moreira, A.F., Soares, J.B., 2007. Physiological, pathological and potential therapeutic roles of ghrelin. Drug Discov. Today 12, 276–288.
Lim, C.T., et al., 2011. The ghrelin/GOAT/GHS-R system and energy metabolism. Rev. Endocr. Metab. Disord. 12, 173–186.
Lopez, M., et al., 2008. Hypothalamic fatty acid metabolism mediates the orexigenic action of ghrelin. Cell Metab. 7, 389–399.
Minokoshi, Y., et al., 2004. AMP-kinase regulates food intake by responding to hormonal and nutrient signals in the hypothalamus. Nature 428, 569–574.
Ning, J., Xi, G., Clemmons, D.R., 2011. Suppression of AMPK activation via S485 phosphorylation by IGF-I during hyperglycemia is mediated by AKT activation in vascular smooth muscle cells. Endocrinology 152, 3143–3154.

Rigault, C., et al., 2007. Ghrelin reduces hepatic mitochondrial fatty acid beta oxidation. J. Endocrinol. Invest. 30, RC4–RC8.
Sangiao-Alvarellos, S., et al., 2009. Central ghrelin regulates peripheral lipid metabolism in a growth hormone-independent fashion. Endocrinology 150, 4562–4574.
Sangiao-Alvarellos, S., et al., 2011. Influence of ghrelin and growth hormone deficiency on AMP-activated protein kinase and hypothalamic lipid metabolism. J. Neuroendocrinol. 22, 543–556.
Sleeman, M.W., Latres, E., 2008. The CAMplexities of central ghrelin. Cell Metab. 7, 361–362.
Smith, P.K., et al., 1985. Measurement of protein using bicinchoninic acid. Anal. Biochem. 150, 76–85.
Stevanovic, D., et al., 2012. Immunomodulatory actions of central ghrelin in diet-induced energy imbalance. Brain Behav. Immun. 26, 150–158.
Suter, M., et al., 2006. Dissecting the role of 5′-AMP for allosteric stimulation, activation, and deactivation of AMP-activated protein kinase. J. Biol. Chem. 281, 32207–32216.
Tschop, M., et al., 2000. Ghrelin induces adiposity in rodents. Nature 407, 908–913.
Van Der Lely, A.J., et al., 2004. Biological, physiological, pathophysiological, and pharmacological aspects of ghrelin. Endocr. Rev. 25, 426–457.
Van Thuijl, H., et al., 2008. Appetite and metabolic effects of ghrelin and cannabinoids: involvement of AMP-activated protein kinase. Vitam. Horm. 77, 121–148.
Wang, W., et al., 2010. Ghrelin inhibits cell apoptosis induced by lipotoxicity in pancreatic beta-cell line. Regul. Pept. 161, 43–50.
Wren, A.M., et al., 2000. The novel hypothalamic peptide ghrelin stimulates food intake and growth hormone secretion. Endocrinology 141, 4325–4328.
Wren, A.M., et al., 2001. Ghrelin enhances appetite and increases food intake in humans. J. Clin. Endocrinol. Metab. 86, 5992.
Xu, X., et al., 2008. Molecular mechanisms of ghrelin-mediated endothelial nitric oxide synthase activation. Endocrinology 149, 4183–4192.
Xu, P., et al., 2011. Genetic selection for body weight in chickens has altered responses of the brain's AMPK system to food intake regulation effect of ghrelin, but not obestatin. Behav. Brain Res. 221, 216–226.

CHAPTER EIGHTEEN

Ghrelin and Gastrointestinal Movement

Naoki Fujitsuka*[,†], Akihiro Asakawa*, Haruka Amitani*, Mineko Fujimiya[‡], Akio Inui*[,1]

*Department of Psychosomatic Internal Medicine, Kagoshima University Graduate School of Medical and Dental Sciences, Kagoshima, Japan
[†]Tsumura Research Laboratories, Ibaraki, Japan
[‡]Department of Anatomy, Sapporo Medical University School of Medicine, Hokkaido, Japan
[1]Corresponding author: e-mail address: inui@m.kufm.kagoshima-u.ac.jp

Contents

Abstract

Ghrelin is a potent stimulant for gastric emptying and gastrointestinal (GI) movement. Clinically, it has been reported that the intravenous administration of ghrelin accelerates the rate of gastric emptying and induces gastric phase III contractions of the migrating motor complex in healthy volunteers. Recent technical advances in the measurement of GI motility in conscious small animals, including rats, mice, and the house musk shrew (*Suncus murinus*), have helped to elucidate the precise mechanism of action of ghrelin. Intravenous administration of ghrelin induces fasted motor activities with phase III-like contractions of the migrating motor complex in the antrum and duodenum in animals. These effects of ghrelin are mediated by activating the hypothalamic orexigenic neuropeptide Y neuron through ghrelin receptors located at the vagal afferent terminal. Stress hormone and anorexigenic peptides cause the disruption of fasted motor activity and induce fed-like motor activity. Ghrelin and

Methods in Enzymology, Volume 514
ISSN 0076-6879
http://dx.doi.org/10.1016/B978-0-12-381272-8.00018-0

the ghrelin signal potentiator rikkunshito successfully restore fed-like motor activities to fasted activities in fenfluramine-treated rats and in a cancer anorexia–cachexia animal model. These findings suggest that ghrelin can be expected to be a therapeutic target for GI disorders.

1. INTRODUCTION

Ghrelin is a potent stimulant of gastric emptying and gastrointestinal (GI) motility. In humans in the fasted state, cyclic changes in contraction waves known as the migrating motor complex (MMC) are observed in the GI tract (Vantrappen et al., 1977). The MMC consists of three phases: a period of motor quiescence (phase I), a period of irregular contractions (phase II), and a period of clustered potent contractions (phase III). These phases are observed at regular intervals of 90–120 min in humans. Clinically, ghrelin accelerates the rate of gastric emptying (Levin et al., 2006) and induces gastric phase III contractions in healthy volunteers (Bisschops, 2008; Tack et al., 2006). The same findings regarding GI motility in animal models have been reported following free-moving, conscious animal experiments. These experiments provide more physiological information than other approaches to estimating GI motility, which is regulated by the brain–gut interaction (Inui et al., 2004). Notably, the dog is a popular model for GI motility research. An early study (Itoh et al., 1976) showed that motilin induces gastric phase III contractions in dogs. However, there have been few reports showing that ghrelin, which has a structural resemblance to motilin, has an effect on the digestive tract in dogs (Ohno et al., 2010).

Recent technical advances have permitted the measurement of GI motility in conscious small animals, including rats, mice, and house musk shrews (*S. murinus*), using manometric methods (Ataka et al., 2008; Fujino et al., 2003; Tanaka et al., 2009) or force-transducer implantation (Ariga et al., 2007; Fujitsuka et al., 2009; Sakahara et al., 2010; Zheng et al., 2009a). These studies have demonstrated that ghrelin induces a fasted motor pattern and augments the motility of the antrum and duodenum in the fed or fasted state of healthy animals through brain–gut interactions. Moreover, ghrelin and rikkunshito have been shown to improve gastric emptying and GI motility in animal models of GI disorder. Rikkunshito, a traditional Japanese herbal (Kampo) medicine, potentiates ghrelin signaling (Fujitsuka et al., 2011; Takeda et al., 2008, 2010; Yakabi et al., 2010) and is widely prescribed for patients exhibiting functional dyspepsia (Kusunoki et al., 2010). In this section, the role of ghrelin in GI motility and the methods for the measurement of GI motility in experimental animals are introduced.

2. THE ROLE OF GHRELIN IN GASTRODUODENAL MOTILITY

2.1. The strain-gauge force-transducer measurement of gastroduodenal motility in conscious rats and mice

The role of ghrelin in the control of gastroduodenal motility was evaluated in free-moving, conscious rats using a strain-gauge force-transducer method (Fig. 18.1A). In fasted rats, cyclic changes in contraction waves were detected in both the antrum and duodenum; these waves included a quiescent period (phase I-like contractions) followed by a group of contractions (phase III-like contractions) (Fig. 18.1B). Phase III-like contractions of the antrum occur periodically at intervals of approximately 10 min, and most of these contractions appear to occur in conjunction with phase III-like contractions of the duodenum. Circulating ghrelin levels in fasted rats fluctuate; the peaks of these fluctuations are highly associated with phase III-like contractions in the antrum (Ariga et al., 2007; Fujitsuka et al., 2009).

Intravenous administration of ghrelin to fasted rats immediately potentiates the fasted motor activity and increases the motility index (MI) and the frequency of phase III-like contractions in the antrum and duodenum (Fig. 18.1C). The physiological fasted motor activity decreases with the administration of the growth-hormone secretagogue receptor (ghrelin receptor) antagonist (D-Lys^{3}) GHRP-6. Exogenous ghrelin eliminates the fed motor pattern, which is irregular contractions of high frequency caused by feeding, and produces a fasted motor pattern (Fig. 18.1B and D) (Fujitsuka et al., 2009). Gastric motility in the physiological fed and fasted states of conscious mice has also been measured successfully by a method involving the implantation of a transducer in the mouse stomach (Zheng et al., 2009a). The protocol presented below has been used to measure gastroduodenal motility in conscious rats using the strain-gauge force-transducer method (Fig. 18.1A).

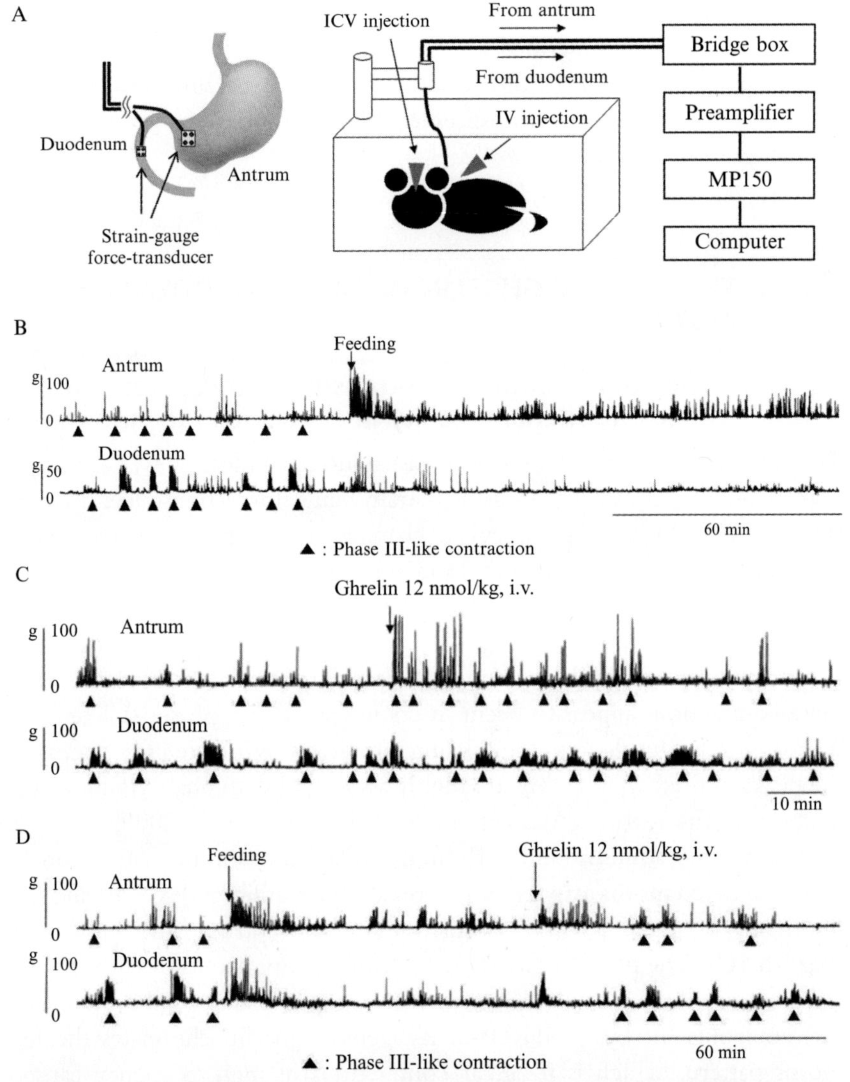

Figure 18.1 Method for strain-gauge force-transducer measurements of gastrointestinal motility in conscious rats. (A) Strain-gauge force transducers were placed on the serosal surface of the antrum and duodenum. The wires of the transducers were drawn out from the back of the neck and connected to a preamplifier via a bridge box. Data were recorded using an MP150. (B) The fasted patterns were replaced by the fed patterns in antrum and duodenum after feeding. (C) Intravenous administration of ghrelin to fasted rats immediately potentiated the frequency of phase III-like contractions in antrum and duodenum. (D) Intravenous administration of ghrelin eliminated the fed motor activities and induced phase III-like contractions in the antrum and duodenum of fed rats (from Fujitsuka et al., 2009). (See Color Insert.)

2.1.1 *Animal preparation involving the strain-gauge force-transducer method*

1. Rats deprived of food overnight and weighing 200–250 g are anesthetized with intraperitoneal injections of pentobarbital sodium (50 mg/kg body weight).
2. After laparotomy, strain-gauge force transducers (F-08IS, Star Medical, Tokyo, Japan) are placed on the serosal surface of the antrum and duodenum.
3. The wires of the transducers are drawn out through a protective coil from the back of the neck via the subcutaneous part of the back.
4. Measurements are made with the animals in a free-moving condition system (Sugiyana-gen Co., LTD, Tokyo, Japan) in individual cages after a 5-day postoperative period for recovery.

2.1.2 *Measurement of gastroduodenal motility*

1. Rats are deprived of food but not water for 16 h before the experiment.
2. The strain-gauge force transducer placed in rats is connected to a preamplifier via a bridge box (Star Medical).
3. Data are recorded using an MP150 (BIOPAC Systems, Goleta, California).
4. The experiment is started when the fasted gastric contraction is stabilized, 2 h after the initial measurement.
5. The frequency of the fasted pattern is obtained from the average of the onset of phase III-like activities for each hour of the experiment.
6. The area under the wave (MI) per minute in the antrum and duodenum is measured and is shown as a percentage (%MI) relative to control data.

2.2. The manometric measurement of gastroduodenal motility in conscious rats and mice

Gastroduodenal motility in the physiological fed and fasted states of conscious rats has also been measured using a manometric method (Fig. 18.2A) (Fujimiya et al., 2000; Fujino et al., 2003). The frequency of phase III-like contractions in the antrum was 5.3 ± 0.5/h and that in the duodenum was 5.6 ± 0.8/h in fasted rats. This fasted pattern was disrupted and replaced by the fed pattern after feeding (Fig. 18.2B). The intravenous injection of ghrelin induced the fasted pattern in the duodenum when rats in the fed state were injected, increasing their %MI in the antrum (Fig. 18.2C). Recent advances in transgenic and knockout technologies have provided tools to investigate the pathogenesis of disease models, and these technologies have typically been applied to mice.

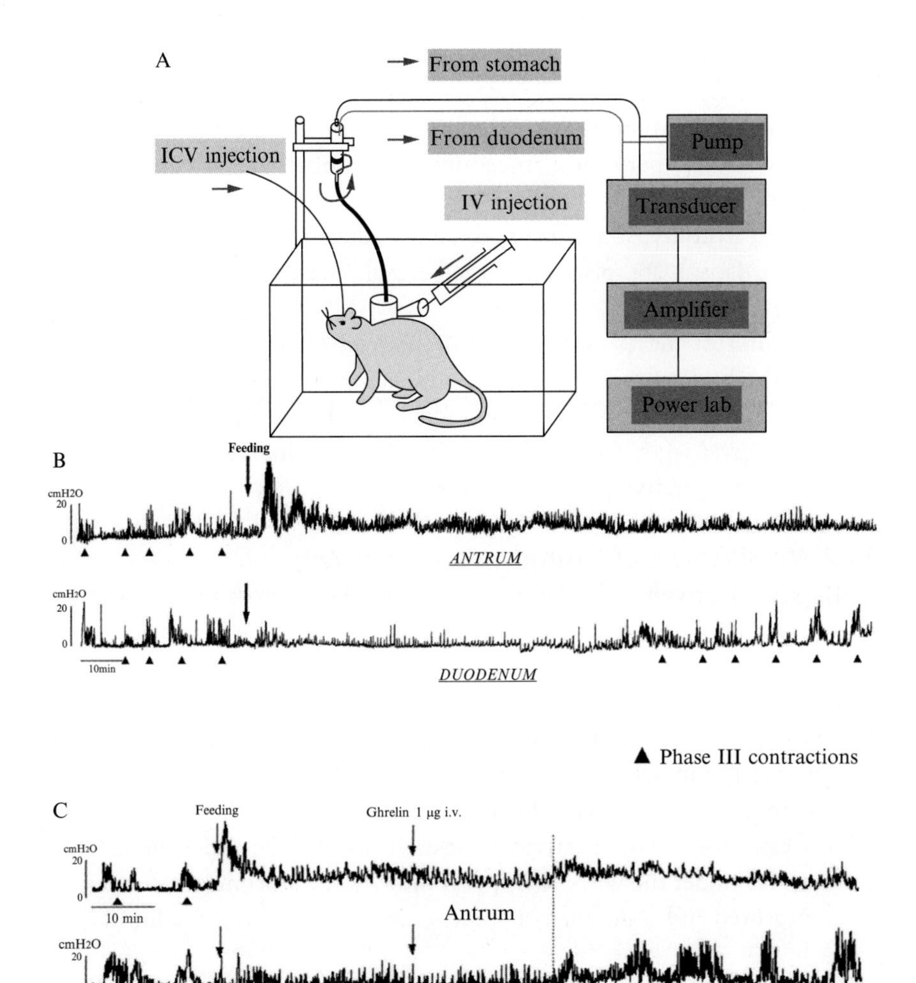

Figure 18.2 Method for manometric measurements of gastrointestinal motility in conscious rats. (A) Catheters for motility recordings are inserted into the antrum and duodenum, connected to the infusion swivel to allow free movement and then connected to a pressure transducer. The data are recorded and stored in a PowerLab. (B) Phase III-like contractions in fasted rats were disrupted and replaced by the fed pattern, which is irregular contraction of high frequency, after feeding. From Kihara et al. (2001). (C) Intravenous administration of ghrelin eliminated the fed motor pattern and produced a fasted motor pattern in duodenum (from Fujino et al., 2003). (See Color Insert.)

Manometric methods allow dual monitoring of the motility of the stomach and duodenum in conscious mice (Tanaka et al., 2009). In fasted mice, the frequency of phase III-like contractions in the antrum was 7.8 ± 0.5/h and that in the duodenum was 6.6 ± 0.7/h. However, the frequency of phase

III-like contractions was lower in the antrum of neuropeptide Y (NPY) Y2 receptor knockout mice than in wild-type mice (Tanaka et al., 2009). The manometric method for the measurement of gastroduodenal motility in conscious rats and mice reported by Fujino et al. (2003) and Tanaka et al. (2009) is described below.

2.2.1 Animal preparation (manometric method)

1. Rats weighing 200–250 g or mice weighing 20–25 g are anesthetized by the intraperitoneal injection of pentobarbital sodium (50 mg/kg body weight).
2. Two polyurethane tubes (3-Fr, 1 mm diameter for rats or ID 0.30 × OD 0.84 mm for mice) are used as a manometric catheter for the motility recordings.
3. One tube is inserted into the stomach through a small incision to the gastric body with the tip placed at the gastric antrum. The other is inserted through the duodenal wall, and the tip is placed 3 cm for rats or 7 mm for mice from the pylorus.
4. The tubes are fixed on the gastric wall and duodenal wall by purse-string sutures, which run subcutaneously to emerge at the top of the neck, and are then secured on the neck skin.
5. Animals are allowed to recover for 1 week before the experiments.

2.2.2 Measurement of gastroduodenal motility (Fig. 18.2A)

1. Mice are deprived of food but not water for 16 h before the experiment.
2. The manometric catheters from the stomach and duodenum are connected to the infusion swivel on a single-axis counter-weighted swivel mount to allow free movement and then connected to a pressure transducer.
3. The catheters are continuously infused with bubble-free saline or distilled water from an infusion pump at the rate of 1.5 ml/h for rats or 0.15 ml/h for mice.
4. The data are recorded and stored in a PowerLab.
5. The mice are placed in a black box (150 × 200 × 300 mm) with the top open.
6. Motor activity is analyzed as described in Section 2.1.2.

2.3. Measurement of gastroduodenal motility in conscious house musk shrews (*S. murinus*)

Ghrelin has a structural resemblance to motilin, and the ghrelin receptor exhibits a 50% identity with the motilin receptor (Asakawa et al., 2001). Ghrelin induces premature phase III contractions in the human stomach

(Tack et al., 2006). Motilin also induces phase III contractions through the cholinergic pathway in humans and dogs (Itoh et al., 1976; Luiking et al., 1998; Suzuki et al., 1998). Ghrelin and motilin are expected to have additive or synergistic effects on the induction of GI contraction. The motilin gene is inactivated in rodents, and mice and rats are therefore not suitable animals for the study of motilin–ghrelin interactions (He et al., 2010).

A recent report (Sakahara et al., 2010) has demonstrated that both physiological ghrelin and motilin are produced and stimulate gastric motility in the house musk shrew (laboratory name: suncus). Strain-gauge force transducers were implanted on the serosa of the gastric body and duodenum in the free-moving, conscious suncus. As a result, clear fasted contractions similar to those observed in humans (Vantrappen et al., 1977) and dogs (Szurszewski, 1969) were observed in the suncus stomach (gastric body). These coordinated contractions consist of three phases: phase I (a period of motor quiescence), phase II (a period of preceding irregular contractions), and phase III (a period of clustered potent contractions). These phases were clearly recognized at regular intervals (every 80–150 min). In addition, ghrelin and/or motilin stimulated suncus gastric motility (Sakahara et al., 2010). Suncus, a small laboratory animal, may be useful as an alternative to humans and dogs for studying the physiological relationships between ghrelin and motilin on GI motility.

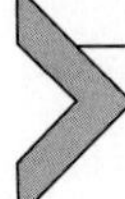

3. THE GI MOTOR EFFECT OF GHRELIN MEDIATED BY THE GUT–BRAIN AXIS

3.1. The brain mechanism responsible for mediating GI motility

The fasted pattern in GI motility is disrupted and replaced by the fed pattern after feeding. Intracerebroventricular injection of NPY, a powerful orexigenic peptide in the brain, induces fasted motor activity in fed rats (Fujimiya et al., 2000). The frequency of phase III-like contractions was lower in the antrum of NPY-Y2 knockout mice than in wild-type mice (Tanaka et al., 2009). However, anorexigenic peptides such as the corticotrophin-releasing factor (CRF) (Bueno et al., 1986), urocortin (Kihara et al., 2001), and cholecystokinin (Rodriguez-Membrilla and Vergara, 1997) cause the disruption of fasted motor activity in animals. These findings may represent an integrated mechanism linking the feeding behavior and GI motor activity through the gut–brain axis.

The effects of intravenous injection of ghrelin were blocked by the immunoneutralization of NPY in the brain, suggesting that peripheral ghrelin induces fasted motor activity by activating the NPY neurons in the brain, probably through ghrelin receptors on vagal afferent neurons (Fujino et al., 2003). Various recent studies have demonstrated the brain mechanism responsible for mediating GI motility. Central and peripheral administration of des-acyl ghrelin has been shown to significantly decrease food intake in food-deprived mice and to decrease gastric emptying (Asakawa et al., 2005). Des-acyl ghrelin exerts inhibitory effects on antrum motility but not on duodenal motility in fasted animals (Chen et al., 2005). Obestatin exerts inhibitory effects on the motility of the antrum and duodenum in the fed state but not in the fasted state (Ataka et al., 2008). CRF receptors in the brain may mediate the actions of des-acyl ghrelin and obestatin. Central administration of nesfatin-1, which has been identified as a hypothalamic anorexigenic peptide, has been shown to decrease food intake and inhibit gastroduodenal motility in mice (Atsuchi et al., 2010). In the experiments that measure gastroduodenal motility, the peptide should be injected through a catheter to avoid the effect of handling stress. The methodology for catheter implantation in rats is described below.

3.1.1 Vessel catheter (Figs. 18.1A and 18.2A)

A vessel catheter (ID 0.36 × OD 0.84 mm, Eicom, Kyoto, Japan) is inserted into the right jugular vein in rats and also led out from the back of the neck. The catheter is filled with heparinized saline (100 units/ml) to avoid blood coagulation. The operation can be performed at the same time as the implantation of a strain-gauge force transducer.

3.1.2 Intracerebroventricular catheter (Figs. 18.1A and 18.2A)

1. The implantation of an intracerebroventricular catheter is performed 4 days before the implantation of a strain-gauge force transducer.
2. The anesthetized rats are placed in a stereotaxic apparatus and implanted with a guide cannula (25 gauge; Eicom, Kyoto, Japan), which reaches the right lateral ventricle.
3. The stereotaxic coordinates are 0.8 mm posterior to bregma, 1.4 mm right lateral to the midline, and 3.4 mm below the outer surface of the skull, when using a Kopf stereotaxic frame (Tujunga, CA, USA), with the incisor bar set at the horizontal plane passing through bregma and lambda.

4. The guide cannula is secured with dental cement anchored by two stainless steel screws that are fixed on the dorsal surface of the skull.
5. After surgery, a dummy cannula (Eicom) is inserted into each guide cannula and a screw cap (Eicom) is placed on the guide cannula to prevent blockade.
6. The correct placement of the intracerebroventricular catheter is verified by the administration of a dye (e.g., 0.05% cresyl violet) into the right lateral ventricle by the brain sections at the end of the experiments.

3.2. Ghrelin and GI disorders

Ghrelin and its receptor agonists possess strong prokinetic properties and therefore have the potential to serve in the treatment of diabetic, neurogenic, or idiopathic gastroparesis as well as for chemotherapy-associated dyspepsia; postoperative, septic, or postburn ileus; opiate-induced bowel dysfunction; and chronic idiopathic constipation (Sallam and Chen, 2010). Abnormalities in gastroduodenal motility are considered key players in the pathogenesis of upper-GI symptoms in certain disorders such as functional dyspepsia and gastroparesis (Suzuki et al., 2006). Zheng et al. (2009b) reported that acute restraint stress inhibits solid gastric emptying and abolishes gastric phase III-like contractions via central CRF in rats. During subsequent chronic stress, the impaired gastric phase III-like contractions were restored by an adaptation mechanism that involves the upregulation of ghrelin expression. Recent work has shown that the central serotonin (5-HT) 2c receptor pathway decreases the peripheral levels of ghrelin, resulting in a shift from fasted to fed-like motor activity. Intravenous administration of ghrelin was shown to replace fed with fasted motor activity in rats treated with fenfluramine, which stimulated 5-HT2cR signaling in the central nervous system. Rikkunshito is widely prescribed for patients exhibiting functional dyspepsia (Kusunoki et al., 2010; Suzuki et al., 2009). Oral administration of rikkunshito has been shown to reduce the incidence of anorexia and improve gastric emptying in animals through increased peripheral plasma ghrelin concentrations (Fujitsuka et al., 2009; Sadakane et al., 2011; Saegusa et al., 2011; Takeda et al., 2008; Yakabi et al., 2011), stimulated central ghrelin secretion (Yakabi et al., 2010), or increased hypothalamic ghrelin receptor activity (Takeda et al., 2010). Recent studies have demonstrated that oral administration of rikkunshito improves gastroduodenal dysmotility in a rat model of cancer anorexia–cachexia by the potentiation of ghrelin receptor signaling

(Fujitsuka et al., 2011). These findings suggest that stimulation of ghrelin signaling may be an attractive approach for the treatment of upper-GI motor dysfunction.

4. SUMMARY

Recent technical advances have permitted the measurement of GI motility in conscious small animals, including rats, mice, and house musk shrews (*S. murinus*). Transgenic and knockout mice are tools to investigate the pathogenesis of disease models. The suncus may be useful as an alternative to humans and dogs for studying the physiological relationships between ghrelin and motilin in the context of GI motility. Recent experiments on free-moving, conscious animal have demonstrated that ghrelin regulates physiological fasted motor activity in the antrum and duodenum. Intravenous injection of ghrelin increases the MI and the frequency of phase III-like contractions, both of which are mediated by hypothalamic NPY neuron activation though ghrelin receptors at the vagal afferent terminal.

Stress hormone and anorexigenic peptides cause the disruption of fasted motor activity through a brain–gut interaction, which is involved in the pathogenesis of upper-GI symptoms in disorders such as functional dyspepsia and gastroparesis. Ghrelin is a signal potentiator that promotes GI motility and could be a good therapeutic target for GI disorders.

ACKNOWLEDGMENTS

This work was supported by Grants-in-Aid for Scientific Research from the Ministry of Education, Culture, Sports, Science, and Technology and the Ministry of Health, Labour, and Welfare of Japan. A. I. has received grant support from Tsumura & Co.

REFERENCES

Ariga, H., et al., 2007. Endogenous acyl ghrelin is involved in mediating spontaneous phase III-like contractions of the rat stomach. Neurogastroenterol. Motil. 19, 675–680.

Asakawa, A., et al., 2001. Ghrelin is an appetite-stimulatory signal from stomach with structural resemblance to motilin. Gastroenterology 120, 337–345.

Asakawa, A., et al., 2005. Stomach regulates energy balance via acylated ghrelin and desacyl ghrelin. Gut 54, 18–24.

Ataka, K., et al., 2008. Obestatin inhibits motor activity in the antrum and duodenum in the fed state of conscious rats. Am. J. Physiol. Gastrointest. Liver Physiol. 294, G1210–G1218.

Atsuchi, K., et al., 2010. Centrally administered nesfatin-1 inhibits feeding behaviour and gastroduodenal motility in mice. Neuroreport 21, 1008–1011.

Bisschops, R., 2008. Ligand and electrically induced activation patterns in myenteric neuronal networks. Confocal calcium imaging as a bridge between basic and human physiology. Verh. K. Acad. Geneeskd. Belg. 70, 105–145.

Bueno, L., et al., 1986. Effects of corticotropin-releasing factor on plasma motilin and somatostatin levels and gastrointestinal motility in dogs. Gastroenterology 91, 884–889.

Chen, C.Y., et al., 2005. Des-acyl ghrelin acts by CRF type 2 receptors to disrupt fasted stomach motility in conscious rats. Gastroenterology 129, 8–25.

Fujimiya, M., et al., 2000. Neuropeptide Y induces fasted pattern of duodenal motility via Y(2) receptors in conscious fed rats. Am. J. Physiol. Gastrointest. Liver Physiol. 278, G32–G38.

Fujino, K., et al., 2003. Ghrelin induces fasted motor activity of the gastrointestinal tract in conscious fed rats. J. Physiol. 550, 227–240.

Fujitsuka, N., et al., 2009. Selective serotonin reuptake inhibitors modify physiological gastrointestinal motor activities via 5-HT2c receptor and acyl ghrelin. Biol. Psychiatry 65, 748–759.

Fujitsuka, N., et al., 2011. Potentiation of ghrelin signaling attenuates cancer anorexia-cachexia and prolongs survival. Transl. Psychiatry 1, e23.

He, J., et al., 2010. Stepwise loss of motilin and its specific receptor genes in rodents. J. Mol. Endocrinol. 44, 37–44.

Inui, A., et al., 2004. Ghrelin, appetite, and gastric motility: the emerging role of the stomach as an endocrine organ. FASEB J. 18, 439–456.

Itoh, Z., et al., 1976. Motilin-induced mechanical activity in the canine alimentary tract. Scand. J. Gastroenterol. 39 (Suppl.), 93–110.

Kihara, N., et al., 2001. Effects of central and peripheral urocortin on fed and fasted gastroduodenal motor activity in conscious rats. Am. J. Physiol. Gastrointest. Liver Physiol. 280, G406–G419.

Kusunoki, H., et al., 2010. Efficacy of Rikkunshito, a traditional Japanese medicine (Kampo), in treating functional dyspepsia. Intern. Med. 49, 2195–2202.

Levin, F., et al., 2006. Ghrelin stimulates gastric emptying and hunger in normal-weight humans. J. Clin. Endocrinol. Metabol. 91, 3296–3302.

Luiking, Y.C., et al., 1998. Motilin induces gall bladder emptying and antral contractions in the fasted state in humans. Gut 42, 830–835.

Ohno, T., et al., 2010. The roles of motilin and ghrelin in gastrointestinal motility. Int. J. Pept. article ID 820794

Rodriguez-Membrilla, A., Vergara, P., 1997. Endogenous CCK disrupts the MMC pattern via capsaicin-sensitive vagal afferent fibers in the rat. Am. J. Physiol. 272, G100–G105.

Sadakane, C., et al., 2011. 10-Gingerol, a component of rikkunshito, improves cisplatin-induced anorexia by inhibiting acylated ghrelin degradation. Biochem. Biophys. Res. Commun. 412, 506–511.

Saegusa, Y., et al., 2011. Decreased plasma ghrelin contributes to anorexia following novelty stress. Am. J. Physiol. Endocrinol. Metab. 301, E685–E696.

Sakahara, S., et al., 2010. Physiological characteristics of gastric contractions and circadian gastric motility in the free-moving conscious house musk shrew (Suncus murinus). Am. J. Physiol. Regul. Integr. Comp. Physiol. 299, R1106–R1113.

Sallam, H.S., Chen, J.D., 2010. The prokinetic face of ghrelin. Int. J. Pept. article ID 493614

Suzuki, H., et al., 1998. Motilin controls cyclic release of insulin through vagal cholinergic muscarinic pathways in fasted dogs. Am. J. Physiol. 274, G87–G95.

Suzuki, H., et al., 2006. Therapeutic strategies for functional dyspepsia and the introduction of the Rome III classification. J. Gastroenterol. 41, 513–523.

Suzuki, H., et al., 2009. Japanese herbal medicine in functional gastrointestinal disorders. Neurogastroenterol. Motil. 21, 688–696.

Szurszewski, J.H., 1969. A migrating electric complex of canine small intestine. Am. J. Physiol. 217, 1757–1763.

Tack, J., et al., 2006. Influence of ghrelin on interdigestive gastrointestinal motility in humans. Gut 55, 327–333.

Takeda, H., et al., 2008. Rikkunshito, an herbal medicine, suppresses cisplatin-induced anorexia in rats via 5-HT2 receptor antagonism. Gastroenterology 134, 2004–2013.

Takeda, H., et al., 2010. Rikkunshito ameliorates the aging-associated decrease in ghrelin receptor reactivity via phosphodiesterase III inhibition. Endocrinology 151, 244–252.

Tanaka, R., et al., 2009. New method of manometric measurement of gastroduodenal motility in conscious mice: effects of ghrelin and Y2 depletion. Am. J. Physiol. Gastrointest. Liver Physiol. 297, G1028–G1034.

Vantrappen, G., et al., 1977. The interdigestive motor complex of normal subjects and patients with bacterial overgrowth of the small intestine. J. Clin. Invest. 59, 1158–1166.

Yakabi, K., et al., 2010. Rikkunshito and 5-HT2C receptor antagonist improve cisplatin-induced anorexia via hypothalamic ghrelin interaction. Regul. Pept. 161, 97–105.

Yakabi, K., et al., 2011. Urocortin 1 reduces food intake and ghrelin secretion via CRF(2) receptors. Am. J. Physiol. Endocrinol. Metab. 301, E72–E82.

Zheng, J., et al., 2009a. Ghrelin regulates gastric phase III-like contractions in freely moving conscious mice. Neurogastroenterol. Motil. 21, 78–84.

Zheng, J., et al., 2009b. Effects of repeated restraint stress on gastric motility in rats. Am. J. Physiol. Regul. Integr. Comp. Physiol. 296, R1358–R1365.

CHAPTER NINETEEN

Ghrelin Acylation by Ingestion of Medium-Chain Fatty Acids

Yoshihiro Nishi*[,1], Hiroharu Mifune[†], Masayasu Kojima[‡]

*Department of Physiology, Kurume University School of Medicine, Asahi-machi, Kurume, Japan
[†]Institute of Animal Experimentation, Kurume University School of Medicine, Asahi-machi, Kurume, Japan
[‡]Molecular Genetics, Institute of Life Science, Kurume University, Hyakunenkouen, Kurume, Japan
[1]Corresponding author: e-mail address: nishiy@med.kurume-u.ac.jp

Contents

Abstract

We found in a primary study that ingestion of medium-chain fatty acids (MCFAs) or medium-chain triacylglycerols (MCTs) increased the stomach contents of acyl ghrelin, and we further showed that the carbon-chain length of the acyl groups that modified the nascent ghrelin peptides corresponded to that of the ingested MCFAs or MCTs. These findings clearly demonstrated that the ingested MCFAs are directly used for the acyl-modification of ghrelin. Before the discovery of ghrelin-*O*-acyltransferase (GOAT), our *in vivo* study suggested that the putative GOAT preferred MCTs (composed of C6:0 to C10:0 FFAs) to either short- or long-chain triglycerides. In another study, we suggested that MCFAs or MCTs might represent a potential therapeutic modality for the clinical manipulation of energy metabolism through the modulation of ghrelin activity. After the discovery of GOAT, many studies have been done on the acylation of ghrelin using MCFAs, MCTs, or

Methods in Enzymology, Volume 514
ISSN 0076-6879
http://dx.doi.org/10.1016/B978-0-12-381272-8.00019-2

their derivatives; however, results and interpretations have been inconsistent, largely due to the differences in experimental conditions. This chapter describes detailed methods for the analysis of ghrelin acylation *in vivo* to facilitate future research in this field.

1. INTRODUCTION

In a primary study, we detected *n*-heptanoyl ghrelin, an unnatural form of acyl ghrelin, in stomachs and plasma of mice after the ingestion of glyceryl triheptanoate, which cannot be synthesized by mammalian cells except under very rare conditions (Nishi et al., 2005a). This suggested that ingested medium-chain fatty acids (MCFAs) or medium-chain triglycerides (MCTs) were directly utilized for the ghrelin acylation. This finding was confirmed by subsequent studies both before (Nishi et al., 2005b; Yamato et al., 2005) and after (Gutierrez et al., 2008; Kirchner et al., 2009) the discovery of ghrelin-*O*-acyltransferase (GOAT) (Yang et al., 2008), which catalyzes the acylation of ghrelin. MCFAs and MCTs are commonly found in milk, vegetable oils (coconut or palm), and butter (Babayan, 1968; Fernando-Warnakulasuriya et al., 1981; Greenberger and Skillman, 1969). Their derivatives, such as medium-chain acyl CoAs, are also produced in the body by metabolic processes involving long-chain fatty acids, a main component of lipid nutrients (Coleman et al., 2000; Eaton et al., 1996). The ability of acyl ghrelins to stimulate their receptor (GHS-R) is highly dependent on the length of the fatty acid that modifies the serine 3 residue (Ser^3) of acyl ghrelins (Matsumoto et al., 2001). These findings confirmed that the availability and carbon-chain length of MCFAs, MCTs, or their derivatives are rate-limiting steps for the acylation and the activation of ghrelin. This chapter describes methods for *in vivo* experiments on ghrelin acylation based on previous studies by us and by other groups of specialists in this field.

2. PREPARING FOOD AND WATER CONTAINING MCFAs OR MCTs

2.1. Preparation of drinking water with MCFAs

MCFAs (*n*-hexanoic acid, *n*-octanoic acid, or *n*-decanoic acid (Sigma-Aldrich, Japan)) were dissolved in distilled water at a concentration of 5 mg/ml (Nishi et al., 2005a). At this concentration, no precipitates were formed at room temperature (25 °C).

2.2. Preparation of chow with MCTs

Standard laboratory chow (CLEA Rodent Diet: CE-2, CLEA Japan Inc., Osaka, Japan) was used. The CE-2 chow contained 50.3% carbohydrate, 25.4% protein, and 4.4% fat. MCTs consisting of glyceryl trihexanoate (C6-MCT) (Tokyo-Kasei, Tokyo, Japan), glyceryl trioctanoate (C8-MCT) (Wako Pure Chemical, Osaka, Japan), glyceryl tridecanoate (C10-MCT) (Wako Pure Chemical), or glyceryl triheptanoate (Fluka Chemie, Switzerland) were mixed into the CE-2 chow at a concentration of 5% (w/w) as described below (Nishi et al., 2005a).

1. Weigh out a portion of the CE-2 chow and sprinkle an appropriate amount of the liquid MCT (5% w/w of the CE-2) onto it (e.g., 5.0 g of liquid MCT onto 95 g of CE-2 chow).
2. Allow the MCT to soak into the CE-2 for at least 2 h at room temperature.
3. In the case of C10-MCT, which is a solid at room temperature, dissolve the C10-MCT by heating in a thermostat chamber at 60 °C.
4. Sprinkle the appropriate amount of liquid C10-MCT (5% w/w to CE-2) onto the CE-2, which is prewarmed to 60 °C.
5. Heat the C10-MCT-sprinkled CE-2 in a thermostat chamber at 60 °C for at least 2 h to allow the C10-MCT to soak into the CE-2.
6. Store the C6-, C8-, or C10-MCT-soaked CE-2 at 4 °C until use.

2.3. Feeding conditions

Male C57BL/6JJcl mice (CLEA Japan, Inc., Osaka, Japan) weighing 20–25 g were used. Mice were given *ad libitum* access to either the MCT-mixed chow (5% w/w) or the MCFA-mixed water (5 mg/ml) for 0–14 days (Nishi et al., 2005a). To minimize the denaturing of MCFAs or MCTs, drinking water and chow were changed twice a week.

3. SAMPLES FOR GHRELIN MEASUREMENT

3.1. Preparation of stomach samples

All mouse tissue samples were collected during a fed state between 9:00 and 10:00 a.m. (Nishi et al., 2005a). Tissue sampling was conducted in accordance with the Guidelines for Animal Experimentation, Kurume University.

1. Dissect the whole mouse stomach immediately after induction of deep anesthesia (intraperitoneal injection of 50 mg/kg pentobarbital).

2. Wash each stomach sample twice in 50 m*M* sodium phosphate buffer Saline (pH 7.4) (PBS) and measure the wet weight of each sample.
3. Boil each stomach sample for 5 min in a 10-fold volume of water to inactivate intrinsic proteases.
4. After cooling on ice, adjust the boiled samples in water to 1 *M* acetic acid (AcOH)–20 m*M* HCl.
5. Homogenize the boiled stomach samples in the 1 *M* AcOH–20 m*M* HCl solution and isolate the extracted peptides by centrifugation (e.g., 12,000 × *g* for 15 min).
6. Transfer the extracted peptides in supernatant (1 mg wet weight tissue equivalent) to stock tubes, lyophilize, and store at −80 °C until assay.
7. Dissolve the lyophilized peptide sample in RIA buffer and measure the levels of ghrelin by ghrelin RIA (the precise procedure is described in chapter 8 in this issue).

3.2. Preparation of plasma samples

A detailed procedure for the measurement of plasma ghrelin level in the mouse (or rat) has been reported (Hosoda et al., 2000) and is also described in chapter 8 in this issue. Therefore, this chapter will focus on the method to semiquantify the levels of *n*-heptanoyl ghrelin (C7-ghrelin) in mouse plasma after the ingestion of glyceryl triheptanoate, a C7-MCT composed of C7-FFA, which is rarely synthesized by mammalian cells under normal conditions.

1. Aspirate blood (~1.0 ml per mouse) by heart puncture after induction of deep anesthesia.
2. Immediately transfer the mixture of blood samples (from 7 to 10 mice) to a chilled polypropylene tube on ice containing EDTA-2Na (1 mg/ml) and aprotinin (500–1000 kallikrein inactivator units per milliliter).
3. Isolate the plasma after centrifugation at 4 °C and immediately transfer it to the polypropylene tube on ice.
4. Add HCl to the plasma to a final concentration of 0.1 *M* and dilute it with an equal volume of ice-cold 0.9% NaCl solution (saline). To minimize the degradation of acyl ghrelins in plasma, steps 1–4 above must be completed within 5 min.
5. Centrifuge the acidified and diluted plasma sample for 5 min at 4 °C and transfer the supernatant to a Sep-Pak C18 cartridge pre-equilibrated in 0.1% trifluoroacetic acid (TFA) and saline.
6. After a serial washing with saline and 5% acetonitrile (CH_3CN)/0.1% TFA, elute the plasma sample with 60% CH_3CN/0.1% TFA from the Sep-Pak C18 cartridge.

7. Collect the eluate from the cartridge into a polypropylene tube and lyophilize.
8. Dissolve the residual material (peptides) in 1 *M* AcOH (~2.0 ml) and adsorb it onto an SP-Sephadex C25 column (H^+-form, GE Healthcare, Uppsala, Sweden) (bed volume: ~1.0 ml) pre-equilibrated in 1 *M* AcOH. Successive elutions in 3.0 ml of 1 *M* AcOH, 2 *M* pyridine, and 2 *M* pyridine–AcOH (pH 5.0) will generate three fractions: SP-I, SP-II, and SP-III.
9. Evaporate the SP-III fraction by heating (~55 °C) and dissolve the residual material in 1 *M* AcOH.
10. Separate the acyl- or des-acyl ghrelins in 1 *M* AcOH using reverse-phase HPLC (RP-HPLC), as described below.

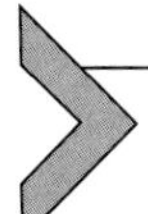

4. Measurement of Ghrelins Modified with or without the *N*-Octanoyl Group

4.1. RP-HPLC plus ghrelin C-terminal RIA

Peptide samples from stomachs or plasma of mice dissolved in 1.0 *M* AcOH are fractionated by RP-HPLC and then quantified by ghrelin RIA (Hosoda et al., 2000). The precise procedure for the measurement of ghrelin, including C-terminal or N-terminal ghrelin RIA, is described in chapter 8 in this issue. A significant increase was detected in stomach levels of *n*-hexanoyl ghrelin (C6-ghrelin), *n*-octanoyl ghrelin (C8-ghrelin), or *n*-decanoyl ghrelin (C10-ghrelin) after the ingestion of C6-MCT, C8-MCT, or C10-MCT, respectively (Table 19.1), by means of the assay system described below:

1. Before the RP-HPLC analyses of stomach ghrelins, the peptide sample (from 1.0 mg wet weight tissue of stomach) is semipurified by a Sep-Pak C18 cartridge (Waters, Milford, MA) and lyophilized. Next, the lyophilized sample is dissolved in 1.0 *M* AcOH. For the analysis of plasma ghrelin, the Sep-Pak C18-purified samples from 7 to 10 ml of plasma are lyophilized and redissolved in 1.0 *M* AcOH.
2. The peptide solution in 1.0 *M* AcOH is fractionated by RP-HPLC in a C18 cartridge (Symmetry 300, 3.9 × 150, Waters) (C18-RP-HPLC). The peptide sample from 0.2 to 0.5 mg wet weight of stomach tissue or that from 7 to 10 ml plasma is injected into the C18-RP-HPLC column.
3. The injected sample (in 1.0 *M* AcOH) is fractionated by the C18-RP-HPLC using a linear gradient from 10% to 60% CH_3CN/0.1% TFA at a flow rate of 1.0 ml/min for 40 min, collecting 500-μl fractions.

Table 19.1 Stomach contents of *n*-hexanoyl, *n*-octanoyl, or *n*-decanoyl ghrelin after the ingestion of medium-chain (C6, C8, and C10) triglycerides

	C6-ghrelin	C8-ghrelin	C10-ghrelin
Control	28.6±3.5	531.1±27.3	30.8±5.5
C6-MCT	237.8±34.8a	360.8±33.3[b]	25.8±6.0
C8-MCT	12.3±4.5	788.8±82.6[b]	8.8±5.7[c]
C10-MCT	24.6±4.4	516.9±42.3	108.4±12.0[c]

Male C57BL/6J mice were fed standard laboratory chow (control) or chow containing 5% (w/w) glyceryl trihexanoate (C6-MCT), glyceryl trioctanoate (C8-MCT), or glyceryl tridecanoate (C10-MCT) for 14 days. The contents of *n*-hexanoyl ghrelin (C6-ghrelin), *n*-octanoyl ghrelin (C8-ghrelin), or *n*-decanoyl ghrelin (C10-ghrelin) in stomach samples (fmol of ghrelins in 0.2 mg wet weight tissue) were measured by ghrelin C-terminal RIA after C18-RP-HPLC fractionation.

[a] $p < 0.001$ versus C6-ghrelin level in control.

[b] $p < 0.001$ versus C8-ghrelin level in control.

[c] $p < 0.001$ versus C10-ghrelin level in control.

4. Peptide samples in each resultant HPLC fraction are lyophilized and dissolved in RIA buffer (50 m*M* sodium phosphate buffer (pH 7.4), 0.5% bovine serum albumin, 0.5% Triton X-100, 80 m*M* NaCl, 25 m*M* EDTA-2Na, and 0.05% NaN_3) containing 0.5% normal rabbit serum.
5. Each sample solution is quantified by ghrelin C-terminal RIA, which recognizes the C-terminal sequence of the ghrelin peptide irrespective of the N-terminal acylation (Hosoda et al., 2000).
6. By this assay system combining C18-RP-HPLC and ghrelin C-terminal RIA, both C6-ghrelin and C7-ghrelin are eluted with shorter retention times (RTs) than C8-ghrelin, while C8-ghrelin is eluted with a shorter retention time than C10-ghrelin. (Fig. 19.1) (Nishi et al., 2005a).

4.2. *n*-Decanoyl ghrelin RIA

Stomach or plasma levels of *n*-decanoyl ghrelin, another acyl from of ghrelin, can be directly measured by an *n*-decanoyl ghrelin RIA (Hiejima et al., 2009; Yoh et al., 2011), which does not cross-react with *n*-octanoyl ghrelin or des-acyl ghrelin (Figs. 19.2 and 19.3). The procedure for the measurement of *n*-decanoyl ghrelin is exactly the same as that of ghrelin C-terminal RIA, except that an antibody raised against the *n*-terminal sequence of the *n*-decanoyl ghrelin is used.

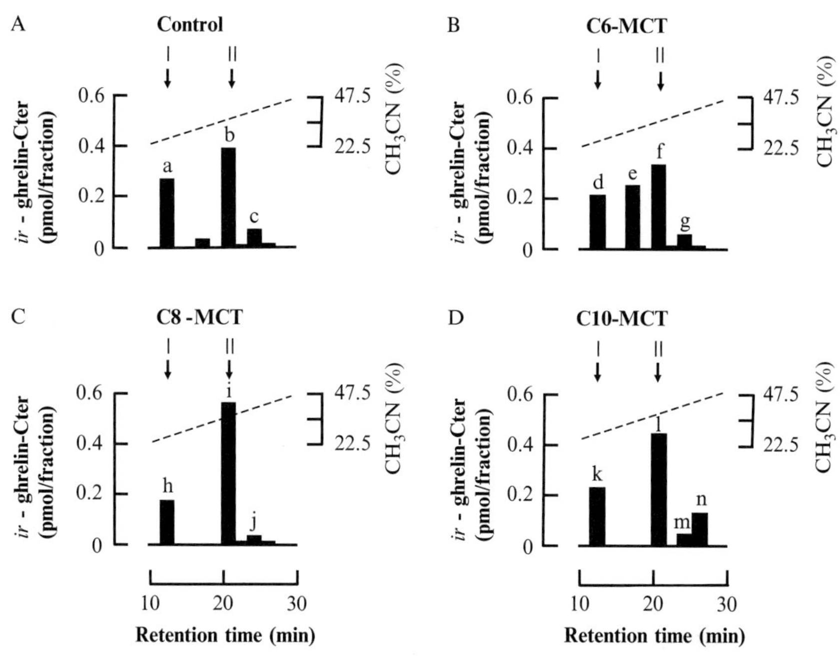

Figure 19.1 RP-HPLC profiles of ghrelin C-terminal immunoreactivity in the stomachs of mice fed standard laboratory chow (control, A) or chow containing glyceryl trihexanoate (C6-MCT, B), glyceryl trioctanoate (C8-MCT, C), or glyceryl tridecanoate (C10-MCT, D). A linear gradient of 10–60% CH_3CN containing 0.1% TFA was run for 40 min. Ghrelin C-terminal immunoreactivity (*ir*-ghrelin-Cter) in the peptides extracted from 0.2 mg of stomach tissue and fractionated by C18-RP-HPLC was quantified by ghrelin C-terminal RIA. Black bars represent levels of *ir*-ghrelin-Cter. Arrows indicate the elution positions of des-acyl ghrelin (I) and *n*-octanoyl ghrelin (II). Based on the retention times of the respective ghrelin forms, peaks a, d, h, and k correspond to des-acyl ghrelin, whereas peaks b, f, i, and l correspond to *n*-octanoyl ghrelin. Peak e corresponds to *n*-hexanoyl ghrelin, whereas peak n corresponds to *n*-decanoyl ghrelin. Peaks c, g, j, and m are considered to correspond to *n*-decenoyl (C10:1) ghrelin and other acyl ghrelins, based on previous studies (Hiejima et al., 2009; Nishi et al., 2005a).

5. PURIFICATION AND CHARACTERIZATION OF ACYL GHRELINS FROM STOMACHS

5.1. Purification of *n*-heptanoyl and other acyl forms of ghrelin

The method for the purification of acyl ghrelins from the stomach is similar to that used for the purification of *n*-heptanoyl ghrelin (C7-ghrelin). Therefore, in this section, we will describe how to purify *n*-heptanoyl ghrelin from

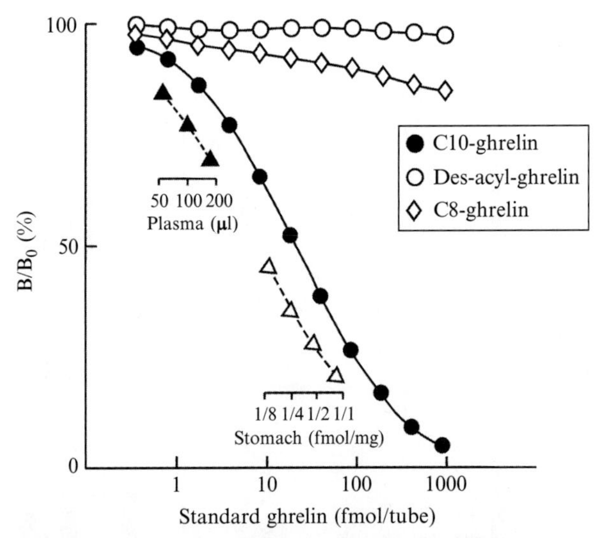

Figure 19.2 Competition curves of *n*-decanoyl ghrelin (C10-ghrelin) and other forms of ghrelin for the binding of [^{125}I]Tyr^{12}-C10-ghrelin [1–11] to the anti-C10-ghrelin antiserum. Three independent experiments were done in duplicate and the results are expressed as a percentage of specific binding (B) in the absence of peptides (B_0) (mean, $n = 3$). Parallel dilution curves to the C10-ghrelin standard curve were drawn from the stomach extract and plasma extract of mice fed *ad libitum*. The antibody (within antiserum) used in this assay exhibited little ($<2\%$) or no cross-reactivity to C8-ghrelin or des-acyl ghrelin, respectively. C10-ghrelin, *n*-decanoyl ghrelin [1–28]; C8-ghrelin, *n*-octanoyl ghrelin [1–28]; des-acyl ghrelin, des-acyl ghrelin [1–28].

stomachs of mice fed glyceryl triheptanoate (C7-MCT) by means of immune-affinity chromatography using rabbit IgG raised against rat ghrelin(1–11) IgG (Kaiya et al., 2001; Nishi et al., 2005a).

1. Prepare the stomach sample (total 1000 mg) from 7 to 10 mice fed C7-MCT for 4 days, as described in steps 1–6 of Section 3.1.
2. Dissolve the residual material (peptides) in 1 *M* AcOH, adsorb it onto an SP-Sephadex C25 column, fractionate it as described in steps 8 and 9 of Section 3.2., and collect the SP-III fraction.
3. Evaporate the SP-III by heating (~55 °C), and dissolve the residual material in 0.5–1.0 ml of 1.0 *M* AcOH.
4. Apply the SP-III fraction in 1.0 *M* AcOH to a Sephadex G-50 fine gel-filtration column (1.9 × 145 cm) (GE Healthcare Japan, Tokyo), fractionate it at a flow rate of 5.0 ml/h, and collect the 5.0-ml fractions.
5. Subject a portion of each fraction (15–50 μl from 5 ml) to a ghrelin calcium-mobilization assay (described in Section 5.2) using a fluorometric imaging plate reader system (Molecular Devices, CA, USA),

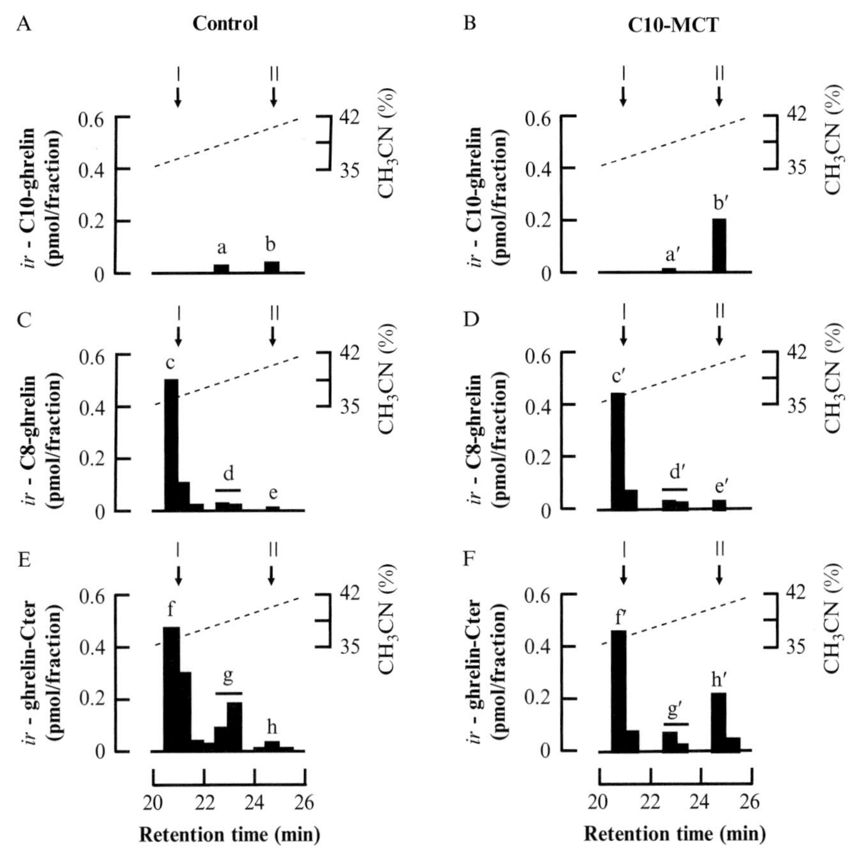

Figure 19.3 RP-HPLC profile of the immunoreactivity for *n*-decanoyl ghrelin (*ir*-C10-ghrelin, A, B), *n*-octanoyl ghrelin (*ir*-C8-ghrelin, C, D), or the C-terminal sequence of ghrelin (*ir*-ghrelin-Cter, E, F) in the stomachs of mice freely fed a standard laboratory chow (Control, A, C, E) or a chow containing 5% (w/w) glyceryl tridecanoate (C10-MCT, B, D, F). *Arrows* I or II indicate the elution point of synthetic *n*-octanoyl ghrelin (C8-ghrelin) or *n*-decanoyl ghrelin (C10-Gghrelin), respectively. Based on the retention times of these synthetic ghrelin peptides (I, II) and that of purified ghrelin immunoreactivities, peaks c, c′, f, and f′ corresponded to C8-ghrelin and peaks b, b′, e, e′, h, and h′ corresponded to C10-ghrelin. Peaks a, a′, d, d′, g, and g′, eluted between peaks of C8- and C10-ghrelin, were considered to represent *n*-decenoyl (C10:1) and other acyl ghrelin fractions.

in a cell line stably expressing the rat ghrelin receptor (Hosoda et al., 2000; Kojima et al., 1999).

6. Collect half (2.5 ml each) of each active fraction (around fraction numbers 47–52) and condense it using a Sep-Pak C18 light cartridge (Waters). Then lyophilize the eluate in 60% CH_3CN/0.1% TFA.

7. Dissolve the lyophilized eluate in 100 m*M* sodium phosphate buffer (pH 7.4) and purify by anti-rat ghrelin (1–11) IgG immuno-affinity chromatography. Elute the adsorbed substances (purified ghrelin peptides) in 500 μl of 10% CH_3CN/0.1% TFA.
8. Lyophilize the eluate from the immune-affinity chromatography, dissolve it in 500 μl of 1.0 *M* AcOH, and load onto the C18-RP-HPLC column (Symmetry 300, 3.9 × 150, Waters) working on a linear gradient from 10% to 60% CH_3CN/0.1% TFA at a flow rate of 1.0 ml/min for 40 min.
9. Collect the eluate of C7-ghrelin in the HPLC peak at a RT around 18.4 min, which comes before the peak of C8-ghrelin (RT around 20.4 min) (Fig. 19.4). In the same fashion, C6-ghrelin or C10-ghrelin is eluted by C18-RP-HPLC at an RT around 17.0 or 24.0 min, respectively.
10. Analyze the peptide sequences of the purified C7-ghrelin or other acyl ghrelins with a protein sequencer (model 494; Applied Biosystems, CA, USA). Molecular weight of each purified peptide is determined by matrix-assisted laser desorption/ionization time of flight (MALDI-TOF) mass spectrometry (Hillenkamp and Karas, 1990) with a Voyager-DE PRO instrument (Applied Biosystems) (Ida et al., 2007; Nishi et al., 2005a).

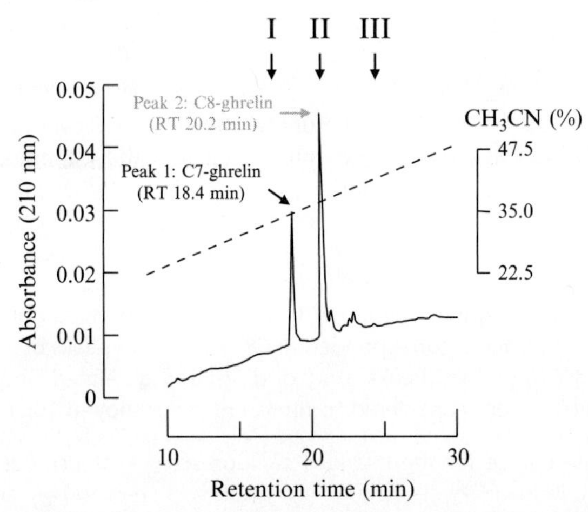

Figure 19.4 RP-HPLC profile of the purified *n*-heptanoyl ghrelin (C7-ghrelin). Ghrelin peptides were purified from the stomachs of mice fed glyceryl triheptanoate. Elutes from anti-rat ghrelin immuno-affinity column were subjected to C18-RP-HPLC. Based on the retention times of control samples in HPLC and MALDI-TOF-MS analysis, peaks 1 and 2 corresponded to C7- and C8-ghrelin, respectively. Arrows indicate the elution points of *n*-hexanoyl (I), *n*-octanoyl (II), and *n*-decanoyl (III) ghrelin. RT, retention time.

5.2. Bioactivities of acyl ghrelins *in vitro*

Bioactivities of acyl ghrelins *in vitro* were estimated by measuring changes in intracellular calcium concentrations ($[Ca^{2+}]_i$) with a fluorometric plate reader (FLIPR) system (Molecular Devices) in a cell line stably expressing rat ghrelin receptor (GHS-R1a) (Chinese hamster ovary (CHO)-GHSR62) (Hosoda et al., 2000; Kojima et al., 1999) as follows (Fig. 19.5):

1. Plate the CHO-GHSR62 cells on flat-bottom, black-walled 96-well plates at a density of 4.0×10^4 cells/well, and incubate at 37 °C for 12–15 h before the assay of ghrelin activity.
2. Incubate the cells for 1 h with 4 m*M* of Fluo-4-AM fluorescent indicator dye (molecular probes, Eugene, OR) in the assay buffer (Hanks' balanced salts solution, 20 m*M* Hepes, 2.5 m*M* probenicid) containing 1% fetal calf serum.
3. Wash the incubated cells four times with the assay buffer.
4. Dissolve the assay samples (e.g., purified peptide samples from stomachs) in the assay buffer containing 0.1% bovine serum albumin.
5. Measure the changes of $[Ca^{2+}]_i$ in CHO-GHSR62 cells through the changes of fluorescence intensity using the FLIPR system (Fig. 19.5).

The *in vitro* bioactivity of *n*-hexanoyl or *n*-heptanoyl ghrelin to increase $[Ca^{2+}]_i$ in CHO-GHSR62 cells is approximately 5% or 60% that of

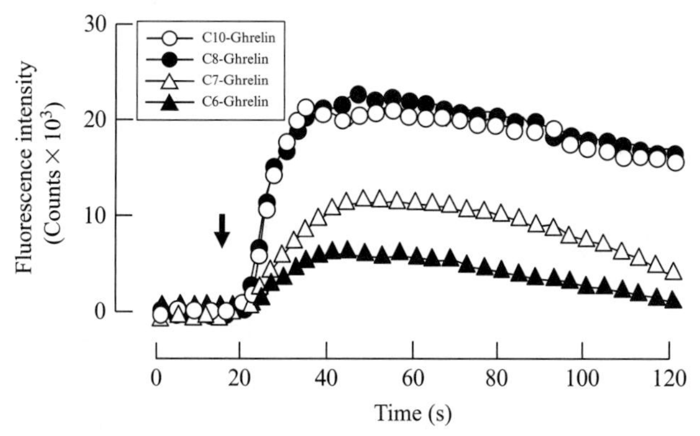

Figure 19.5 Time course for the changes of fluorescence intensity as a measure of $[Ca^{2+}]_i$ changes induced by *n*-hexanoyl, *n*-heptanoyl, *n*-octanoyl, and *n*-decanoyl ghrelin in ghrelin receptor (GHS-R1a)-expressing cells (CHO-GHSR62). Ghrelin peptides (1.0×10^{-8} *M*) were added at the time indicated by the arrow. The *in vitro* bioactivity of *n*-hexanoyl or *n*-heptanoyl ghrelin to increase $[Ca^{2+}]_i$ in CHO-GHSR62 cells is approximately 5% or 60% that of *n*-octanoyl ghrelin, respectively. In contrast, the activity of *n*-decanoylghrelin is almost equipotent to that of *n*-octanoyl ghrelin in this assay system (Matsumoto et al., 2001; Nishi et al., 2005a).

n-octanoyl ghrelin, respectively. In contrast, the activity of *n*-decanoyl ghrelin is almost equipotent to that of *n*-octanoyl ghrelin in this assay system (Matsumoto et al., 2001; Nishi et al., 2005a).

6. SUMMARY

One of the main physiological roles of the ghrelin–GOAT system is to serve as a nutrient sensor, sending the brain information about lipids within the gastrointestinal tract. The bioactivity of ghrelin is regulated by the type of medium-chain fatty acyl group modifying the ghrelin peptides. In fact, the bioactivity of C6-ghrelin measured by the CHO-GHSR62 system is far smaller (~5%) than that of C8-ghrelin. Since the medium-chain acyl molecules modifying the stomach ghrelin peptides are easily displaced by ingested MCFAs or MCTs, it might be possible to manipulate appetite and control obesity, to some degree, simply by ingesting MCFAs or MCTs that weaken the bioactivity of the acylated ghrelins. *In vivo* tracing of ghrelin acylation processes by unnatural MCFAs or MCTs (e.g., tracing the C7-ghrelin produced by the ingestion of C7-MCT) is a simple and useful system to investigate the production, secretion, and degradation of acyl ghrelins. However, to maximize the accuracy of this system, great care must be taken to prepare samples properly before the ghrelin assay, since acyl ghrelins are rapidly degraded by intrinsic esterases that catalyze their deacylation. Precise examinations of *in vivo* acylation of ghrelin by ingested MCFAs using this method, in concert with *in vitro* enzymatic studies on GOAT, will help to clarify the ghrelin–GOAT system and stimulate further research in this important field.

ACKNOWLEDGMENTS

This work was supported in part by a Japanese Grant-in-Aid for Scientific Research (C; KAKENHI Nos. 22591145 and 22500361), a grant from the Kurume University Millennium Box Foundation for the promotion of Science, and a grant from the Morinaga Foundation for Health and Nutrition. We are grateful to Dr. H. Hosoda, Dr. H. Kaiya, Dr. K. Kangawa (Department of Biochemistry, National cerebral and Cardiovascular Center Research Institute, Osaka, Japan), and Dr. G. Wyckoff (Guest Professor, Kurume University) for their valuable comments on this chapter. We also thank Dr. J. Yoh and Dr. H. Hiejima (Kurume University School of Medicine) for technical assistance.

REFERENCES

Babayan, V.K., 1968. Medium-chain triglycerides—their composition, preparation, and application. J. Am. Oil Chem. Soc. 45, 23–25.

Coleman, R.A., et al., 2000. Physiological and nutritional regulation of enzymes of triacylglycerol synthesis. Annu. Rev. Nutr. 20, 77–103.
Eaton, S., et al., 1996. Mammalian mitochondrial beta-oxidation. Biochem. J. 320 (Pt. 2), 345–357.
Fernando-Warnakulasuriya, G.J., et al., 1981. Studies on fat digestion, absorption, and transport in the suckling rat. I. Fatty acid composition and concentrations of major lipid components. J. Lipid Res. 22, 668–674.
Greenberger, N.J., Skillman, T.G., 1969. Medium-chain triglycerides. N. Engl. J. Med. 280, 1045–1058.
Gutierrez, J.A., et al., 2008. Ghrelin octanoylation mediated by an orphan lipid transferase. Proc. Natl. Acad. Sci. U.S.A. 105, 6320–6325.
Hiejima, H., et al., 2009. Regional distribution and the dynamics of n-decanoyl ghrelin, another acyl-form of ghrelin, upon fasting in rodents. Regul. Pept. 156, 47–56.
Hillenkamp, F., Karas, M., 1990. Mass spectrometry of peptides and proteins by matrix-assisted ultraviolet laser desorption/ionization. Methods Enzymol. 193, 280–295.
Hosoda, H., et al., 2000. Ghrelin and des-acyl ghrelin: two major forms of rat ghrelin peptide in gastrointestinal tissue. Biochem. Biophys. Res. Commun. 279, 909–913.
Ida, T., et al., 2007. Purification and characterization of feline ghrelin and its possible role. Domest. Anim. Endocrinol. 32, 93–105.
Kaiya, H., et al., 2001. Bullfrog ghrelin is modified by n-octanoic acid at its third threonine residue. J. Biol. Chem. 276, 40441–40448.
Kirchner, H., et al., 2009. GOAT links dietary lipids with the endocrine control of energy balance. Nat. Med. 15, 741–745.
Kojima, M., et al., 1999. Ghrelin is a growth-hormone-releasing acylated peptide from stomach. Nature 402, 656–660.
Matsumoto, M., et al., 2001. Structure-activity relationship of ghrelin: pharmacological study of ghrelin peptides. Biochem. Biophys. Res. Commun. 287, 142–146.
Nishi, Y., et al., 2005a. Ingested medium-chain fatty acids are directly utilized for the acyl modification of ghrelin. Endocrinology 146, 2255–2264.
Nishi, Y., et al., 2005b. Developmental changes in the pattern of ghrelin's acyl modification and the levels of acyl-modified ghrelins in murine stomach. Endocrinology 146, 2709–2715.
Yamato, M., et al., 2005. Exogenous administration of octanoic acid accelerates octanoylated ghrelin production in the proventriculus of neonatal chicks. Biochem. Biophys. Res. Commun. 333, 583–589.
Yang, J., et al., 2008. Identification of the acyltransferase that octanoylates ghrelin, an appetite-stimulating peptide hormone. Cell 132, 387–396.
Yoh, J., et al., 2011. Plasma levels of n-decanoyl ghrelin, another acyl- and active-form of ghrelin, in human subjects and the effect of glucose- or meal-ingestion on its dynamics. Regul. Pept. 167, 140–148.

CHAPTER TWENTY

Islet β-Cell Ghrelin Signaling for Inhibition of Insulin Secretion

Katsuya Dezaki[1], Toshihiko Yada[1]

Division of Integrative Physiology, Department of Physiology, Jichi Medical University School of Medicine, Shimotsuke, Japan

[1]Corresponding author: e-mail address: dezaki@jichi.ac.jp, tyada@jichi.ac.jp

Contents

Abstract

Ghrelin, an acylated 28-amino acid peptide, was isolated from the stomach, where circulating ghrelin is produced predominantly. In addition to its unique role in regulating growth-hormone release, mealtime hunger, lipid metabolism, and the cardiovascular system, ghrelin is involved in the regulation of glucose metabolism. Ghrelin is expressed in pancreatic islets and released into pancreatic microcirculations. Ghrelin inhibits insulin release in mice, rats, and humans. Pharmacological and genetic blockades of islet-derived ghrelin markedly augment glucose-induced insulin release. The signal transduction mechanisms of ghrelin in islet β-cells are very unique, being distinct from those utilized for growth-hormone release. Ghrelin attenuates the glucose-induced cAMP production and PKA activation, which drives activation of Kv channels and suppression of the glucose-induced $[Ca^{2+}]_i$ increase and insulin release in β-cells. Insulinostatic function of the ghrelin–GHS-R system in islets is a potential therapeutic target for type 2 diabetes.

Methods in Enzymology, Volume 514
ISSN 0076-6879
http://dx.doi.org/10.1016/B978-0-12-381272-8.00020-9

1. INTRODUCTION

Given ghrelin's wide spectrum of biological activities such as growth-hormone (GH) release (Kojima et al., 1999), feeding stimulation (Nakazato et al., 2001), adiposity (Tschöp et al., 2000), and cardiovascular actions (Nagaya et al., 2001), its discovery has opened up many new perspectives within neuroendocrine, metabolic, and cardiovascular research, thus suggesting its possible clinical application (Kojima and Kangawa, 2006). Circulating ghrelin is produced predominantly in the stomach, and its receptor GH-secretagogue receptor (GHS-R) is expressed in a variety of central and peripheral tissues. Ghrelin and GHS-R are also located in pancreatic islets. Multiple experimental systems have shown ghrelin immunoreactivity in α-cells (Date et al., 2002; Dezaki et al., 2004; Heller et al., 2005; Wang et al., 2007), β-cells (Granata et al., 2007; Volante et al., 2002; Wang et al., 2007), and novel islet cells (Wierup et al., 2002, 2004) including those named ε-cells (Prado et al., 2004). Regarding receptors for ghrelin in islets, double immunohistochemistry revealed that GHS-R-like immunoreactivity colocalized extensively with glucagon immunoreactivity and occasionally with insulin immunoreactivity in rat pancreatic islets (Kageyama et al., 2005), indicating expression of GHS-R in α- and β-cells. Ghrelin *O*-acyltransferase (GOAT), which has been identified as the enzyme that promotes the acylation of the third serine residue of ghrelin, is highly expressed in the pancreatic islets and β-cell line (An et al., 2010; Gutierrez et al., 2008; Yang et al., 2008). Recent evidence highlights an important role of ghrelin in glucose homeostasis. Ghrelin inhibits insulin release in mice, rats, and humans (Broglio et al., 2001; Dezaki et al., 2004; Egido et al., 2002; Reimer et al., 2003). Conversely, ghrelin receptor antagonists and GOAT inhibitor could increase insulin release and decrease blood glucose in glucose tolerance tests, thereby acting as antidiabetics (Barnett et al., 2010; Dezaki et al., 2004; Esler et al., 2007). Here we review the physiological role of ghrelin and its signaling mechanisms in the regulation of insulin release.

2. GHRELIN IS RELEASED IN THE PANCREATIC ISLETS

Ghrelin is expressed in the pancreatic islets. Release of ghrelin from pancreatic islets was assessed by comparing the ghrelin level in the pancreatic vein with that in the pancreatic artery in anesthetized rats.

1. Blood samples are collected from the pancreatic arteries (celiac artery) and veins (splenic vein), and portal veins of anesthetized rats.
2. To avoid inflow of ghrelin from the intestine and stomach to splenic vein, the inferior mesenteric vein and the spleen side of splenic vein, including short gastric and left gastro-omental veins, are ligated.
3. Plasma concentrations of ghrelin and des-acyl ghrelin are measured using ELISA kits (Mitsubishi Kagaku Iatron, Tokyo, Japan) according to the manufacturer's instructions.

The concentrations of both ghrelin and des-acyl ghrelin in the pancreatic vein were significantly higher (approximately eight and three times, respectively) than those in the pancreatic artery in rats, indicating that ghrelin is released from the pancreas (Dezaki et al., 2006).

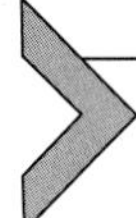

3. INSULINOSTATIC FUNCTION OF ISLET-DERIVED GHRELIN

3.1. *In vitro* perfusion of the pancreas

To examine physiological roles of the pancreatic islet-produced ghrelin, we employed insulin release from the perfused rat pancreas, an *in vitro* system that retains the intact circulation in pancreatic islets while excluding the influence of other organs.

1. Male Wistar rats aged 8–12 weeks are anesthetized by intraperitoneal injection of pentobarbitone at 80 mg/kg.
2. Pancreas with segments of the duodenum and spleen is isolated from anesthetized rats.
3. An arterial cannula is introduced into the celiac artery, and a venous cannula is inserted into the portal vein.
4. The pancreas is perfused for 10 min with the baseline perfusate of HEPES-added Krebs–Ringer bicarbonate buffer (HKRB) solution ((in m*M*): NaCl 129, $NaHCO_3$ 5.0, KCl 4.7, KH_2PO_4 1.2, $CaCl_2$ 2.0, $MgSO_4$ 1.2, and HEPES 10 at pH 7.4) containing 2.8 m*M* glucose, 0.5% bovine serum albumin, and 4% dextran T70, at 37 °C, oxygenated in an atmosphere of 95% O_2 and 5% CO_2.
5. Perfusion is continued for 30 min with 8.3 m*M* glucose-containing HKRB solution with or without test reagents.
6. Fractions collected in chilled tubes at 1-min intervals are assayed for immunoreactive insulin.

A rise in the perfusate glucose concentration from 2.8 to 8.3 m*M* evoked insulin release in a biphasic manner. Both the first and second phases of glucose-induced insulin release were significantly enhanced by the blockade of GHS-R with a GHS-R antagonist [D-Lys^3]-GHRP-6 and by immunoneutralization of endogenous ghrelin with antighrelin antiserum (Dezaki et al., 2006). Conversely, administration of exogenous ghrelin (10 n*M*) suppressed both phases of glucose-induced insulin release (Dezaki et al., 2006; Egido et al., 2002). These findings indicate that both endogenous and exogenous ghrelin suppress glucose-induced insulin secretion within islets.

In perfused pancreas from rats pretreated with pertussis toxin (PTX), a specific inhibitor of G_i and G_o subtypes of trimeric GTP-binding proteins, both phases of glucose-induced insulin release were markedly enhanced and ghrelin failed to affect them (Dezaki et al., 2011). The membrane-permeable cAMP analogue dibutyryl-cAMP markedly enhanced both phases of glucose-induced insulin release, and in the presence of dibutyryl-cAMP, ghrelin failed to suppress both phases of insulin release. In the pancreas preincubated with an irreversible adenylate cyclase inhibitor MDL-12330A, both phases of glucose-induced insulin release were attenuated and were not further altered by the administration of ghrelin (Dezaki et al., 2011). These results suggest that ghrelin attenuates glucose-induced insulin release via PTX-sensitive G-protein that is coupled to modulation of cAMP signaling.

3.2. Insulin release from isolated islets

1. Male Wistar rats as well as ghrelin-knockout (KO) and wild-type mice are anesthetized by intraperitoneal injection of pentobarbitone at 80 mg/kg.
2. Collagenase (Sigma, 1.05 mg/ml) dissolved in HKRB with 0.1% bovine serum albumin is injected into the common bile duct.
3. The pancreas is dissected out and incubated at 37 °C for 16 min.
4. Islets are collected and groups of 10–15 islets are incubated for 1 h in HKRB at 37 °C with 2.8 m*M* glucose for stabilization, followed by test incubation for 1 h in HKRB with various concentrations of glucose and test agents.
5. Insulin concentrations are determined by ELISA kit (Morinaga Institute of Biological Science) according to the manufacturer's instructions.

In isolated rat islets, GHS-R antagonists [D-Lys^3]-GHRP-6 and [D-Arg^1, D-Phe^5, D-Trp^{7,9}, Leu^{11}]-substance P markedly increased insulin release in the presence of glucose at 5.6 m*M* or higher. Furthermore, antiserum against ghrelin, but not nonimmune serum, increased insulin release. These results suggested that endogenous ghrelin suppressed glucose-induced insulin release. Administration of exogenous ghrelin at the relatively high concentration of 10 n*M*, but not 0.1 n*M* and 1 p*M*, attenuated 8.3 m*M* glucose-induced insulin release in islets, while it had no effect on basal insulin release at 2.8 m*M* glucose (Dezaki et al., 2004). Similar concentration- and glucose-related inhibitory effects of exogenous ghrelin on insulin release *in vitro* have been reported in the rat (Colombo et al., 2003) and mouse islets (Reimer et al., 2003). These effects of endogenous and exogenous ghrelin were blunted in islets isolated from PTX-treated rats, whereas addition of 25 m*M* KCl enhanced insulin release from these islets (Dezaki et al., 2007) (Fig. 20.1).

The role of islet-derived ghrelin in insulin release has also been demonstrated by the analysis in ghrelin-deficient mice. Glucose (8.3 and 16.7 m*M*)-induced insulin release from isolated islets of KO mice was significantly greater than that of wild-type mice, while basal levels of insulin release at 2.8 m*M* glucose were not altered. No difference was observed between KO and wild-type mice in insulin content per islet (Dezaki et al., 2006).

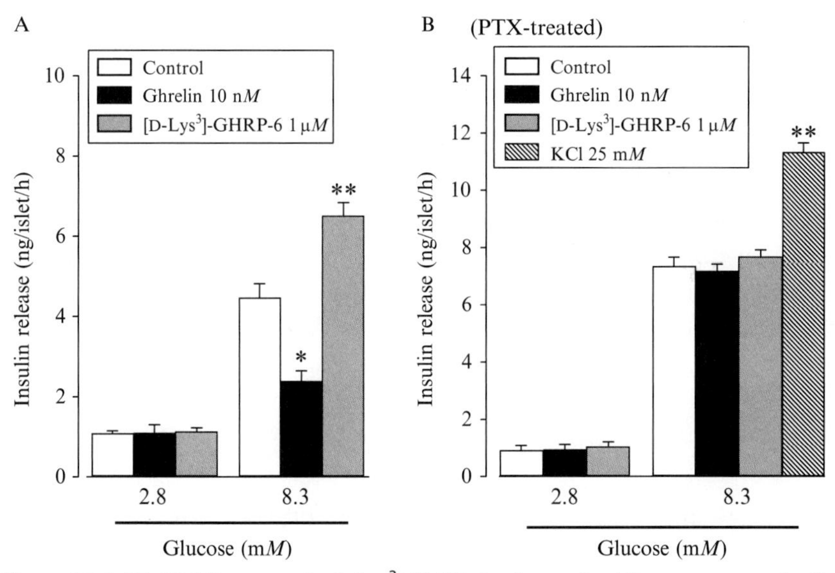

Figure 20.1 (A) GHS-R antagonist [D-Lys^3]-GHRP-6 enhanced, while exogenous ghrelin inhibited, glucose (8.3 m*M*)-induced insulin release from rat isolated islets. (B) Neither ghrelin nor [D-Lys^3]-GHRP-6 altered the insulin release in islets isolated from PTX-treated rats. Addition of 25 m*M* KCl enhanced insulin release as a positive control. *$P < 0.05$; **$P < 0.01$ versus control. The figures were modified from Dezaki et al. (2007).

4. GHRELIN SIGNALING IN ISLET β-CELLS

It is known that PTX-sensitive G_i-proteins inhibit adenylate cyclase that produces cAMP in the cells. In pancreatic β-cells, intracellular cAMP signals are generated by nutrient secretagogues and play a critical role in regulating insulin secretion (Dyachok et al., 2008; Henquin, 2000; Prentki and Matschinsky, 1987). We aimed to clarify whether ghrelin regulates cAMP pathway in islet β-cells and whether this regulation leads to insulinostatic cascade in islet β-cells.

4.1. Measurements of cAMP productions in rat islets

1. Islets are isolated from male Wistar rats by collagenase digestion as mentioned above.
2. Groups of 10 isolated islets are incubated for 1 h at 37 °C in HKRB with 500 μ*M* 3-isobutyl-1-methylxanthine, a phosphodiesterase inhibitor, to avoid degradation of cAMP in the samples.
3. Total cellular cAMP content in islets is determined by EIA kit (GE Healthcare, Buckinghamshire, UK) according to the manufacturer's instructions.

Static incubation of islets with 8.3 m*M* glucose induced modest cAMP productions in islets compared with those with 2.8 m*M* glucose ($P < 0.05$). The glucose-induced cAMP responses were significantly inhibited by exogenous ghrelin (10 n*M*) and augmented by a GHS-R antagonist [D-Lys^3]-GHRP-6 (1 μ*M*) (Dezaki et al., 2011). These effects of exogenous and endogenous ghrelin were blunted in islets isolated from PTX-treated rats: neither exogenous ghrelin (10 n*M*) nor [D-Lys^3]-GHRP-6 (1 μ*M*) affected 8.3 m*M* glucose-induced cAMP productions. Moreover, immunoneutralization of endogenous ghrelin with antighrelin antiserum (0.1%) significantly enhanced the glucose-induced cAMP productions compared with the control with normal rabbit serum (0.1%) (Dezaki et al., 2011).

4.2. Real-time monitoring of cytosolic cAMP concentration in β-cell line

To determine the direct effect of ghrelin on the glucose-induced cAMP production, cytosolic cAMP concentrations ($[cAMP]_i$) were monitored in mouse β-cell line transfected with a fluorescent-translocation biosensor using evanescent-wave microscopy.

1. Mouse MIN6 β-cells are transfected with a fluorescent-translocation biosensor comprising a truncated and membrane-anchored protein kinase A (PKA) regulatory subunit tagged with cyan-fluorescent protein (ΔRIIβ-CFP-CAAX) and a catalytic subunit labeled with yellow-fluorescent protein (Cα-YFP).
2. Emission wavelengths are detected with interference filters (485/25 nm for cyan-fluorescent protein (CFP) excited by 442 nm, and 560/40 nm for yellow-fluorescent protein (YFP) excited by 514 nm line).
3. Holoenzyme dissociation caused by elevation of cytosolic cAMP concentrations ($[cAMP]_i$) results in the translocation of Cα-YFP to the cytoplasm, recorded with evanescent-wave microscopy as selective loss of YFP fluorescence with rise of the CFP-to-YFP fluorescence ratio.

Raising the glucose concentration from 3 to 11 m*M* induced an oscillatory rise in $[cAMP]_i$ in MIN6 cells. Administration of ghrelin markedly suppressed the glucose-induced oscillatory rise in $[cAMP]_i$. Furthermore, in a MIN6 cell with pronounced oscillations of $[cAMP]_i$ during 11 m*M* glucose challenge, the $[cAMP]_i$ oscillations were inhibited by administration of ghrelin, indicating that ghrelin directly inhibits glucose-induced cAMP signaling in β-cells (Dezaki et al., 2011).

4.3. Measurements of $[PKA]_i$ activity in single β-cells

The activity of PKA, a downstream effector of cAMP elevations, was measured in rat single β-cells using PKA-sensitive DR-II fluorescence with confocal microscopy.

1. Isolated islets are collected and dispersed into single cells in Ca^{2+}-free HKRB.
2. The single cells are plated sparsely on coverslips.
3. The single cells are maintained for 1 day at 37 °C in an atmosphere of 5% CO_2 and 95% air in Eagle's minimal essential medium supplemented with 10% fetal bovine serum, 100 μg/ml streptomycin, and 100 U/ml penicillin.
4. A PKA-sensitive dye DR-II (Dojindo, Kumamoto, Japan) is loaded into the cells for 1 h.
5. The DR-II-loaded cells on coverslips are mounted in an open chamber and superfused in HKRB.
6. DR-II fluorescence was excited at 780 nm by a mode-locked Ti:sapphire laser every 10 s, and the emission signal at 475 nm is detected with a photomultiplier tube of the multiphoton laser-scanning confocal microscope (FluoView FV300-TP, Olympus).

7. After scanning, cells are fixed with 4% paraformaldehyde, and then β-cells were identified by insulin immunostaining.

A rise in the perfusate glucose concentration from 2.8 to 8.3 m*M* decreased DR-II fluorescence, reflecting an increase in PKA activity in rat single β-cells since DR-II fluorescence is inversely related to PKA activity. The glucose (8.3 m*M*)-induced increase in PKA activity was significantly inhibited by treatment with ghrelin in the majority of β-cells examined (Dezaki et al., 2011).

4.4. Patch-clamp experiments in rat single β-cells

At substimulatory glucose concentrations, β-cells maintain the resting membrane potential at a hyperpolarized level of around −70 mV. Elevation of the blood glucose concentration increases glucose uptake and metabolism by β-cells, resulting in closure of the ATP-sensitive K^+ (K_{ATP}) channels. When K^+ efflux is reduced, the membrane depolarizes and action potentials are produced by orchestrated openings of voltage-dependent Ca^{2+} channels and voltage-gated K^+ channels. We measured electrophysiological activities in β-cells using patch-clamp analysis.

1. Patch pipettes are pulled from glass tubings (Narishige, Tokyo, Japan); the resistance of the pipettes ranges from 4 to 7 MΩ when filled with pipette solution which contained (in m*M*) K_2SO_4 40, KCl 50, $MgCl_2$ 5, EGTA 0.5, and HEPES 10 at pH 7.2 with KOH.
2. Perforated whole-cell currents are recorded using a pipette solution containing nystatin (150 μg/ml).
3. β-Cells are voltage-clamped at a holding potential of −70 mV, and then shifted to test potentials from −60 to +40 mV in 10-mV steps with pulses of 100-ms durations at 5–8-s intervals at room temperature (25 °C).
4. The membrane potentials are recorded by switching to the current-clamp mode from the whole-cell clamp mode on the amplifier (Axopatch, 200B, Foster, CA).
5. Outward K^+ currents in rat β-cells under nystatin-perforated whole-cell clamp are measured in the presence of 100 μ*M* tolbutamide to inhibit the K_{ATP} channel and thereby exclude involvement of this channel in the currents.
6. For measurement of K_{ATP} channel currents, the β-cells are voltage-clamped to the holding potential of −70 mV, and then applied with a voltage ramp from −100 to −50 mV at a rate of 25 mV/100 ms and voltage step-back to −70 mV at 10-s intervals.

7. For conventional whole-cell clamp experiments measuring Ca^{2+}-channel currents, the pipette solution contains (in m*M*) 90 aspartate, 10 HEPES, 5 $MgCl_2$, 10 EGTA, 5 ATP-Mg, 20 TEA-Cl, and 90 CsOH at pH 7.2 with CsOH.
8. Cells are superfused with 10 m*M* Ca^{2+}-containing HKRB, voltage-clamped at a holding potential of -70 mV, and then shifted to the test potentials from -60 to $+40$ mV in 10-mV steps with pulses of 50-ms duration.
9. Data are stored online in a computer using pCLAMP9 software.

4.4.1 Membrane potentials

Under the condition of nystatin-perforated whole-cell current-clamp mode, glucose (8.3 m*M*) elicited firings of action potentials in rat β-cells. The firings were characterized by spike-like and repetitively occurring action potentials on top of the plateau phase of slow waves (Dezaki et al., 2007). These electrical firings were attenuated by ghrelin (10 n*M*) in a reversible manner. Ghrelin decreased both the frequency and amplitude of the firings, whereas the mean membrane potentials measured at most repolarized levels between slow-wave potentials at 8.3 m*M* glucose were not significantly altered. These findings suggest that ghrelin does not hyperpolarize the membrane potential but decreases the activity of action potentials in β-cells.

4.4.2 Voltage-dependent Kv channel

In pancreatic β-cells, activation of voltage-dependent delayed rectifier K^+ (Kv) channels repolarizes cells and attenuates glucose-stimulated action potentials, thereby limiting Ca^{2+} entry through voltage-dependent Ca^{2+} channels to suppress insulin secretion (MacDonald and Wheeler, 2003). In the presence of 8.3 m*M* glucose, outward K^+ currents evoked by depolarizing pulses from a holding potential of -70 to $+20$ mV were increased by exposure to 10 n*M* ghrelin (Dezaki et al., 2004). The current–voltage relationship demonstrated that ghrelin increased the amplitude of current densities through Kv channels at potentials positive to -30 mV. The ghrelin enhancement of Kv currents in the entire range of potentials was blunted in β-cells treated with PTX (Dezaki et al., 2007). Moreover, ghrelin did not potentiate the Kv currents in the presence of the membrane-permeable cAMP analogues dibutyryl-cAMP and 6-Phe-cAMP (Dezaki et al., 2007, 2011). These data suggest that ghrelin activates Kv channels by counteracting the cAMP signaling.

4.4.3 ATP-sensitive K^+ channel

K_{ATP} channel is the key molecule that determines the resting membrane potentials and converts the glucose metabolism to the membrane excitation in β-cells. Effects of ghrelin on the K_{ATP} channel currents in rat β-cell under nystatin-perforated whole-cell clamp were investigated. The currents evoked by the ramp pulses during superfusion with 2.8 m*M* glucose were substantially inhibited by exposure to tolbutamide or by increasing the glucose concentration to 8.3 m*M*. The conductance density of the K_{ATP} channel current measured at a slope during ramp pulses was not altered by ghrelin, indicating that K_{ATP} channels may not contribute to the inhibition of β-cell membrane excitability by ghrelin (Dezaki et al., 2007).

4.4.4 Voltage-dependent Ca^{2+} channel

In the control external solution containing 8.3 m*M* glucose, a depolarizing pulse from a holding potential of −70 to 0 mV evoked a long-lasting inward current in rat β-cells, and this current was markedly inhibited by an L-type Ca^{2+} channel blocker nifedipine. Exposure of cells to ghrelin (10 n*M*) altered neither the current amplitude nor current–voltage relationships, indicative of no direct effect of ghrelin on L-type Ca^{2+} channels in β-cells (Dezaki et al., 2011).

4.5. Calcium signaling

In islet β-cells, $[Ca^{2+}]_i$ is considered the major regulator of insulin secretion. The role of ghrelin in the regulation of $[Ca^{2+}]_i$ in islet β-cells was explored by measuring $[Ca^{2+}]_i$ in whole islets or single β-cells by fura-2 fluorescence imaging.

1. Isolated single islets and single β-cells on coverslips are incubated with 1 μ*M* fura-2/acetoxymethylester for 30 min at 37 °C.
2. Fura-2-loaded islets and β-cells on coverslips are mounted in an open chamber and superfused in HKRB.
3. $[Ca^{2+}]_i$ in single islets and single β-cells is measured by dual-wavelength fura-2 microfluorometry with excitation at 340/380 nm and emission at 510 nm using a cooled charge-coupled device camera.
4. The ratio image is produced on an Aquacosmos system (Hamamatsu Photonics, Hamamatsu, Japan).

$[Ca^{2+}]_i$ in a whole islet was elevated mildly by increasing the glucose concentration from 2.8 to 5.6 m*M*. In the presence of [D-Lys^3]-GHRP-6, which antagonizes the effect of endogenous ghrelin, the peak of the first phase $[Ca^{2+}]_i$ response was enhanced and, in some islets, oscillations of $[Ca^{2+}]_i$ were induced (Dezaki et al., 2004). Similarly, antiserum against ghrelin enhanced the $[Ca^{2+}]_i$ response to 8.3 m*M* glucose. These results indicate that the endogenous ghrelin in islets restricts glucose-induced $[Ca^{2+}]_i$ increases in β-cells and thereby insulin secretion, presumably via a paracrine and/or autocrine route.

In rat single β-cells, ghrelin at a relatively high concentration of 10 n*M*, but not 0.1 n*M*, markedly suppressed the peak of the first phase $[Ca^{2+}]_i$ responses to 8.3 m*M* glucose and this effect was blocked by GHS-R antagonist (Dezaki et al., 2004). Ghrelin preincubated with antiserum against ghrelin failed to affect $[Ca^{2+}]_i$ responses, confirming that the antiserum employed in our study neutralizes the activity of ghrelin. Suppressions of $[Ca^{2+}]_i$ increases by ghrelin were blunted in β-cells treated with PTX and with antisense oligonucleotide (AS) specific for G-protein $G\alpha_{i2}$ subunit (Fig. 20.2). Furthermore, in the β-cells treated with AS for $G\alpha_{i2}$, ghrelin failed to suppress glucose (8.3 m*M*)-induced insulin release (Dezaki et al., 2007). The action of ghrelin to suppress glucose-induced $[Ca^{2+}]_i$ increases was also blunted in the β-cells whose cAMP levels were clamped upward by a cAMP analogue and downward by an adenylate cyclase inhibitor (Dezaki et al., 2011). These results suggest that $G\alpha_{i2}$-mediated cAMP signaling is crucial for the action of ghrelin to suppress glucose-induced $[Ca^{2+}]_i$ increase and insulin release.

5. CONCLUSION

Ghrelin, including that produced by pancreatic islets, plays a pivotal role in the regulation of insulin release and blood glucose in rodents and humans. Ghrelin attenuates the glucose-induced cAMP production and PKA activation, which drives activation of Kv channels and suppression of the glucose-induced $[Ca^{2+}]_i$ increase and insulin release in islet β-cells (Fig. 20.3). This novel ghrelin signaling in β-cells is distinct from that reported in other tissues including pituitary GH cells, in which GHS-R signaling is primarily coupled to G_{11}-phospholipase C signaling (Howard et al., 1996). When the systemic demand for insulin exceeds the physiological range, including insulin resistance and obesity, antagonism of ghrelin function can promote insulin secretion and thereby prevent glucose intolerance, providing a potential therapeutic avenue to counteract the progression of type 2 diabetes (Dezaki et al., 2008; Yada et al., 2008).

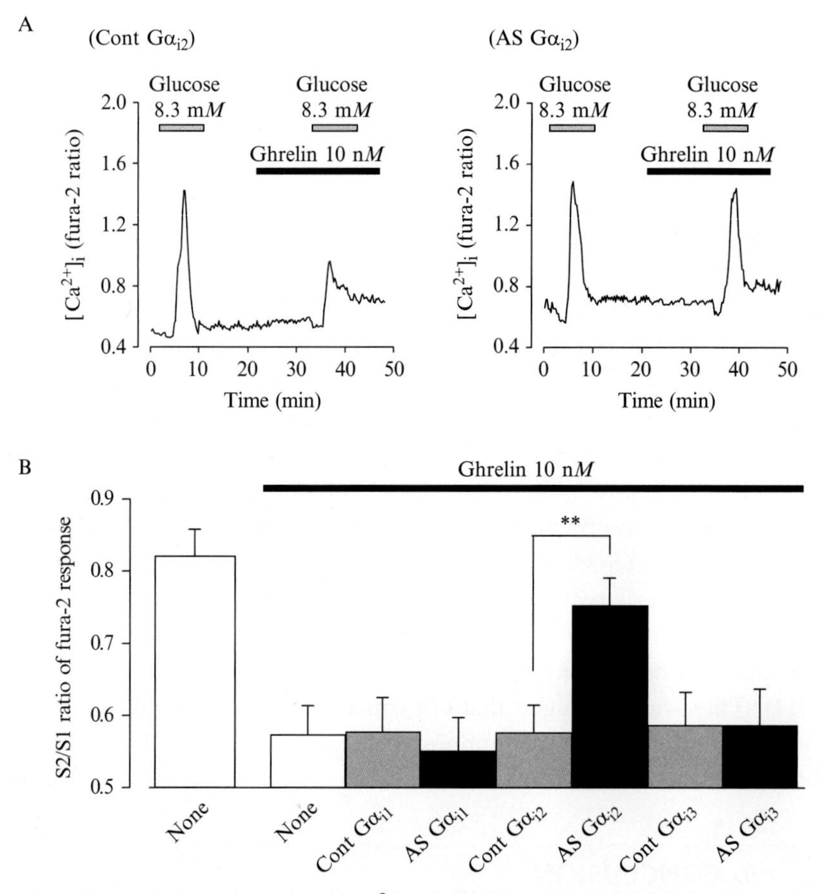

Figure 20.2 (A) Effects of ghrelin on $[Ca^{2+}]_i$ responses to 8.3 m*M* glucose in β-cells following treatment with an antisense oligonucleotide (AS) specific for the G-protein $G\alpha_{i2}$ subunit and control oligonucleotide (Cont). (B) Ghrelin decreased the S2/S1 ratio of the peak $[Ca^{2+}]_i$ responses to the second glucose (8.3 m*M*) stimulation (S2) over those to the first glucose stimulation (S1), indicating inhibition of $[Ca^{2+}]_i$ responses to glucose. AS-induced knockdown of $G\alpha_{i2}$, but not $G\alpha_{i1}$ and $G\alpha_{i3}$, blocked ghrelin attenuation of $[Ca^{2+}]_i$ responses. $^{**}P < 0.01$ versus control oligonucleotide. The figures were modified from Dezaki et al. (2007).

ACKNOWLEDGMENTS

This publication was subsidized by JKA through its promotion funds from KEIRIN RACE. This work was supported by Grants-in-Aid for Scientific Research (to K. D. and T. Y.) and that on Priority Areas (to K. D.) from the Japan Society for the Promotion of Science; by grants from the Salt Science Research Foundation (to K. D.), the Pharmacological Research Foundation, Tokyo (to K. D.), and Takeda Science Foundation (to K. D. and T. Y.); by grants from the Support Program for Strategic Research Program for Brain Sciences by the Ministry of Education, Culture, Sports, Science and Technology of Japan

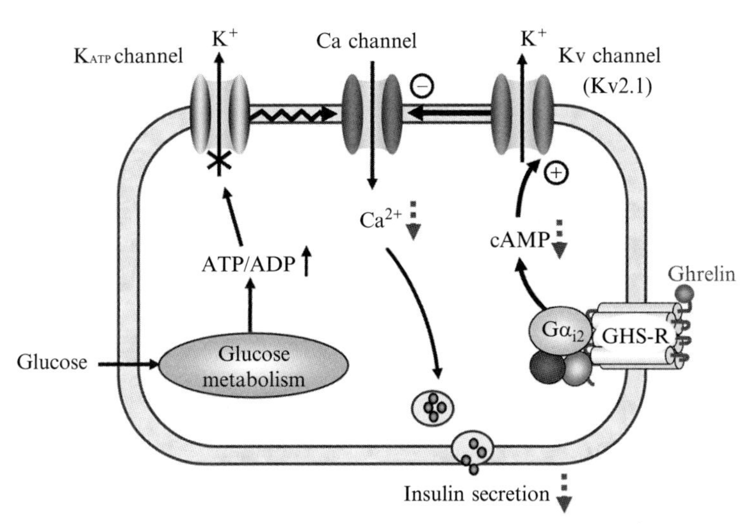

Figure 20.3 Ghrelin signaling in islet β-cells. Closure of ATP-sensitive K^+ (K_{ATP}) channels by increases in ATP/ADP ratio following glucose metabolism induces membrane depolarization and increase in $[Ca^{2+}]_i$ via voltage-dependent Ca^{2+} channels, leading to insulin secretion in β-cells. Ghrelin activates β-cell GHS-R that is coupled with PTX-sensitive G-protein $G\alpha_{i2}$, decreases cAMP production, and attenuates membrane excitability via activation of voltage-dependent K^+ channels (Kv2.1 subtype), and consequently suppresses Ca^{2+} influx and insulin release. (See Color Insert.)

(MEXT), MEXT-Supported Program for the Strategic Research Foundation at Private Universities, the Japan Diabetes Foundation, and Uehara Memorial Foundation; and by the Insulin Research Award from Novo Nordisk and the Basic Science Research Award from Sumitomo Foundation (to T. Y.). The authors thank S. Ookuma, M. Warashina, and M. Motoshima for technical assistance at Jichi Medical University.

REFERENCES

An, W., et al., 2010. Modulation of ghrelin *O*-acyltransferase expression in pancreatic islets. Cell. Physiol. Biochem. 26, 707–716.

Barnett, B.P., et al., 2010. Glucose and weight control in mice with a designed ghrelin *O*-acyltransferase inhibitor. Science 330, 1689–1692.

Broglio, F., et al., 2001. Ghrelin, a natural GH secretagogue produced by the stomach, induces hyperglycemia and reduces insulin secretion in humans. J. Clin. Endocrinol. Metab. 86, 5083–5086.

Colombo, M., et al., 2003. Effects of ghrelin and other neuropeptides (CART, MCH, orexin A and B, and GLP-1) on the release of insulin from isolated rat islets. Pancreas 27, 161–166.

Date, Y., et al., 2002. Ghrelin is present in pancreatic β-cells of humans and rats and stimulates insulin secretion. Diabetes 51, 124–129.

Dezaki, K., et al., 2004. Endogenous ghrelin in pancreatic islets restricts insulin release by attenuating Ca^{2+} signaling in β-cells: implication in the glycemic control in rodents. Diabetes 53, 3142–3151.

Dezaki, K., et al., 2006. Blockade of pancreatic islet-derived ghrelin enhances insulin secretion to prevent high-fat diet-induced glucose intolerance. Diabetes 55, 3486–3493.
Dezaki, K., et al., 2007. Ghrelin uses $G\alpha_{i2}$ and activates Kv channels to attenuate glucose-induced Ca^{2+} signaling and insulin release in islet β-cells: novel signal transduction of ghrelin. Diabetes 56, 2319–2327.
Dezaki, K.T., et al., 2008. Ghrelin is a physiological regulator of insulin release in pancreatic islets and glucose homeostasis. Pharmacol. Ther. 118, 239–249.
Dezaki, K., et al., 2011. Ghrelin attenuates cAMP-PKA signaling to evoke insulinostatic cascade in islet β-cells. Diabetes 60, 2315–2324.
Dyachok, O., et al., 2008. Glucose-induced cyclic AMP oscillations regulate pulsatile insulin secretion. Cell Metab. 8, 26–37.
Egido, E.M., et al., 2002. Inhibitory effect of ghrelin on insulin and pancreatic somatostatin secretion. Eur. J. Endocrinol. 146, 241–244.
Esler, W.P., et al., 2007. Small-molecule ghrelin receptor antagonists improve glucose tolerance, suppress appetite, and promote weight loss. Endocrinology 148, 5175–5185.
Granata, R., et al., 2007. Acylated and unacylated ghrelin promote proliferation and inhibit apoptosis of pancreatic β-cells and human islets: involvement of 3',5'-cyclic adenosine monophosphate/protein kinase A, extracellular signal-regulated kinase 1/2, and phosphatidyl inositol 3-Kinase/Akt signaling. Endocrinology 148, 512–529.
Gutierrez, J.A., et al., 2008. Ghrelin octanoylation mediated by an orphan lipid transferase. Proc. Natl. Acad. Sci. U.S.A. 105, 6320–6325.
Heller, R.S., et al., 2005. Genetic determinants of pancreatic β-cell development. Dev. Biol. 286, 217–224.
Henquin, J.C., 2000. Triggering and amplifying pathways of regulation of insulin secretion by glucose. Diabetes 49, 1751–1760.
Howard, A.D., et al., 1996. A receptor in pituitary and hypothalamus that functions in growth hormone release. Science 273, 974–977.
Kageyama, H., et al., 2005. Morphological analysis of ghrelin and its receptor distribution in the rat pancreas. Regul. Pept. 126, 67–71.
Kojima, M., Kangawa, K., 2006. Drug insight: the functions of ghrelin and its potential as a multitherapeutic hormone. Nat. Clin. Pract. Endocrinol. Metab. 2, 80–88.
Kojima, M., et al., 1999. Ghrelin is a growth-hormone-releasing peptide from stomach. Nature 402, 656–660.
MacDonald, P.E., Wheeler, M.B., 2003. Voltage-dependent K^+ channels in pancreatic beta cells: role, regulation and potential as therapeutic targets. Diabetologia 46, 1046–1062.
Nagaya, N., et al., 2001. Hemodynamic and hormonal effects of human ghrelin in healthy volunteers. Am. J. Physiol. 280, R1483–R1487.
Nakazato, M., et al., 2001. A role for ghrelin in the central regulation of feeding. Nature 409, 194–198.
Prado, C.L., et al., 2004. Ghrelin cells replace insulin-producing β cells in two mouse models of pancreas development. Proc. Natl. Acad. Sci. U.S.A. 101, 2924–2929.
Prentki, M., Matschinsky, F.M., 1987. Ca^{2+}, cAMP, and phospholipids-derived messengers in coupling mechanisms of insulin secretion. Physiol. Rev. 67, 1185–1248.
Reimer, M.K., et al., 2003. Dose-dependent inhibition by ghrelin of insulin secretion in the mouse. Endocrinology 144, 916–921.
Tschöp, M., et al., 2000. Ghrelin induces adiposity in rodents. Nature 407, 908–913.
Volante, M., et al., 2002. Expression of ghrelin and of the GH secretagogue receptor by pancreatic islet cells and related endocrine tumors. J. Clin. Endocrinol. Metab. 87, 1300–1308.
Wang, X., et al., 2007. The effects of intrauterine undernutrition on pancreas ghrelin and insulin expression in neonate rats. J. Endocrinol. 194, 121–129.

Wierup, N., et al., 2002. The ghrelin cell: a novel developmentally regulated islet cell in the human pancreas. Regul. Pept. 107, 63–69.
Wierup, N., et al., 2004. Ghrelin is expressed in a novel endocrine cell type in developing rat islets and inhibits insulin secretion from INS-1 (832/13) cells. J. Histochem. Cytochem. 52, 301–310.
Yada, T., et al., 2008. Ghrelin regulates insulin release and glycemia: physiological role and therapeutic potential. Curr. Diabetes Rev. 4, 18–23.
Yang, J., et al., 2008. Identification of the acyltransferase that octanoylates ghrelin, an appetite-stimulating peptide hormone. Cell 132, 387–396.

CHAPTER TWENTY-ONE

Rikkunshito as a Ghrelin Enhancer

Hiroshi Takeda*,†,1, Shuichi Muto*,‡, Koji Nakagawa*, Shunsuke Ohnishi‡, Chiharu Sadakane*,§, Yayoi Saegusa*,§, Miwa Nahata§, Tomohisa Hattori§, Masahiro Asaka¶

*Pathophysiology and Therapeutics, Hokkaido University Faculty of Pharmaceutical Sciences, Sapporo, Hokkaido, Japan
†Gastroenterology and Hematology, Hokkaido University Graduate School of Medicine, Sapporo, Hokkaido, Japan
‡Gastroenterology, Tomakomai City General Hospital, Tomakomai, Hokkaido, Japan
§Tsumura & Co., Tsumura Research Laboratories, Ibaraki, Japan
¶Cancer Preventive Medicine, Hokkaido University Graduate School of Medicine, Sapporo, Hokkaido, Japan
[1]Corresponding author: e-mail address: h_takeda@pharm.hokudai.ac.jp

Contents

Abstract

Rikkunshito is a kampo herbal medicine which is widely used in Japan for the treatment of the upper gastrointestinal symptoms of patients with functional dyspepsia, gastroesophageal reflux disease, dyspeptic symptoms of postgastrointestinal surgery patients, and chemotherapy-induced dyspepsia in cancer patients. Recently, very unique characteristics of rikkunshito have been unveiled; oral administration of rikkunshito potentiates orexigenic action of ghrelin through several different mechanisms. In addition, several lines of evidence obtained from both animal and human studies indicate that

Methods in Enzymology, Volume 514
ISSN 0076-6879
http://dx.doi.org/10.1016/B978-0-12-381272-8.00021-0

rikkunshito can be an attractive and promising therapeutic option for the anorectic conditions including cisplatin-induced dyspepsia, anorexia of aging, stress-induced hypophagia, and cancer cachexia–anorexia syndrome. In this chapter, we highlight the orexigenic effect of rikkunshito with a special focus on its interaction with ghrelin signaling system.

1. INTRODUCTION

Rikkunshito is a kampo herbal medicine which is prepared by compounding eight herbal medicines: *Atractylodis Lanceae Rhizoma, Ginseng Radix, Pinelliae Tuber, Hoelen, Zizyphi Fructus, Aurantii Nobilis Pericarpium, Glycyrrhizae Radix*, and *Zingiberis Rhizoma*. Rikkunshito is widely used in Japan for the treatment of the upper gastrointestinal symptoms of patients with functional dyspepsia and gastroesophageal reflux disease, chemotherapy-induced dyspepsia in cancer patients, and dyspeptic symptoms of postgastrointestinal surgery patients (Mochiki et al., 2010; Suzuki et al., 2009; Takeda et al., 2012).

Ghrelin is a peripherally active orexigenic gut hormone consisting of 28 amino acids, and the third N-terminal amino-acid serine (Ser) residue is octanoylated (Kojima and Kangawa, 2005; Kojima et al, 1999). Ghrelin is involved in the hypothalamic regulation of energy homeostasis by increasing food intake and reducing fat utilization (Nakazato et al., 2001; Tschop et al., 2000). Plasma levels of ghrelin rise during fasting, and fall upon eating, which has led to the suggestion that ghrelin is a meal-initiating hormone (Cummings et al., 2001). Plasma levels of ghrelin are inversely correlated with body weight in humans and rise after weight loss (Cummings et al., 2002). Besides the regulation of energy homeostasis, ghrelin mediates an increase in gastric motility, induces a positive inotropic effect on the heart, and causes vasodilatation (Chen et al., 2009; Kojima and Kangawa, 2005; Leite-Moreira and Soares, 2007).

Recently, it was shown that oral administration of rikkunshito stimulates secretion of ghrelin in rodents and humans (Hattori, 2010; Matsumura et al., 2010; Takeda et al., 2008). More recent evidence suggests that rikkunshito enhances ghrelin's orexigenic effect by several additional mechanisms (Fujitsuka et al., 2011; Sadakane et al., 2011; Yakabi et al., 2010a,b). In this chapter, we discuss the currently available evidence on the orexigenic effect of rikkunshito with a special attention to its interaction with the ghrelin signaling system.

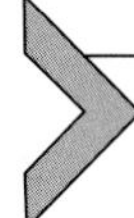

2. CISPLATIN-INDUCED ANOREXIA

2.1. Cisplatin-induced anorexia and ghrelin

Patients with cancer being treated with cytotoxic drugs such as cisplatin often experience a number of undesirable side effects which include acute and delayed nausea, vomiting, anorexia, dyspepsia, and disrupted gastrointestinal function.

Recent evidence has demonstrated the relationship between chemotherapy-induced gastrointestinal disorders and ghrelin in both clinical and animal studies. In human studies, one report has demonstrated that an increase in plasma ghrelin concentrations was observed after the start of anticancer chemotherapy (Shimizu et al., 2003), but more recent studies revealed that the plasma concentration of acylated ghrelin was decreased during the treatment with anticancer drugs (Hiura et al., 2011; Ohno et al., 2011). In animal studies, we and others reported that circulating ghrelin concentrations were reduced in cisplatin-treated rats until 6 h during the early stage of anorexia (Takeda et al., 2008; Yakabi et al., 2010a).

Intraperitoneal injection of 5-HT decreased 24-h food intake as well as plasma acylated-ghrelin level in a dose-dependent manner (Takeda et al., 2008). This result suggests that the cisplatin-induced reduction in the plasma level of acylated ghrelin may be mediated via a release of 5-HT from the gastrointestinal tract mucosa triggered by cisplatin. Indeed, a 5-HT2B receptor agonist BW723C86 and a 5-HT2C agonist *m*-chlorophenylpiperazine HCl (mCPP) markedly decreased the acylated-ghrelin levels in plasma and increased intragastric ghrelin content, suggesting that 5-HT2B/2C receptor stimulation inhibits the release of gastric ghrelin into the circulation (Takeda et al., 2008). In contrast, 5-HT3 and 5-HT4 agonists had no effect on ghrelin dynamics. 5-HT2B and 5-HT2C antagonists suppressed the cisplatin-induced decrease of plasma acylated-ghrelin level and food intake. These results strongly imply that activation of 5-HT2B and 5-HT2C receptors, but not 5-HT3 and 5-HT4 receptors, plays an important role in the decrease in plasma ghrelin level in cisplatin-induced anorexia. Of note, granisetron used in this study clearly inhibited delayed gastric emptying after cisplatin treatment, but it failed to improve cisplatin-induced anorexia (Takeda et al., 2008; Yakabi et al., 2010a).

Peripheral administration of exogenous ghrelin ameliorates anorexia (Liu et al., 2006; Takeda et al., 2008) and vomiting (Rudd et al., 2006) induced by cisplatin. Administration of exogenous ghrelin has been shown to have the potential to reduce each of these symptoms in relevant animal models treated with cisplatin as a cytotoxic agent: emesis in the ferret (Rudd et al., 2006) and anorexia in the rat and mouse (Liu et al., 2006).

Yakabi et al. (2010a) examined the changes of hypothalamic ghrelin secretion in cisplatin-treated rats to elucidate the mechanism underlying chemotherapy-induced delayed anorexia. Although ghrelin secretion in the hypothalamus did not decrease within 24 h of cisplatin administration, it started to decline significantly after 24 h and continued to decrease at least until 48 h, while their plasma ghrelin levels were comparable (Liu et al., 2006; Takeda et al., 2008). Yakabi et al. (2010a) also showed that hypothalamic 5-HT2C receptor gene expression increased significantly in cisplatin-treated rats and the administration of mCPP inhibited hypothalamic ghrelin secretion. Intracerebroventricularly (icv) administered 5-HT2C antagonist SB242084 prevented a decrease in secretion of hypothalamic ghrelin in cisplatin-treated rats, but granisetron, a 5-HT3 antagonist, did not (Yakabi et al., 2010a). These results indicate that the reduced ghrelin secretion in the hypothalamus secondary to 5-HT2C receptor activation may be involved in cisplatin-induced anorexia.

In another study, Yakabi et al. (2010b) demonstrated that hypothalamic GHS-R1a gene expression was significantly reduced after cisplatin or mCPP treatment and this change was reversed by the treatment with the 5-HT2C receptor antagonist SB242084 but not with 5-HT3 receptor antagonists. 5-HT2C receptor antagonist also suppressed cisplatin-induced delayed anorexia. Injection (icv) of GHS-R1a antagonist to saline- or cisplatin-treated rats significantly reduced food intake compared with those injected with saline alone, and this inhibitory effect was abolished by the coadministration of the 5-HT2C receptor antagonist. From these results, it was suggested that delayed-onset anorexia induced by cisplatin may be partly mediated by the activation of the hypothalamic 5-HT2C receptor and the resultant suppression of hypothalamic GHS-R1a gene expression as well as decreased ghrelin secretion in the hypothalamus.

2.2. Effect of rikkunshito on cisplatin-induced anorexia

Rikkunshito ameliorated the decrease in circulating ghrelin concentration and this effect was abolished by coadministration of a GHS-R1a antagonist [D-Lys^3]-GHRP-6 (Takeda et al., 2008). This finding suggests that the mechanism of improvement of anorexia by rikkunshito may involve ghrelin receptor activation. Moreover, Yakabi et al. (2010a) found that rikkunshito reversed the decrease in hypothalamic ghrelin secretion and the decrease in GHS-R1a gene expression 24 h after cisplatin treatment. Injection (icv) of the GHS-R1a antagonist impedes the rikkunshito-mediated improvement in cisplatin-induced anorexia (Yakabi et al., 2010a). Hence, it seems likely that rikkunshito ameliorates cisplatin-induced anorexia by restoring ghrelin secretion and GHS-R1a expression in the hypothalamus. Collectively, rikkunshito suppressed cisplatin-induced anorexia by improving ghrelin signal transduction system by both the peripheral and the central mechanisms.

The induction of ghrelin secretion by rikkunshito is supposed to be based on the 5-HT2B/2C receptor antagonism due to multiple active ingredients. We screened 33 compounds contained in rikkunshito and found that 13 showed antagonistic activity against binding to any of 5-HT 1A, 1B/D, 2A, 2B, 2C, 3, 4, 6, and 7 receptors (Takeda et al., 2008). Among them, 3,3′,4′,5,6,7,8-heptamethoxyflavone (HMF), nobiletin, and tangeretin contained in *Aurantii nobilis pericarpium* had potent 5-HT2B receptor antagonistic activity. The inhibitory activity of hesperidin against the 5-HT2B receptor was weak, but the concentration of hesperidin in rikkunshito is the highest among the ingredients tested. In addition, isoliquiritigenin, which is an ingredient of *Glycyrrhizae radix*, had the most potent activity against the 5-HT2C receptor binding. Our study indicated that the administration of HMF, isoliquiritigenin, and hesperidin attenuated the decrease in plasma ghrelin level, while tangeretin, nobiletin, and 8-shogaol did not. This suggested that the ingredients that inhibit 5-HT2B/5-HT2C-receptor binding are likely to be effective *in vivo*.

2.3. Experimental methods

2.3.1 Animals

Male SD rats aged 7 weeks (Charles River Laboratories Japan, Inc, Tokyo, Japan) were used. During the study period, the rats were kept in an animal room under the following conditions: controlled temperature and humidity, four to five rats in each cage, with a 12-h light/darkness cycle (7:00 a.m.–7:00p.m.), and food and water given *ad libitum*. All experiments were performed between 9:00 a.m. and 6:00 p.m..

2.3.2 Effects of cisplatin, various 5-HT receptor agonists, and antagonists on plasma ghrelin level

Rats were administered cisplatin (2, 6, 10, or 14 mg/kg) and 5-HT (1, 2, 4, or 8 mg/kg) and subjected to blood collection by decapitation at 120 min (cisplatin) or 30 min (5-HT) after administration. Other groups of rats were fasted for 24 h and administered BW723C86 (4 or 16 mg/kg), mCPP (3 or 9 mg/kg), 1-(3-chlorophenyl) biguanide HCl (CPB, 1.25 or 6 mg/kg), 2-methylserotonin (1 or 4 mg/kg), cisapride (1.25 or 5 mg/kg), SB215505 (6 mg/kg), or SB242084HCl (10 mg/kg) and subjected to blood collection by decapitation at 60 or 120 min after administration. The stomach was removed from the animals of the treated groups in which the plasma ghrelin level was decreased. All drugs were given intraperitoneally as solutions in physiologic saline.

2.3.3 Effects of rikkunshito and active herbal components on plasma acylated-ghrelin level in rats administered cisplatin

The rats were orally administered rikkunshito (500 or 1000 mg/kg) dissolved in 1 ml of distilled water per 100 g of body weight. After 16 h, the rats were administered cisplatin intraperitoneally and at the same time a second dose of rikkunshito (which was the same size as the first dose) was administered orally. A single dose of HMF (0.8, 4.0, or 20 mg/kg), hesperidin (4.0 or 20 mg/kg), or isoliquiritigenin (4.0 mg/kg) was given orally simultaneously with the administration of cisplatin. At 2 h after the last dose, the rats were subjected to blood collection by decapitation, and plasma samples were obtained by the method detailed later.

2.3.4 Determination of ghrelin level

The plasma samples were promptly centrifuged at 4 °C, and the supernatant was acidified with 1 mol/l HCl (1/10 volume). The tissue samples were boiled in water for 7 min and acidified with 1 mol/l HCl after cooling. The samples then were homogenized and centrifuged at 10,000 rpm for 15 min. The supernatants were stored at −80 °C until use. The ghrelin level was determined using Active Ghrelin or Desacyl Ghrelin Enzyme-Linked Immunoassay Kit (Mitsubishi Chemical Medience Corporation, Tokyo, Japan).

2.3.5 Effects of rat ghrelin, rikkunshito, and 5-HT2B/2C receptor antagonists on food intake in rats administered with cisplatin

Cisplatin was administered intraperitoneally, and after 2 h acylated ghrelin (1 or 5 nmol/rat) was administered intravenously via the tail vein of rats. Rikkunshito (500 or 1000 mg/kg) was administered orally, and after 16 h cisplatin was administered intraperitoneally, and at the same time, a second dose of rikkunshito was administered orally (which was of the same size as the first dose). SB215505 or SB242084HCl was administered intraperitoneally 30 min before the administration of cisplatin. The food intake and body weight gain were recorded in the subsequent 24 h. In the second experiment, after following the procedures described earlier, physiologic saline or a solution of [D-Lys^{3}]-GHRP-6 (0.4 μmol/rat) was administered intraperitoneally simultaneously with cisplatin and the second dose of rikkunshito (1000 mg/kg); the food intake in the subsequent 6 h was recorded.

2.3.6 Radioligand binding assays

CHO-K1 cells expressing human recombinant 5-HT receptor were used to prepare membranes in modified Tris–HCl buffer. A membrane protein was incubated with 1.2 nmol/l [^{3}H]LSD for 60 min at 37 °C. Nonspecific binding was estimated in the presence of 10 μmol/l 5-HT. The filters were then counted for radioactivity to determine the amount of specifically bound [^{3}H] LSD (Wolf and Schuts, 1997). CHO-K1 cells stably transfected with a plasmid encoding the human 5-HT2C receptor were used to prepare membranes in modified Tris–HCl buffer. A membrane protein was incubated with 1.0 nmol/ml [^{3}H]-Mesulergine for 60 min at 25 °C. Nonspecific binding was estimated in the presence of 1 μmol/l Mianserin. The filters then were counted for radioactivity to determine the amount of specifically bound [^{3}H]-Mesulergine. Likewise, membranes prepared from cells stably transfected with human recombinant 5-HT3, 5-HT6 receptors, and guinea pig striatum (5-HT4) were incubated with radiolabeled ligands with a high affinity for the given receptor, that is, [^{3}H]LSD, [^{3}H]-GR-65630, and [^{3}H]-GR-113808, respectively. Nonspecific radioligand binding was defined by 5-HT or MDL-72222.

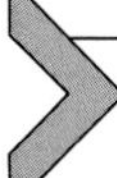

3. ANOREXIA OF AGING

3.1. Anorexia of aging and ghrelin

In elderly subjects, the reduction in energy intake often exceeds energy expenditure, resulting in weight loss and protein energy malnutrition (Morley, 2001). Protein energy malnutrition in the elderly is a frequent and clinically important problem, which leads to increased morbidity, mortality, disability, and health costs in this growing population. One of the most important causes of the reduction in energy intake is anorexia. The causes of the anorexia of aging have not yet been fully defined; they are probably multifactorial and include sensory impairment, social isolation, and psychological and physiologic factors, in addition to the presence of disease (Di Francesco et al., 2007; Morley, 2001).

Although many peripheral anorexigenic hormones including cholecystokinin, leptin, and insulin have been found to rise with increased age, findings for ghrelin are controversial (Takeda et al., 2012). Several lines of animal studies also have revealed mixed results. The reason for these conflicting data seems to be due to the differences in their experimental conditions under which the plasma ghrelin concentration was measured. Indeed, our group found that plasma ghrelin in aged C57BL/6 mice does not increase under fasted conditions but is higher than that in young mice under freely fed conditions (Takeda et al., 2010). This suggests that regulation of ghrelin secretion from the stomach may be disturbed in older mice. In agreement with this, recent clinical studies have suggested that disturbance of regulation of ghrelin secretion and reduced production during hunger and satiety may cause "anorexia of aging" in elderly people (Takeda et al., 2012).

In a previous study, Ariyasu et al. (2008) reported that subcutaneous injection of ghrelin (360 μg/kg twice a day) enhanced food intake in 72-h-fasted and aged mice and restored the decrease in body weight caused by fasting. Contrary to their data, we recently found that much lower dose of ghrelin (33 μg/kg) failed to increase food intake in 75-week-old mice, whereas the same dose of ghrelin had an orexigenic effect in young mice (Takeda et al., 2010), suggesting that aging is associated with attenuation of endogenous ghrelin signaling. Collectively, it seems that dysregulation of ghrelin secretion as well as ghrelin resistance in the appetite control system is occurring in aged mice.

Although the detailed mechanisms of disturbed ghrelin dynamics remain unclear, one of the possible causes seems to be leptin. We have found that plasma leptin and insulin levels in aged mice are significantly higher than in young ones (Takeda et al., 2010). Leptin and insulin are reported to inhibit ghrelin secretion from the stomach into the circulation (Barazzoni et al., 2003); hence elevated leptin and insulin in the elderly may contribute to the inhibition of secretion of ghrelin during fasting, resulting in prolonged satiety and inhibition of hunger sensation. Moreover, the activation of the phosphoinositide 3-kinase (PI3K)–phosphodiesterase 3 (PDE3) pathway was recently proposed as a mechanism by which leptin blocks ghrelin signaling in neuropeptide Y (NPY) neurons, and it may counteract the adenylate cyclase–cAMP–protein kinase A system implicated in the effect of ghrelin (Kohno et al., 2007). Other studies showed that the effect of leptin was abolished by the administration of either PDE3 inhibitor (Niswender et al., 2001) or PI3K inhibitor (Zhao et al., 2002). We demonstrated that the plasma leptin level in aged mice was greatly increased under both feeding and fasting conditions. Furthermore, we found that administration of either a PI3K inhibitor LY-294002 or the PDE3 inhibitor cilostamide improved anorexia in aged mice (Takeda et al., 2010). These results suggest that plasma leptin, which increases with age, may induce resistance to ghrelin reactivity via camp downregulation.

3.2. Effect of rikkunshito on hypophagia of aged mice

Recently, we demonstrated that the administration of rikkunshito improves anorexia of aging. In addition, we found that rikkunshito increased the reactivity of ghrelin by inhibiting PDE3 activity. The components of rikkunshito (nobiletin, isoliquiritigenin, and HMF) had inhibitory effects against PDE3 activity (Takeda et al., 2012). These results suggest that dysregulation of ghrelin secretion and ghrelin resistance in the appetite control system occurred in aged mice and that rikkunshito ameliorated aging-associated anorexia via inhibition of PDE3.

3.3. Experimental methods

3.3.1 Animals

Male C57BL/6J mice aged 6 weeks were obtained from Japan SLC, Inc. (Shizuoka, Japan) and were maintained until 75 weeks of age in Tsumura Research Laboratories (Tokyo, Japan). Male B6.V-Lepob/J homozygous, ob/ob and heterozygous, ob/+ mice aged 6 weeks were purchased from Charles River Japan (Tsukuba, Japan). All animals were fed a normal chow diet and had free access to water and food. Mice were housed five animals per cage in plastic cages on a 12-h light (07:00–19:00 h), 12-h dark cycle. All experiments were carried out between 09:00 and 18:00 h.

3.3.2 Food intake

Food intake was measured by subtracting uneaten food from the initially premeasured food 1 and 24 h after administration. Rat ghrelin (Peptide Institute, Inc., Osaka, Japan) was dissolved in physiological saline, and 33 μg/kg was administered intravenously (i.v.) to conscious mice from the tail vain. Cilostamide (PDE3 inhibitor) or LY-294002 (PI3K inhibitor) was dissolved in distilled water and administered orally at 30 mg/kg. Rikkunshito was dissolved in 10 ml of DW and orally administered at 1000 mg/kg.

3.3.3 Extraction of total RNA, reverse transcription, and real-time RT-PCR analysis

Mice were decapitated and their brains and stomachs removed immediately. The hypothalamus and gastric bodies were dissected out, frozen in liquid nitrogen, and kept at −80 °C until they were processed for RNA extraction.

Isolated tissue homogenization and total RNA extraction were performed according to the manufacturer's protocol using an RNeasy universal tissue kit (QIAGEN Co., Valencia, CA). The OD of the total RNA solution was determined using an ND-1000 UV/vis spectrophotometer (Thermo Fisher Scientific, Inc., Waltham, MA), and each sample was diluted to100 ng/μl. The diluted total RNA was incubated at 70 °C for 5 min and then cooled on ice. Total RNA (1000 ng) was reverse-transcribed using TaqMan RT reagents (Applied Biosystems Co., Foster City, CA) according to the manufacturer's protocol. Quantitative PCR assays were performed using the TaqMan universal PCR master mix (Applied Biosystems) on a Prism 7900HT sequence detection system (Applied Biosystems). mRNA expression was normalized to ribosomal protein S29 expression, which served as an endogenous control to correct for differences in the amount of total RNA added to each reaction as expressed by the delta threshold cycle (dCt) value: dCt value $= 2^{(-|A-B|)}$, where A is the number of cycles of the control gene and B is the number of cycles of the target gene. All oligonucleotide primers and fluorogenic probe sets for TaqMan real-time PCR were obtained from Applied Biosystems: ribosomal protein S29, Mm02342448_gH; preproghrelin, Mm00445450_m1; GHS-R, Mm00616415_m1; NPY, Mm00445771_m1; AGRP, Mm00475829_g1; POMC-α, Mm00435874_m1.

3.3.4 Hormone assay

Blood was collected from the caudal vena cava in tubes containing aprotinin and EDTA-2Na. The blood samples were promptly centrifuged at 4 °C, and the plasma was acidified with 1 mol/l HCl (1:10 plasma volume). The ghrelin level was determined using an active ghrelin or des-acyl ghrelin ELISA kit. The leptin and GH levels were determined using the mouse leptin ELISA kit and rat/mouse GH ELISA kit (Millipore Corp., Billerica, MA). The insulin level was measured by Bio-plex using a Milliplex mouse serum adipokine panel (Millipore).

3.3.5 Assay for PDE3 activity

PDE3 activity was measured using 1 μmol cyclic nucleotides as a substrate in a two-step radioassay procedure adapted from Hidaka and Asano (Hidaka and Asano, 1976). The 5′-[^{3}H]AMP formed by PDE was converted to [^{3}H]adenosine by the activation of nucleotidase. The radioactivity of the product that was isolated by cation exchange resin was counted in a liquid scintillation counter. Appropriate dilutions of the enzyme were incubated in 50 m*M* Tris–HCl (pH 7.5), 5 m*M* $MgCl_2$, 50 μg BSA, and 1 μmol [^{3}H] AMP. After 15 min at 25 °C, the reaction was terminated by boiling for 5 min. Next, snake venom was added, and an additional incubation was performed for 20 min at 25 °C. Water was then added, and the mixture was applied to a small ion exchange resin column. After the column was washed with DW, adenosine or guanosine was eluted with NH_4OH. The components of rikkunshito, including hesperetin, naringenin, HMF, nobiletin, tangeretin, isoliquiritigenin, liquiritigenin, glycycoumarin, 8-shogaol, 10-shogaol, 10-dehydrogingerdion, 10-gingerdion, and 8-gingerol, were also subjected to a PDE assay.

3.3.6 Assay for PI3K activity

The test compound was preincubated for 5 min at room temperature with the PI3K enzyme (6.5 ng) in a buffer containing 50 m*M* HEPES/NaOH (pH 7.4), 5 m*M* $MgCl_2$, 1 m*M* dithiothreitol, 40 μ*M* Na_3VO_4, and 0.005% Tween 20. The reaction was then initiated by the addition of 35 n*M* substrate, biotin-KEAKEKRQEQIAKRRRLSSLRASTSKSGGSQK, and 11 μ*M* ATP. The mixture was then incubated for 90 min at room temperature. After incubation, the reaction was stopped by the addition of 33 m*M* EDTA. The fluorescence acceptor (XL665-labeled streptavidin) and the fluorescence donor (antiphospho-S6-rib antibody labeled with europium cryptate) were then added. After 60 min, the fluorescence transfer was measured at $\lambda_{ex}=337$ nm, $\lambda_{em}=620$ and 665 nm using a microplate reader (Zhao et al., 2005).

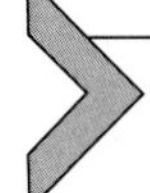

4. STRESS

4.1. Stress and ghrelin

Stress and negative emotions have been associated with both increased and decreased food intake. The mechanism underlying this opposed behavioral responses to similar stressors has not been determined, but high stress levels appear to lead to decreased eating (Stone and Brownell, 1994).

Conflicting data are available regarding the effect of stress on ghrelin secretion (Takeda et al., 2012). In animal studies, elevations in plasma ghrelin have been observed in response to various psychological/environmental stressors, including tail-pinch stress, water-avoidance stress, chronic exposure to cold, repeated restraint stress, and chronic social defeat stress. In contrast, exposure to immune, visceral, or strenuous physical stressors causes reduction of plasma ghrelin level. In humans, acute psychosocial stress or cold exposure increased plasma ghrelin levels. However, there are several reports showing that plasma ghrelin level did not change or even decreased by an exposure to stresses. Collectively, these findings support the idea that acute or severe stress causes a reduction of circulating ghrelin level, resulting in the suppression of appetite, whereas mild or chronic repeated stress causes an upregulation of ghrelin secretion as an adaptation to stress (Takeda et al., 2012). In support of this notion, Lutter et al. (Lutter et al., 1998) found that increased ghrelin levels produced anxiolytic and antidepressant responses in mice, suggesting that increased ghrelin in response to stress protects against depressive reactions to stress and helps them cope with stress.

Corticotropin-releasing factor (CRF) and its family peptides urocortin1 (Ucn1), urocortin2 (Ucn2), and urocortin3 (Ucn3) play an important role in the control of food intake (Richard et al., 2002). Among the CRF family peptides, Ucn1 was shown to have the most potent inhibitory effect on the food intake in mice. Ucn1 has been identified in the brain and has a higher affinity for CRF2 receptors (CRFR2) than for CRF1 receptors (CRFR1), and hence it is believed that CRFR2 plays the major role in satiety.

Activation of CRFR1 in the brain can suppress feeding independently of CRFR2-mediated mechanisms. CRF1 and CRF2 receptor-mediated anorexia appear to exhibit different time courses; in rats, i.c.v. administration of CRFR1 agonists elicited rapid-onset anorexia with short duration, while CRFR2 agonists caused delayed-onset, prolonged anorexia (Richard et al., 2002).

There are several reports showing that administration of Ucn1 to humans and rodents reduces plasma ghrelin concentrations (Tanaka et al., 2009; Yakabi et al., 2011). In addition, Ucn1-induced reduction of plasma ghrelin and food intake was restored by CRFR2 but not CRFR1 (Yakabi et al., 2011). However, much less information is available on the relationship between ghrelin and CRFR1.

The novelty stress model, where animals are removed from their home cage and placed somewhere they have never been before, has been used to estimate the levels of anxiety and depression (Fone et al., 1996; Miura et al., 2002). Using this stress model, we found that 3 h after the novelty stress, appetite reduction was associated with decreased plasma ghrelin level, reduced levels of NPY/agouti-related peptide mRNA, and increased levels of pro-opiomelanocortin mRNA in the hypothalamus (Saegusa et al., 2011). Administering a CRFR1 selective antagonist, but not a

CRFR2 antagonist, resolved the reduction in food intake 3 h after the novelty stress by enhancing circulating ghrelin concentrations. Interestingly, 5-HT1B/2CR antagonist and melanocortin-4 receptor (MC4R) antagonist alleviated the novelty-stress-induced hypophagia and the reduction in the circulating ghrelin level (Saegusa et al., 2011). We hypothesized that the acute appetite suppression due to CRFR1 activation after a novelty stress is caused by a chain reaction of appetite control mechanisms mediated by 5-HT1B/2CR in ARC to MC4R system in paraventricular nucleus, causing lowered peripheral ghrelin secretion.

4.2. Effect of rikkunshito on the novelty stress model

Oral administration of rikkunshito inhibited the reduction of food intake at 1 and 3 h in mice exposed to the novelty stress, and coadministration of the ghrelin receptor antagonist [D-Lys^{3}]-GHRP-6 with rikkunshito abolished this effect (Saegusa et al., 2011). Rikkunshito also increased plasma acyl-ghrelin concentrations at 1 and 3 h after the novelty stress, suggesting that blocking the decrease in endogenous peripheral ghrelin in mice exposed to the novelty stress also acts to sustain the feeding behavior. We found that the oral administration of glycycoumarin and isoliquiritigenin inhibited the reduction in food intake in mice exposed to novelty stress (Saegusa et al., 2011). We have previously shown that glycycoumarin and isoliquiritigenin potently inhibit 5-HT2C receptor ligand binding and that orally administering rikkunshito abolishes the decrease in food intake in mCPP-treated rats (Takeda et al., 2008). These findings support the notion that rikkunshito improved hypophagia and decreased plasma ghrelin levels via 5-HT2C receptor antagonism-like action in mice exposed to the novelty stress.

4.3. Experimental methods

4.3.1 Animals

Male C57BL/6J mice aged 6 weeks (Japan Charles River, Tokyo, Japan) were used. Before the experiment, five mice per cage were maintained in a room with controlled temperature and humidity under a 07:00–19:00 h light cycle with free access to food and water. For the novelty stress, each mouse was transferred from group-housed cages to individual cages. Control mice were housed in individual cages for 7 days before the experiment was initiated. Mice in each group were handled in the same way. All experiments were performed between 09:00 and 18:00 h.

4.3.2 Intracerebroventricular infusion

A 26-gauge stainless steel indwelling cannula was implanted 2.6 mm below the skull surface into the lateral ventricle (1.1 mm lateral to the bregma) of the mice. Injections were performed using a 26-gauge stainless steel injector attached to PE-10 tubing fitted to a 10-μl microsyringe.

4.3.3 Food intake

All protocols were performed under a 24-h-fasted condition. The time course evaluation of the effect of the novelty stress on food intake in 24-h-fasted mice was undertaken 1, 3, and 6 h after the novelty stress and was calculated as the difference between the food weights before and after the feeding period at each time interval. Subsequently, to clarify the role of the CRFR or 5-HT1B/2CR on food intake in control or stressed mice, we investigated the effects of the icv administration of the CRFR1 antagonist NBI27914 (10 μg/mouse), the CRFR2 antagonist astressin2B (10 μg/mouse), the 5-HT2CR antagonist SB-242084 (0.5 or 1.5 μg/mouse), or the 5-HT1BR antagonist SB-224289 (0.5 or 1.5 μg/mouse) on the novelty stress-induced decrease in food intake during a 24-h-fasted condition. Each drug was immediately administered after the novelty stress, and subsequently, each mouse was placed in a single housing cage with access to preweighed mouse chow from the group housing cage.

In another experiment, icv administration of CRF, a CRFR1 agonist, was performed in 24-h-fasted mice (0.1, 0.5, and 1.0 μg/mouse). In addition, the 5-HT1B/2CR agonist mCPP (2, 10, and 50 μg/mouse) in 24-h-fasted mice was administered (icv) 3 h after the novelty stress. This is because icv administration of the 5-HT1B/2CR antagonist significantly suppressed decreased food intake for 3 h after the novelty stress was introduced in the present results. CRF (0.1 μg/mouse) was administered icv to mice 15 min before the mCPP (10 μg/mouse icv), and food intake was evaluated 3 h after the mCPP administration. Administration (icv) of the MC4R antagonist HS014 (0.15 μg/mouse) was conducted immediately in 24-h-fasted mice after the novelty stress.

Rat ghrelin (3.3 and 33 μg/mouse) was administered iv in 24-h-fasted mice, and food intake was measured to clarify the contribution of peripheral ghrelin. Rikkunshito (250 or 500 mg/kg) was administered orally in 24-h-fasted mice 1 h before the novelty stress, in combination with the GHS-R antagonist [D-Lys^3]-GHRP-6 (0.12 mg/mouse i.v.) or saline. In another experiment, two components contained in rikkunshito, glycycoumarin (4 mg/kg po) and isoliquiritigenin (4 mg/kg p.o.), which have 5-HT2CR antagonistic-like activity, were administered.

4.3.4 Determining plasma levels of ghrelin or serum levels of corticosterone

Blood was collected from mice by ether anesthesia 0.5, 1, 3, and 6 h after the novelty stress to investigate changes in plasma corticosterone during the novelty stress. The blood collection to determine plasma ghrelin was performed between 10 a.m. and 12 p.m.. We first investigated the effects of NBI27914 (10 μg/mouse), astressin2B (10 μg/mouse), SB-242084 (1.5 μg/mouse), SB-224289 (1.5 μg/mouse), and HS014 (0.15 μg/mouse) on plasma ghrelin concentration in mice exposed to the novelty stress to clarify the role of 5-HT1B/2CR or MC4R on ghrelin secretion 3 h after the novelty stress. Each test drug was administered, the mice were isolated simultaneously, and blood was collected 3 h after the novelty stress. The results of our evaluation of the post-novelty stress time course revealed that plasma ghrelin decreased significantly after 3 h. We collected blood samples 3 h after administration to clarify the relationship between this change in plasma ghrelin level and improved food intake.

CRF (0.2 μg/mouse) or mCPP (50 μg/mouse) was administered alone, CRF was coadministered with SB-242084 (1.5 μg/mouse) or SB-224289 (1.5 μg/mouse), and blood was collected 1 h later. To clarify MC4R for the decrease in food intake after the novelty stress, icv administration of THIQ (1.5 μg/mouse), an MC4R agonist, or α-MSH (1.5 μg/mouse), an MC3R agonist, was performed, and blood was collected 1 h later. RKT (500 mg/kg) was administered orally 1 h before the novelty stress, and blood was collected 1 or 3 h after the stress.

To determine the serum corticosterone concentration after the novelty stress, blood was collected in a laboratory dish by severing the carotid artery. Ghrelin levels were measured using the Active Ghrelin ELISA kit/Desacyl-Ghrelin ELISA kit (Mitsubishi Chemical Medience), and the corticosterone levels were measured using the Correlate-EIA Corticosterone kit (Enzo Biochem, New York, NY).

4.3.5 Extraction of total RNA for RT-PCR

The hypothalamus and stomach were rapidly removed and immediately frozen by placing them in a tube on dry ice. Homogenization of the isolated tissue and total RNA extraction were performed according to the protocol from the RNeasy Universal Tissue Kit (QIAGEN), following which each sample was diluted to 100 ng/μl. The diluted total RNA was incubated at 70 °C for 5 min and then cooled on ice. Total RNA (1000 ng) was reverse-transcribed using TaqMan Reverse Transcription Reagents

(Applied Biosystems) according to the manufacturer's protocol. Quantitative PCR assays were performed using TaqMan Universal PCR Master Mix (Applied Biosystems), using a Prism 7900HT Sequence Detection System (Applied Biosystems). To correct the differences in the amount of total RNA added to each reaction, mRNA expression was normalized using ribosomal protein S29 (RPS29) as an endogenous control. These differences were expressed by the dCT value: $\mathrm{dCT} = 2^{(-|A-B|)}$, where A is the number of cycles needed to reach the threshold for the housekeeping gene (CT: threshold cycle) and B is the number of cycles needed for the target gene. All oligonucleotide primers and fluorogenic probe sets for TaqMan real-time PCR were manufactured by Applied Biosystems (Rps29: Mm02342448_gH; Npy: Mm00445771_m1; Agrp: Mm00475829_g1; Pomc: Mm00435874_m1; ghrelin: Mm00445450_m1).

5. GHRELIN-DEGRADING ENZYME

5.1. Ghrelin-degrading enzyme and rikkunshito

The acylated-ghrelin level in plasma is regulated by both the secretion from stomach and the elimination from circulation which includes degradation of acylghrelin by deacylating enzymes. Acylated ghrelin in plasma is rapidly deacylated in a process that is believed to be primarily mediated by carboxylesterase (CES) in rodents and butyrylcholinesterase (BuChE) in humans (De Vriese et al., 2004).

Rikkunshito administration has been shown to stimulate food intake and peripheral and central ghrelin secretion in rodents (Matsumura et al., 2010; Saegusa et al., 2011; Takeda et al., 2008; Yakabi et al., 2010a) and humans (Matsumura et al., 2010). Interestingly, it was shown that oral administration of rikkunshito enhanced circulating acyl-ghrelin level without a significant effect on the plasma level of des-acyl ghrelin in both normal-fed and cisplatin-treated rats, leading to the increase in the acyl- to des-acyl-ghrelin (A/D) ratio (Takeda et al., 2008). Furthermore, in a human study, it was demonstrated that plasma acyl-ghrelin level and A/D ratio increased significantly after rikkunshito administration, whereas des-acyl-ghrelin level showed a decreasing trend (Matsumura et al., 2010). These results raise the possibility that rikkunshito increases circulating acyl-ghrelin level by inhibiting its degradation.

To test the hypothesis that some components of rikkunshito contribute to the inhibition of ghrelin degradation, 48 components in rikkunshito were screened for inhibitory activity against ghrelin-degrading enzymes such as

CES and BuChE (Sadakane et al., 2011). It was found that eight compounds exhibited inhibitory activity against CES or BuChE. Among them, 10-gingerol exhibited the highest inhibitory activity against CES (5.2 μ*M* inhibition constant) in a competitive manner. In addition, pachymic acid and glycycoumarin were shown to be competitive inhibitors of BuChE. Furthermore, rikkunshito and its component 10-gingerol inhibited the decrease in plasma acyl-ghrelin level of exogenously administered ghrelin, and the CES inhibitor BNPP inhibited cisplatin-induced decreases in food intake (Sadakane et al., 2011). On the basis of these findings, it is conceivable that rikkunshito may enhance plasma acyl-ghrelin level, at least in part, by inhibiting the circulating ghrelin-degrading enzyme.

5.2. Experimental methods

5.2.1 Animals

During the experimental period, 7-week-old male Sprague–Dawley rats (Charles River Japan, Yokohama, Japan) were housed in each cage in a room with controlled temperature and humidity under a 12-h light cycle (07:00–19:00 h) with *ad libitum* access to food and water. All experiments were performed between 09:00 and 18:00 h.

5.2.2 Food intake

The rats were administered saline or cisplatin 1 h after administration of BNPP (20 mg/kg) to 24-h-fasted rats. Food intake was defined as the difference between the weight of food before examination and that recovered subsequent to the test session at 2 and 24 h after cisplatin administration. Food spilled from the food cage was collected, combined with the remaining food, and added to the total weight.

5.2.3 Enzyme assay

CES activity was determined by measuring the hydrolysis of α-naphthyl acetate (Duysen et al., 2001). After preincubating sample solutions for 20 min with 0.01 U CES (Sigma–Aldrich) in 100 m*M* phosphate buffer (pH 7.0), 10 μl of 20 m*M* α-naphthyl acetate was added. Absorbance was measured at 321 nm every 15 s for 10 min. The rate of the absorbance increase (Δ321 nm/s/ml) was calculated as the amount of enzyme hydrolysis activity. BuChE activity was measured by the method developed by Ellman et al. (1961). Briefly, after preincubating the samples with purified human BuChE (0.01 U, Sigma–Aldrich) in 50 m*M* Tris–HCl (pH 7.4), 100 n*M* butyrylthiocholine iodide and 0.25 m*M* 5,5′-dithiobis-2-nitrobenzoate in

Tris buffer were added. The absorbance was read at 405 nm every 15 s for 10 min. The rate of the absorbance increase (Δ405 nm/s/ml) was calculated as the amount of enzyme hydrolysis activity.

REFERENCES

Ariyasu, H., et al., 2008. Efficacy of ghrelin as a therapeutic approach for age-related physiological changes. Endocrinology 149, 3722–3728.

Barazzoni, R., et al., 2003. Hyperleptinemia prevents increased plasma ghrelin concentration during short-term moderate caloric restriction in rats. Gastroenterology 124, 1188–1192.

Chen, C.Y., et al., 2009. Ghrelin gene products and the regulation of food intake and gut motility. Pharmacol. Rev. 61, 430–481.

Cummings, D.E., et al., 2001. A preprandial rise in plasma ghrelin levels suggests a role in meal initiation in humans. Diabetes 50, 1714–1719.

Cummings, D.E., et al., 2002. Plasma ghrelin levels after diet-induced weight loss or gastric bypass surgery. N. Engl. J. Med. 346, 1623–1630.

De Vriese, C., et al., 2004. Ghrelin degradation by serum and tissue homogenates: identification of the cleavage sites. Endocrinology 145, 4997–5005.

Di Francesco, V., et al., 2007. The anorexia of aging. Dig. Dis. 25, 129–137.

Duysen, E.G., et al., 2001. Evidence for nonacetylcholinesterase targets of organophosphorus nerve agent: supersensitivity of acetylcholinesterase knockout mouse to VX lethality. J. Pharmacol. Exp. Ther. 299, 528–535.

Ellman, G.L., et al., 1961. A new and rapid colorimetric determination of acetylcholinesterase activity. Biochem. Pharmacol. 7, 88–95.

Fone, K.C., et al., 1996. Increased 5-HT2C receptor responsiveness occurs on rearing rats in social isolation. Psychopharmacology (Berl) 123, 346–352.

Fujitsuka, N., et al., 2011. Potentiation of ghrelin signaling attenuates cancer anorexia–cachexia and prolongs survival. Transl. Psychiatry 1, e23.

Hattori, T., 2010. Rikkunshito and ghrelin. Int. J. Pept. 2010, pii: 283549.

Hidaka, H., Asano, T., 1976. Human blood platelet 3′: 5′-cyclic nucleotide phosphodiesterase. Isolation of low-Km and high-Km phosphodiesterase. Biochim. Biophys. Acta 429, 485–497.

Hiura, Y., et al., 2011. Fall in plasma ghrelin concentrations after cisplatin-based chemotherapy in esophageal cancer patients. Int. J. Clin. Oncol. http://dx.doi.org/10.1007/s10147-011-0289-0.

Kohno, D., et al., 2007. Leptin suppresses ghrelin-induced activation of neuropeptide Y neurons in the arcuate nucleus via phosphatidylinositol 3-kinase-andphosphodiesterase 3-mediated pathway. Endocrinology 148, 2251–2263.

Kojima, M., Kangawa, K., 2005. Ghrelin: structure and function. Physiol. Rev. 85, 495–522.

Kojima, M., et al., 1999. Ghrelin is a growth-hormone-releasing acylated peptide from stomach. Nature 402, 656–660.

Leite-Moreira, A.F., Soares, J.B., 2007. Physiological, pathological and potential therapeutic roles of ghrelin. Drug Discov. Today 12, 276–288.

Liu, Y.L., et al., 2006. Ghrelin alleviates cancer chemotherapy-associated dyspepsia in rodents. Cancer Chemother. Pharmacol. 58, 326–333.

Lutter, M., et al., 1998. The orexigenic hormone ghrelin defends against depressive symptoms of chronic stress. Nat. Neurosci. 11, 752–753.

Matsumura, T., et al., 2010. The traditional Japanese medicine Rikkunshito increases the plasma level of ghrelin in humans and mice. J. Gastroenterol. 45, 300–307.

Miura, H., et al., 2002. Influence of aging and social isolation on changes in brain monoamine turnover and biosynthesis of rats elicited by novelty stress. Synapse 46, 116–124.

Mochiki, E., et al., 2010. The effect of traditional Japanese medicine (Kampo) on gastrointestinal function. Surg. Today 40, 1105–1111.
Morley, J.E., 2001. Anorexia, sarcopenia, and aging. Nutrition 17, 660–663.
Nakazato, M., et al., 2001. A role for ghrelin in the central regulation of feeding. Nature 409, 194–198.
Niswender, K.D., et al., 2001. Intracellular signalling. Key enzyme in leptininduced anorexia. Nature 413, 794–795.
Ohno, T., et al., 2011. Rikkunshito, a traditional Japanese medicine, suppresses cisplatin-induced anorexia in humans. Clin. Exp. Gastroenterol. 4, 291–296.
Richard, D., et al., 2002. The corticotropin-releasing factor family of peptides and CRF receptors: their roles in the regulation of energy balance. Eur. J. Pharmacol. 440, 189–197.
Rudd, J.A., et al., 2006. Anti-emetic activity of ghrelin in ferrets exposed to the cytotoxic anti-cancer agent cisplatin. Neurosci. Lett. 392, 79–83.
Sadakane, C., et al., 2011. 10-Gingerol, a component of rikkunshito, improves cisplatin-induced anorexia by inhibiting acylated ghrelin degradation. Biochem. Biophys. Res. Commun. 412, 506–511.
Saegusa, Y., et al., 2011. Decreased plasma ghrelin contributes to anorexia following novelty stress. Am. J. Physiol. Endocrinol. Metab. 301, E685–E696.
Shimizu, Y., et al., 2003. Increased plasma ghrelin level in lung cancer cachexia. Clin. Cancer Res. 9, 774–778.
Stone, A., Brownell, K., 1994. The stress-eating paradox: multiple daily measurements in adult males and females. Psychol. Health 9, 425–436.
Suzuki, H., et al., 2009. Japanese herbal medicine in functional gastrointestinal disorders. Neurogastroenterol. Motil. 21, 688–696.
Takeda, H., et al., 2008. Rikkunshito, an herbal medicine, suppresses cisplatin-induced anorexia in rats via 5-HT2 receptor antagonism. Gastroenterology 134, 2004–2013.
Takeda, H., et al., 2010. Rikkunshito ameliorates the aging-associated decrease in ghrelin receptor reactivity via phosphodiesterase III inhibition. Endocrinology 151, 244–252.
Takeda, H., et al., 2012. Rikkunshito and ghrelin secretion. Curr. Pharm. Des. (in press).
Tanaka, C., et al., 2009. Comparison of the anorexigenic activity of CRF family peptides. Biochem. Biophys. Res. Commun. 390, 887–891.
Tschop, M., et al., 2000. Ghrelin induces adiposity in rodents. Nature 407, 908–913.
Wolf, W.A., Schuts, J.S., 1997. The serotonin 5-HT2C receptor is a prominent serotonin receptor in basal ganglia; evidence from functional studies on serotonin-mediated phosphoinositide hydrolysis. J. Neurochem. 69, 1449–1458.
Yakabi, K., et al., 2010a. Reduced ghrelin secretion in the hypothalamus of rats due to cisplatin-induced anorexia. Endocrinology 151, 3773–3782.
Yakabi, K., et al., 2010b. Rikkunshito and 5-HT2C receptor antagonist improve cisplatin-induced anorexia via hypothalamic ghrelin interaction. Regul. Pept. 161, 97–105.
Yakabi, K., et al., 2011. Urocortin 1 reduces food intake and ghrelin secretion via CRF(2) receptors. Am. J. Physiol. Endocrinol. Metab. 301, E72–E82.
Zhao, A.Z., et al., 2002. A phosphatidylinositol 3-kinase phosphodiesterase 3B-cyclic AMP pathway in hypothalamic action of leptin on feeding. Nat. Neurosci. 5, 727–728.
Zhao, J.J., et al., 2005. The oncogenic properties of mutant p110α and p110β phosphatidylinositol 3-kinases in human mammary epithelial cells. Proc. Natl. Acad. Sci. U.S.A. 102, 18443–18448.

SECTION 6

Tg and KO Mice of Ghrelin

CHAPTER TWENTY-TWO

Thermogenic Characterization of Ghrelin Receptor Null Mice

Ligen Lin, Yuxiang Sun[1]

Department of Pediatrics, USDA/ARS Children's Nutrition Research Center, Baylor College of Medicine, Houston, Texas, USA

[1]Corresponding author: e-mail address: yuxiangs@bcm.edu

Contents

Abstract

Ghrelin is the only known circulating orexigenic hormone that increases food intake and promotes adiposity, and these physiological functions of ghrelin are mediated through its receptor growth hormone secretagogue receptor (GHS-R). Ghrelin/GHS-R signaling plays a crucial role in energy homeostasis. Old GHS-R null mice exhibit a healthy phenotype—lean and insulin sensitive. Interestingly, the GHS-R null mice have

Methods in Enzymology, Volume 514
ISSN 0076-6879
http://dx.doi.org/10.1016/B978-0-12-381272-8.00022-2

increased energy expenditure, yet exhibit no difference in food intake or locomotor activity compared to wild-type mice. We have found that GHS-R is expressed in brown adipose tissue (BAT) of old mice. Ablation of GHS-R attenuates age-associated decline in thermogenesis, exhibiting a higher core body temperature. Indeed, the BAT of old GHS-R null mice reveals enhanced thermogenic capacity, which is consistent with the gene expression profile of increases in glucose/lipid uptake, lipogenesis, and lipolysis in BAT. The data collectively suggest that ghrelin/GHS-R signaling has important roles in thermogenesis. The recent discovery that BAT also regulates energy homeostasis in adult humans makes the BAT a new antiobesity target. Understanding the roles and molecular mechanisms of ghrelin/GHS-R in thermogenesis is of great significance. GHS-R antagonists might be a novel means of combating obesity by shifting adiposity balance from obesogenesis to thermogenesis.

1. INTRODUCTION

There are two types of adipose tissues that control whole-body energy metabolism: energy-storing white adipose tissue (WAT) and energy-burning brown adipose tissue (BAT). In contrast to WAT, BAT consists of small adipocytes containing reduced amounts of triglyceride (in multilocular lipid droplets) and has a high density of mitochondria (Cannon and Nedergaard, 2004). BAT activity correlates positively with energy expenditure (EE) and negatively with fat mass (Cannon and Nedergaard, 2004; Nedergaard and Cannon, 2010). BAT is a key organ of nonshivering thermogenesis, which plays an important role in EE in rodents and in human neonates; thermogenesis is responsible for more than half the total oxygen consumption in small animals (Cannon and Nedergaard, 2004). Nonshivering thermogenesis can be regulated directly in BAT, and/or through activation of central sympathetic nerve activity. In rodents, cold exposure or diet causes release of norepinephrine at sympathetic nerve endings in the BAT, which in turn stimulates BAT proliferation and activation (Cannon and Nedergaard, 2004). BAT activity can be also activated by β-adrenergic activators such as catecholamines (Bartness and Song, 2007; Cannon and Nedergaard, 2004). BAT thermogenesis is dependent on the β-adrenergically mediated activation of lipolysis and subsequent degradation of fatty acids, via uncoupling protein 1 (UCP1), to produce heat (Inokuma et al., 2005; Lowell and Spiegelman, 2000). Recent evidence has shown that BAT is also present in adult humans; further, thermogenic function is severely impaired in obese individuals and the elderly (Frontini and Cinti, 2010; Nedergaard and Cannon, 2010;

Pfannenberg et al., 2010). Interventions that activate thermogenesis may represent an attractive, novel option for combating obesity.

Ghrelin is the only circulating orexigenic hormone known to stimulate appetite and promote obesity (Cowley et al., 2003; Kojima and Kangawa, 2005; Shimbara et al., 2004; Sun et al., 2004; Tschop et al., 2000). We and others have shown that ghrelin's effect on appetite and adiposity is mediated through the activation of growth hormone secretagogue receptor (GHS-R) (Davies et al., 2009; Sun et al., 2004). Ghrelin is ubiquitously expressed in all tissues, with the highest expression present in the stomach and intestines (Gnanapavan et al., 2002). In contrast, the expression of GHS-R is much more limited. GHS-R is primarily expressed in hypothalamus and pituitary; lower levels of expression have been detected only in some peripheral tissues (Gnanapavan et al., 2002; Sun et al., 2007). We and others have shown that GHS-R knockout mice ($Ghsr^{-/-}$) have modestly reduced body weight and adiposity, but have a normal appetite and activity (Longo et al., 2008; Mano-Otagiri et al., 2010; Sun et al., 2004; Zigman et al., 2005). While these studies support the notion that the ghrelin receptor plays an important role in energy homeostasis, it is still unknown what metabolic changes result in the lean phenotype of $Ghsr^{-/-}$ mice. GHS-R is not detectable in BAT of young mice (Sun et al., 2007). Surprisingly, we have recently found that GHS-R is detectable in BAT of old mice (Lin et al., 2011), and others have found GHS-R is expressed in BAT of old rats (Davies et al., 2009). This unique expression pattern of GHS-R suggests that GHS-R may play a role in thermogenic impairment during aging. It has been shown that ghrelin, given pharmacologically, suppresses the release of norepinephrine in BAT of rats (Mano-Otagiri et al., 2009; Yasuda et al., 2003). To elucidate whether GHS-R regulates thermogenesis in BAT during aging, we studied thermogenesis of old wild-type (WT) and $Ghsr^{-/-}$ mice. Old $Ghsr^{-/-}$ mice have increased EE, but no change in food intake or activity; this suggests that the lean phenotype of the mice may be the result of increased thermogenesis. Indeed, while thermogenic capacity in BAT is decreased during aging, the ablation of GHS-R attenuates this age-associated decline of thermogenesis, resulting in higher core temperature (Lin et al., 2011). BAT of old $Ghsr^{-/-}$ mice shows enhanced thermogenic capacity, consistent with gene expression profile of increased glucose/lipid uptake, lipogenesis, and lipolysis in BAT. The data collectively suggest that ghrelin signaling plays an important role in thermogenesis, and GHS-R is a key regulator of thermogenic function.

GHS-R antagonists may represent a paradigm-shifting new antiobesity drug by activating thermogenesis.

2. *IN VIVO* METABOLIC AND THERMOGENIC CHARACTERIZATIONS

Metabolic parameters of WT and *Ghsr*$^{-/-}$ mice have been obtained using an Oxymax open-circuit indirect calorimetry system (Columbus Instruments, Columbus, OH), as previously described (Lin et al., 2011; Ma et al., 2011). The telemetric thermometry recoding has been carried out as described (Rudaya et al., 2005).

2.1. EE evaluation by indirect calorimetry

1. To reduce the impact of stress on metabolic profiles, mice are individually caged in metabolic chambers, and given free access to regular chow and water for 1 week prior to the calorimetry tests.
2. Before the calorimetry tests, measurements of body weight and analysis of body composition (fat and lean mass) are carried out using an Echo MRI-100 analyzer (Echo Medical Systems, Houston, TX).
3. Mice are caged individually in the metabolic chambers. The oxygen consumption (VO_2), carbon dioxide production (VCO_2), locomotor activity, and food intake are monitored for a 72-h period. The first 24 h are considered the acclimation phase, and the data of second and third 24 h are analyzed.
4. After the calorimetry test, body weight is remeasured and body composition is reanalyzed.
5. Respiratory exchange ratio (RER) is calculated as $RER = VCO_2/VO_2$. EE (or heat) is calculated as $EE = (3.815 + 1.232 \times RER) \times VO_2$ (Obici et al., 2002). EE is then normalized by body weight (mean body weights, before and after the tests) or lean body mass (mean lean mass, before and after the tests). Resting metabolic rate is determined by selecting the three lowest points of the EE curve during each light cycle, as previously described (Nuotio-Antar et al., 2007).
6. The activity of each mouse is measured using the break counts of infrared beams. The locomotor activity is measured on the *x*- and *z*-axes during the recording period. Horizontal activity (X_{tot}) = every time a beam (new or same) is broken in the *x*-axis; total count is accumulated as X_{tot}. Vertical activity (Z_{tot}) = every time a beam is broken in the *z*-axis, and

total count is accumulated as Z_{tot}. Total activity is calculated as $X_{tot} + Z_{tot}$.

7. The system measures food intake in real time. This can be used to evaluate food intake at different periods of time (e.g., dark and light periods) and meal sizes for each feeding period.

2.2. Core body temperature measurement by rectal probe thermometer or telemetric thermometer

1. Rectal telemetry: Rectal temperature of the mice is measured using a TH-8 temperature monitor system (Physitemp, Clifton, NJ). The probe is lubricated with petroleum jelly and then gently inserted into the rectum of the mice about 2 cm deep. Once temperature reading is stable, the temperature is recorded and the probe is removed. Thermogenic responses are often enhanced when mice are exposed to cold temperature. When rectal telemetry is carried out on mice kept in a cold room (4 °C), the rectal temperature should be monitored hourly, and the mice that experience hypothermia should be returned to normal housing temperature for recovery.
2. Telemetric thermometry: If the rectal temperature is not sensitive enough, a temperature transmitter system (Mini-Mitter Company, Bend, OR) can be used to record the core body temperature of freely moving mice. To implant a temperature transmitter system, the mouse is first anesthetized. Then, a miniature temperature transmitter (4000 E-mitter) is surgically implanted into the peritoneal cavity. The mice are allowed to recover in warm (30 °C) housing temperature for 3 days and then subjected to telemetric thermometry testing. The temperature of the mice is captured by an ER-4000 Energizer-Receiving system, which is positioned inside a climatic chamber and connected to a computer. The mouse is kept in an enclosed metabolic chamber and placed directly on top of the receiver.

3. HORMONAL CHARACTERIZATION: THYROID HORMONES AND CATECHOLAMINES

Nonshivering thermogenesis can be obligatory or facultative (adaptive). Obligatory thermogenesis is the heat production automatically caused by basal metabolism, which is largely determined by the basal metabolic rate and regulated by thyroid hormones (T3 and T4) (Cannon and Nedergaard, 2004). Facultative thermogenesis is activated in response to environmental changes (cold exposure or diet) and is primarily mediated by β3-adrenoreceptors and regulated by sympathetic nervous system and catecholamine (norepinephrine) (Argyropoulos and Harper, 2002; van Marken Lichtenbelt and Schrauwen, 2011).

3.1. Serum thyroid hormone assays

Enzyme-linked immunosorbent assay (ELISA) has been used to measure mice plasma total T3 (Mouse/Rat Triiodothyronine ELISA, Calbiotech, Spring Valley, CA) and T4 (Mouse/Rat Thyroxine ELISA, Calbiotech, Spring Valley, CA) (Lin et al., 2011).

1. Collect 100 μl blood specimens from each mouse under either fed or overnight fasting conditions. Spin down the blood in EDTA-coated tubes, and transfer plasma to new tubes. Refrigerate at -20 °C immediately.
2. Pipet 25 μl (for T3) or 10 μl (for T4) of samples into the designed wells for serum reference, control, or plasma specimen, respectively.
3. Add 100 μl of T3-enzyme conjugate solution or T4-enzyme conjugate solution to all wells; incubate with shaking for 60 min at room temperature.
4. Remove liquid from all wells. Wash wells three times with 300 μl of 1 × wash buffer. Drain the excess liquid with paper towels.
5. Add 100 μl of TMB substrate solution and incubate with shaking at room temperature for 15 min.
6. Add 50 μl of stop solution and gently mix for 15–20 s and read the absorbance on an ELISA reader (PerkinElmer Life Sciences, Waltham, MA) at 450 nm within 15 min after adding the stop solution.
7. To construct the standard curve, plot the absorbance for T3/T4 standards (vertical axis) versus T3/T4 standard concentrations (horizontal axis) in a linear graph. Read the absorbance of controls from the curve to validate the accuracy of the assay. Then the absorbances of all unknown samples are plotted on the standard curve to determine the concentrations of T3 and T4, respectively.

3.2. Urinary catecholamine assays

It has been shown that urinary catecholamine levels are less stress prone than those in plasma (Lee et al., 2010). The urine epinephrine and norepinephrine levels of the mice have been determined by ELISA (IBL Inc., Minneapolis, MN). The urine creatinine level (Quidel Corporation, San Diego, CA) is used as an internal control.

1. Collect urine samples between 08:00 a.m. and 10:00 a.m. to minimize the influence of circadian rhythm.
2. Grab a mouse, and let it urinate voluntarily on a foil. Pipet 200 μl urine in a tube, with 2 μl of 6N HCl as a preservative for epinephrine and norepinephrine assays. Pipet 10 μl urine into another tube for creatinine assay. Save all samples in a refrigerator (−20 °C). If a mouse does not urinate right after the grab, urine can be collected on a second day. It is important to avoid repeated handling because it will stress the mice and affect their catecholamine levels.
3. Measure epinephrine, norepinephrine, and creatinine concentrations with ELISA assays, following manufacturer's instructions. Creatinine-normalized epinephrine/norepinephrine levels are presented as catecholamine levels.

3.3. Microdialysis of norepinephrine in BAT

Mano-Otagiri and colleagues have directly measured norepinephrine levels in BAT of rats, using the microdialysis method (Mano-Otagiri et al., 2009).

1. Four days before the experiment, with the rats under anesthesia (50 mg/kg body weight of sodium pentobarbital), insert an i.v. catheter into the right jugular vein.
2. Anesthetize rat on the day of the experiment. Shave the skin in the interscapular area, and clean with 70% ethanol. Cut a small incision along the midline, exposing a deposit of white fat and the underlying interscapular brown fat.
3. Insert a microdialysis probe (OP-100-05; EicomCorp., Kyoto, Japan) with 5-mm dialyzable membrane into either lobe of the brown fat and close the skin incision with sutures. The end of the probe is exteriorized through a midscapular skin incision.
4. After recovery from anesthesia, connect the probe with tubing for microdialysis and perform microdialysis under free-moving conditions. Perfuse the probe continuously with Ringer's solution (147 m*M* NaCl, 4 m*M* KCl, and 2.3 m*M* $CaCl_2$, pH 7.0) at a flow rate of 2 μl/min, and collect the dialysate every 20 min.
5. Determine norepinephrine concentrations in dialysates by a combination of HPLC and electrochemical detection, using an Eicompak CA-5ODS column (2.1 mm i.d. × 150 mm; Eicom Corp.) and aWE-3G graphite electrode (Eicom Corp.) set at +450 mV against an Ag/AgCl reference electrode. The current sensitivity is 0.1 nA. The mobile phase in the

HPLC column is 0.1 *M* sodium phosphate buffer (pH 6.0) containing 1.85 m*M* sodium octanesulfonic acid, 0.17 m*M* EDTA, and 5.0% (v/v) methanol.
6. After a 3-h stabilization period, the average release of norepinephrine in three consecutive fractions is defined as baseline norepinephrine levels.
7. Give an i.v. injection of 30 nmol of ghrelin to rats. Collect dialysates 2 h after injection, and measure norepinephrine levels using HPLC.

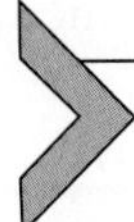

4. GENE/PROTEIN EXPRESSION PROFILES AND *EX VIVO* LIPOLYSIS OF BAT

The BAT of old *Ghsr*$^{-/-}$ mice reveals enhanced thermogenic capacity (increased UCP1 and PPAR-γ coactivator 1α (PGC1α) expression), as well as a gene expression profile of increased glucose/lipid uptake, lipogenesis, and lipogenesis (Lin et al., 2011), supporting higher substrate reserves. It is possible that GHS-R ablation increases lipolysis of BAT, thus promoting heat production by BAT.

4.1. Expression profiles of thermogenic and glucose/lipid metabolic genes

To characterize the expression profiles of thermogenic (Table 22.1) and glucose/lipid metabolic (Table 22.2) genes, total RNA is isolated from the BAT of old WT and *Ghsr*$^{-/-}$ mice, and real-time RT-PCR is used to quantitate expression levels of various genes involved in thermogenic and metabolic regulations (Lin et al., 2011). All primer and probe information is listed in Tables 22.1 and 22.2.

Table 22.1 Thermogenic and adrenergic receptor genes in BAT

Transcripts	Probe/primers
Thermogenic genes	
UCP1[a]	Applied Biosystems (Mm00494070_m1)
PGC-1α[a]	Applied Biosystems (Mm00447183_m1)
Lipid recycling	
β3-AR	Forward: TGC CAA CTC TGC CTT CAA CCC GCT C
	Reverse: CGC TCA CCT TCA TAG CCA TCA AAC C

[a]Using the ABI TaqMan Master Mix.
UCP1, uncoupling protein 1; PGC1α, PPAR-γ coactivator 1α; β3-AR, β3 adrenergic receptor.

Table 22.2 Glucose and lipid metabolic genes in BAT

Transcripts	**Probe/primers**
Adipogenic	
PPAR-γ2	Forward: GCC TAT GAG CAC TTC ACA AGA AAT T
	Reverse: TGC GAG TGG TCT TCC ATC AC
C/EBPα[a]	Applied Biosystems (Mm00514283_s1)
Glucose/lipid uptake	
GLUT4	Forward: GCC TTG GGA ACA CTC AAC CA
	Reverse: CAC CTG GGC AAC CAG AAT G
Lipoprotein lipase	Forward: GGC CAG ATT CAT CAA CTG GAT
	Reverse: GCT CCA AGG CTG TAC CCT AAG
CD36	Forward: CCT GCA AAT GTC AGA GGA AA
	Reverse: GCG ACA TGA TTA ATG GCA CA
Lipogenic	
aP2	Forward: AGT GAA AAC TTC GAT GAT TAC ATG AA
	Reverse: GCC TGC CAC TTT CCT TGT G
Fatty acid synthase[a]	Applied Biosystems (Mm00662319_m1)
Lipin1	Forward: GGT CCC CCA GCC CCA GTC CTT
	REVERSE: GCA GCC TGT GGC AAT TCA
Lipid utilization	
UCP2	Forward: TCA CTG TGC CCT TAC CAT GCT
	Reverse: AGG CAT GAA CCC CTT GTA GAA G
Lipolytic	
Perilipin	Forward: GAC ACC ACC TGC ATG GCT
	Reverse: TGA AGC AGG GCC ACT CTC
ATGL	Forward: CAG CAC ATT TAT CCC GGT GTA C
	Reverse: AAA TGC CGC CAT CCA CAT AG
HSL	Forward: GCT GGG CTG TCA AGC ACT GT
	Reverse: GTA ACT GGG TAG GCT GCC AT

Continued

Table 22.2 Glucose and lipid metabolic genes in BAT—cont'd

Transcripts	Probe/primers
Lipid recycling	
PEPCK	Forward: TCC TGG CAC CTC AGT GAA GAC AAA
	Reverse: TGT CCT TCC GGA ACC AGT TGA CAT

[a]Using the ABI TaqMan Master Mix.
PPAR-γ2, peroxisome proliferator-activated receptor-γ2; C/EBPα, CCAAT-enhancer-binding protein α; GLUT4, glucose transporter 4; CD36, cluster of differentiation 36; aP2, adipocyte protein 2; UCP2, uncoupling protein 2; ATGL, adipose triglyceride lipase; HSL, hormone-sensitive lipase; PEPCK, phosphoenolpyruvate carboxykinase.

1. Extract total RNA from BAT using TRIzol reagent (Invitrogen Corp., Carlsbad, CA), following the manufacturer's instructions.
2. Treat RNA with DNaseI (Ambion Inc., Austin, TX) to remove any remaining trace amount of DNA, per manufacturer's instructions.
3. Run RNA samples on RNA gel to validate the quality of the RNA.
4. Synthesize cDNA from 1 μg RNA, using the SuperScript III First-Strand Synthesis System for RT-PCR (Invitrogen Corp., Carlsbad, CA). The cDNA is diluted 10 times with DEPC water before PCR amplification.
5. SYBR Green PCR Master Mix or TaqMan Gene expression Master Mix (Applied Biosystems, Carlsbad, CA) is used for the PCR cDNA amplification. 18s RNA, β-actin, and 36B4 are used as internal controls. Thermal cycling is carried out with an ABI prism 7900 sequence detection system (Applied Biosystems) under factory default settings (50 °C for 2 min, 95 °C for 10 min, and 40 cycles at 95 °C for 15 s and 60 °C for 1 min). Relative quantitation between the WT and *Ghsr*$^{-/-}$ mice is calculated using the ddCt method.

4.2. Western blotting analysis of key thermogenic regulators UCP1 and PGC1α

1. Homogenize BAT completely in 1 ml RIPA buffer (Millipore, Billerica, MA), supplied with complete protease inhibitor cocktail (Roche Inc., Mannheim, Germany).
2. Centrifuge protein lysate at 20,000 × *g* for 15 min at 4 °C, yielding three layers: lipid layer on top, clear protein layer in the middle, and pellet at the bottom. Transfer the middle clear protein layer to a new ice-chilled tube. If it is needed, repeat this step to remove remained lipid.

3. Measure the protein concentration with a BCA protein assay kit (Thermo Fisher Scientific, Rockford, IL).
4. Separate 20 μg protein by SDS-PAGE and then transfer to nitrocellulose membrane for immunoblotting analysis.
5. Block the nitrocellulose membrane in 10 ml blocking solution with constant shaking for 1 h at room temperature. Blocking solution consists of 5% nonfat dry milk in TBST (Tris-buffered saline with Tween 20: 0.1% Tween 20, 137 m*M* NaCl, 2.7 m*M* KCl, 25 m*M* Tris base, pH 7.4).
6. Wash the nitrocellulose membrane with 10 ml TBST twice, for 10 min each time.
7. Dilute the primary antibody in TBST with 3% BSA. Place the nitrocellulose membrane in 10 ml primary antibody solution, and incubate overnight with constant shaking at 4 °C. The following antibodies are used: rabbit anti-UCP1 (Millipore, Billerica, MA, 1:1000), rabbit anti-PGC1α (Cell signaling, Danvers, MA, 1:1000), and mouse anti-β-actin (Santa Cruz Biotechnology, Santa Cruz, CA, 1:10,000).
8. Wash the nitrocellulose membrane with TBST five times, for 5 min each.
9. Incubate the nitrocellulose membranes in the secondary antibody solutions composed of HRP-conjugated anti-mouse or anti-rabbit antibodies (GE Healthcare UK Limited, Bucks, UK, 1:10,000) in TBST with 1% BSA, for 2 h at room temperature with shaking.
10. Wash the nitrocellulose membrane with TBST five times for 5 min each.
11. Pour off TBST, and then perform detection of proteins using the SuperSignal West Pico Chemiluminescent kit (Pierce, Rockford, IL).

4.3. *Ex vivo* lipolysis

The lipolysis activity of BAT of *Ghsr*$^{-/-}$ mice has been measured using *ex vivo* lipolysis assay, as described (Qatanani et al., 2009).

1. Isolate interscapular BAT from mice, and carefully remove the white fat layer on top of the BAT.
2. Weigh two pieces of BAT, each at about 20 mg. Put each piece into culture medium (DMEM with 0.5% fatty-acid-free BSA (Sigma, St. Louis, MO)), and then cut tissue into tiny pieces with a pair of scissors in the culture medium.
3. The BAT tissue from each mouse is divided into two wells: add 10 μ*M* CL316243 (Sigma, St. Louis, MO) into one well as stimulated condition and add the same amount of DMSO into another well as basal condition.

4. Incubate samples at 37 °C. Then collect medium after 2 h, 4 h, and overnight.
5. Measure free glycerol in the media using the Free Glycerol Reagent (Sigma, St. Louis, MO), following the manufacturer's instructions.
6. *Ex vivo* lipolysis is represented by glycerol levels, normalized by the weight of the tissue.

5. ISOLATION AND CHARACTERIZATION OF BROWN ADIPOCYTES

Higher mitochondrial density is a hallmark of BAT and determines the thermogenic capacity of BAT. We have characterized the mitochondria of BAT by histological, molecular, and functional approaches (Lin et al., 2011; Ma et al., 2011). The isolation method of brown adipocytes has been described in the literature (Fasshauer et al., 2000, 2001).

5.1. Histological characterization of morphology and mitochondrial content of BAT

1. Fix BAT in 10% formalin at room temperature overnight; dehydrate, and then embed in paraffin. The tissue block is sectioned at 5 μm.
2. H&E staining is carried out following the standard protocols. BAT has variably sized small lipid droplets and an abundance of blood-filled capillaries.
3. Immunohistochemical analysis of mitochondrial protein optic atrophy 1 (OPA1) is carried out as described (Dali-Youcef et al., 2007). The OPA1 antibody (BD Biosciences, San Jose, CA) is diluted 1:1000. The intensity of OPA1 staining is correlated with the number of mitochondria.

5.2. Quantification of mitochondrial density of BAT using ratio of mitochondrial DNA/total nuclear DNA

Mitochondrial DNA (mtDNA) has been extracted and quantified as described, with modifications (Justo et al., 2005).

1. Dissect the interscapular BAT, and homogenize the tissue in 1 ml isolation buffer (300 m*M* sucrose, 1 m*M* EDTA, 5 m*M* MOPS, 5 m*M* KH_2PO_4, 0.01% BSA, pH 7.4), in a glass homogenizer with Teflon pestle. Six to eight strokes are required.
2. Filter the homogenate through a layer of gauze, and centrifuge at $8000 \times g$ for 10 min at 4 °C.

3. Discard the fat layer and supernatant, resuspend the pellet (containing cell debris, nuclei, and mitochondria) in a 300-μl isolation buffer, and transfer to a new tube.
4. Centrifuge at 800 × *g* for 10 min at 4 °C. Transfer the supernatant (containing mitochondria) carefully to a new tube, and save the pellet as nuclei (including cell debris).
5. Centrifuge the supernatant at 8000 × *g* for 10 min at 4 °C. Save the resulting pellet as mitochondria.
6. Extract nuclear DNA with the phenol/chloroform method (Chomczynski and Sacchi, 1987).
7. Digest an aliquot of the mitochondrial fraction overnight in 100-μl lysis buffer (10 m*M* Tris, pH 8.0, 10 m*M* EDTA, 10 m*M* NaCl, 0.5% SDS, 100 mg/ml proteinase K) at 37 °C, and then boil for 5 min.
8. Linearize mtDNA by digestion with Bcl-I (New England Biolabs, Ipswich, MA) for 3 h at 50 °C, and then boil the samples for 5 min.
9. Centrifuge the samples at 7000 × *g* for 5 min, and use the resulting supernatant for subsequent PCR amplification.
10. Perform PCR to amplify the 162-nt region of the mitochondrial NADH dehydrogenase subunit 4 gene. The primer sequences are 5′-TACACGATGAGGCAACCAAA-3′ and 5′-GGTAGGGGGT GTGTGTTGTGAG-3′. Purify the PCR product with the high-pure PCR template preparation kit (Roche, Indianapolis, IN).
11. Quantify nuclear DNA and the amplified PCR product of mtDNA with NanoDrop (ND-1000 Thermo Scientific, Waltham, MA), and calculate the ratio of mtDNA:total nuclear DNA.

5.3. Isolation of primary brown adipocytes

1. Remove the interscapular brown fat pad from mice. Place the tissue in a Petri dish with 500 μl PBS (sterile); then cut the tissue into tiny pieces in PBS.
2. Transfer the tissue into an Eppendorf tube, and add 500 μl isolation buffer (123 m*M* NaCl, 5 m*M* KCl, 1.3 m*M* $CaCl_2$, 5 m*M* glucose, 100 n*M* HEPES, 1% penicillin/streptomycin, and 4% BSA) with 1.5 mg/ml collagenase I (Worthington Biochemical Corporation, Lakewood, NJ).
3. Vortex for 10 s, and then place the Eppendorf tube in a shaking water bath (37 °C) for 40 min; vortex for 10 s every 5 min.

4. Filter the digested tissue through a 125-μm nylon mesh (VWR Corp., Radnor, PA) into a new Eppendorf tube, and then centrifuge at room temperature at 1500 rpm for 5 min.
5. The top fat layer is collected as mature brown adipocytes, and the pellet is collected as brown preadipocytes.

5.4. Measurement of oxygen consumption of brown adipocytes

Mitochondrial oxygen consumption can be measured with a Clark-type oxygen electrode (Hansatech Instruments, Norfolk, England) as described previously (Shi et al., 2005; Wu et al., 1999).

1. To calibrate the Clark-type oxygen electrode, fill the electrode chamber with 37 °C DMEM without cover and allow the output to stabilize (about 217 nmol O_2/ml).
2. Remove the calibration solution, and carefully wash the electrode chamber with distilled water. Connect the chamber to a circulating water bath at 37 °C.
3. Add 1 ml DMEM with about 10,000 brown adipocytes to the electrode chamber with a magnetic stir bar.
4. Tightly seal the chamber with paraffin film, and allow the system to stabilize.
5. Record the oxygen consumption as the basal respiration rate.
6. Using a Hamilton microsyringe, add the 2.5 μg/ml oligomycin (an ATP synthase inhibitor) through the small hole in the cover of the chamber. Record the O_2 consumption rate for 5 min.
7. Wait until the respiration slows down and returns to a rate comparable to basal level.
8. Using a Hamilton microsyringe, add 2 μ*M* cyanide *p*-trifluoromethoxyphenylhydrazone (FCCP) into the chamber. Record the O_2 consumption rate for 5 min as the maximum respiration rate.
9. After finishing all the measurements, collect all cells from the chamber. Determine the protein concentration using the BCA kit (Thermo Fisher Scientific, Rockford, IL) to normalize the O_2 consumption rate.

6. SUMMARY

BAT plays an important role in energy homeostasis by regulating thermogenesis. Aging is associated with severe declines of BAT mass and activity (Pfannenberg et al., 2010). Ablation of GHS-R attenuates age-associated thermogenic impairment in BAT, leading to increased EE and reduced adiposity (Lin et al., 2011). This new finding suggests that GHS-R plays an important role in thermogenic regulation. Ghrelin signaling may regulate energy homeostasis through a thermogenic mechanism, in addition to an orexigenic mechanism. GHS-R antagonists might prove to be new class of drug combating obesity by "turning up the heat."

ACKNOWLEDGMENTS

This study was supported by NIH/NIA grant 1R03AG029641-01 (Y. S.) and USDA/CRIS grant ARS 6250-51000-055 (Y. S.) and partly supported by the NIH Diabetes and Endocrinology Research Center (P30DK079638) at Baylor College of Medicine.

REFERENCES

Argyropoulos, G., Harper, M.E., 2002. Uncoupling proteins and thermoregulation. J. Appl. Physiol. 92, 2187–2198.

Bartness, T.J., Song, C.K., 2007. Brain-adipose tissue neural crosstalk. Physiol. Behav. 91, 343–351.

Cannon, B., Nedergaard, J., 2004. Brown adipose tissue: function and physiological significance. Physiol. Rev. 84, 277–359.

Chomczynski, P., Sacchi, N., 1987. Single-step method of RNA isolation by acid guanidinium thiocyanate-phenol-chloroform extraction. Anal. Biochem. 162, 156–159.

Cowley, M.A., et al., 2003. The distribution and mechanism of action of ghrelin in the CNS demonstrates a novel hypothalamic circuit regulating energy homeostasis. Neuron 37, 649–661.

Dali-Youcef, N., et al., 2007. Adipose tissue-specific inactivation of the retinoblastoma protein protects against diabesity because of increased energy expenditure. Proc. Natl. Acad. Sci. USA 104, 10703–10708.

Davies, J.S., et al., 2009. Ghrelin induces abdominal obesity via GHS-R-dependent lipid retention. Mol. Endocrinol. 23, 914–924.

Fasshauer, M., et al., 2000. Essential role of insulin receptor substrate-2 in insulin stimulation of Glut4 translocation and glucose uptake in brown adipocytes. J. Biol. Chem. 275, 25494–25501.

Fasshauer, M., et al., 2001. Essential role of insulin receptor substrate 1 in differentiation of brown adipocytes. Mol. Cell. Biol. 21, 319–329.

Frontini, A., Cinti, S., 2010. Distribution and development of brown adipocytes in the murine and human adipose organ. Cell Metab. 11, 253–256.

Gnanapavan, S., et al., 2002. The tissue distribution of the mRNA of ghrelin and subtypes of its receptor, GHS-R, in humans. J. Clin. Endocrinol. Metab. 87, 2988.

Inokuma, K., et al., 2005. Uncoupling protein 1 is necessary for norepinephrine-induced glucose utilization in brown adipose tissue. Diabetes 54, 1385–1391.

Justo, R., et al., 2005. Brown adipose tissue mitochondrial subpopulations show different morphological and thermogenic characteristics. Mitochondrion 5, 45–53.

Kojima, M., Kangawa, K., 2005. Ghrelin: structure and function. Physiol. Rev. 85, 495–522.

Lee, S., et al., 2010. Disrupting circadian homeostasis of sympathetic signaling promotes tumor development in mice. PLoS One 5, e10995.

Lin, L., et al., 2011. Ablation of ghrelin receptor reduces adiposity and improves insulin sensitivity during aging by regulating fat metabolism in white and brown adipose tissues. Aging Cell 10, 996–1010.

Longo, K.A., et al., 2008. Improved insulin sensitivity and metabolic flexibility in ghrelin receptor knockout mice. Regul. Pept. 150, 55–61.
Lowell, B.B., Spiegelman, B.M., 2000. Towards a molecular understanding of adaptive thermogenesis. Nature 404, 652–660.
Ma, X., et al., 2011. Ablations of ghrelin and ghrelin receptor exhibit differential metabolic phenotypes and thermogenic capacity during aging. PLoS One 6, e16391.
Mano-Otagiri, A., et al., 2009. Ghrelin suppresses noradrenaline release in the brown adipose tissue of rats. J. Endocrinol. 201, 341–349.
Mano-Otagiri, A., et al., 2010. Genetic suppression of ghrelin receptors activates brown adipocyte function and decreases fat storage in rats. Regul. Pept. 160, 81–90.
Nedergaard, J., Cannon, B., 2010. The changed metabolic world with human brown adipose tissue: therapeutic visions. Cell Metab. 11, 268–272.
Nuotio-Antar, A.M., et al., 2007. Carbenoxolone treatment attenuates symptoms of metabolic syndrome and atherogenesis in obese, hyperlipidemic mice. Am. J. Physiol. Endocrinol. Metab. 293, E1517–E1528.
Obici, S., et al., 2002. Identification of a biochemical link between energy intake and energy expenditure. J. Clin. Invest. 109, 1599–1605.
Pfannenberg, C., et al., 2010. Impact of age on the relationships of brown adipose tissue with sex and adiposity in humans. Diabetes 59, 1789–1793.
Qatanani, M., et al., 2009. Macrophage-derived human resistin exacerbates adipose tissue inflammation and insulin resistance in mice. J. Clin. Invest. 119, 531–539.
Rudaya, A.Y., et al., 2005. Thermoregulatory responses to lipopolysaccharide in the mouse: dependence on the dose and ambient temperature. Am. J. Physiol. Regul. Integr. Comp. Physiol. 289, R1244–R1252.
Shi, T., et al., 2005. SIRT3, a mitochondrial sirtuin deacetylase, regulates mitochondrial function and thermogenesis in brown adipocytes. J. Biol. Chem. 280, 13560–13567.
Shimbara, T., et al., 2004. Central administration of ghrelin preferentially enhances fat ingestion. Neurosci. Lett. 369, 75–79.
Sun, Y., et al., 2004. Ghrelin stimulation of growth hormone release and appetite is mediated through the growth hormone secretagogue receptor. Proc. Natl. Acad. Sci. USA 101, 4679–4684.
Sun, Y., et al., 2007. Ghrelin and growth hormone secretagogue receptor expression in mice during aging. Endocrinology 148, 1323–1329.
Tschop, M., et al., 2000. Ghrelin induces adiposity in rodents. Nature 407, 908–913.
van Marken Lichtenbelt, W.D., Schrauwen, P., 2011. Implications of nonshivering thermogenesis for energy balance regulation in humans. Am. J. Physiol. Regul. Integr. Comp. Physiol. 301, R285–R296.
Wu, Z., et al., 1999. Mechanisms controlling mitochondrial biogenesis and respiration through the thermogenic coactivator PGC-1. Cell 98, 115–124.
Yasuda, T., et al., 2003. Centrally administered ghrelin suppresses sympathetic nerve activity in brown adipose tissue of rats. Neurosci. Lett. 349, 75–78.
Zigman, J.M., et al., 2005. Mice lacking ghrelin receptors resist the development of diet-induced obesity. J. Clin. Invest. 115, 3564–3572.

CHAPTER TWENTY-THREE

Transgenic Mice Overexpressing Ghrelin or Ghrelin Analog

Hiroyuki Ariyasu*,†,1, Go Yamada*, Hiroshi Iwakura*,†,‡, Takashi Akamizu†,§, Kenji Kangawa†,¶, Kazuwa Nakao*

*Department of Endocrinology and Metabolism, Kyoto University Graduate School of Medicine, Kyoto, Japan
†Ghrelin Research Project of Translational Research Center, Kyoto University Graduate School of Medicine, Kyoto, Japan
‡Medial Innovation Center, Kyoto University Graduate School of Medicine, Kyoto, Japan
§The First Department of Medicine, Wakayama Medical University, Wakayama, Japan
¶National Cerebral and Cardiovascular Center Research Institute, Suita, Osaka, Japan
[1]Corresponding author: e-mail address: ariyasu@kuhp.kyoto-u.ac.jp

Contents

Abstract

To understand the chronic effects of ghrelin, genetically engineered mouse models would be useful. Early studies, however, suggested that it was challenging to generate ghrelin gain-of-activity models by standard procedures. Although several groups have been trying to generate transgenic (Tg) mice overexpressing ghrelin, almost all these animals produced only des-acyl ghrelin rather than acylated ghrelin. Therefore, to elucidate the mechanism for the fatty acid modification in ghrelin, many researchers have been seeking an enzyme that would catalyze the acylation of ghrelin with an octanoic acid. In 2008, ghrelin *O*-acyltransferase (GOAT) was identified at last, and thereafter double-Tg mice overexpressing ghrelin and GOAT were generated by Kirchner et al.

Methods in Enzymology, Volume 514
ISSN 0076-6879
http://dx.doi.org/10.1016/B978-0-12-381272-8.00023-4

On the other hand, we have succeeded in generating Tg mice overexpressing Trp^3-ghrelin, a ghrelin analog that does not require posttranscriptional modification with GOAT for activity. These ghrelin gain-of-activity models are useful tools for evaluating the long-term pathophysiological and/or pharmacological effects of ghrelin or ghrelin analogs and provide insight into the physiological roles of ghrelin/GHS-R systems.

1. INTRODUCTION

Genetically engineered mouse models are now widely utilized for investigating the long-term effects of hormones and are used as preclinical tools for the investigation of novel therapeutic approaches for the prevention and treatment of human disease. In order to understand the chronic effects of ghrelin, several groups have been trying to generate transgenic (Tg) mice overexpressing ghrelin under the control of different promoters (Ariyasu et al., 2005; Asakawa et al., 2005; Iwakura et al., 2005; Reed et al., 2008; Wei et al., 2006; Zhang et al., 2008). It was, however, very difficult to generate Tg mice overexpressing ghrelin by standard procedures because the mechanism of acylation of ghrelin had been unclear until the identification of ghrelin *O*-acyltransferase (GOAT) (Gutierrez et al., 2008; Yang et al., 2008). Therefore, almost all these animals produced only des-acyl ghrelin rather than acylated ghrelin. After the identification of GOAT, the human GOAT and ghrelin Tg mice were generated and reported by Kirchner et al. (2009). In this chapter, we introduce three lines of Tg mice: the first is des-acyl ghrelin Tg mice; the second is Tg mice overexpressing a ghrelin analog possessing ghrelin-like activity without Ser^3 acylation; the last is the human GOAT and ghrelin Tg mice.

2. DES-ACYL GHRELIN TG MICE

2.1. Concept of the study

We generated Tg mice bearing the preproghrelin gene under the control of a cytomegalovirus immediate early enhancer and a modified chicken beta-actin promoter, designated the *CAG* promoter (Ariyasu et al., 2005; Niwa et al., 1991). This Tg mouse overexpressed des-acyl ghrelin in plasma and a wide variety of tissues.

2.2. Generation of des-acyl ghrelin Tg mice

1. Mouse stomach cDNA library was constructed from 1 μg of mouse stomach poly$(A)^+$ RNA with a cDNA synthesis kit (Amersham Biosciences). Mouse *ghrelin* cDNA (Accession No. NM_001126314) was isolated from this library using rat *ghrelin* cDNA (Accession No. NM_021669) as a probe.
2. A fusion gene comprising *pCAGGS* expression vector including the *CAG* promoter (Niwa et al., 1991) and mouse *ghrelin* cDNA coding sequences was designed (plasmid *pCAGGS–ghrelin*). It was constructed by inserting the mouse preproghrelin cDNA into the unique *Eco*RI site between the *CAG* promoter and 3′-flanking sequence of the rabbit beta-globin gene of the *pCAGGS* expression vector.
3. The DNA fragment was excised from its plasmid by digestion with *Sal*I and *Hin*dIII.
4. The purified fragment (10 μg/ml) was microinjected into the pronucleus of fertilized C57/B6J mice (SLC, Shizuoka, Japan) eggs. The viable eggs were transferred into the oviducts of pseudopregnant female ICR mice (SLC) using standard techniques.
5. Founder Tg mice were identified by PCR analysis and bred with C57BL/6 mice.

2.3. Phenotypes of des-acyl ghrelin Tg mice

We originally intended to generate mice overexpressing biologically active ghrelin. However, there were no differences in plasma acylated ghrelin levels between non-Tg and Tg mice (non-Tg vs. Tg: 83.7 ± 11.9 and 86.3 ± 21.1 fmol/ml), and acylated ghrelin levels in all tissues examined were not elevated. On the other hand, plasma des-acyl ghrelin levels in Tg mice were markedly elevated (non-Tg vs. Tg: 1104.5 ± 94.4 vs. $48{,}565.5 \pm 9291.5$ fmol/ml). Although the Tg mice overexpressed only des-acyl ghrelin, they showed small phenotype due to suppressed GH/IGF-1 axis.

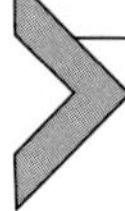

3. TG MICE OVEREXPRESSING A GHRELIN ANALOG

3.1. Concept of the study

Before the identification of GOAT, we initiated a plan to generate Tg mice overexpressing a ghrelin analog that possessed ghrelin-like activity in the absence of acylation at Ser^3 (Yamada et al., 2010). As the replacement of Ser^3 of ghrelin with Trp^3 (Trp^3-ghrelin) preserves a low level of ghrelin activity and

Trp^3-ghrelin can be synthesized *in vivo* (Matsumoto et al., 2001), we generated mice overexpressing Trp^3-ghrelin by using the *hSAP* (human serum-amyloid-P) promoter.

3.2. Generation of Tg mice overexpressing a ghrelin analog

1. A fusion gene of the *hSAP* promoter and mouse preproghrelin cDNA coding sequences was designed (plasmid *hSAP–ghrelin*) (Kojima et al., 1999; Ogawa et al., 1999). It was constructed by inserting the mouse preproghrelin cDNA into the unique *Eco*RI site between the *hSAP* promoter and 3′-flanking sequence of the rabbit beta-globin gene.
2. Mutations were created using a QuikChange Site-Directed Mutagenesis kit, according to the manufacturer's instruction. The hSAP–ghrelin plasmid was used as the template for PCR amplification. To replace the AGC codon encoding Ser to a TGG codon encoding Trp, we used two oligonucleotide primers:
   5′-GGACATGGCCATGGCAGGCTCCTGGTTCCTGAGCC CAGAGC-3′ and 5′-GCTCTGGGCTCAGGAACCAGGAGCCTG CCATGGCCATGTCC-3′.
3. The mutated construct was verified by sequencing.
4. The DNA fragment was excised from its plasmid by digestion with *Sal*I and *Hin*dIII, and then purified and microinjected into the pronuclei of fertilized eggs.
5. Founder Tg mice were identified by PCR analysis and bred against C57BL/6 mice.

3.3. Phenotypes of Tg mice overexpressing a ghrelin analog

Hepatic transgene expression in Tg mice was 3.02 ± 1.15 in arbitrary units after normalization to preproghrelin mRNA expression levels seen in the stomachs of non-Tg mice (1.00 ± 0.18). No expression of preproghrelin mRNA was seen in the livers of non-Tg mice. There were no differences in plasma ghrelin and des-acyl ghrelin concentrations between non-Tg and Tg mice (ghrelin; non-Tg vs. Tg: 40.5 ± 10.2 vs. 36.6 ± 4.4 fmol/ml, des-acyl ghrelin; non-Tg vs. Tg: 167.5 ± 51.8 vs. 235.7 ± 44.8 fmol/ml).

Plasma Trp^3-ghrelin concentrations in Tg mice was 3437.8 ± 571.6 (2546.4–5101.7) fmol/ml, which was approximately 85-fold ($3437.8/40.5 = 84.9$-fold) higher than plasma ghrelin (acylated ghrelin) concentrations seen in non-Tg mice. Because Trp^3-ghrelin is approximately 1/10–1/20 less potent than ghrelin *in vivo* (Matsumoto et al., 2001), plasma

Trp^3-ghrelin concentrations in Tg mice were calculated to have an activity approximately six times greater than that of ghrelin (acylated ghrelin) seen in non-Tg mice (84.9-fold $\times$ 1/10–1/20 = 4.2–8.5-fold). Tg mice exhibited normal growth in their early life stage. There were no significant differences in serum GH and IGF-I concentrations and the average food intake between Trp^3-ghrelin-Tg mice and non-Tg mice. In addition, glucose metabolism in Trp^3-ghrelin-Tg mice did not differ from that seen in non-Tg mice in early life. However, 1-year-old Tg mice demonstrated impaired glucose tolerance and reduced insulin sensitivity. Although there were no differences between Trp^3-ghrelin-Tg and non-Tg mice in anthropometric parameters, including body weight, total body fat percentage, and lean body mass, blood glucose levels after glucose injection were significantly higher than those in non-Tg mice and the acute phase of insulin secretion typically seen in response to glucose tended to be suppressed in Trp^3-ghrelin-Tg mice. In addition, the hypoglycemic response after the injection with insulin was blunted in Trp^3-ghrelin-Tg mice. There were no differences, however, in pancreatic insulin mRNA levels between 1-year-old Trp^3-ghrelin-Tg and non-Tg mice.

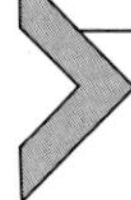

4. TG MICE OVEREXPRESSING BOTH HUMAN GOAT AND GHRELIN

4.1. Concept of the study

After the identification of GOAT, the human GOAT and ghrelin Tg mice were generated by Kirchner et al. (2009). They generated Tg mice designed to express the human *GHRL* (Accession No. AB029434) and human *MBOAT4* (membrane-bound *O*-acyltransferase domain containing 4; Accession No. NM_001100916) genes in the liver under control of the human *APO-E* (encoding apolipoprotein E) promoter.

4.2. Generation of Tg mice overexpressing both human GOAT and ghrelin

1. The open reading frames for the human preproghrelin and a C-terminal Flag-tagged version of the human *GOAT* were independently cloned with the addition of a 5′-Kozak sequence into the multiple cloning site of a *plivhHL1*-derived vector (Fan et al., 1998). The *plivhHL1* vector has the constitutive human *apo-E* gene promoter and its hepatic control region for specific liver tissue expression.

2. Both transgenes were microinjected at equimolar amounts into fertilized eggs to generate Tg mice.
3. The mutated construct was verified by sequencing.
4. DNA PCR analyses specific for the Tg sequences were performed to identify founder mice harboring both transgenes.
5. Functional expression of both genes was determined by blood ghrelin immunoprecipitation matrix-assisted laser desorption/ionization time-of-flight mass spectrometry (IPMS) analyses (Gutierrez et al., 2008) and used to establish this line of human ghrelin and GOAT Tg mice.

4.3. Phenotypes of the Tg mice overexpressing both human GOAT and ghrelin

Tg mice have been reported to have high circulating levels of des-acyl ghrelin under normal dietary conditions. However, when the Tg mice were fed on a diet containing MCTs (medium-chain triglycerides), these mice exhibited elevated concentrations of fatty-acid-modified forms of ghrelin; human octanoyl-modified ghrelin concentrations were approximately 32 ng/ml in Tg mice but were undetectable in WT mice. Although it is well known that pharmacological administration of ghrelin stimulates food intake (Nakazato et al., 2001; Shintani et al., 2001; Tschöp et al., 2000; Wren et al., 2000), mice overproducing ghrelin did not increase food consumption. It was also reported that ghrelin Tg mice showed decreased energy expenditure and exhibited a significantly increased rate of body weight gain. These data suggest that ghrelin overproduction may lead to obesity without increasing food intake.

5. SUMMARY

It is challenging to generate ghrelin gain-of-activity models because ghrelin requires posttranscriptional modification, an octanoylation of Ser^3. GOAT is responsible for this octanoylation of ghrelin and confers its biological activity (Gutierrez et al., 2008; Yang et al., 2008).

Kirchner et al. (2009) generated Tg mice simultaneously expressing human ghrelin and GOAT in the liver under the control of the human *APO-E* promoter. Surprisingly, these mice exhibited elevated concentrations of fatty-acid-modified forms of ghrelin only when given a diet rich in MCTs. They reported that ghrelin Tg mice exhibited obese phenotypes without increasing food intake. This Tg mouse generated by Kirchner et al. is a genuine ghrelin Tg mice, but it may be difficult to characterize the phenotype of the mice precisely, especially the metabolic phenotype, because plasma ghrelin concentrations in Tg mice are elevated only when given a diet of MCTs. We have succeeded in generating Tg mice overexpressing Trp^3-ghrelin, which is a ghrelin analog that does not require posttranscriptional modification with GOAT for activity. We believe that this unique mouse model is a useful tool for evaluating the long-term pathophysiological and/or pharmacological effects of ghrelin or ghrelin analogs and would provide insights into the physiological roles of ghrelin/GHS-R systems.

REFERENCES

Ariyasu, H., et al., 2005. Transgenic mice overexpressing des-acyl ghrelin show small phenotype. Endocrinology 146, 355–364.

Asakawa, A., et al., 2005. Stomach regulates energy balance via acylated ghrelin and desacyl ghrelin. Gut 54, 18–24.

Fan, J., et al., 1998. Increased expression of apolipoprotein E in transgenic rabbits results in reduced levels of very low density lipoproteins and an accumulation of low density lipoproteins in plasma. J. Clin. Invest. 101, 2151–2164.

Gutierrez, J.A., et al., 2008. Ghrelin octanoylation mediated by an orphan lipid transferase. Proc. Natl. Acad. Sci. USA 105, 6320–6325.

Iwakura, H., et al., 2005. Analysis of rat insulin II promoter-ghrelin transgenic mice and rat glucagon promoter-ghrelin transgenic mice. J. Biol. Chem. 280, 15247–15256.

Kirchner, H., et al., 2009. GOAT links dietary lipids with the endocrine control of energy balance. Nat. Med. 15, 741–745.

Kojima, M., et al., 1999. Ghrelin is a growth-hormone-releasing acylated peptide from stomach. Nature 402, 656–660.

Matsumoto, M., et al., 2001. Structural similarity of ghrelin derivatives to peptidyl growth hormone secretagogues. Biochem. Biophys. Res. Commun. 284, 655–659.

Nakazato, M., et al., 2001. A role for ghrelin in the central regulation of feeding. Nature 409, 194–198.

Niwa, H., et al., 1991. Efficient selection for high-expression transfectants with a novel eukaryotic vector. Gene 108, 193–199.

Ogawa, Y., et al., 1999. Increased glucose metabolism and insulin sensitivity in transgenic skinny mice overexpressing leptin. Diabetes 48, 1822–1829.

Reed, J.A., et al., 2008. Mice with chronically increased circulating ghrelin develop age-related glucose intolerance. Am. J. Physiol. Endocrinol. Metab. 294, E752–E760.

Shintani, M., et al., 2001. Ghrelin, an endogenous growth hormone secretagogue, is a novel orexigenic peptide that antagonizes leptin action through the activation of hypothalamic neuropeptide Y/Y1 receptor pathway. Diabetes 50, 227–232.

Tschöp, M., et al., 2000. Ghrelin induces adiposity in rodents. Nature 407, 908–913.

Wei, W., et al., 2006. Effect of chronic hyperghrelinemia on ingestive action of ghrelin. Am. J. Physiol. Regul. Integr. Comp. Physiol. 290, R803–R808.

Wren, A.M., et al., 2000. The novel hypothalamic peptide ghrelin stimulates food intake and growth hormone secretion. Endocrinology 141, 4325–4328.

Yamada, G., et al., 2010. Generation of transgenic mice overexpressing a ghrelin analog. Endocrinology 151, 5935–5940.

Yang, J., et al., 2008. Identification of the acyltransferase that octanoylates ghrelin, an appetite-stimulating peptide hormone. Cell 132, 387–396.

Zhang, W., et al., 2008. Effect of des-acyl ghrelin on adiposity and glucose metabolism. Endocrinology 149, 4710–4716.

SECTION 7

Clinical Application of Ghrelin

CHAPTER TWENTY-FOUR

Therapeutic Potential of Ghrelin in Restricting-Type Anorexia Nervosa

Mari Hotta[*,†,1], **Rina Ohwada**[†], **Takashi Akamizu**[‡], **Tamotsu Shibasaki**[§], **Kenji Kangawa**[¶]

[*]Health Services Center, National Graduate Institute for Policy Studies, Tokyo, Japan
[†]Department of Medicine, Tokyo Women's Medical University, Tokyo, Japan
[‡]Department of Medicine, Wakayama Medical University, Wakayama, Japan
[§]Department of Physiology, Nippon Medical School, Tokyo, Japan
[¶]National Cerebral and Cardiovascular Center Research Institute, Suita, Osaka, Japan
[1]Corresponding author: e-mail address: marihs@grips.ac.jp

Contents

Abstract

Anorexia nervosa (AN) is an eating disorder characterized by a decrease in caloric intake and malnutrition. It is associated with a variety of medical morbidities as well as significant mortality. Nutritional support is of paramount importance to prevent impaired quality of life later in life in affected patients. Some patients with restricting-type AN who are fully motivated to gain body weight cannot increase their food intake because of malnutrition-induced gastrointestinal dysfunction. Chronicity of AN prevents

Methods in Enzymology, Volume 514
ISSN 0076-6879
http://dx.doi.org/10.1016/B978-0-12-381272-8.00024-6

participation in social activities and leads to increased medical expenses. Therefore, there is a pressing need for effective appetite-stimulating therapies for patients with AN.

Ghrelin is the only orexigenic hormone that can be given intravenously. Intravenous infusion of ghrelin is reported to increase food intake and body weight in healthy subjects as well as in patients with poor nutritional status. Here, we introduce the results of a pilot study that investigated the effects of ghrelin on appetite, energy intake, and nutritional parameters in five patients with restricting-type AN, who are fully motivated to gain body weight but could not increase their food intake because of malnutrition-induced gastrointestinal dysfunction.

1. INTRODUCTION

Since the 1960s, eating disorders, including anorexia nervosa (AN), began to be recognized as an important health problem among adolescent girls and young women in Western societies (Bruch, 1985). Although few Japanese patients with AN were recognized early on, according to the 2010 Survey Committee for Eating Disorders of the Japanese Ministry of Health, Labor and Welfare, the prevalence rate of AN in Japanese high-school girls is 0.25%.

AN is a psychosomatic disorder characterized by obsessive dieting in spite of thinness, fear of weight gain, abnormal eating attitudes, and starvation-induced psychological symptoms. Cultural, social, familial, psychological, and biological factors are involved in the pathogenesis of this disorder. There are two types of dieting in AN: restricting type, in which patients severely restrict food intake and binge eating/purging type, in which patients engage in self-induced vomiting or the misuse of laxatives after eating. Many patients with AN have a history of stressful events or traumatic experiences and develop AN as a way of coping with difficult circumstances, such as focusing on tallying calories or thinking about food, or binge eating and purging that distracts patients from their stress or emotional pain. In addition, patients may develop AN because they can control food and body weight but cannot manage other aspects of their life (Birmingham and Treasure, 2010).

AN is associated with extensive morbidity due to malnutrition as well as significant mortality (Neumarker, 1997). Some of the complications of this disorder remain even after recovery. Thus, nutritional support is of paramount importance to prevent impaired quality of life later on in patients with AN. Some patients with restricting-type AN who are fully motivated to gain body weight cannot increase their food intake because of malnutrition-induced gastrointestinal dysfunction, which delays recovery. Chronicity of AN prevents patients from participating in social activities and increases medical expenses. It is reported that the cost of long-term disability for patients with AN is up to 30 times the yearly cost for care service and treatment (Su and Birmingham, 2003). Therefore, there is a pressing need for effective appetite-stimulating therapies for patients with AN.

Ghrelin is mainly secreted by the stomach during starvation and exerts a potent stimulatory effect on food intake and growth hormone (GH) secretion (Ariyasu et al., 2001; Kojima et al., 1999). Ghrelin is the only orexigenic hormone that can be given intravenously. Intravenous infusion of ghrelin is reported to increase food intake and body weight in healthy subjects (Wren et al., 2001) and to stimulate appetite and food intake in patients with congestive heart failure (Nagaya et al., 2004), chronic obstructive pulmonary disease (Nagaya et al., 2005), cancer (Neary et al., 2004), and functional dyspepsia (Akamizu et al., 2008).

2. PATHOPHYSIOLOGY OF AN

Patients with AN show diverse symptoms that affect multiple organ systems. Gains in body weight improve most of the medical complications and prevent deterioration of long-term sequelae.

2.1. Medical complications and sequelae due to malnutrition in AN

Routine laboratory examination reveals pancytopenia, decreased serum levels of total protein and albumin or rapid turnover of proteins, liver dysfunction, or serum cholesterol abnormalities. Hypoglycemic coma, dehydration-induced renal failure, rhabdomyolysis, pseudo-Bartter syndrome, and arrhythmia due to hypokalemia are serious complications and can cause death (Neumarker, 1997). Adolescence is a time of accelerated physical growth and maturity, which are affected by the nutritional status. Children with AN often show reduced height, and short stature is a sequelae of this disorder. AN is frequently accompanied by osteoporosis, which involves a reduction of bone formation and an increase in bone resorption (Hotta et al., 1998). We previously reported that duration of body mass index (BMI) < 16 kg/m^2 is a potent risk factor for bone loss because catabolic bone metabolism is improved in patients with BMI greater than 16 kg/m^2 with increased serum levels of insulin-like growth factor-I (IGF-I) as an osteogenic growth factor and estradiol as a strong inhibitor of bone resorption (Hotta et al., 2000). The recovery of spinal bone mineral density also positively correlates with body weight. We find a variety of endocrinological abnormalities, including euthyroid sick syndrome, increased plasma levels of GH, hypogonadotropic hypogonadism, increased plasma levels of ACTH, and hypercortisolemia with the loss of diurnal rhythm in patients with AN (Hotta et al., 1986). Plasma levels of leptin, which positively correlate with body fat, are suppressed (Grinspoon et al., 1996), while plasma levels of adiponectin increase (Delporte et al., 2003) in patients with AN.

2.2. Gastrointestinal symptoms and complications

Chronic malnutrition induces both functional and organic changes in the gastrointestinal tract (Abell et al., 1987; Heather et al., 2002). Most patients with AN complain of early satiety, postprandial abdominal discomfort and fullness, and constipation, which are usually chronic or recurrent. Decreased and impaired motility of the stomach are common. Laboratory examinations of the stomach reveal atrophy of the mucosa, alteration of peristalsis, and delayed emptying time (Benini et al., 2004; Domstad et al., 1987; McCallum et al., 1985). Because acetaminophen is absorbed in the duodenum, plasma concentrations of acetaminophen can be used to measure gastric excretion (Fig. 24.1). Gastric excretion estimated by plasma acetaminophen concentrations in patients with AN were delayed (Heading et al., 1973), which can result in gastric stasis and early satiety and predispose patients to esophageal reflux. Currently prescribed appetite-stimulating drugs such as metoclopramide, cyproheptadine, and sulpiride are not always effective, and any increase in appetite may be minor. Therefore, there is a pressing need for effective appetite-stimulating therapies for patients with AN. In addition, even after becoming fully motivated to gain body weight, patients with AN may succumb to the fear of gastrointestinal discomfort and often cannot increase their food intake. Emaciation induces a vicious downward spiral of malnutrition and resistance to psychotherapy.

3. PLASMA GHRELIN IN AN

Intact ghrelin, which comprises 28 amino acid residues with an *n*-octanoyl ester at Ser^3, is unstable and rapidly degrades to inactive des-octanoyl form or smaller fragments (degraded ghrelin). Although it has been reported that plasma levels of total ghrelin (intact and degraded ghrelin) increase in patients with AN, we have to pay special attention to which form of ghrelin increases in patients with AN.

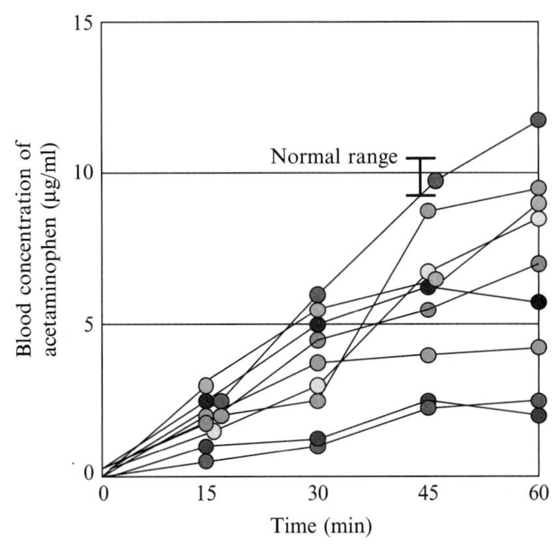

Figure 24.1 Plasma concentrations of acetaminophen in patients with AN. At 08:00 h after overnight fasting for longer than 12 h, blood samples were drawn from female patients with AN (age: 18.2–31.4 year, BMI: 11.48–17.08 kg/m^2) for the determination of plasma acetaminophen concentrations immediately before and every 15 min after the ingestion of 1.5 g acetaminophen and 200 kcal liquid diet for 60 min. The plasma level of acetaminophen in age-matched young women was 9.4 $\pm$ 0.5 μg/ml. Gastric excretion was delayed in 9 out of 10 patients with AN. (See Color Insert.)

3.1. Plasma levels of intact and degraded ghrelin in patients with AN

It was previously reported that plasma levels of ghrelin negatively correlate with body fat and are higher in patients with AN than in healthy controls. However, the antibodies used for measuring plasma ghrelin levels in those studies were against C-terminus ghrelin (13–28) (Shiiya et al., 2002), ghrelin (1–11) (Nakai et al., 2003), or full-length ghrelin with or without octanoylation at Ser^3 (Otto et al., 2001; Tolle et al., 2003), all of which might detect intact as well as degraded ghrelin. Concentrations of intact ghrelin actually represent $<10\%$ of total circulating ghrelin levels (Yoshimoto et al., 2002). Because only intact ghrelin, not degraded ghrelin, has an orexigenic effect, plasma levels of intact ghrelin should be measured in patients with AN. Evaluation of plasma ghrelin levels depends on the specificity of the ghrelin antibody.

To determine differences in the profile of plasma levels of intact and degraded ghrelin between patients with AN and healthy women, plasma levels

Table 24.1 Five assays used to determine differences in the profile of plasma levels of intact and degraded ghrelin between female patients with AN and healthy women

Name of assay or kit	Ghrelin and its fragments against which antibodies are raised	
ICT-EIA	Cys^{12} ghrelin (1–11)	Cys^{29}ghrelin (1–28)
Active ghrelin ELISA	Ghrelin (1–10)	Ghrelin (13–28)
Des-acyl ghrelin ELISA	Des-octanoyl ghrelin (1–10)	Ghrelin (13–28)
Ghrelin active RIA	Ghrelin (1–10)	
Ghrelin total RIA		Ghrelin (14–28)

of intact, des-octanoyl-, N-terminus, and C-terminus ghrelin were measured using five assays (Table 24.1) (Hotta et al., 2004). The immunocomplex transfer-enzyme immunoassay (ICT-EIA) is designed to measure intact human ghrelin based on the principle of two-site sandwich enzyme-linked immunosorbent assay (ELISA) using two different polyclonal antibodies and specifically detects intact human ghrelin but does not detect shorter fragments or des-octanoyl ghrelin. The active ghrelin ELISA kit (Mitsubishi Kagaku Iatron, Tokyo, Japan) detects intact human or murine ghrelin specifically (Hosoda et al., 2000). Plasma levels of ghrelin from ICT-EIA were significantly correlated with values from the active ghrelin ELISA kit in healthy women ($r=0.876$, $p=0.0007$) and patients with AN ($r=0.796$, $p<0.0001$) (Hotta et al., 2004). Plasma levels of des-octanoyl ghrelin were measured using the des-acyl ghrelin ELISA kit (Mitsubishi Kagaku Iatron) (Hosoda et al., 2000). Plasma levels of N- and C-terminus ghrelin were measured using the ghrelin active RIA and the ghrelin total RIA kits (Linco Research, St. Charles, MO, USA), respectively. The antibody used in the ghrelin active RIA kit recognizes intact ghrelin and octanoyl ghrelin (1–10), but not des-octanoyl ghrelin, whereas the antibody used in the ghrelin total RIA kit recognizes intact and des-octanoyl ghrelin, and ghrelin (14–28). After an overnight fasting of longer than 12 h, blood was taken from subjects at 08:00 h and transferred into tubes with 1 mg/ml EDTA-2Na and 500 U/ml aprotinin. Blood samples were immediately centrifuged at 4 °C. Plasma samples were then acidified with 1 N HCl and stored at −80 °C until assay.

Mean plasma levels of intact ghrelin in 30 female patients with AN (BMI, 8.81–22.4 kg/m^2) obtained by ICT-EIA or the active ghrelin ELISA kit were lower than or similar to those of 16 age-matched healthy women, whereas levels of degraded forms of ghrelin, such as des-octanoyl ghrelin,

Table 24.2 Plasma levels of ghrelin in controls and patients with anorexia nervosa

Subject	Controls	AN
n	16	30
ICT-EIA (pmol/l)	65.0 ± 4.9	49.2 ± 2.9*
Active ghrelin ELISA (pmol/l)	29.9 ± 3.1	34.7 ± 3.2
Des-acyl ghrelin ELISA (pmol/l)	94.1 ± 7.5	223.5 ± 37.3*
Ratio of des-acyl to active ghrelin ELISA	3.34 ± 0.24	6.14 ± 0.44*
Ghrelin active RIA (pmol/l)	104.1 ± 9.5	136.7 ± 12.9*
Ghrelin total RIA (nmol/l)	1.85 ± 0.13	2.87 ± 0.25*
Ratio of ghrelin total to Active RIA	19.9 ± 2.1	21.9 ± 1.2

Data are expressed as mean ± SEM. *$p < 0.05$ compared to values of controls.

octanoyl N-terminus ghrelin, and C-terminus ghrelin, were significantly elevated in patients with AN compared with normal women (Table 24.2). Results also showed that no correlation existed between BMI and intact ghrelin in patients with AN. In contrast, plasma levels of degraded ghrelin or the ratio of des-octanoyl ghrelin to intact ghrelin correlated with BMI in patients and controls (Fig. 24.2) (Hotta et al., 2004).

Of note is the fact that plasma levels of intact ghrelin in patients with AN were not higher than in controls. Plasma ghrelin levels reportedly increase after cure of *Helicobacter pylori* infection (Nwokolo et al., 2003). In addition, the gastric banding procedure strongly suppresses plasma ghrelin levels despite a massive and permanent reduction in body weight (Leonetti et al., 2003), as gastric banding reduces plasma levels of motilin. These results indicate that injury to the gastric mucosa or impaired gastric peristalsis could decrease ghrelin secretion. Therefore, the reason patients with AN did not show higher plasma levels of intact ghrelin than controls seems likely due to a decrease in ghrelin secretion from gastric mucosa, induced by malnutrition.

Plasma levels of degraded ghrelin were much higher in patients with AN than in controls, which indicates that the profiles of intact and degraded forms of ghrelin in plasma of patients with AN differ from those of healthy women. Plasma levels of total ghrelin, but not intact ghrelin, are significantly correlated with renal function (Akamizu et al., 2004). The kidney represents an important site for the clearance and/or degradation of ghrelin. In patients with end-stage renal disease, plasma levels of C-terminus ghrelin are significantly correlated with serum creatinine levels (Yoshimoto et al., 2002).

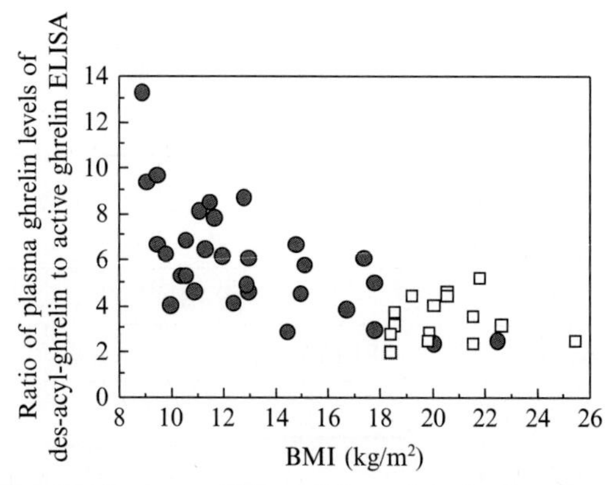

Figure 24.2 Relationship between BMI and the ratio of values from des-acyl ghrelin ELISA to those from Active Ghrelin ELISA. BMI significantly correlated with plasma ghrelin levels from degraded ghrelin and the ratio of values from des-acyl ghrelin ELISA to those from active ghrelin ELISA ($r = -0.737$, $p < 0.0001$). Open squares and closed circles represent healthy women and patients with AN, respectively.

Elevated plasma levels of C-terminus ghrelin have recently been demonstrated in hepatic cachexia (Tacke et al., 2003) with deterioration of the clinical status as determined by signs such as ascites or reduced renal clearance. AN is also usually complicated by dehydration, reduced glomerular filtration rate, and decreased creatinine clearance (Aperia et al., 1978). Elevation of plasma levels of degraded ghrelin in patients with AN may therefore result from decreased renal clearance related to decreased BMI.

3.2. Effects of glucose on plasma levels of intact and degraded ghrelin

Intravenous infusion of 50 g glucose or oral administration of 75 g glucose suppressed the secretion of C-terminus ghrelin in healthy subjects (Nakagawa et al., 2002; Shiiya et al., 2002), whereas meal-induced decrease in plasma ghrelin levels was not found in patients with AN (Nedvídkova et al., 2003). Since gastric excretion time is delayed in these patients, changes in plasma glucose levels and insulin secretion are extremely variable after oral glucose tolerance testing or after eating food in AN (Nozaki et al., 1994). We therefore used the 50-g glucose infusion test to investigate the effects of hyperglycemia on plasma ghrelin levels. When glucose was infused in six female patients with AN and six age-matched healthy women, plasma glucose levels increased significantly (controls: from 92.3 ± 2.3 to 182.0 ± 15.1 mg/dl, AN: from 68.7 ± 6.5 to

227.0 ± 29.7 mg/dl), and plasma levels of intact ghrelin promptly decreased significantly in both groups (controls: 58.8 ± 3.3% vs. AN: 63.2 ± 9.8% of the basal levels, $p = 0.206$). After glucose infusion in controls, plasma levels of degraded ghrelin significantly decreased. Conversely, plasma levels of degraded ghrelin displayed no significant changes after glucose infusion in patients with AN. These results may suggest that acute elevation of plasma glucose inhibits secretion of intact ghrelin from the stomach and that the substantial increase in fragments of degraded ghrelin in plasma would mask the response of degraded ghrelin.

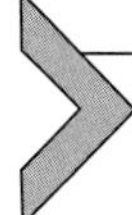

4. CLINICAL APPLICATION OF GHRELIN IN PATIENTS WITH AN

There are two reports about the effects of ghrelin on appetite in patients with AN. In one study, 5 pmol/kg/min ghrelin infusion for 300 min had little effect on appetite in severely emaciated as well as weight-recovered patients with AN (Miljic et al., 2006). Since those patients with AN refused to eat, food intake was not investigated in the study. In another study, 1 μg/kg ghrelin infusion made patients with AN feel hunger sensations, although food intake was not evaluated (Broglio et al., 2004). We therefore believe that studies aiming to investigate ghrelin as an appetite-stimulating therapy should recruit only those patients with AN who are fully motivated by psychoeducational therapy to gain body weight.

4.1. Study design

Five Japanese female patients with restricting-type AN were included in the present study, who met the Diagnostic and Statistical Manual IV criteria for AN, in addition to those of the Survey Committee for Eating Disorders of the Japanese Ministry of Health, Labor and Welfare (Hotta et al., 1986) (Table 24.3). All subjects tested negative for *H. pylori*, and none of the patients had started medication prior to the trial. They had already taken intense counseling and supervision by a dietitian as well as total parenteral nutrition during a previous hospital admission but then had not been able to increase weight or lost weight again. All patients had been motivated to gain weight but could not increase their food intake for several years in four out of five in part because of gastrointestinal discomfort.

For ethical reasons, involving a nontreated group was not possible and a randomized controlled design or blinding methods were not applied to this study. The study protocol was approved by the Institutional Review Board

Table 24.3 Clinical profile of patients with anorexia nervosa in the present study

Case no.	**1**	**2**	**3**	**4**	**5**
Age on entry	27	31	25	35	14
Height (cm)	161	157	156.2	154	149.6
Weight before illness (kg) (BMI kg/m^2)	48 (18.5)	48 (19.5)	44.2 (18.1)	50 (21.0)	43 (19.2)
Age of onset (years)	16	24	17	20	13
Duration of illness (years)	12	6	8	15	1.25
The minimal weight (kg) (BMI kg/m^2)	29 (11.2)	30 (12.2)	32 (13.11)	23 (9.70)	27.4 (12.2)
Weight on entry (kg) (BMI kg/m^2)	37.9 (14.6)	32.5 (13.2)	35.0 (14.4)	24.2 (10.2)	28.2 (12.6)
The increment of total calorie (kcal)	12%	36%	16%	33%	14%
Weight at the end of study (kg) (BMI kg/m^2)	36.4 (14.0)	31.5 (12.8)	35.7 (14.6)	26.6 (11.2)	28.4 (12.7)
Weight at 6 months after discharge (kg) (BMI kg/m^2)	43 (16.6)	38.5 (15.6)	38.2 (15.7)	28 8 (11.8)	34.5 (15.4)

of Tokyo Women's Medical University. All patients provided written informed consent to participate in this study.

Patients were hospitalized for 26 days (day −6 to day 20) in Tokyo Women's Medical University Hospital. Food intake and subjective hunger sensation were measured for 24 days (day −5 to day 19). The pretreatment period was defined as the 5 days before ghrelin injection (day −5 to day −1). Patients received an intravenous infusion of ghrelin (3 μg/kg body weight) for 5 min twice a day (before breakfast and dinner) for 14 days (day 1 to day 14). After ghrelin infusion, subjects were monitored for clinical efficacy and safety of ghrelin for 5 days (day 15 to day 19) as the posttreatment period (Fig. 24.3). Since ghrelin at doses of 1 and 5 μg/kg tended to increase appetite dose dependently without severe adverse effects (Akamizu et al., 2004), we employed 3 μg/kg of ghrelin in the present study.

The ghrelin used in the study was an acylated peptide that was dissolved in 3.75% D-mannitol to yield a final concentration of 180 μg/ml, as described previously (Akamizu et al., 2008). The solutions were filtered and stored at −20 °C in sterile vials. Examination by the Japan Food Research

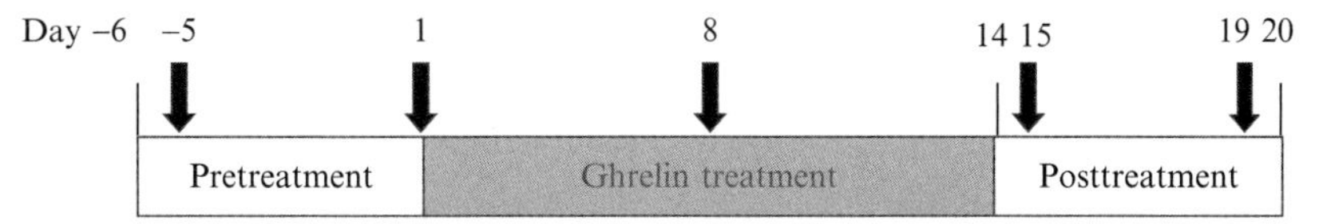

Figure 24.3 Timeline of ghrelin study. Subjects were hospitalized for 26 days (day −6 to day 20). The pretreatment period was defined as the 5 days before ghrelin injection (day −5 to day −1). Subjects received an intravenous infusion of ghrelin (3 μg/kg body weight) for 5 min twice a day (before breakfast and dinner) for 14 days (day 1 to day 14). After ghrelin infusion, subjects were monitored for the clinical efficacy and safety of ghrelin for 5 days (day 15 to day 19) as the posttreatment period. (For color version of this figure, the reader is referred to the online version of this chapter.)

Laboratories (Tokyo, Japan) did not find any traces of endotoxin in the ghrelin solutions. A pyrogen test based on the Pharmacopeia of Japan was also negative.

The primary end point of this study was energy intake. Since patients with AN lose appetite when too large an amount of food to eat is served, they were initially served an amount of food equivalent to their meals at home before hospitalization plus an additional 200 kcal.

Each dish was weighed before and after eating and was photographed by a digital camera. Energy intake was calculated by dieticians as the total energy, carbohydrate, fat, and protein intakes. When subjects ate all the food served and wanted more, they were allowed to eat self-prepared foods yielding approximately 200 kcal such as fruits or other snacks. Attitudes toward food were evaluated by a questionnaire incorporating visual analogue scales (VAS) rating hunger, satiety, prospective food consumption, fullness, desire for some meat or fish, desire for something salty, desire for something sweet, and desire for something fatty. It had been demonstrated that food intake correlated with perceptions of hunger and fullness as assessed by VAS in healthy volunteers (Parker et al., 2004). During ghrelin treatment, patients with AN answered the VAS questionnaire at 15 min before ghrelin infusion and breakfast or dinner, 15 min after ghrelin infusion before breakfast or dinner, and after those meals. They also answered the questionnaire before and after every meal without ghrelin treatment (Fig. 24.4).

Blood and urine samples for biochemical and endocrinological parameters, including complete blood count (CBC), total protein, albumin, rapid turnover proteins, liver function, lipid profile, immunoreactive insulin, leptin, GH, IGF-I, prolactin, ACTH, active ghrelin, and des-acyl ghrelin were taken in the morning after overnight fasting longer than 10 h on day −5, day 1, day 8, day 15, and day 19.

Psychological states were evaluated using the Japanese versions of the self-rating depression scale (Zung, 1965), state-trait anxiety inventory

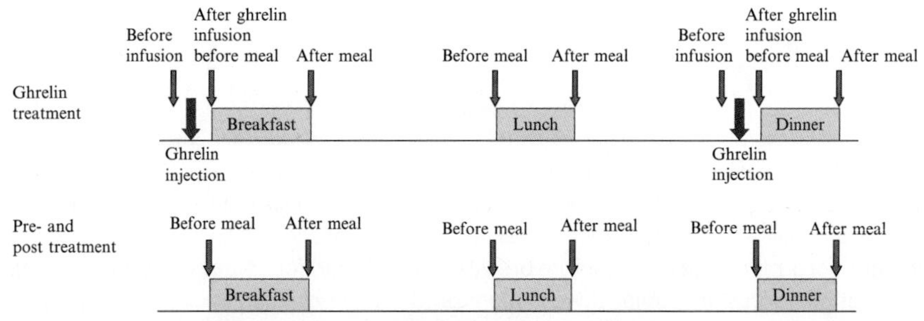

Figure 24.4 Schedule of visual analogue scales. Attitudes toward food were evaluated by a questionnaire incorporating visual analogue scales (VAS). During ghrelin treatment, patients with AN answered the VAS questionnaire at 15 min before ghrelin infusion and breakfast or dinner, 15 min after ghrelin infusion before breakfast or dinner, and after those meals. They answered the questionnaire also before and after every meal without ghrelin treatment. (See Color Insert.)

(Iwata et al., 1998), and eating disorder inventory (Garner and Garfinkel, 1979) on day −5, day 1, day 8, day 15, and day 19.

Data were expressed as mean ± SE. Two-way analysis of variance was used for energy and nutrient intakes and for biochemical and endocrinologic data. Data were examined by Student's two-tailed paired *t*-test when appropriate. Appetite scores were analyzed by a paired *t*-test comparing the changes in VAS. Statistical analyses were performed using the computer statistical package SPSS (version 11.0.1; SPSS Inc., Chicago, IL). Levels of significance were determined at $p < 0.05$.

4.2. Effects of ghrelin infusion on hunger sensation and gastrointestinal symptoms

All patients reported that they had sensations of stomach activity or that their upper abdominal fullness disappeared 5 min after ghrelin injection, which continued for 30–60 min. Borborygmi were also audible within 30 min just after each ghrelin infusion, and no patient reported constipation during ghrelin treatment. Hunger sensation assessed by VAS was higher after ghrelin infusion than before ghrelin infusion in four patients. The stimulatory effects of ghrelin on hunger sensation disappeared after eating and did not last until the next meal.

The sensation of hunger is usually correlated with gastric emptying in humans (Sepple and Read, 1989). Ghrelin plays a role in the regulation of gastrointestinal motility and acid secretion in rats (Edholm et al., 2004; Masuda et al., 2000) and increases the gastric emptying rate in normal-weight humans (Levin et al., 2006). Because recognition of hunger and satiety in patients with AN is generally impaired, appetite cannot be always analyzed correctly by VAS alone. However, hunger sensation was higher just after ghrelin infusion than before ghrelin infusion in four patients, and ghrelin improved epigastric discomfort in all patients. This was probably mediated through increased gastric peristalsis as shown in other diseases with gastrointestinal dysfunction (Binn et al., 2006; Murray et al., 2005; Strasser et al., 2008).

4.3. Effects of ghrelin infusion on food intake and body weight

The daily energy intake of the five patients during the pretreatment period ranged from 825 to 1426 kcal. During ghrelin infusion, four patients showed a statistically significant increase in daily energy intake (Fig. 24.5). Mean increase in daily energy intake during ghrelin infusion was 20±4% when compared with the pretreatment period. Analysis of nutrients revealed significant increases in daily intakes of carbohydrate in three patients, fat in one

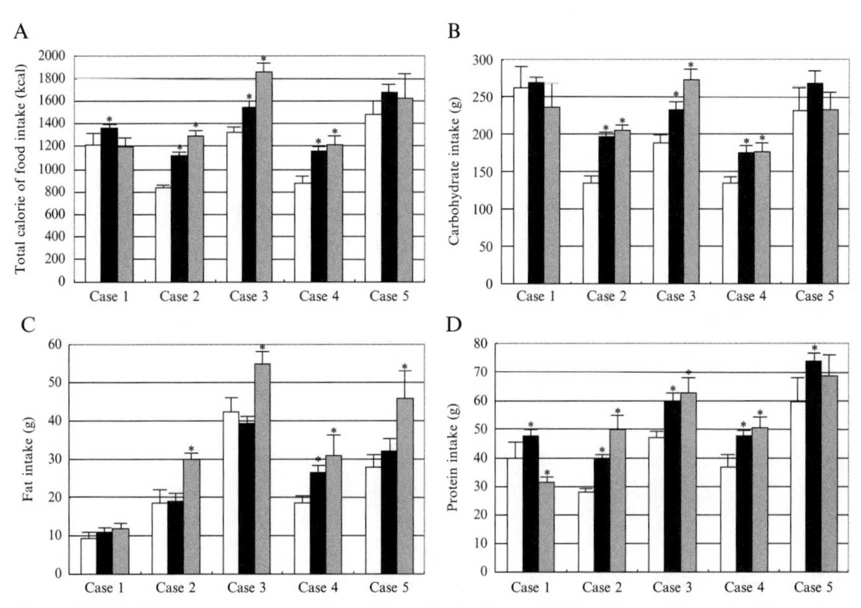

Figure 24.5 Changes in total energy (A), carbohydrate (B), fat (C), and protein (D) intakes of AN patients. During ghrelin infusion, four patients showed a statistically significant increase in daily energy intake. Mean increase in daily energy intake during ghrelin infusion was 20±4% when compared with the pretreatment period. Open, closed, and gray bars represent intake during pretreatment, ghrelin treatment, and posttreatment periods, respectively. Data are expressed as mean±SE. *$p < 0.05$ versus prescreening period.

patient, and protein in all patients (Hotta et al., 2009). Interestingly, daily energy intake during postscreening remained higher than in the prescreening period in three patients. Those residual effects continued for several days. The increments of body weight in five patients ranged from −1.5 to 2.4 kg during the study. Because a 1-kg weight gain requires 7000–8000 kcal, the increase in energy intake achieved for 14 days in this study was not enough to lead to any considerable weight gain. Although a patient was able to gain 2.4 kg and showed remarkable improvement in nutritional parameters and malnutrition-related liver dysfunction, we believe that water retention during the refeeding period contributed to this weight gain (Yucel et al., 2005). Although two patients lost body weight during the study, this effect was thought to be due to a decrease in malnutrition-induced fluid retention or improvement in bowel movements. The improvement effects of ghrelin on gastrointestinal symptom disappeared after the study ended. However, all patients gained weight after discharge and one resumed menstruation 6 months after discharge. As the patients told us that they were happy to eat free from uncomfortable gastrointestinal symptoms after ghrelin treatment, it is speculated that ghrelin triggered an improvement in gastrointestinal function, which ameliorated the fear of gastrointestinal discomfort after eating.

4.4. Effects of ghrelin infusion on biochemically nutritional markers

Serum total protein and triglyceride levels increased significantly between before and after ghrelin treatment. Other nutritional markers including serum levels of transferrin and glucose showed a tendency to increase after ghrelin infusion but did not reach statistical significance. Mean plasma levels of insulin and leptin did not increase significantly during ghrelin treatment. The mean level of serum IGF-I during ghrelin treatment did not significantly change. Mean plasma levels of prolactin, ACTH, and cortisol, which were measured early in the morning, before ghrelin injection, did not change significantly during ghrelin treatment.

4.5. Adverse effects of ghrelin infusion

No serious adverse events occurred during ghrelin treatment. We did not detect any changes in vital signs or biochemical and endocrinologic data after ghrelin treatment. Adverse effects such as abdominal discomfort, diarrhea, transient flushing, truncal perspiration (Akamizu et al., 2004), and somnolence have been reported after ghrelin injection (Miljic et al., 2006). The only noted event was loose stools on day 6 in case 5, whose dose of ghrelin was reduced to 1.5 μg/body weight from day 7 to day 14, resulting in an improvement of symptoms. An occasionally warm sensation in the trunk or mild sweating was noted in two subjects. No patient developed somnolence during ghrelin treatment. Although we were concerned that the increase in appetite induced by ghrelin might aggravate mental status in patients with AN, we did not observe increased fear of weight gain, abnormal behavior, anxiety, or depression due to an increase in appetite during ghrelin treatment, and psychological tests did not demonstrate any significant changes.

5. CONCLUSIONS

We performed a pilot study of patients with restricting-type AN to investigate the effects of ghrelin on appetite, caloric intake, and nutritional parameters. Results indicated that ghrelin may have therapeutic potential in restricting-type AN patients who cannot gain weight because of gastrointestinal dysfunction. However, a major limitation of the present study relates to the lack of a randomized, placebo-controlled group and the small number of patients recruited.

Clinicians need to carefully consider whether patients may benefit from ghrelin therapy. Based on this study, appropriate patients for ghrelin therapy may include those with restricting-type AN who are fully motivated to gain weight through intense psychotherapy, but who cannot increase food intake because of persistent gastrointestinal discomfort. Clinicians also need to remember than even patients with AN who are motivated to gain weight may get scared when they feel hunger sensations after ghrelin therapy and may still refrain from eating even though they show an increase in borborygmi after each ghrelin infusion and experience increased hunger.

ACKNOWLEDGMENTS

We would like to thank SRL (Tachikawa, Tokyo, Japan) for technical assistance with the ELISA assay for ghrelin. This study was supported by a fund from the Ministry of Health, Labor and Welfare of Japan. There is no conflict of interest that would prejudice the impartiality of the research.

REFERENCES

Abell, T.L., et al., 1987. Gastric electromechanical and neurohormonal function in anorexia nervosa. Gastroenterology 93, 958–965.

Akamizu, T., et al., 2004. Pharmacokinetics, safety, and endocrine and appetite effects of ghrelin administration in young healthy subjects. Eur. J. Endocrinol. 150, 447–455.

Akamizu, T., et al., 2008. Repeated administration of ghrelin to patients with functional dyspepsia: its effects on food intake and appetite. Eur. J. Endocrinol. 158, 491–498.
Aperia, A., et al., 1978. Renal function in anorexia nervosa. Acta Paediatr. Scand. 67, 219–224.
Ariyasu, H., et al., 2001. Stomach is a major source of circulating ghrelin, and feeding state determines plasma ghrelin-like immunoreactivity levels in humans. J. Clin. Endocrinol. Metab. 86, 4753–4758.
Benini, L., et al., 2004. Gastric emptying in patients with restricting and binge/purging subtypes of anorexia nervosa. Am. J. Gastroenterol. 99, 1448–1454.
Binn, M., et al., 2006. Ghrelin gastrokinetic action in patients with neurogenic gastroparesis. Peptides 27, 1603–1606.
Birmingham, C.L., Treasure, J., 2010. Medical Management of Eating Disorders, 2nd ed. Cambridge University Press, Cambridge 5–13.
Broglio, F., et al., 2004. The endocrine response to acute ghrelin administration is blunted in patients with anorexia nervosa, a ghrelin hypersecretory state. Clin. Endocrinol. (Oxf) 60, 592–599.
Bruch, H., 1985. Four decades of eating disorders. In: Garner, D.M., Garfinkel, P.E. (Eds.), Handbook of Psychotherapy for Anorexia Nervosa and Bulimia. Guilford, New York, USA, 7–18.
Delporte, M.L., et al., 2003. Hyperadiponectinaemia in anorexia nervosa. Clin. Endocrinol. (Oxf) 58, 22–29.
Domstad, P.A., et al., 1987. Radionuclide gastric emptying studies in patients with anorexia nervosa. J. Nucl. Med. 28, 816–819.
Edholm, T., et al., 2004. Ghrelin stimulates motility in the small intestine of rats through intrinsic cholinergic neurons. Regul. Pept. 121, 25–30.
Garner, D.M., Garfinkel, P.E., 1979. The eating attitude test: an incex of symptoms of anorexia nervosa. Psychol. Med. 9, 273–279.
Grinspoon, S., et al., 1996. Serum leptin levels in women with anorexia nervosa. J. Clin. Endocrinol. Metab. 81, 3861–3863.
Heading, R.C., et al., 1973. The dependence of paracetamol absorption on the rate of gastric emptying. Br. J. Pharmacol. 47, 415–421.
Heather, J.C., et al., 2002. Anorexia nervosa: manifestation and management for the gastroenterologist. Gastroenterology 97, 255–269.
Hosoda, H., et al., 2000. Ghrelin and des-acyl ghrelin: two major forms of rat ghrelin peptide in gastrointestinal tissue. Biochem. Biophys. Res. Commun. 279, 909–913.
Hotta, M., et al., 1986. The responses of plasma adrenocorticotropin and cortisol to corticotropin-releasing hormone (CRH) and cerebrospinal fluid immunoreactive CRH in anorexia nervosa patients. J. Clin. Endocrinol. Metab. 62, 319–324.
Hotta, M., et al., 1998. The importance of body weight history in the occurrence and recovery of osteoporosis in patients with anorexia nervosa: evaluation by dual X-ray absorptiometry and bone metabolic markers. Eur. J. Endocrinol. 139, 276–283.
Hotta, M., et al., 2000. The relationship between bone turnover and body weight, serum insulin-like growth factor (IGF) I, and serum IGF-binding protein levels in patients with anorexia nervosa. J. Clin. Endocrinol. Metab. 85, 200–206.
Hotta, M., et al., 2004. Plasma levels of intact and degraded ghrelin and their responses to glucose infusion in anorexia nervosa. J. Clin. Endocrinol. Metab. 89, 5707–5712.
Hotta, M., et al., 2009. Ghrelin increases hunger and food intake in patients with restricting-type anorexia nervosa: a pilot study. Endocr. J. 56, 1119–1128.
Iwata, N., et al., 1998. The Japanese adaptation of the STAI Form Y in Japanese working adults—the presence or absence of anxiety. Ind. Health 36, 8–13.
Kojima, M., et al., 1999. Ghrelin is a growth-hormone-releasing acylated peptide from stomach. Nature 402, 656–660.

Leonetti, F., et al., 2003. Different plasma ghrelin levels after laparoscopic gastric bypass and adjustable gastric banding in morbid obese subjects. J. Clin. Endocrinol. Metab. 88, 4227–4231.

Levin, F., et al., 2006. Ghrelin stimulates gastric emptying and hunger in normal-weight humans. J. Clin. Endocrinol. Metab. 91, 3296–3302.

Masuda, Y., et al., 2000. Ghrelin stimulates gastric acid secretion and motility in rats. Biochem. Biophys. Res. Commun. 276, 905–908.

McCallum, R.W., et al., 1985. Definition of a gastric emptying abnormality in patients with anorexia nervosa. Dig. Dis. Sci. 30, 713–722.

Miljic, D., et al., 2006. Ghrelin has partial or no effect on appetite, growth hormone, prolactin, and cortisol release in patients with anorexia nervosa. J. Clin. Endocrinol. Metab. 91, 1491–1495.

Murray, C.D., et al., 2005. Ghrelin enhances gastric emptying in diabetic gastroparesis: a double-blind, placebo-controlled, cross-over study. Gut 54, 1693–1698.

Nagaya, N., et al., 2004. Effects of ghrelin administration on left ventricular function, exercise capacity, and muscle wasting in patients with chronic heart failure. Circulation 110, 3674–3679.

Nagaya, N., et al., 2005. Treatment of cachexia with ghrelin in patients with COPD. Chest 128, 1187–1193.

Nakagawa, E., et al., 2002. Hyperglycaemia suppresses the secretion of ghrelin, a novel, growth-hormone-releasing peptides: responses to the intravenous and oral administration of glucose. Clin. Sci. (Lond.) 103, 325–328.

Nakai, Y., et al., 2003. Plasma levels of active form of ghrelin during oral glucose tolerance test in patients with anorexia nervosa. Eur. J. Endocrinol. 149, R1–R3.

Neary, N.M., et al., 2004. Ghrelin increases energy intake in cancer patients with impaired appetite: acute, randomized, placebo-controlled trial. J. Clin. Endocrinol. Metab. 89, 2832–2836.

Nedvídkova, J., et al., 2003. Loss of meal-induced decrease in plasma ghrelin levels in patients with anorexia nervosa. J. Clin. Endocrinol. Metab. 88, 1678–1682.

Neumarker, K.-J., 1997. Mortality and sudden death in anorexia nervosa. Int. J. Eat. Disord. 21, 205–212.

Nozaki, T., et al., 1994. Insulin response to intravenous glucose in patients with anorexia nervosa showing low insulin response to oral glucose. J. Clin. Endocrinol. Metab. 79, 217–222.

Nwokolo, C.U., et al., 2003. Plasma ghrelin following cure of *Helicobacter pylori*. Gut 52, 637–640.

Otto, B., et al., 2001. Weight gain decreases elevated plasma ghrelin concentrations of patients with anorexia nervosa. Eur. J. Endocrinol. 145, 669–673.

Parker, B.A., et al., 2004. Relation between food intake and visual analogue scale ratings of appetite and other sensations in healthy older and young subjects. Eur. J. Clin. Nutr. 58, 212–218.

Sepple, C.P., Read, N.W., 1989. Gastrointestinal correlates of the development of hunger in man. Appetite 13, 183–191.

Shiiya, T., et al., 2002. Plasma ghrelin levels in lean and obese humans and the effect of glucose on ghrelin secretion. J. Clin. Endocrinol. Metab. 87, 240–244.

Strasser, F., et al., 2008. Safety, tolerability and pharmacokinetics of intravenous ghrelin for cancer-related anorexia/cachexia: a randomised, placebo-controlled, double-blind, double-crossover study. Br. J. Cancer 98, 300–308.

Su, J.C., Birmingham, C.L., 2003. Anorexia nervosa: the cost of long-term disability. Eat. Weight Disord. 8, 76–79.

Tacke, F., et al., 2003. Ghrelin in chronic liver disease. J. Hepatol. 38, 447–454.

Tolle, V., et al., 2003. Balance in ghrelin and leptin plasma levels in anorexia nervosa patients and constitutionally thin women. J. Clin. Endocrinol. Metab. 88, 109–116.
Wren, A.M., et al., 2001. Ghrelin enhances appetite and increases food intake in humans. J. Clin. Endocrinol. Metab. 86, 5992–5995.
Yoshimoto, A., et al., 2002. Plasma ghrelin and des-acyl ghrelin concentrations in renal failure. J. Am. Soc. Nephrol. 13, 2748–2752.
Yucel, B., et al., 2005. Weight fluctuations during early refeeding period in anorexia nervosa: case reports. Int. J. Eat. Disord. 37, 175–177.
Zung, W.W., 1965. A self-rating depression scale. Arch. Gen. Psychiatry 12, 63–70.

CHAPTER TWENTY-FIVE

Clinical Application of Ghrelin for Chronic Respiratory Diseases

Nobuhiro Matsumoto[1], **Masamitsu Nakazato**

Division of Neurology, Respirology, Endocrinology, and Metabolism, Department of Internal Medicine, Faculty of Medicine, University of Miyazaki, Miyazaki, Japan

[1]Corresponding author: e-mail address: nobu@fc.miyazaki-u.ac.jp

Contents

Abstract

The discovery of ghrelin has resulted in the development of potential therapeutics for cachexia caused by multiple underlying diseases. When chronic respiratory diseases progress to their advanced stages, cachexia often occurs, thereby worsening the patient's prognosis. A small clinical trial that enrolled cachectic patients with chronic respiratory disease revealed that administration of ghrelin improved their nutritional status and exercise tolerance. Short-term administration of ghrelin was found to be safe and tolerated with adverse events, including suffusion, sleepiness, peristalsis, hunger, and sweating. Further large-scale and long-term clinical trials are needed.

1. INTRODUCTION

Chronic respiratory failure is a condition that results in the inability to effectively exchange carbon dioxide and oxygen, and induces chronically low oxygen levels or chronically high carbon dioxide levels. The condition is usually caused by chronic respiratory diseases, such as chronic obstructive

Methods in Enzymology, Volume 514
ISSN 0076-6879
http://dx.doi.org/10.1016/B978-0-12-381272-8.00025-8

pulmonary disease (COPD), chronic bronchitis, and idiopathic interstitial pneumonia. If diseases with chronic respiratory failure progress to their advanced stages, cachexia commonly occurs (Schols, 2002). Cachexia is characterized by physical and muscle wasting, increased metabolic rate, and decreased appetite. The presence of cachexia worsens the patients' quality of life and prognosis (Anker et al., 1997; Schols, 2002). Although cachectic status is an important therapeutic potential target, there are no adequate treatments for cachexia in clinical use. The discovery of ghrelin has resulted in the development of approaches to treat cachexia. The various effects of ghrelin, which are to increase food intake and body weight, anti-inflammatory effects, and energy metabolism regulation, are ideal targets for the treatment of cachexia. In this chapter, we focus on the clinical application of ghrelin for the treatment of cachectic chronic respiratory disease.

2. CLINICAL APPLICATION OF GHRELIN FOR CHRONIC RESPIRATORY DISEASE

COPD has been defined by the Global Initiative for Chronic Obstructive Lung Disease as a disease state characterized by airflow limitation that is not fully reversible (Pauwels et al., 2001). The airflow limitation is usually progressive and causes significant respiratory symptoms and extrapulmonary symptoms. Patients with COPD often show a certain degree of cachexia, which is a catabolic state characterized by skeletal muscle wasting, appetite loss, and weakness as a result of increased apoptosis and muscle disuse (Agustí et al., 2002). Increased production of inflammatory mediators, such as tumor necrosis factor (TNF)-α, interleukin, and oxygen-derived free radicals, may mediate this state (Schols et al., 1996). Cachexia is commonly seen in patients with severe COPD and is an independent risk factor for mortality in such patients (Wilson et al., 1989). Various physiological effects of ghrelin, such as GH release, appetite regulation, and energy metabolism regulation, are expected to improve the clinical condition of COPD patients with cachexia.

2.1. Preparation of synthetic human ghrelin

Synthesis of ghrelin is detailed in Chapter 12. Synthetic human ghrelin was dissolved in distilled water with 4% D-mannitol and sterilized by passage through a 0.22-μm filter. Ghrelin was stored in 2-ml volumes, each containing 200 μg ghrelin. The chemical nature and content of the human ghrelin in vials were verified by high-performance liquid chromatography and radioimmunoassay. To ensure the safety of clinical application, a bacterial endotoxin test and pyrogen test were performed according to the standards of the United States Pharmacopeial Convention. All vials were stored frozen at −80 °C from the time of dispensing until the time of preparation for administration.

2.2. Administration of synthetic human ghrelin

Human ghrelin (2 μg/kg) was dissolved in saline and the total amount of the solution was adjusted to 10 ml. The ghrelin solution was administered intravenously for 30 min at a constant rate. The infusion was repeated twice a day (before breakfast and before dinner) for 3 weeks.

As synthetic ghrelin dissolved in distilled water with 4% D-mannitol is stable at room temperature at least for 12 h, there is no problem about the stability of the solution during the administration.

The appropriate dose and period of ghrelin administration are still unclear. In previous exploratory studies, 2 μg/kg of human ghrelin administrated intravenously twice a day for 3 weeks improved exercise capacity and weight loss in patients with cachectic COPD (Nagaya et al., 2005).

2.3. Exercise training

Exercise training has been proved to be an essential component of pulmonary rehabilitation (Gosselink et al., 1997) and was conducted in three sets daily, every weekday for 3 weeks using a cycle ergometer. The initial workload of each exercise set was for 6 min and 60% of the maximal oxygen intake (VO_2max) measured on the baseline cardiopulmonary exercise testing (CPET). Each daily exercise set was increased to 10 min by 2-min increments if the patient tolerated it. After that, the training workload of each set was increased daily by 5 W and then extended to a workload corresponding to 80% of the baseline VO_2max. If the patient could not tolerate the load, the training workload was reduced to its previous setting. Supplemental oxygen was used if necessary to maintain an oxygen saturation of more than 90% during exercise training.

2.3.1 Cardiopulmonary exercise testing

While breathing room air with a mask, symptom-limited CPET was conducted on a cycle ergometer using an incremental workload protocol (continuous ramp rate of 5 W/min). Expired gas data were measured breath by breath and collected as 30-s averages at rest and during exercise. The CPET was done until subject exhaustion.

2.4. Outcome measures

The main purpose of the clinical application of ghrelin for respiratory cachexia is to improve cachexia, exercise tolerance, and patients' health-related quality of life.

Cachexia is characterized by physical wasting, involving loss of weight and muscle mass. Body weight, food intake, and peripheral and respiratory muscle strength are appropriate endpoints of this purpose. Food intake for three consecutive days was assessed before ghrelin administration and during the last week of ghrelin therapy. The food intake was semiquantitatively assessed by staff nurses using a calorie count based on a 10-point scale method (0 = null intake to 10 = full intake, 1800 kcal), which was averaged for 3 days. Peripheral muscle strength was measured by the maximal voluntary handgrip maneuver. The patients performed four maneuvers on each side, with at least a 1-min interval between each of the maneuvers. The average of the best values on the left and right sides was reported. Respiratory muscle strength was examined during maximal voluntary efforts against an occluded airway. The maximum inspiratory pressure and the maximum expiratory pressure were measured from functional residual capacity. The patients performed four maneuvers, and the highest value was reported.

The most popular clinical exercise tests are stair-climbing, a 6-min walk test, a shuttle-walk test, and a CPET (Wasserman et al., 1999; Weisman and Zeballos, 1994). The 6-min walk test is easy to administer, better tolerated, and more reflective of activities of daily living than the other walk tests (Solway et al., 2001). It evaluates the global and integrated responses of all the systems involved during exercise, including the pulmonary and cardiovascular systems, systemic circulation, peripheral circulation, blood, neuromuscular units, and muscle metabolism. The 6-min walk test was performed according to the guideline of the American Thoracic Society (ATS Committee on Proficiency Standards for Clinical Pulmonary Function Laboratories, 2002).

Health-related quality of life was evaluated by the Medical Outcome Study Short Form 36 and the St. George Respiratory Questionnaire. The former is a questionnaire that evaluates the general medical quality of life; on the other hand, the latter is a questionnaire that evaluates the chronic respiratory disease-specific quality of life (Jones et al., 1992; Ware and Sherbourne, 1992). The questionnaires were completed before and after treatment.

2.5. Adverse effects of synthetic human ghrelin

Ghrelin could be administered relatively safely. Fifteen clinical trials of the ghrelin administration were conducted in Japan, and the enrolled subjects received intravenous injection of ghrelin in these trials (Adachi et al., 2010; Akamizu et al., 2004, 2008a,b; Enomoto et al., 2003; Hataya et al., 2001; Hotta et al., 2009; Kodama et al., 2008; Nagaya et al., 2001a,b, 2004, 2005; Takaya et al., 2000; Takeno et al., 2004; Yamamoto et al., 2010).

One or more adverse events have been reported in 42.4% of enrolled subjects, and no serious adverse events have ever been reported in these trials. The most frequently seen adverse effect was transient suffusion, which occurred in 17.6% of subjects. Sleepiness (8.8%), peristalsis (6.4%), hunger (2.4%), sweating (2.4%), diarrhea (1.6%), abdominal discomfort (0.8%), abdominal pain (0.8%), fever (0.8%), urinary sugar (0.8%), sore throat (0.8%), and leukocytosis (0.8%), which were only transient, were reported in these trials. In a study of chronic ghrelin administration, subcutaneous injections of ghrelin for 3 months were reported to be safe and well-tolerated with adverse events generally mild, including hyperhidrosis, dizziness, nausea, back and abdominal pain, insomnia, and headache (Gertner and Levinson, 2010).

2.6. Effects of ghrelin

2.6.1 Chronic obstructive pulmonary disease

Miki et al. conducted a multicenter, randomized, double-blind, placebo-controlled clinical trial. Thirty-three COPD patients with body mass index (BMI) less than 21 were included, and randomly assigned ghrelin (2 μg/kg, twice daily) or saline administration for 3 weeks. All the patients received pulmonary rehabilitation, which was lower extremity exercise training. This study showed significant improvement of symptom scores in the ghrelin treatment group. In addition, the 6-min walk distance was extended significantly in the ghrelin treatment group. The most frequent adverse event related to the ghrelin treatment was a feeling of being warm, and it was mild. No serious side effects occurred (Miki et al., 2012).

2.6.2 Chronic lower respiratory infection

Kodama et al. conducted an open-label intervention study of ghrelin (Kodama et al., 2008). Synthetic human ghrelin (2 μg/kg, twice daily) was intravenously administered to the patients for 30 min at a constant rate before breakfast and dinner for 3 weeks. Inclusion criteria of this study were

(i) persistent productive cough with purulent sputum for more than 6 months, (ii) isolation of multidrug-resistant pathogens from sputum, and (iii) cachectic status, which was defined as the loss of greater than 7.5% of weight over a period of more than 6 months and BMI less than 21. Three-week ghrelin administration decreased neutrophil density and inflammatory cytokine levels in the sputum; reduced plasma norepinephrine levels; and increased body weight, serum protein levels, and 6-min walk distance. The ghrelin treatment improved not only the patients' nutritional condition but also airway inflammation. Previous studies suggested that the ghrelin decreased TNF-α-induced adhesion molecule expression on vascular endothelial cells *in vitro* (Li et al., 2004) and that ghrelin suppressed the biosynthesis of TNF-α in T cells and mononuclear cells (Dixit et al., 2004). In their study, Kodama et al. also showed that ghrelin administration decreased the serum levels of soluble ICAM-1, which is an adhesion molecule expressed on vascular endothelial cells that regulates neutrophil adhesion (Kodama et al., 2008).

2.6.3 Effects of octanoic acid-rich nutrition formula

Acyl ghrelin, an active form of ghrelin, is synthesized in the stomach and inactivated as des-acyl ghrelin by deacylation. Acyl modification of ghrelin is indispensable for various physiological actions of ghrelin (Kojima et al., 1999). Octanoic acids are one of the omega-3 polyunsaturated acids and are essential for the acyl modification of ghrelin, which is vital for biological activity. In an animal experiment, oral administration of octanoic acids significantly increased plasma levels of ghrelin and appetite (Yamato et al., 2005). In humans, it is still unclear whether increased intake of octanoic acids increases plasma acyl ghrelin levels and leads to significant clinical effects for cachectic patients.

Based on the hypothesis that octanoic acids are necessary for acylation in the biosynthesis of acyl ghrelin, Ashitani et al. investigated whether oral administration of an octanoic acid-rich nutrition formula would increase plasma acyl ghrelin levels in cachectic patients with chronic respiratory disease (Ashitani et al., 2009). Twenty-three cachectic patients, including those with bronchiectasis, COPD, and old pulmonary tuberculosis, were enrolled this open-label study. First, changes in 24-h plasma ghrelin profiles after single oral administration of the nutrition formula were investigated. Five hours after oral intake of the formula, plasma acyl ghrelin levels were significantly elevated compared with the control. On the other hand, no significant changes were found in the plasma des-acyl ghrelin levels. Second, the effects of 2-week administration of the formula on plasma ghrelin levels and nutritional status were evaluated. BMI, appetite score, and rapid-turnover proteins (including prealbumin, transferrin, and retinol binding protein, which represent the short-term nutritional status) were significantly elevated by the 2-week treatment. Moreover, plasma acyl ghrelin levels were remarkably elevated whereas those of des-acyl ghrelin were not. Oral intake of an octanoic acid-rich nutrient increased plasma levels of acyl ghrelin and showed its clinical effectiveness.

3. SUMMARY

Cachexia is commonly seen in patients with progressive chronic respiratory disease, especially in patients with COPD. Cachexia is an independent risk factor for mortality in patients with the disease. Based on small clinical trials, there appears to be a benefit in administering ghrelin and oral octanoic acid-rich nutrition formula for the treatment of cachexia due to chronic respiratory disease. No serious adverse events associated with the use of ghrelin have been reported, and the clinical application of the ghrelin seems to be safe. Clearly, more large-scale and long-term clinical trials are necessary.

REFERENCES

Adachi, S., et al., 2010. Effects of ghrelin administration after total gastrectomy: a prospective, randomized, placebo-controlled phase II study. Gastroenterology 138, 1312–1320.

Agustí, A.G., et al., 2002. Skeletal muscle apoptosis and weight loss in chronic obstructive pulmonary disease. Am. J. Respir. Crit. Care Med. 166, 485–489.

Akamizu, T., et al., 2004. Pharmacokinetics, safety, and endocrine and appetite effects of ghrelin administration in young healthy subjects. Eur. J. Endocrinol. 150, 447–455.

Akamizu, T., et al., 2008a. Repeated administration of ghrelin to patients with functional dyspepsia: its effects on food intake and appetite. Eur. J. Endocrinol. 158, 491–498.

Akamizu, T., et al., 2008b. Effects of ghrelin treatment on patients undergoing total hip replacement for osteoarthritis: different outcomes from studies in patients with cardiac and pulmonary cachexia. J. Am. Geriatr. Soc. 56, 2363–2365.

Anker, S.D., et al., 1997. Wasting as independent risk factor for mortality in chronic heart failure. Lancet 349, 1050–1053.

Ashitani, J., et al., 2009. Effect of octanoic acid-rich formula on plasma ghrelin levels in cachectic patients with chronic respiratory disease. Nutr. J. 8, 25.

ATS Committee on Proficiency Standards for Clinical Pulmonary Function Laboratories, 2002. ATS statement: guidelines for the six-minute walk test. Am. J. Respir. Crit. Care Med. 166, 111–117.

Dixit, V.D., et al., 2004. Ghrelin inhibits leptin- and activation-induced proinflammatory cytokine expression by human monocytes and T cells. J. Clin. Invest. 114, 57–66.

Enomoto, M., et al., 2003. Cardiovascular and hormonal effects of subcutaneous administration of ghrelin, a novel growth hormone-releasing peptide, in healthy humans. Clin. Sci. (Lond.) 105, 431–435.

Gertner, J., Levinson, B., 2010. SUN11031 (synthetic human Ghrelin) improves lean body mass and function in advanced COPD cachexia in a placebo controlled trial. Growth Horm. IGF Res. 20, S1–S38.

Gosselink, R., et al., 1997. Exercise training in COPD patients: the basic questions. Eur. Respir. J. 10, 2884–2891.

Hataya, Y., et al., 2001. A low dose of ghrelin stimulates growth hormone (GH) release synergistically with GH-releasing hormone in humans. J. Clin. Endocrinol. Metab. 86, 4552.

Hotta, M., et al., 2009. Ghrelin increases hunger and food intake in patients with restricting-type anorexia nervosa: a pilot study. Endocr. J. 56, 1119–1128.

Jones, P.W., et al., 1992. A self-complete measure of health status for chronic airflow limitation. The St. George's Respiratory Questionnaire. Am. Rev. Respir. Dis. 145, 1321–1327.

Kodama, T., et al., 2008. Ghrelin treatment suppresses neutrophil-dominant inflammation in airways of patients with chronic respiratory infection. Pulm. Pharmacol. Ther. 21, 774–779.

Kojima, M., et al., 1999. Ghrelin is a growth-hormone-releasing acylated peptide from stomach. Nature 402, 656–660.

Li, W.G., et al., 2004. Ghrelin inhibits proinflammatory responses and nuclear factor-kappaB activation in human endothelial cells. Circulation 109, 2221–2226.

Miki, K., et al., 2012. Ghrelin treatment of cachectic patients with chronic obstructive pulmonary disease: A multicenter, randomized, double-blind, placebo-controlled trial. PLoS One 7, e35708.

Nagaya, N., et al., 2001a. Hemodynamic and hormonal effects of human ghrelin in healthy volunteers. Am. J. Physiol. Regul. Integr. Comp. Physiol. 280, R1483–1487.

Nagaya, N., et al., 2001b. Hemodynamic, renal, and hormonal effects of ghrelin infusion in patients with chronic heart failure. J. Clin. Endocrinol. Metab. 86, 5854–5859.

Nagaya, N., et al., 2004. Effects of ghrelin administration on left ventricular function, exercise capacity, and muscle wasting in patients with chronic heart failure. Circulation 110, 3674–3679.

Nagaya, N., et al., 2005. Treatment of cachexia with ghrelin in patients with COPD. Chest 128, 1187–1193.

Pauwels, R.A., et al., 2001. Global strategy for the diagnosis, management, and prevention of chronic obstructive pulmonary disease. NHLBI/WHO Global Initiative for Chronic Obstructive Lung Disease (GOLD) Workshop summary. Am. J. Respir. Crit. Care Med. 163, 1256–1276.

Schols, A.M., 2002. Pulmonary cachexia. Int. J. Cardiol. 85, 101–110.

Schols, A.M., et al., 1996. Evidence for a relation between metabolic derangements and increased levels of inflammatory mediators in a subgroup of patients with chronic obstructive pulmonary disease. Thorax 51, 819–824.

Solway, S., et al., 2001. A qualitative systematic overview of the measurement properties of functional walk tests used in the cardiorespiratory domain. Chest 119, 256–270.

Takaya, K., et al., 2000. Ghrelin strongly stimulates growth hormone release in humans. J. Clin. Endocrinol. Metab. 85, 4908–4911.

Takeno, R., et al., 2004. Intravenous administration of ghrelin stimulates growth hormone secretion in vagotomized patients as well as normal subjects. Eur. J. Endocrinol. 151, 447–450.

Ware, J.E., Sherbourne, C.D., 1992. The MOS 36-item short-form health survey (SF-36). I. Conceptual framework and item selection. Med. Care 30, 473–483.

Wasserman, K., et al., 1999. Principles of Exercise Testing and Interpretation, third ed. Lippincott, Williams & Wilkins, Philadelphia.
Weisman, I.M., Zeballos, R.J., 1994. An integrated approach to the interpretation of cardiopulmonary exercise testing. Clin. Chest Med. 15, 421–445.
Wilson, D.O., et al., 1989. Body weight in chronic obstructive pulmonary disease. The National Institutes of Health Intermittent Positive-Pressure Breathing Trial. Am. Rev. Respir. Dis. 139, 1435–1438.
Yamamoto, K., et al., 2010. Randomized phase II study of clinical effects of ghrelin after esophagectomy with gastric tube reconstruction. Surgery 148, 31–38.
Yamato, M., et al., 2005. Exogenous administration of octanoic acid accelerates octanoylated ghrelin production in the proventriculus of neonatal chicks. Biochem. Biophys. Res. Commun. 333, 583–589.

CHAPTER TWENTY-SIX

Clinical Trial of Ghrelin Synthesis Administration for Upper GI Surgery

Shuji Takiguchi, Yuichiro Hiura, Yasuhiro Miyazaki, Akihiro Takata, Kohei Murakami, Yuichiro Doki

Division of Gastroenterological Surgery, Department of Surgery, Graduate School of Medicine, Osaka University, Osaka, Japan

Contents

Abstract

Appetite and weight loss following gastrectomy or esophagectomy is one of the major problems that affect the postoperative QoL. Ghrelin, mainly secreted from the stomach, is related to appetite, weight gain, and positive energy balance. This hormone level had been shown to become low for a long time after upper GI surgery. The efficacy of ghrelin synthesis administration for postoperative weight loss was investigated from a clinical trial to develop a new strategy for weight gain. In addition to this treatment for appetite and weight loss, we focused on the anti-inflammatory role of ghrelin. For the purpose of controlling postoperative cytokine storm after esophagectomy, this

Methods in Enzymology, Volume 514
ISSN 0076-6879
http://dx.doi.org/10.1016/B978-0-12-381272-8.00026-X

hormone was introduced in the clinical trial. Finally, ghrelin replacement therapy during chemotherapy in patients with esophageal cancer is also presented. Our clinical trials and their results are presented in this chapter.

1. INTRODUCTION

Ghrelin was first identified as an endogenous ligand for the growth hormone (GH) secretagogue receptor and is predominantly secreted by gastric endocrine cells (Kojima et al., 1999). One of the most important functions of ghrelin is to stimulate the appetite signal in the hypothalamus (Date et al., 2000; Leite-Moreira and Soares, 2007) and to gain body weight. The majority of ghrelin is produced in the stomach and a smaller amount is secreted from other organs, such as the intestine, pancreas, kidney, and hypothalamus. Ghrelin signal is transmitted from the stomach to the hypothalamus by the vagal nerve. Therefore, circulating ghrelin levels decreased to 10–20% of the preoperative level immediately after total gastrectomy (TG) and to approximately 50% after distal gastrectomy or esophagectomy with vagotomy as we already reported (Doki et al., 2006; Takachi et al., 2006). Not surprisingly, patients suffer from appetite and body weight (BW) loss after operative procedures in gastroenterology.

Recent articles reported that patients with lower levels of ghrelin are also observed under chemotherapy with cisplatin-based regimen (Liu et al., 2006) and systemic inflammatory response syndrome (SIRS) (Koch et al., 2010). The former is noted for inducing high toxicity related to gastrointestinal symptoms and the latter during postoperative course after traditional invasive surgery associated with high morbidity and mortality such as esophagectomy in the gastroenterology field.

We hypothesized that treatment for hypoghrelinemia led to improvement in patients' quality of life (QoL) and amelioration of a variety of symptoms. Based on this hypothesis, several randomized clinical trials as to replacing ghrelin hormone were performed in our institute. First, we planned the trials for patients after gastrectomy and esophagectomy. Next, investigations for patients with SIRS and under chemotherapy were accomplished.

In this chapter, we introduce the designs and results of our clinical trials and discuss the significance of ghrelin replacement therapy.

2. GHRELIN REPLACEMENT THERAPY AFTER TG AND ESOPHAGECTOMY

2.1. Ghrelin reduction after TG and esophagectomy

Doki et al. studied ghrelin reduction after gastroenterological surgery as described below (Doki et al., 2006) and stated that BW loss was observed commonly after TG and esophagectomy because the functional and anatomical alterations to the stomach by these surgeries affected ghrelin secretion.

2.1.1 Methods

We measured early phase postoperative alteration of serum ghrelin before surgery and at day 3 and day 7 after surgery (esphagectomy, TG, distal gastrectomy, and colectomy).

2.1.2 Results

After esophageal replacement by the stomach, serum ghrelin decreased to approximately half the preoperative level at day 3 (45.5 fmol/ml) and day 7 (44.1 fmol/ml) after surgery (Fig. 26.1). TG showed an extreme decline in serum ghrelin (12% of the preoperative level) immediately after surgery. Distal gastrectomy, which removed two-thirds of the stomach, showed an approximately 50% reduction of serum ghrelin at days 3 and 7. Colectomy, which did not include gastric manipulation, did not show any ghrelin alteration on either day 3 or day 7 after surgery.

2.2. Administration of ghrelin for TG patient

BW loss and reduction of blood ghrelin level are commonly observed after TG. Our prospective study was designed to elucidate whether exogenous ghrelin administration prevented postoperative BW loss by improving appetite and oral food intake in patients with gastric cancer after undergoing TG (Adachi et al., 2010).

2.2.1 Protocol design

The study design is summarized in Fig. 26.2A and B. In this randomized phase II study, 21 patients undergoing TG were assigned to the ghrelin (11 patients) or the placebo group (10 patients). The patient usually started oral food intake of rice porridge between postoperative days 5 and 7. In the following 10 days after starting oral food intake, intravenous drip infusion of

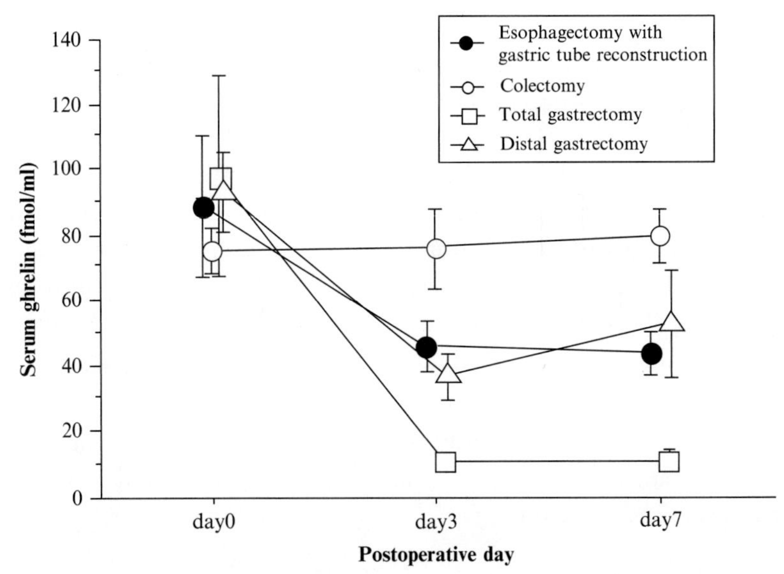

Figure 26.1 Early phase effect of esophagectomy and gastrectomy on serum ghrelin. Serum ghrelin was measured before surgery and on days 3 and 7 after surgery, including esophagectomy with gastric tube reconstruction (closed circles, nine patients), total gastrectomy (open squares, eight patients), distal gastrectomy (open triangles, five patients), and colectomy (open circles, five patients). The mean of each time point is indicated with a *standard error bar*. Significant decline of serum ghrelin after surgery by Student's *t*-test ($P=0.05$) was observed at day 7 for esophagectomy, at days 3 and 7 for total gastrectomy, and at day 3 for distal gastrectomy patients.

synthetic human ghrelin (3 g/kg) or placebo (pure saline) was administered twice daily (before breakfast and before dinner).

During the study period, the same protocol of intravenous infusion and the same menu of meals were provided for the two groups.

2.2.2 End points

The primary end point of this study was an increase in orally ingested calories following ghrelin administration. The secondary end points included changes in BW, appetite, body composition, basal metabolism, and blood tests.

2.2.3 Methodology (measurements)

1. *Appetite*: Preprandial appetite at every meal was scored by the visual analog scale (VAS, possible scores 0–10 cm) recorded in the account sheet by each patient.

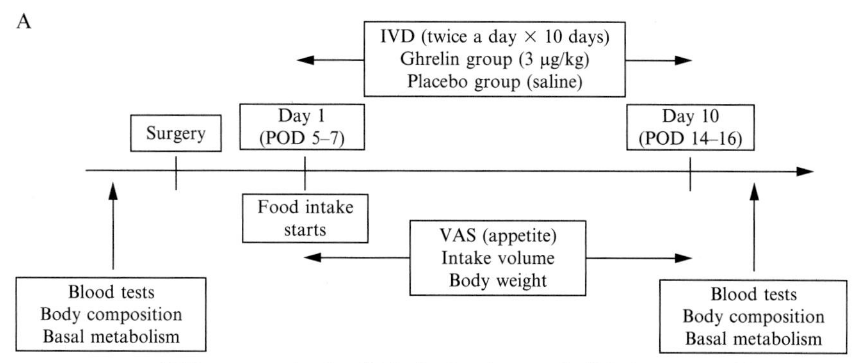

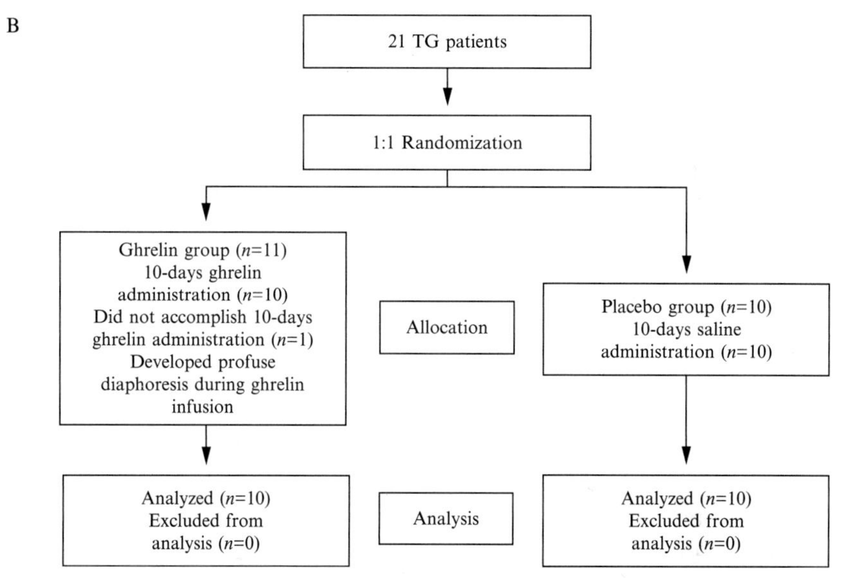

Figure 26.2 Study protocol and flow diagram. (A) Study design (B) The study flow diagram as to patients enrollment.

2. *Food intake*: Food intake calories based on the food weight measured by the patient, including standard meal and extra foods, were calculated by dietitians using a calorimeter.
3. *Body weight*: BW was measured with a beam scale to the nearest 0.1 kg, with patients standing barefoot and in light clothing.

2.2.4 Results

Appetite, oral food intake, and BW were recorded by the patients throughout the 10 days of ghrelin/saline administration. The mean appetite VAS score was significantly higher in the ghrelin group than in the placebo group during the 10-day period (Fig. 26.3A; repeated measures ANOVA, $P=0.032$).

Food intake calories (kcal/kg/day) during the 10-day period were significantly higher in the ghrelin group than in the placebo group (Fig. 26.3B; repeated measures ANOVA, $P=0.030$). Food intake gradually increased at an earlier period of food intake and then remained unchanged thereafter; both groups showed a similar difference throughout the 10-day period. The mean intake over the 10-day period was 13.8 and 10.4 kcal/kg/day for the ghrelin and placebo groups, and ghrelin administration accounted for about 32.7% of the increase.

BW loss was calculated with reference to the first day of oral food intake. During this period, BW gradually decreased in both groups, although the loss was more evident in the placebo group. At the end of the intravenous drip protocol (day 10), BW loss was −3.7% for the placebo group but only −1.4% for the ghrelin group. For the 10-day period, BW loss of the ghrelin group was less than that of the placebo group (Fig. 26.3C; repeated measures ANOVA, $P=0.044$).

2.3. Administration of ghrelin for esophagectomy patient

Ghrelin is a peptide hormone with pleiotropic functions including stimulation of GH secretion and appetite, and its levels decrease after esophagectomy. The aim of this study was to evaluate whether exogenous ghrelin administration could ameliorate the postoperative decrease of oral food intake and BW, which are serious complications after esophagectomy (Yamamoto et al., 2010).

2.3.1 Protocol design

The study was performed in a single-blind manner.

This prospective, randomized, placebo-controlled clinical trial assigned a total of 20 patients with thoracic esophageal cancer who underwent radical operation into either a ghrelin ($n=10$) or a placebo ($n=10$) group.

Only the person responsible for mixing study drugs knew the allocation, but not the subjects and the individuals who collected data. Patients assigned to the ghrelin group received ghrelin treatment at a dose of 3 mg/kg BW diluted in 50 ml saline given over 30 min and repeated twice daily (before breakfast and before dinner) for 10 consecutive days (days 1–10), beginning the day after the start of food intake (day 0). Patients in the placebo group

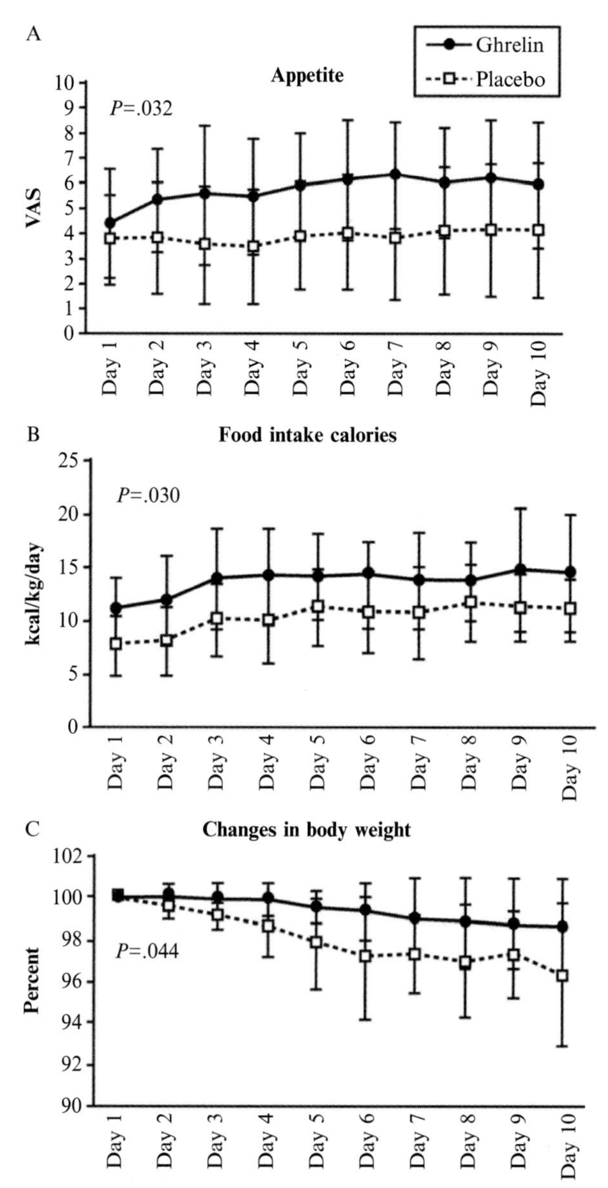

Figure 26.3 Serial changes in appetite, food intake, and body weight during the 10-day study in the ghrelin and placebo groups. Data are expressed as the mean ± SD of visual analog scale scores of (A) preprandial appetite at every meal, (B) daily total food intake calories per body weight (kcal/kg/day), and (C) percent body weight relative to the first day of oral intake in the ghrelin and placebo groups. The visual analog scale score throughout the study period, which was evaluated by repeated measures ANOVA, was significantly higher in the ghrelin group than in the placebo group (5.7 vs. 3.9 cm; $P = 0.032$). Likewise, food intake calories were significantly higher in the ghrelin group than in the placebo group (average, 13.8 vs. 10.4 kcal/kg/day; repeated measures ANOVA, $P = 0.030$). Body weight loss in the ghrelin group was significantly lower than in the placebo group (−1.4% vs. −3.7%; repeated-measures ANOVA, $P = 0.044$).

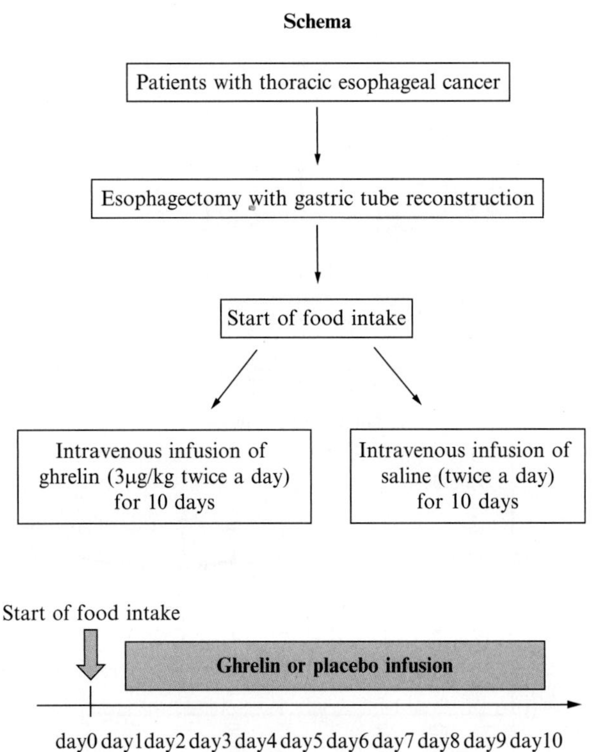

Figure 26.4 Study protocol.

received a corresponding placebo (saline) infusion in a similar fashion (Fig. 26.4).

All subjects received similar nutritional therapy during the study period with the exception of the administration of test treatment.

2.3.2 End points

Alteration of food intake was the primary end point of this trial.

2.3.3 Methodology (measurements)

1. *Food intake calories*: Calories of food intake beginning on day 0 (start of food intake) and for the next 10 consecutive days (days 1–10) were assessed by a national registered dietitian by measuring the weight of each dish containing diet before and after every meal.
2. *Appetite*: To evaluate the change in appetite, a VAS rating of hunger (possible scores 0–10 cm) was recorded by subjects immediately after administration of ghrelin or placebo before every meal during the study period

(days 0–10). The average score was calculated each day. To determine the change in appetite score, the average appetite score of day 0 was considered 100%, and all values measured during the administration of the test drug (days 1–10) were expressed as a percent of the score of day 0.

3. *Body weight*: BW measurement was performed before the start of food intake (day 0) and 10 days after the start of food intake (day 10).

2.3.4 Results

Changes in calories of food intake showed a similar pattern of increased intake in both groups; however, the increase in the ghrelin group was greater than that of the placebo group (repeated measure ANOVA; $P=0.015$). On average, the intake of food calories was 44% greater in the ghrelin group compared with the placebo group (874 vs. 605 kcal/day; Fig. 26.5).

At the beginning of oral food intake (days 0–2), the VAS (0–10 cm) score for appetite was similar in the two groups, but patients in the ghrelin group tended to show a greater increase in hunger sensation compared to patients of the placebo group ($P=0.094$; Fig. 26.6).

During the study period, changes in BW were monitored as percent of the weight of day 0 (Fig. 26.7A). We compared BW measured on day 0 (before start of the food intake) and day 10 (after completion of the study period; Fig. 26.7B).

The percent decrease in total weight was $-3\% \pm 2\%$ in the placebo group but only $-1\% \pm 2\%$ in the ghrelin group ($P=0.019$).

2.3.5 Conclusion of this section

These prospective randomized studies provide convincing data on the beneficial effects of ghrelin on BW and dietary activity after TG and esophagectomy.

3. ANTI-INFLAMMATORY ROLE OF GHRELIN AFTER ESOPHAGEAL SURGERY

3.1. Relationship between ghrelin and SIRS on esophagectomy

3.1.1 Surgical stress and ghrelin

As pointed out previously, ghrelin has a variety of biological functions including anabolic effects via stimulation of GH secretion from the pituitary gland, control of energy expenditure, stimulation of appetite, and improvement of cardiopulmonary functions (Akamizu and Kangawa, 2011).

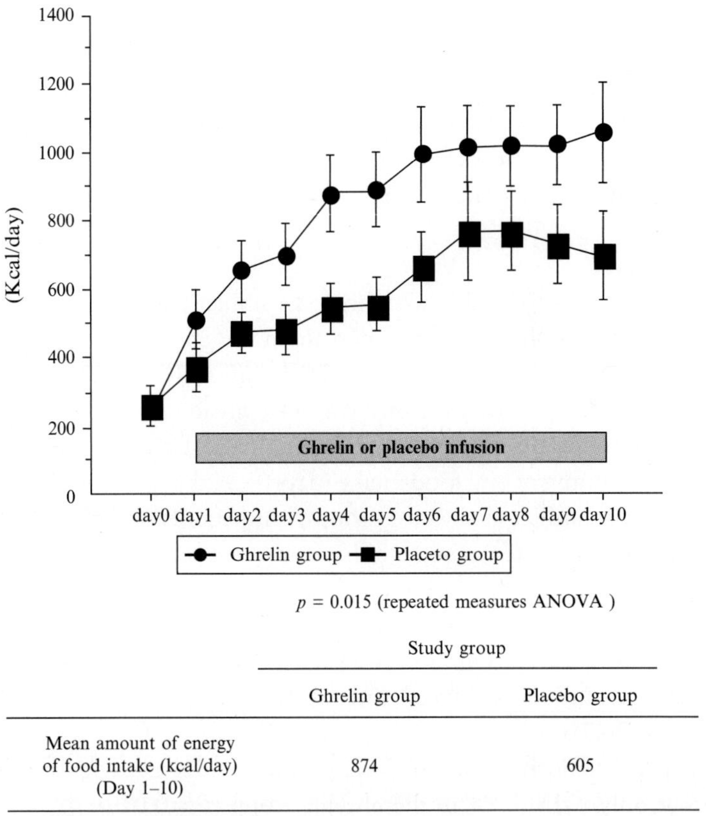

	Study group	
	Ghrelin group	Placebo group
Mean amount of energy of food intake (kcal/day) (Day 1–10)	874	605

Figure 26.5 Serial changes in food intake calories after esophagectomy. Changes in daily intake of food calories were compared between ghrelin group (closed circles) and placebo group (closed squares). In both groups, food intake calories increased with time, but the increase was greater in the ghrelin group than in the placebo group (repeated measure ANOVA; $P=0.015$). Data are presented as mean values ± standard error of the mean.

Recently, evidence that ghrelin exerts anti-inflammatory actions has been accumulating. Ghrelin suppresses the production of proinflammatory cytokines, including IL-1β, IL-6, and TNF-α in both *in vitro* (Li et al., 2004) and *in vivo* (Gonzalez-Rey et al., 2006).

However, it is not known how plasma ghrelin changes during operation and early postoperative period in any kind of invasive surgery. On the other hand, esophageal cancer is most commonly resected with the use of right thoracotomy combined with laparotomy and this surgical treatment is the most invasive among all gastrointestinal surgeries and correlates with the risk of SIRS, cardiopulmonary complications, and stress-induced organ dysfunction state (Enzinger and Mayer, 2003).

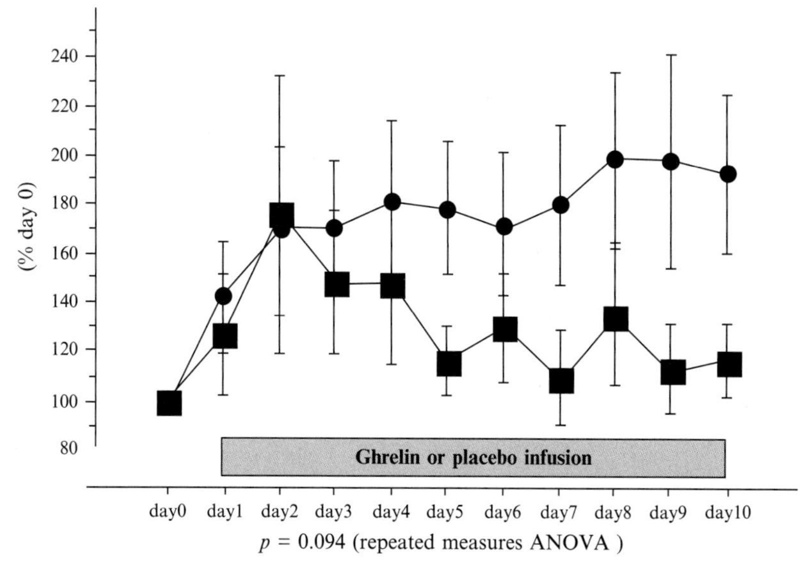

Figure 26.6 Serial changes in subjective appetite measured by VAS score. Percentage changes in VAS score (compared with score of day 0) were compared between ghrelin group (closed circles) and placebo group (closed squares). The VAS score of ghrelin group tended to be greater than placebo group ($P = 0.094$; repeated measure ANOVA). Data presented as mean values ± standard error of the mean.

Therefore, we conducted a prospective cohort study to examine the serial change of plasma ghrelin concentration during perioperative period of esophagectomy and investigated the relationship between the change of plasma ghrelin concentration and the postoperative clinical course.

3.1.2 Material and methods

3.1.2.1 Study patients

Enrolled patients underwent surgery at our institution. All the subjects received same procedure and management during the perioperative period.

The primary end point of this study was the serial change of plasma ghrelin concentration during esophagectomy. The secondary end points included the clinical course after the surgery such as the morbidity and the duration of SIRS. Blood samples were taken at six time points: on preoperative day, 2 h after the start of the surgery (during thoracotomy), 4 h after the start of the surgery (during laparotomy), POD1, 3 days after surgery (POD3), and 10 days after surgery (POD10). SIRS was diagnosed according to the American College of Chest PhysiciansSociety of Critical Care Medicine Consensus.

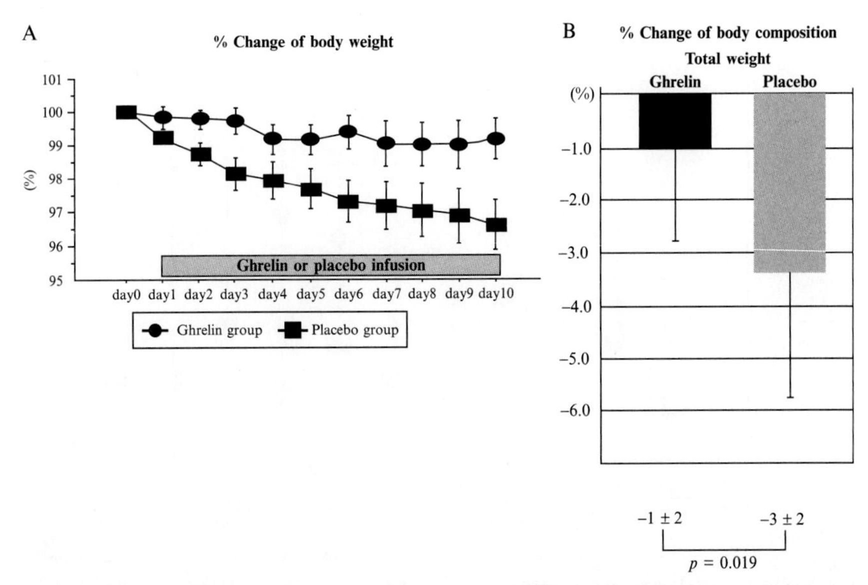

Figure 26.7 Changes in weight and body composition. Changes in postoperative weight (A) and measured before start of food intake (day 0) and after completing protocol treatment (day 10) (B). (A) Percent change in postoperative weight (compared with day 0) in the ghrelin group (closed circles) and placebo group (closed squares). (B) Changes in body composition between the two groups. The decrease in total weight was less in the ghrelin group than in the placebo group (−1% vs. −3%; $P=0.019$). Data are presented as mean values ± standard error of the mean.

3.1.3 Result

3.1.3.1 Patient characteristics and clinical course

A total of 20 patients with pathologically proved squamous cell carcinoma were enrolled in this study. None of the patients withdrew from the study. The patients' characteristics are shown in Table 26.1.

3.1.3.2 The change of plasma ghrelin concentration

The total plasma ghrelin concentration of preoperative day, 2 hours after the start of the surgery (2H), 4 hours after the start of the surgery (4H), the next day of the surgery (POD1), POD3, and POD10 was 100.7 ± 88.4, 66.0 ± 52.3, 50.3 ± 33.8, 38.1 ± 35.3, 50.1 ± 46.6, and 53.4 ± 39.3 fmol/ml, respectively. These data indicated that plasma ghrelin concentration level decreased immediately after the operation and reached the nadir at POD1. After that, it recovered the half level of preoperative state.(Fig. 26.8).

Table 26.1 Patient characteristics

	Characteristics
Age (years)	64.2±7.4
Gender (male/female)	16/4
BMI (kg/m^2)	21.2±3.1
Tumor localization (thoracic) upper/middle/lower	4/9/7
cStage (UICC) I/II/III/IV	1/8/10/1
Preoperative therapy	2
None chemotherapy	17
Chemoradiotherapy	1
Operating time (min)	463.7±53.8
Blood loss (ml)	635±211.1

To examine the serial change of ghrelin concentration of each enrolled subject and search for the specific change of the patients whose duration of SIRS was prolonged, we assigned the percent ghrelin concentration of each time points (2H, 4H, POD1, POD3, POD10), with 100% representing preoperative ghrelin concentration of each subject (Fig. 26.9). When we presented the change of the ghrelin level of the patients whose postoperative duration of SIRS was prolonged for more than 7 days ($n=4$) (solid line) and the others ($n=16$) (dotted line), the former tended to drop especially at POD1.

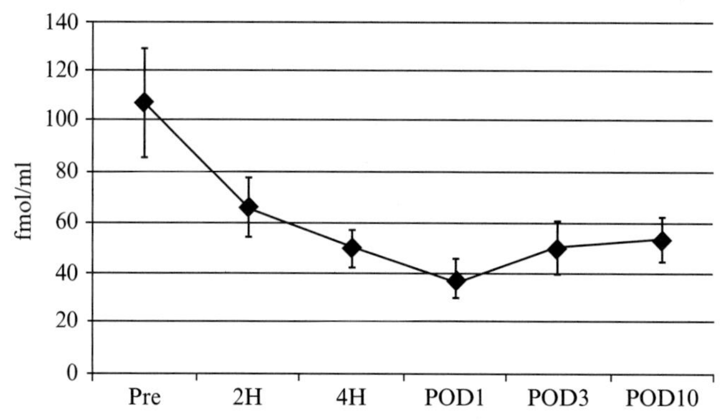

Figure 26.8 Change of concentration of plasma ghrelin postoperation.

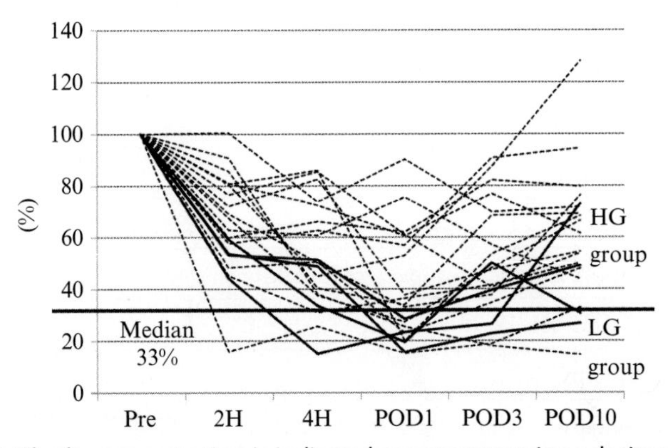

Figure 26.9 Ghrelin concentration is indicated as percentage in each time point compared with its preoperative ghrelin concentration. Solid line: postoperative duration of SIRS was more than 7 days.

3.1.4 Relationship of ghrelin deterioration of POD1 and postoperative clinical course

We assigned the percent plasma ghrelin concentration of POD1 compared with preoperative ghrelin concentration as percent POD1 (median 33%, range 15–90%) and divided the subjects into high ghrelin concentration group (HG group; %POD1 was more than 33%, $n=10$) and low ghrelin concentration group (LG group; %POD1 was less than 33%, $n=10$). (Table 26.2) Patients' perioperative data and blood examination results are as shown in Table 26.3.

Table 26.2 Patient characteristics of the high ghrelin concentration (HG) group and the low ghrelin concentration (LG) group

	HG* (*n* – 10)	**LG** (*n* = 10)**	***P*-value**
Gender (male/female)	8/2	8/2	
Age (years)	64 ± 2	65 ± 3	0.661
Preoperative BMI (kg/m^2)	21.7 ± 0.7	20.1 ± 0.9	0.172
Preoperative complications	1 (DM)	2 (OMI, COPD)	0.531
Localization (Ut/Mt/Lt)	2/6/2	2/3/5	0.309
Clinical stage (I/IIA/IIB/III)	1/4/1/4	0/1/2/7	0.213
Preoperative therapy (none/CT/CRT)	2/7/1	1/9/0	0.453
Preoperative ghrelin (fmol/ml)	94.5 ± 33.3	120.3 ± 28.5	0.562

*HG: high ghrelin concentration group
**LG: low ghrelin concentration group

Table 26.3 Perioperative data and blood examination in the HG group and the LG group

	HG ($n=10$)	LG ($n=10$)	P-value
Operative time (min)	464±19	464±15	0.987
Operative blood loss (ml)	612±175	665±249	0.592
Postoperative complications	2	3	0.605
Anastomotic leakage	1	0	
Pneumonia	1	3	
Duration of SIRS (days)	2.1±0.6	6.0±1.3	0.019
Duration required to reach positive NB[a] (days)	4.0±0.7	5.8±2.6	0.060
Serum IL-6 value of POD1	106.2±36.0	117.8±41.0	0.834
Serum IL-6 value of POD3	62.5±24.1	188.3±48.6	0.037
Serum CRP value of POD1	6.2±0.6	6.7±0.5	0.544
Serum CRP value of POD3	13.4±1.8	17.4±2.7	0.223

[a]NB, nitrogen balance.

The characteristics of two groups were not different statistically. However, duration of SIRS in the LG group was 6.0±1.3 days, which was significantly longer than the corresponding 2.1±0.6 days in HG group ($P=0.019$) and the LG group tended to require longer duration to reach positive nitrogen balance than the HG group (5.8±2.6 vs. 4.0±0.7, $P=0.060$). In terms of acute phase parameters, the degree of postoperative increase in the serum level of IL-6 and C-reactive protein (CRP) of POD1 was similar for the two groups, whereas the serum level of IL-6 of POD3 was significantly higher in the LG group than in the HG group (188.3±48.6 vs. 62.5±24.1, $P=0.037$) and CRP level of POD3 in the LG group tended to be higher than that in the HG group, but this difference did not reach statistical significance (17.4±2.7 vs. 13.4±1.8, $P=0.223$).

Next, we tried to find any parameters that could be obtained at POD1 and correlated with the entire postoperative duration of SIRS. They were not only clinically helpful as predictor of SIRS but might also be useful as therapeutic target, since some of them should be causative for excess inflammatory reaction. We therefore examined the relationship between the duration of SIRS and several blood examinations of POD1 including CRP,

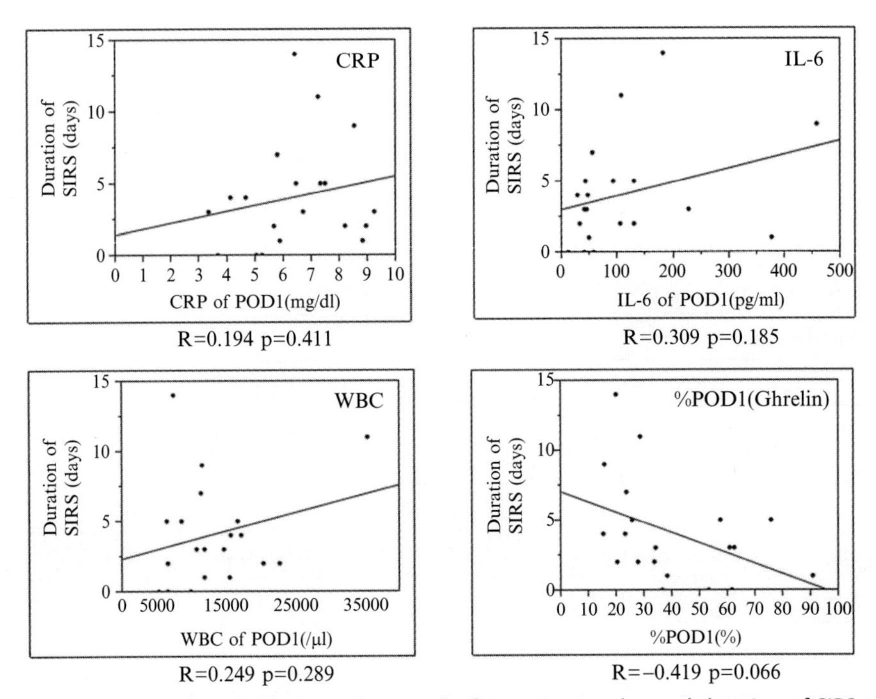

Figure 26.10 Relations between data on the first operative day and duration of SIRS. Patients with more decreased ghrelin concentration were characterized by a longer duration of SIRS ($P=0.066$).

IL-6, white blood cell counts, and %POD1 (ghrelin) by a linear regression model to investigate the most reliable predictor of the duration of SIRS after esophagectomy from the data of POD1. (Fig. 26.10) The correlation between duration of SIRS and %POD1 (ghrelin) indicated the strongest correlation among them, and the patients with more decreased ghrelin concentration were characterized by a longer duration of SIRS ($P=0.066$). Although we sought any parameters that could be obtained preoperatively and during surgery and correlated with duration of SIRS, such as age, weight, preoperative therapies, preoperative complications, tumor stage, extent of lymph node retrieval, operative time, and blood loss, there was no predictive factor that correlated with the duration of SIRS more strongly than %POD1 (ghrelin) among these parameters.

3.2. Administration ghrelin for SIRS after esophagectomy

From the result described above, supplementation of reduced ghrelin could be expected to minimize excess inflammatory response for these patients. We conducted a clinical study to examine the safety and effect of ghrelin in the perioperative period to improve inflammation. The subjects of this study were 20 patients. The subjects included two subgroups: one was administered ghrelin intravenously twice a day for 3 μg/kg (drip infusion); the other was administered ghrelin intermittently for 5 days at 0.5 μg/kg/h (continuous infusion). Since ghrelin synthesis has a short half-life, it is considered to be necessary that this hormone will be administered not as bolus but as continuously for suppressing surgery-related inflammation. This comparative trial has been already completed and the results are under analysis.

4. GHRELIN REPLACEMENT THERAPY DURING CHEMOTHERAPY IN PATIENTS WITH ESOPHAGEAL CANCER

4.1. Serial changes of total ghrelin concentration during chemotherapy in patients with esophageal cancer

It has been reported that, in rodents, ghrelin can greatly alleviate the behaviors associated with chemotherapy-induced dyspepsia. In rats, administration of cisplatin resulted in a marked decrease in plasma ghrelin, and exogenously administered ghrelin improved cisplatin-induced reduction of food intake. There has been no report on the effect of exogenous ghrelin on the efficacy of chemotherapy in humans. To tackle these issues, we conducted a prospective observational study in patients with esophageal cancer scheduled for cisplatin-based neoadjuvant chemotherapy (Hiura et al., 2011).

4.1.1 Methods

Changes in gastrointestinal hormones including ghrelin were measured and correlated with feeding activity including appetite and dietary intake, nutritional status including rapid turnover proteins, and adverse events by chemotherapy.

4.1.2 Results

Total plasma ghrelin significantly decreased at days 3 and 8 of chemotherapy but recovered at day 28 (baseline: 140 ± 54, day 3: 107 ± 46, day 8: 82 ± 32, day 28: 126 ± 43 fmol/ml, $P = 0.023$ for day 3 and $P = 0.034$ for day 8) (Fig. 26.11). Among blood nutritional parameters, transferrin was the only parameter that decreased significantly, and its decline, as well as loss of oral intake and appetite, correlated significantly with plasma ghrelin levels ($P = 0.0013$, 0.0063, and 0.013, respectively) (Fig. 26.12).

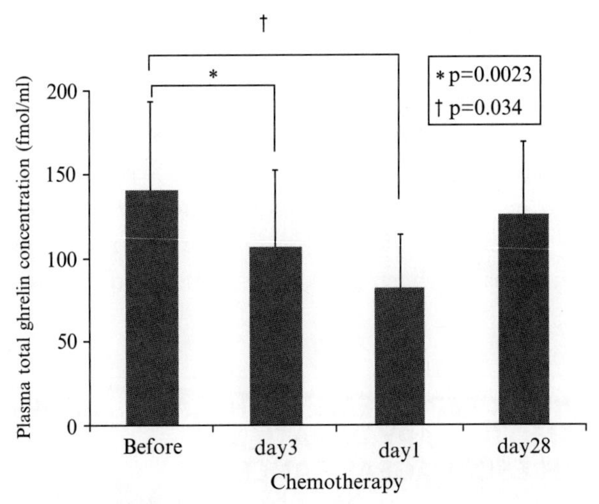

Figure 26.11 Serial changes in plasma total ghrelin concentrations in FAP and DCF regimens. Changes in total ghrelin concentration were compared with that before chemotherapy. Data are mean ± SD.

4.2. Administration of ghrelin for patients with esophageal cancer during chemotherapy

Cisplatin reduces plasma ghrelin levels via the 5-hydroxytryptamine receptor. This may cause cisplatin-induced gastrointestinal disorders and hinder the continuation of chemotherapy. We conducted a prospective, randomized phase II trial to evaluate the effects of exogenous ghrelin during cisplatin-based chemotherapy (Hiura et al., 2012).

4.2.1 Protocol design

The study design is summarized in Fig. 26.13.

Forty-two patients with esophageal cancer undergoing cisplatin-based neoadjuvant chemotherapy were assigned to either the ghrelin ($n=21$) or the placebo ($n=21$) group. They received intravenous infusion of synthetic human ghrelin (3 μg/kg) or saline twice daily for 1 week with cisplatin administration. The primary end point was changes in oral calorie intake, and the secondary end points were chemotherapy-related adverse events, appetite VAS scores, changes in gastrointestinal hormones and nutritional status including rapid turnover proteins, and QoL estimated by QLQ-C30.

4.2.2 End points

The primary end point of this study was alteration in oral calorie intake from day 1 to day 7 of chemotherapy. Patients in this study were served standard meals and were allowed to receive extra food if desired. The secondary end

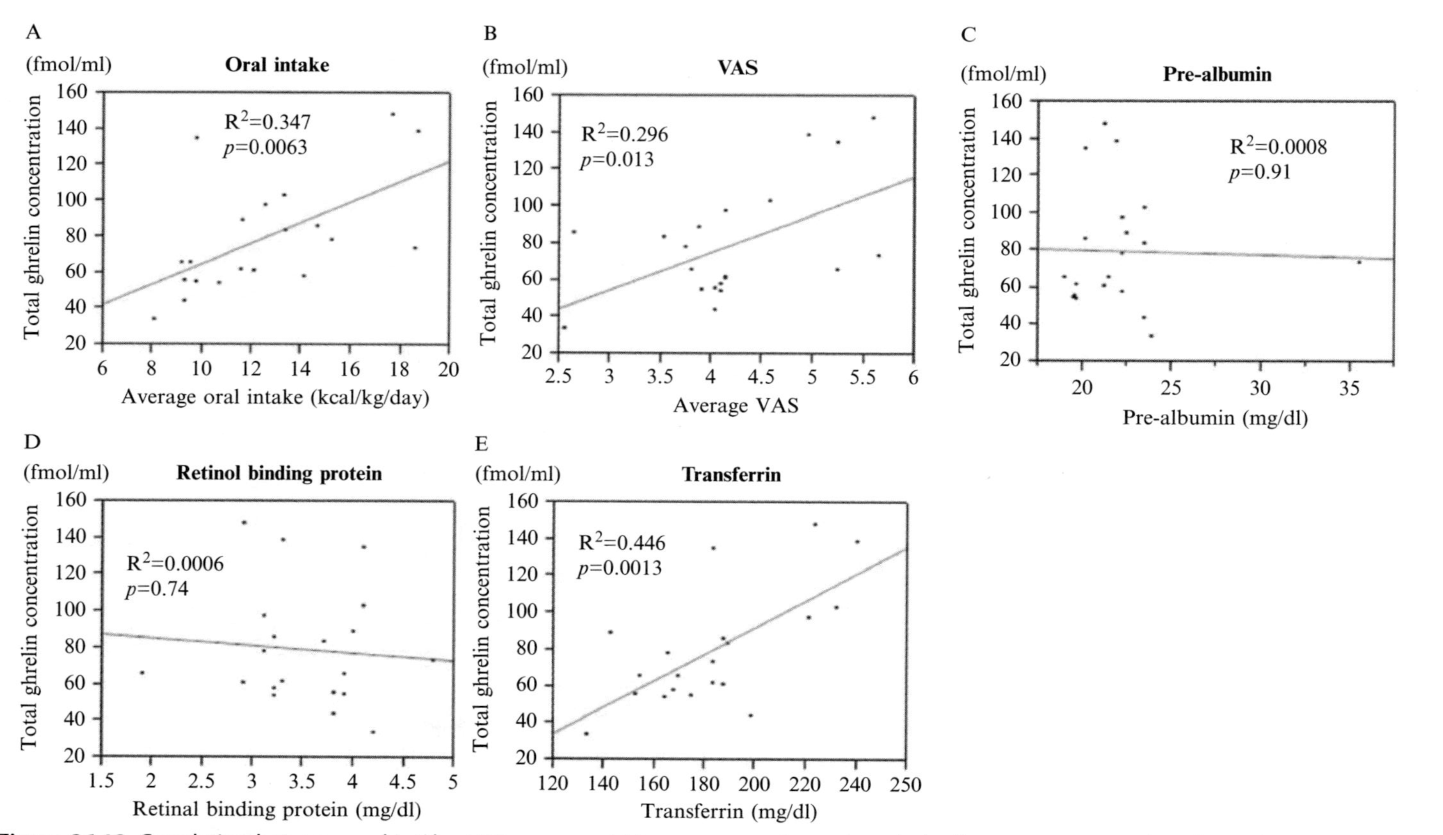

Figure 26.12 Correlation between oral intake, VAS score, rapid turnover protein, and total ghrelin concentration after chemotherapy (day 8). Data are for the mean oral intake, mean VAS score, mean average daily dietary intake (kcal/kg/day), and mean daily VAS score during days 1–7. The relationship among these parameters was analyzed by Pearson correlation analysis.

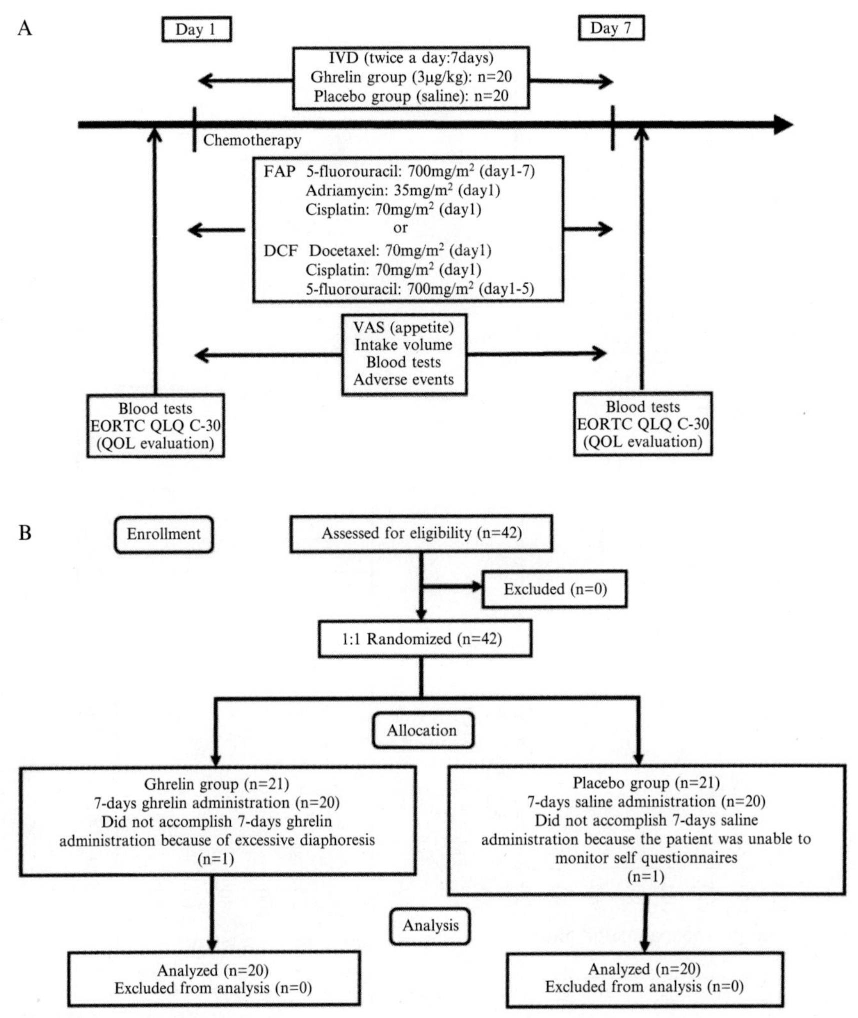

Figure 26.13 (A) Flow diagram of process through the trial. (B) Study protocol. IVD: intravenous drip; VAS: visual analogue scale.

points included changes in appetite, adverse events, QoL, BW, nutritional status, hormonal assays, and blood tests.

4.2.3 Methodology (measurements)

1. *Food intake*: Food intake calories, based on the food weight measured by the patient, including standard meal and extra foods, were calculated by dietitians using a calorimeter.

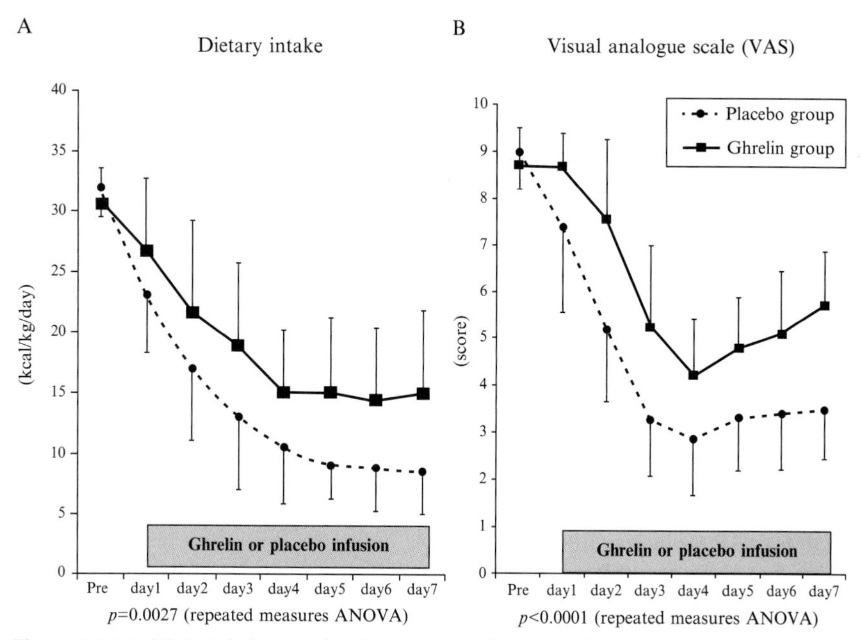

Figure 26.14 (A) Serial changes in dietary calorie intake before and during chemotherapy in the ghrelin group (closed squares) and placebo group (closed circles). (B) The VAS score for appetite was similar in the two groups before chemotherapy. Data are mean $\pm$ SD.

2. *Appetite*: Appetite profile was measured using a 100-mm VAS, with the questions "How hungry are you?" and "How full do you feel?" that were anchored with "0-not at all" and "100-extremely." Patients were instructed to rate themselves by selecting the scale before each meal that was most appropriate to their feeling at that time.
3. *QoL*: Questionnaires included the European Organization for Research and Treatment of Cancer core questionnaire (QLQ-C30) before and after chemotherapy (day8).

4.2.4 Results

Although patients in the ghrelin and placebo groups showed the above trend, the decline in dietary intake with chemotherapy was significantly less in the ghrelin group compared to patients receiving placebo (18.1 vs. 12.7 kcal/kg/day overall), especially at day 1 (26.7 vs. 23.1 kcal/kg/day) compared to day 7 (15.0 vs. 8.5 kcal/kg/day) (Fig. 26.14A). In other words, the improved oral food intake due to ghrelin administration was more significant in the later phase of chemotherapy (repeated measures ANOVA, ghrelin group vs. placebo group, $P = 0.0027$). Changes in the VAS score reflected the changes in dietary intake between the two groups with a significant difference among them (repeated measures ANOVA, ghrelin group vs. placebo group, $P < 0.0001$, Fig. 26.4B). Notably, the appetite scores showed a faster recovery after day 4 of chemotherapy in the ghrelin group than the control.

Patients of the ghrelin group had a lower frequency of adverse events with chemotherapy related to anorexia and nausea than the control. Significant deterioration was noted after chemotherapy in the placebo group in QoL scores, appetite, nausea and vomiting, and global health status.

4.2.5 Conclusion of this section

The present study demonstrated that short-term administration of exogenous ghrelin at the start of cisplatin-based chemotherapy stimulated food intake and minimized adverse events. We believe that ghrelin administration could increase the efficiency of chemotherapy and recommend the use of ghrelin in clinical practice.

REFERENCES

Adachi, S., et al., 2010. Effects of ghrelin administration after total gastrectomy: a prospective, randomized, placebo-controlled phase II study. Gastroenterology 138, 1312–1320.

Akamizu, T., Kangawa, K., 2011. Therapeutic applications of ghrelin to cachexia utilizing its appetite-stimulating effect. Peptides 32, 2295–2300.

Date, Y., et al., 2000. Ghrelin, a novel growth hormone-releasing acylated peptide, is synthesized in a distinct endocrine cell type in the gastrointestinal tracts of rats and humans. Endocrinology 141, 4255–4261.

Doki, Y., et al., 2006. Ghrelin reduction after esophageal substitution and its correlation to postoperative body weight loss in esophageal cancer patients. Surgery 139, 797–805.

Enzinger, P.C., Mayer, R.J., 2003. Esophageal cancer. N. Engl. J. Med. 349, 224.

Gonzalez-Rey, E., et al., 2006. Therapeutic action of ghrelin in a mouse model of colitis. Gastroenterology 130, 1707–1720.

Hiura, Y., et al., 2011. Fall in plasma ghrelin concentrations after cisplatin-based chemotherapy in esophageal cancer patients. Int. J. Clin. Oncol. Jul 20. http://dx.doi.org/10.1007/s10147-011-0289-0 [Epub ahead of print].

Hiura, Y., et al., 2012. Effects of ghrelin administration during chemotherapy with advanced esophageal cancer patients: a prospective, randomized, placebo-controlled phase II study. Cancer Jan 26. http://dx.doi.org/10.1002/cncr.27430 [Epub ahead of print].

Koch, A., et al., 2010. Regulation and prognostic relevance of serum ghrelin concentrations in critical illness and sepsis. Crit. Care 14, R94.

Kojima, M., et al., 1999. Ghrelin is a growth-hormone-releasing acylated peptide from stomach. Nature 402, 656–660.

Leite-Moreira, A.F., Soares, J.B., 2007. Physiological, pathological and potential therapeutic roles of ghrelin. Drug Discov. Today 12, 276–288.

Li, W.G., et al., 2004. Ghrelin inhibits proinflammatory responses and nuclear factor-kappaB activation in human endothelial cells. Circulation 109, 2221–2226.

Liu, Y.L., et al., 2006. Ghrelin alleviates cancer chemotherapy-associated dyspepsia in rodents. Cancer Chemother. Pharmacol. 58, 326–333.

Takachi, K., et al., 2006. Postoperative ghrelin levels and delayed recovery from body weight loss after distal or total gastrectomy. J. Surg. Res. 130, 1–7.

Yamamoto, K., et al., 2010. Randomized phase II study of clinical effects of ghrelin after esophagectomy with gastric tube reconstruction. Surgery 148, 31–38.

AUTHOR INDEX

Note: Page numbers followed by "*f*" indicate figures, and "*t*" indicate tables.

D

E

I

J

K

L

M

N

T

SUBJECT INDEX

Note: Page numbers followed by "*f*" indicate figures, and "*t*" indicate tables.

C

H

I

R

S

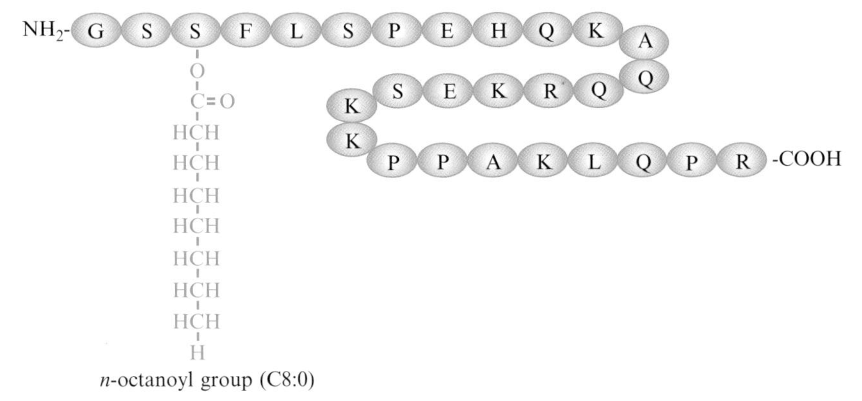

Masayasu Kojima *et al.*, Figure 3.5 Structure of rat ghrelin. The third amino acid, serine, is modified by a fatty acid, *n*-octanoic acid, which is essential for activity.

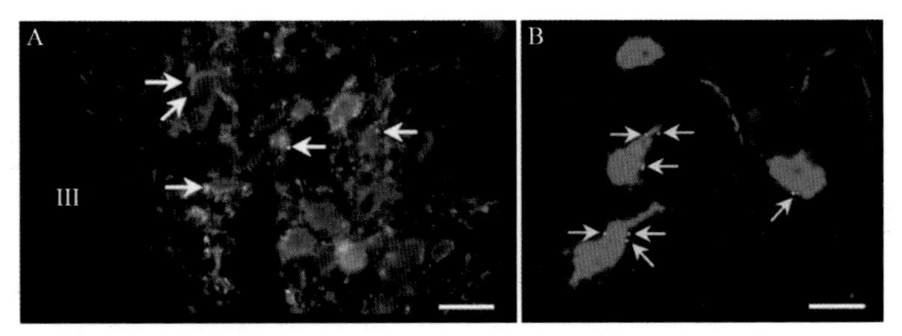

Haruaki Kageyama *et al.*, Figure 6.2 The innervation of ghrelin-immunoreactive axons to neuropeptide Y (NPY) (A) or orexin-immunoreactive cell bodies (B). (A) Ghrelin immunoreactivity is shown in red (Alexa Fluor 546). NPY immunoreactivity is shown in green (Alexa Fluor 488). Ghrelin-immunoreactive fibers can be seen in close proximity to NPY-immunopositive neurons. Arrows show that ghrelin-immunopositive fibers are in close apposition with NPY-immunopositive cell bodies. (B) ghrelin immunoreactivity is shown in red (Alexa Fluor 546) and orexin immunoreactivity in green (Alexa Fluor 488). Ghrelin-immunoreactive fibers are found in close proximity to orexin-immunoreactive neurons. Arrows indicate the apposition of ghrelin fibers to orexin neurons. III; The third ventricle. Scale bar is 10 μm in (A) and (B).

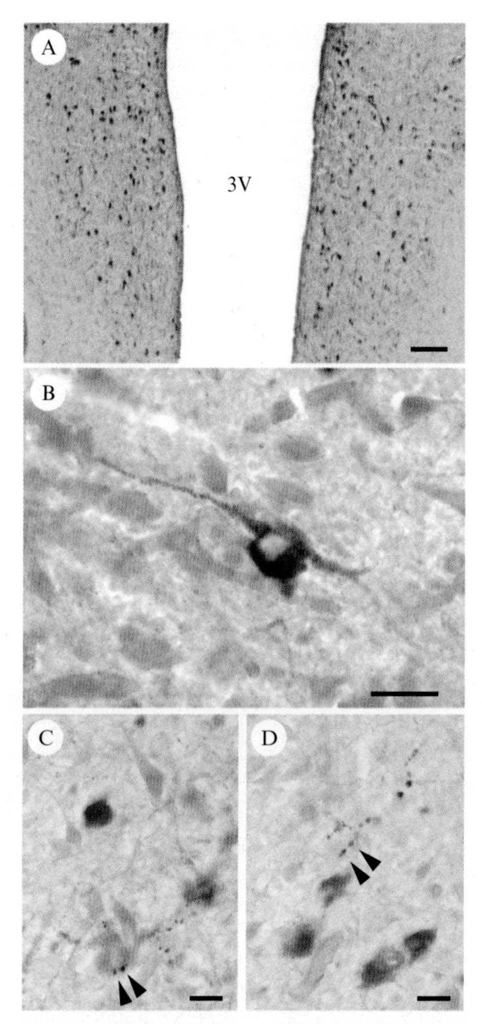

Takahiro Sato and Masayasu Kojima, Figure 7.3 Localization of ghrelin-immunopositive neurons in the porcine hypothalamus. (A) Ghrelin neuron distribution in the paraventricular nucleus. (B) A ghrelin-producing neuron in paraventricular nucleus. A subset of ghrelin-positive neurons projected to cell bodies of either additional ghrelin-positive neurons (C, arrowheads) or ghrelin-negative neurons (D, arrowheads). 3V, Third ventricle. Bar, 200 μm (A), 20 μm (B–D) [Sato et al., 2005].

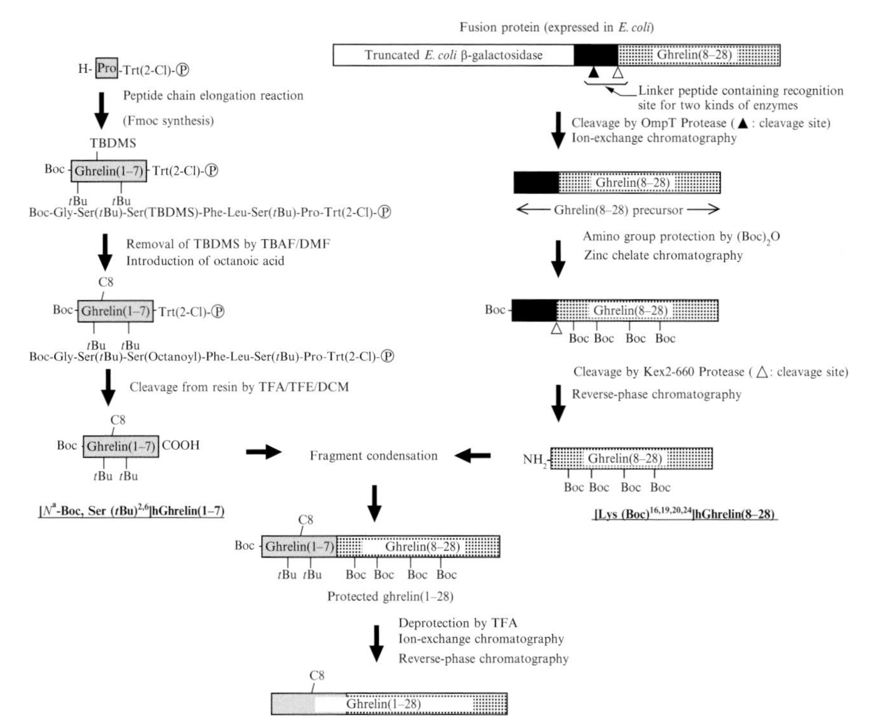

Tomohiro Makino *et al.*, Figure 12.2 Scheme of semisynthesis of ghrelin. Adapted from Makino et al. (2005).

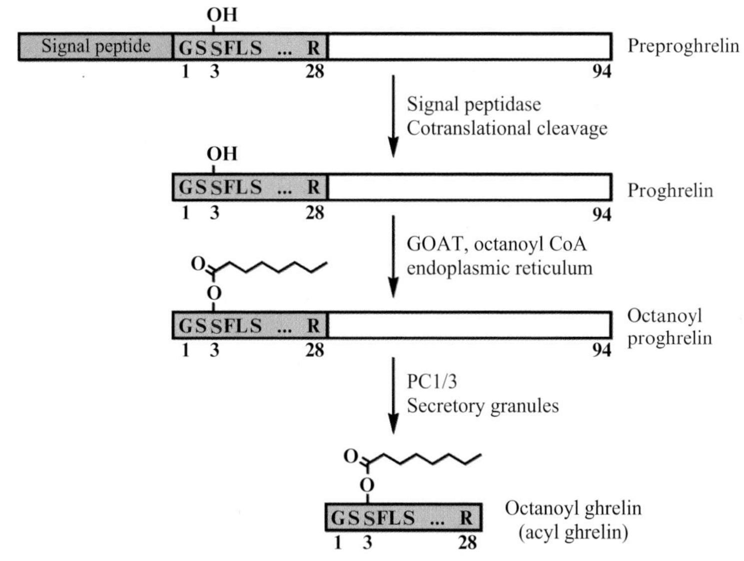

Martin S. Taylor *et al.*, Figure 13.1 Ghrelin biosynthetic pathway. Ghrelin is synthesized as a 117-amino acid precursor, preproghrelin, containing a signal peptide, the 28-amino acid ghrelin sequence, and a 66-amino acid C-terminal peptide. The signal peptide is cotranslationally cleaved, releasing the 94-amino acid proghrelin into the lumen of the endoplasmic reticulum (ER). Attachment of the octanoate group to Ser3 of proghrelin occurs in the ER and is catalyzed by GOAT. In secretory granules, prohormone convertase 1/3 (PC1/3) then cleaves at the C-terminus of acyl proghrelin to give the mature acyl ghrelin.

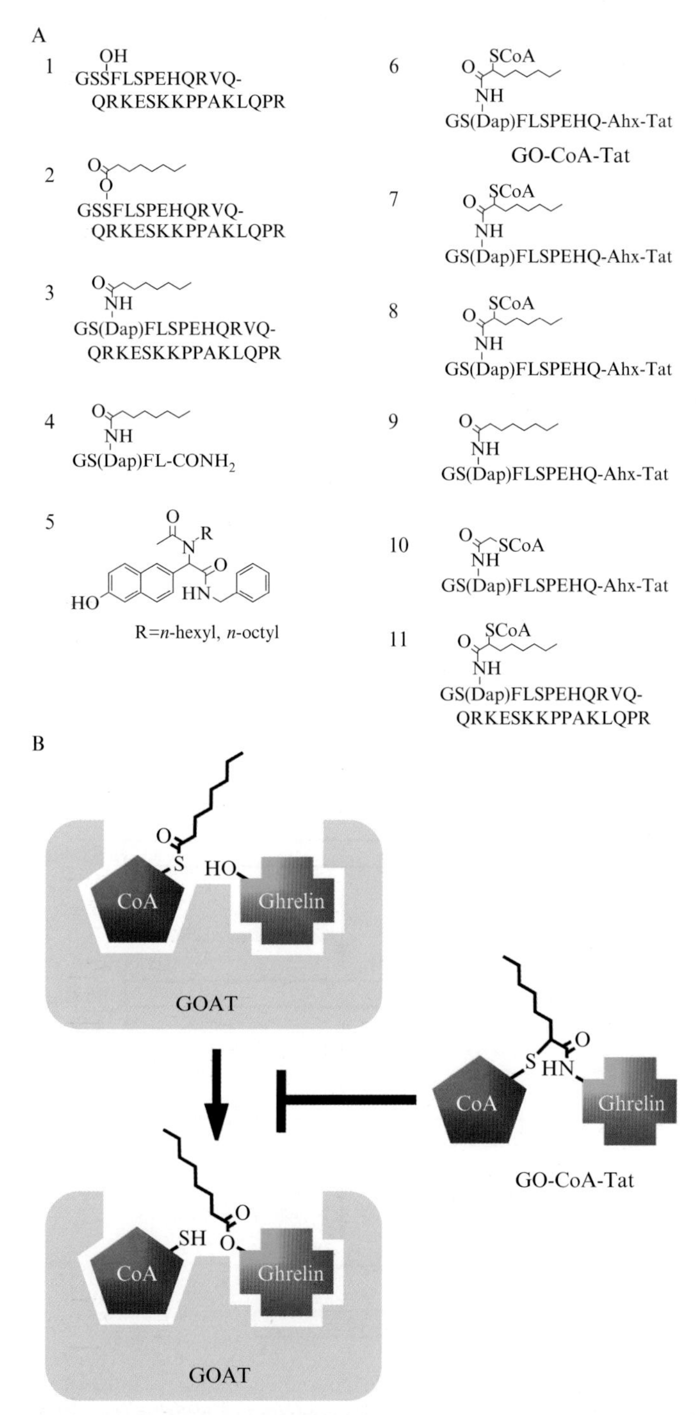

Martin S. Taylor *et al.*, Figure 13.2 GOAT inhibitors. (A) Chemical structures of ghrelin and GOAT inhibitors. 1: Des-acyl ghrelin. 2: Acyl ghrelin. 3: Amide-linked octanoyl ghrelin. 4: Amide-linked 5-mer octanoyl ghrelin with C-terminus amidated. 5: Inhibitors discovered by Garner and Janda (2011). 6: GO-CoA-Tat. 7,8: Bisubstrate compounds with

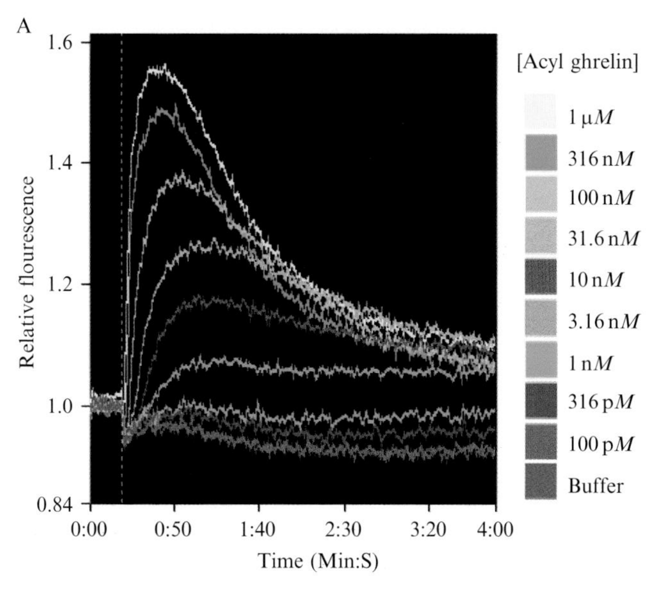

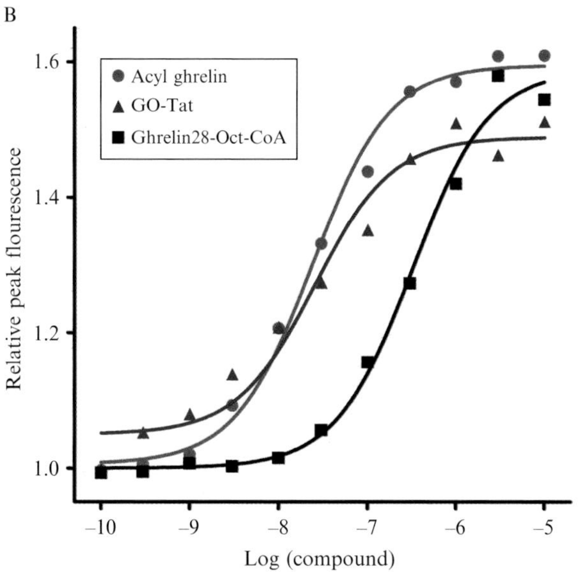

Martin S. Taylor *et al.*, Figure 13.3 GHS-R1a assay in stably transfected HEK-293T-GHS-R1a cells. (A) Typical dose–response traces for acyl ghrelin, with concentrations on half-log scale from 1 μM to 100 pM, with buffer-only control. (B) Agonism for acyl ghrelin, GO-Tat (Fig. 13.2, Compound 9), and the bisubstrate compound Ghrelin28-Oct-CoA (Fig. 13.2, Compound 11). EC_{50} values are reported in Table 13.1.

---

five and three amino acids of ghrelin. 9: GO-Tat: an octanoyl-amide Tat-tagged product analog. 10: Bisubstrate inhibitor with two-carbon acyl group. 11: Ghrelin28-Oct-CoA, a bisubstrate compound. (B) Mechanism-based design strategy of GO-CoA-Tat. The lipid–enzyme interaction is not shown but may be important. Also, the form of ghrelin acylated by GOAT is likely proghrelin; the smaller version is shown for clarity.

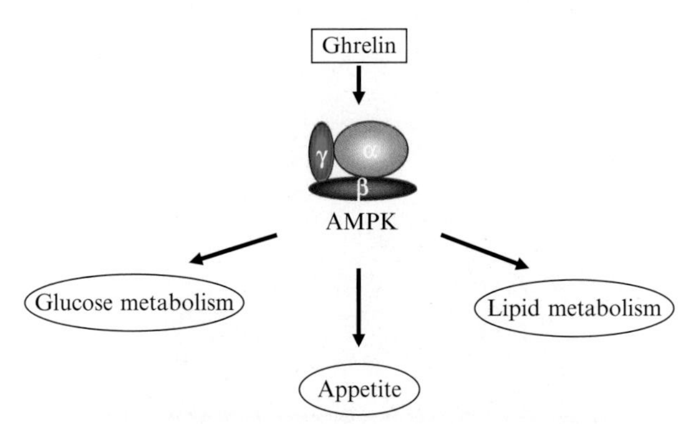

Chung Thong Lim *et al.*, Figure 17.1 The heterotrimeric structure of AMPK, consisting of the α-, β-, and γ-subunits. AMPK has been shown to mediate several metabolic effects of ghrelin, which include stimulating appetite as well as affecting glucose and lipid metabolism.

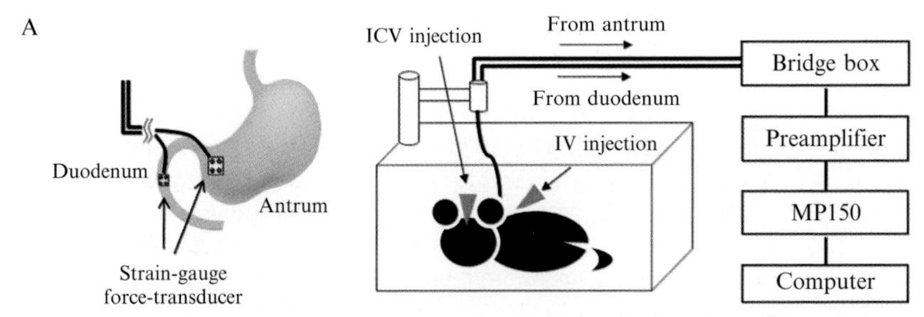

Naoki Fujitsuka *et al.*, Figure 18.1 Method for strain-gauge force-transducer measurements of gastrointestinal motility in conscious rats. (A) Strain-gauge force transducers were placed on the serosal surface of the antrum and duodenum. The wires of the transducers were drawn out from the back of the neck and connected to a preamplifier via a bridge box. Data were recorded using an MP150.

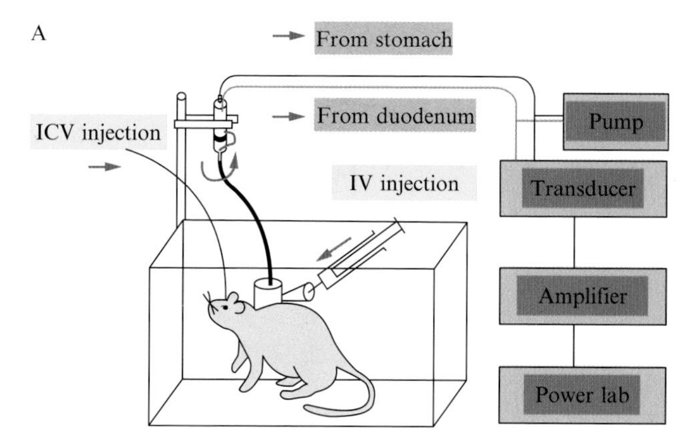

Naoki Fujitsuka *et al.*, Figure 18.2 Method for manometric measurements of gastrointestinal motility in conscious rats. (A) Catheters for motility recordings are inserted into the antrum and duodenum, connected to the infusion swivel to allow free movement and then connected to a pressure transducer. The data are recorded and stored in a PowerLab.

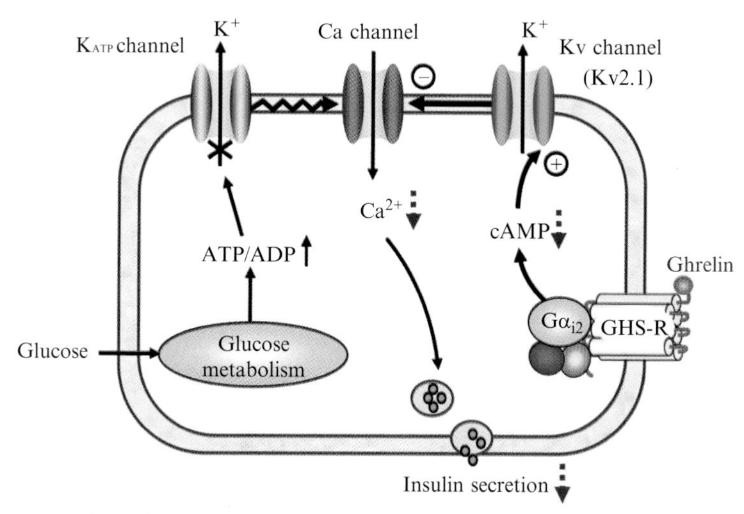

Katsuya Dezaki and Toshihiko Yada, Figure 20.3 Ghrelin signaling in islet β-cells. Closure of ATP-sensitive K^+ (K_{ATP}) channels by increases in ATP/ADP ratio following glucose metabolism induces membrane depolarization and increase in $[Ca^{2+}]_i$ via voltage-dependent Ca^{2+} channels, leading to insulin secretion in β-cells. Ghrelin activates β-cell GHS-R that is coupled with PTX-sensitive G-protein $G\alpha_{i2}$, decreases cAMP production, and attenuates membrane excitability via activation of voltage-dependent K^+ channels (Kv2.1 subtype), and consequently suppresses Ca^{2+} influx and insulin release.

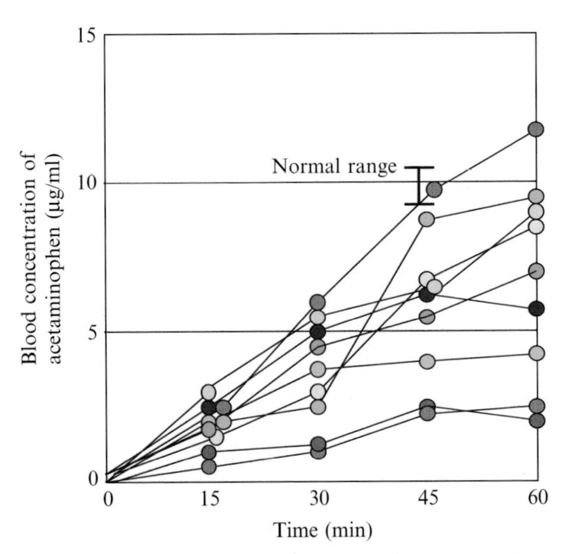

Mari Hotta *et al.*, Figure 24.1 Plasma concentrations of acetaminophen in patients with AN. At 08:00 h after overnight fasting for longer than 12 h, blood samples were drawn from female patients with AN (age: 18.2–31.4 year, BMI: 11.48–17.08 kg/m^2) for the determination of plasma acetaminophen concentrations immediately before and every 15 min after the ingestion of 1.5 g acetaminophen and 200 kcal liquid diet for 60 min. The plasma level of acetaminophen in age-matched young women was 9.4$\pm$0.5 μg/ml. Gastric excretion was delayed in 9 out of 10 patients with AN.

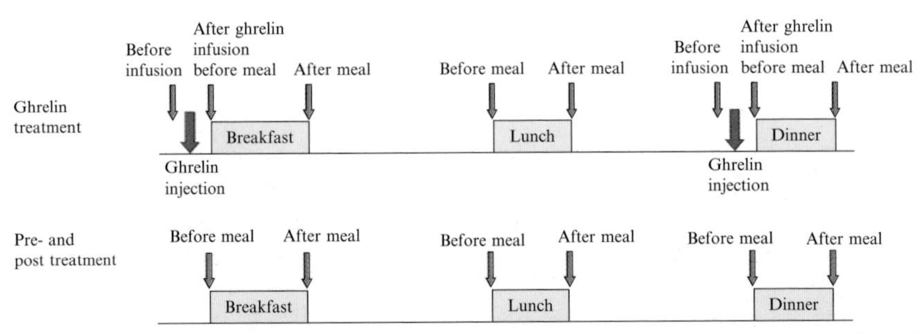

Mari Hotta *et al.*, Figure 24.4 Schedule of visual analogue scales. Attitudes toward food were evaluated by a questionnaire incorporating visual analogue scales (VAS). During ghrelin treatment, patients with AN answered the VAS questionnaire at 15 min before ghrelin infusion and breakfast or dinner, 15 min after ghrelin infusion before breakfast or dinner, and after those meals. They answered the questionnaire also before and after every meal without ghrelin treatment.